Atlas of Axial, Sagittal, and Coronal Anatomy

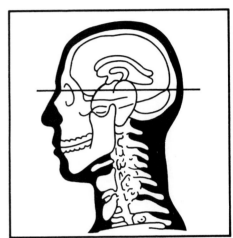

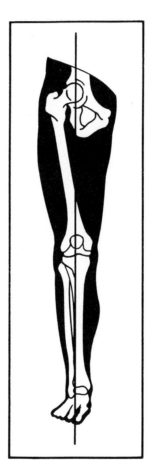

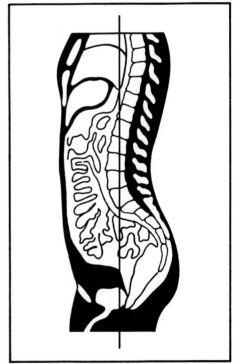

with CT and MRI

A. John Christoforidis, M.D., Ph.D.

Professor & Chairman
Department of Radiology
Ohio State University
Columbus, Ohio

1988 **W. B. SAUNDERS COMPANY**
Harcourt Brace Jovanovich, Inc.

Philadelphia London Toronto Montreal Sydney Tokyo

W. B. SAUNDERS COMPANY
Harcourt Brace Jovanovich, Inc.

West Washington Square
Philadelphia, PA 19105

Library of Congress Cataloging-in-Publication Data

Christoforidis, A. J., 1924-
    Atlas of axial, sagittal, and coronal anatomy with CT and MRI.

    1. Anatomy, Human—Atlases.   2. Tomography—
Atlases.   3. Magnetic resonance imaging—Atlases.
I. Title. [DNLM: 1. Anatomy—atlases.   2. Nuclear
Magnetic Resonance—atlases.   3. Tomography,
X-Ray Computed—atlases.   QS 17 C556a]
QM25.C47   1988   611   86-28041

ISBN 0-7216-1278-4

Editor: Dean Manke
Developmental Editor: Kathleen McCullough
Production Manager: Carolyn Naylor
Manuscript Editor: Susan Thomas
Indexer: Dennis Dolan

Atlas of Axial, Sagittal, and Coronal Anatomy with CT and MRI   ISBN   0-7216-1278-4

Last digit is the print number: 9   8   7   6   5   4   3   2   1

Contributors

Delmas J. Allen, Ph.D.
Professor and Associate Dean of Academic
    Affairs
College of Health Sciences
Georgia State University
Atlanta, Georgia

Javier Beltran, M.D.
Assistant Professor of Radiology
Ohio State University College of Medicine
Head, Division of Musculoskeletal
    Radiology
Ohio State University Hospital
Columbus, Ohio

Donald W. Chakeres, M.D.
Associate Professor of Radiology
Ohio State University College of Medicine
Clinical Director, Magnetic Resonance
    Imaging
Ohio State University Hospital
Columbus, Ohio

A. John Christoforidis, M.D., Ph.D.
Professor and Chairman
Department of Radiology
Ohio State University College of Medicine
Chairman, Department of Radiology
Ohio State University Hospital
Columbus, Ohio

Charles F. Mueller, M.D., F.A.C.R.
Professor of Radiology
Ohio State University College of Medicine
Director of Post-Graduate Education in
    Radiology
Ohio State University Hospital
Columbus, Ohio

John Alexander Negulesco, Ph.D.
Associate Professor of Anatomy and
    Anesthesiology
Chairman, Graduate Studies Committee
Ohio State University College of Medicine
Columbus, Ohio

Preface

Recent extraordinary scientific and technological advances have revolutionized the field of imaging. Their application to medicine led radiologists as well as other physicians and anatomists to utilize the newly discovered approaches and instrumentation to visualize living biological structures.

As life, especially academic life, in my view, is characterized by a sinusoidal curve, the anatomic study of biological structures in vivo revitalizes old interests, seen under different angles or through new prisms. When it seems that a field of inquiry can yield no further avenues for learning or teaching, which proves to be repeatedly an empty, temporary presumption, there is a period of renaissance: As a new Prometheus, anatomy gains new life. The so-called "dead branch of Biology" sprouts and blossoms again. With the successful recent application of electron microscopy to the study of biological structures, anatomy reacquired its ancient and prestigious rank among medical or health sciences. By sharing the same technique to study biological units at the subcellular and at the molecular level, anatomy became again the center of the basic medical sciences, much closer to biochemistry, genetics, physiology, microbiology, and pathology. With x-ray microanalysis the anatomic study was taken to the atomic level. At the opposite extreme of the structural spectrum, macro-anatomy was naturally brought closer to radiology by the advent of computed tomography and nuclear magnetic resonance imaging. The cross or transverse sections, as well as anteroposterior or sagittal sections and frontal or coronal sections of the human body, suddenly became the focus of new attention, regaining the status of indispensable specimens. These anatomic preparations are now critically important and must be studied and learned by teachers, students, and physicians. The almost instantaneous revival of the importance of sectional anatomy was prompted by the need to correlate it with radiographs, computed tomograms, and magnetic resonance images. Such a correlation was essential in order to provide a tridimensional view of each anatomical structure or its parts in vivo and obviously, to learn, practice, and teach medicine.

Toward this multifaceted end, anatomists and radiologists at the Medical College of Ohio, under the leadership of Dr. A. J. Christoforidis, former Chairman and Professor of the Department of Radiology and, presently, occupying the same position at The Ohio State University College of Medicine, set out to use the personal and physical resources then available to prepare a comprehensive atlas, which would include the new developments and advances in imaging. The resulting publication contains the data of comparative study, involving (a) plates of anatomic sections in the major directions to obtain a stereoscopic view of topographical relationships of the structures of the human body, (b) computed tomograms, and (c) magnetic resonance images.

An increasing number of anatomic textbooks have included radiograms of the head, neck, trunk, and members, in what amounts to a volume of radiologic anatomy within a volume. Recently, a few computed tomograms and magnetic resonance images have been added to update the textbooks and to introduce the new developments in imaging to medical and dental students, at least as a reminder of the innovative techniques and at best to stimulate the readers to devote more time to technological advances in modern medicine. To take real advantage of this additional material, however, a much higher number of computed tomograms and magnetic resonance images is necessary, consequently overloading already long anatomic textbooks.

Having the essential resources, we joined the efforts of Dr. A. J. Christoforidis in providing an alternative option, a separate atlas that gives a thorough stereoscopic view, from the clinical standpoint, of the human body and its parts, comparing anatomic sections with computed tomograms and magnetic resonance imaging. The use of such an atlas will serve the task of anatomists and will prepare students to interpret normal and pathological anatomy, seen with new imaging techniques for clinical and surgical purposes.

Once familiarized with these imaging techniques, students, anatomists, radiologists, physicians, and dentists will correlate and properly interpret anatomic sections-paired-with-image to understand the topographic morphology of the whole body, its organs, and its parts.

On the other hand, it should be emphasized that only a thin slice of the human body or of its organs is visualized with a single computed tomogram, which alone does not give a three-dimensional image. In addition, computed tomograms are usually images of transverse or horizontal sections (along and perpendicular to the vertical or longitudinal axis), which until the early seventies have been seldom used by radiologists. Nonetheless, the availability of the whole-body computed tomographic scanner and its numerous clinical applications immediately triggered the renewed interest in cross-sectional anatomy, and later in sections cut along other planes, such as the sagittal and the frontal planes. Computed tomography was soon followed by magnetic resonance imaging, in which magnetic properties of atomic nuclei, such as those of the nucleus of hydrogen,

account for better contrast images of anatomic structures. Thus, the study of anatomic sections of the human body and its parts, once limited to illustration and demonstrations in topographic or regional anatomy, became the focus of interest of radiologists and of a wider variety of specialists for teaching, learning, and clinical as well as surgical purposes.

Ultimately, the preparation of this atlas is another attempt to present a comprehensive and stereoscopic view of the human body by comparing photographs of anatomical sections in different planes with computed tomograms and magnetic resonance imaging.

The objective has been to include in this publication the essential material to make its use worthwhile for anatomic, radiologic, and clinical purposes. This atlas joins a series of others to make possible the updating of knowledge in the field of anatomic structure visualization for medical purposes.

Liberato J. A. DiDio, M.D., D. Sc., Ph.D.
Dean of the Graduate School
Chairman and Professor, Department of Anatomy
Medical College of Ohio
Toledo, Ohio

Foreword

The year 1895 is a great landmark in the history of science in general and of medicine in particular. On November 8 of that year Wilhelm Conrad Roentgen discovered x-rays, giving rise to the discipline of radiology. More appropriately, we should regard this historic event as marking the origin of medical imaging.

The significance of Roentgen's discovery became apparent immediately, as indicated by the more than 1000 scientific papers published within the first year following the discovery. The Nobel Foundation awarded him the first Prize in physics when this most universally recognized distinction was established in 1901.

Important advances took place in the intervening decades, some of which include the introduction of contrast media, particularly barium sulfate for the examination of the alimentary tract, and the iodinated compounds, primarily for intravascular use. Other significant landmarks are represented by the discovery of the hot-cathode tube in 1913 by W. D. Coolidge, and later by the introduction of the image amplification technology of the early 1950s. These discoveries facilitated the development and the routine use of invasive procedures, starting with angiography of large vessels and then selective exploration of the entire vascular system. These procedures were followed by the application of needle biopsy techniques for most organs. Percutaneous entrance into body spaces and cavities for diagnostic and therapeutic purposes became a routine practice in most hospitals.

In 1972, a unique breakthrough revolutionized the field of medical imaging with the introduction of cross-sectional imaging by computed tomography (CT), combining advanced x-ray technology with computer science.

Computers had a profound impact on the world in general, expanding the horizons of science and influencing practically the entire spectrum of human activities. Godfrey Hounsfield received the Nobel Prize in physiology and medicine in 1979 for the discovery and introduction of computed tomography. Hounsfield shared the prize with Allan MacLeod Cormack; his work in applied reconstruction techniques through solution of mathematical problems, published in 1963, was thus recognized.

Today, computed tomography plays a dominant role in medical diagnosis. The practicing physician is keenly aware of the significance of this modality, which contributes importantly to the study and treatment of diseases.

No more than one year had lapsed since the introduction of computed tomography when Lauterbur published the first images of a new epoch-making breakthrough in the journal *Nature;* and R. Damadian made useful suggestions on the potential use of nuclear magnetic resonance in tumor detection. The application of nuclear magnetic resonance in imaging (MRI), the principles of which had been in use by chemists and physicists for more than three decades, gave birth to a most exciting and versatile new diagnostic modality. The statement made by many in regard to the application of nuclear magnetic resonance (NMR) in medical imaging being as important as the discovery of the roentgen rays may well prove in the future to be justified. The discoverers of the (NMR) principles, Alex Block and Edward Purcell, who worked independently, jointly received the Nobel Prize in physics in 1952.

The principles on which magnetic resonance imaging and computer tomography operate are quite different. Computed tomography exploits the differential absorption of the x-ray beam by the body tissues. This is based on a large number of calculations by complex mathematical formulas made possible and practical through computers. CT is primarily an imaging modality of the normal or pathologic anatomy.

Magnetic resonance, on the other hand, uses the magnetic properties of the atoms and their nuclei. In proton MRI, the molecular structure as related to the presence of hydrogen, abundant in human tissues, determines the all-important imaging contrast.

The possibility of combining magnetic resonance imaging with spectroscopic analysis of biochemical happenings at the cellular level is of unique significance. It is anticipated that accurate morphologic study of tissues will be complemented by the understanding of their biochemistry and physiology. The in vivo biochemical studies of human tissues and their imaging is on the horizon of this new frontier.

The objective of this atlas is to present an adequately annotated reference source useful to radiologists, the imaging physician, and the student of anatomy and radiology in general. We have attempted to emphasize the close correlation of the gross anatomy with two important imaging modalities which are most useful in the study of the human body and of great practical significance in patient care.

We hope that this volume will meet these objectives to some extent.

A. John Christoforidis

Acknowledgments

The preparation and completion of this atlas was made possible with the close collaboration and advice from a number of colleagues and associates.

Members of the Anatomy Department of the Medical College of Ohio, where this work started, were very supportive in this effort. Among them is Dr. Liberato DiDio, Chairman of the Department of Anatomy and Dean of the Graduate School at the Medical School of Ohio, whose advice and support to the end have been invaluable. His Department provided a large part of the necessary anatomic material. Dr. Prem Chandnani, Associate Professor of Radiology, and Mrs. Roberta Easter, Chief Technologist, cooperated in the preparation of the computed tomographic examination of the first cadaver. Ms. Faye Keene, Photographic Specialist in the Department of Radiology at the Medical College of Ohio, produced the photographs of the cadaver CT, while Mr. Elmer Schoenrock made the color photographs of the first cadaver. Mr. Craig Cramer and Mr. Kallol Chaudhuri, doctoral students in the Department of Anatomy, as well as Mr. James Mansfield were helpful in the preparation of the anatomic plates.

The members of the Anatomy Department at Ohio State University were most helpful. We acknowledge the help of Dr. James S. King, Chairman of the Department of Anatomy, Professor Margaret Hines, and Professor Emeritus George R. Gaughran. Pathology Specialist Donald Kincaid and Technical Assistant Dwight Jack assisted in the preparation of the anatomic material at Ohio State University. Professor William Hunter was very supportive, particularly in obtaining the MRI images, for some of which he himself was kind enough to volunteer. Greg Christoforidis, medical student, assisted in the annotation of the anatomic plates of the thorax, abdomen, and pelvis.

This atlas would not have been possible without the dedicated work of Mr. John Croyle, BA, Photographic Specialist of the Department of Radiology at Ohio State University. We are grateful for his expert and constant support, for most of the color and black and white anatomic specimens, and for his photographic work of the CT and MRI imaging.

To Ms. Melinda McCalla, Chief of Magnetic Resonance Technology, special thanks for her tireless cooperation. Thanks are also due to Ms. Lyn Vandervort, Technologic Supervisor of the CT section, and to Mrs. Linda Chakeres, who assisted in proofreading the manuscript.

To my secretary, Mrs. Tonda Robinson, and to Mrs. Nancy Wilburn and Ms. Sandy Lewis our sincere appreciation for their continuous support in preparing the manuscript and the legends and performing all the related secretarial work.

Special thanks are due to all the members of the Department of Radiology at Ohio State University who directly or indirectly supported this undertaking.

Finally, I want to express my appreciation to our publishers, particularly to Mr. Dean Manke and Ms. Kitty McCullough for their cooperation and support, to Ms. Susan Thomas for her advice and editorial coordination of this effort, and to Ms. Carolyn Naylor for overseeing the completion of the work.

Introduction

The significance of being familiar with the anatomy of the axial, sagittal, and coronal planes in the everyday clinical practice has been more evident in the last few years than ever before.

The detailed study of the anatomy of the human body used to be primarily the concern of the surgeons and the anatomists, who explored the human body on the surgical or the necrotomic table. Radiologists were concerned mainly with the skeleton, the digestive tract, the cardiorespiratory and circulatory systems, and anatomic areas important to certain special procedures. These studies were performed in the frontal, lateral, and occasionally oblique projections. The radiologically detectable parts had great limitations, particularly in the exploration of soft tissues. Important radiological examinations were possible primarily through the use of contrast media introduced via the alimentary canal or by invading the vascular system by means of needles and/or catheters. Direct visualization of most of the organs and of the adjacent soft tissues was markedly limited.

In the last 15 years, with the advent of ultrasound, followed by the introduction of computed tomography, and more recently with the expanding use of magnetic resonance imaging, one can study organs and tissues of the human body with exquisite detail in vivo and with only limited use of contrast media. The routine diagnostic studies for patients in most hospitals and a great number of outpatient clinics can now be accomplished in great detail, including the exploration of soft tissue organs and of the cardiovascular organs, without resorting to invasive procedures.

The imaging physician, usually a radiologist, responsible for the performance and study of the requested examination should be very familiar with the detailed anatomy of the human body in all planes. The referring physician, on the other hand, should be able to communicate with the radiologist and understand the normal and pathologic anatomy of his or her patients.

The surgeon in particular, can be guided accurately and therefore can plan accordingly for the intended operative procedures. The radiation oncologist can effectively study the area to be treated by radiation with great accuracy in order to calculate the dose to the patient and use an effective plan of therapy.

Never before has the knowledge of anatomy been so important and beneficial in the daily routine of medical practice.

In the past, students and physicians used the frontal and lateral views during their studies, particularly with the conventional radiographic, fluoroscopic, and nuclear imaging techniques. The axial (transverse) view, which was rarely, if ever, used in the past, is now the prevailing one, particularly with the use of computed tomography. The coronal and sagittal planes have also become very important in the use of computed tomography, occasionally for direct scanning of the head or more commonly for reconstruction techniques. All three planes, however, axial, coronal, and sagittal, are routinely used in magnetic resonance imaging. Familiarity, therefore, with these projections for the study of the human body has become a most important prerequisite. Medical students are required to learn the anatomic relationship of the organs in these planes, rarely used before in the daily practice, thus making necessary the introduction of a new concept in the study of anatomy. To the radiographic and computer tomographic images, the Anatomy Department at Ohio State University has added, since 1985, the correlation of magnetic resonance images in the study of gross anatomy for medical students. As for the physician, it has become necessary to refresh his knowledge of anatomy, or to put it more realistically, learn it again. For the imaging physician, this knowledge represents the sine qua non.

In this atlas we have used appropriate anatomic sections from 11 cadavers and corresponding images from two of the established imaging modalities: computed tomography and magnetic resonance. The photographs of these anatomic sections can be studied with the corresponding images of these modalities. Emphasis has been placed on the areas of anatomy that are clinically significant. Joints of practical importance, including the temporomandibular joint, the shoulder, the wrist, the hip, the knee, and the ankle, are the subjects of a separate section.

Nine of the cadavers used were unembalmed. The preparation of the cadavers consisted of freezing the fresh cadaver for 8 to 10 days, depending on the size. Subsequently, sectioning at 2-cm intervals was performed using a fine band saw with minimal loss of tissue during the sectioning. Before photographing the sections, the frozen specimen was cleaned carefully with room temperature tap water to eliminate remnants of tissue left on the surface during the sectioning. This process also facilitated the elimination of tiny ice particles, which are detrimental to photography owing to reflection of light. Twenty to 30 additional minutes were adequate for allowing the specimen to thaw before photographing.

An exception to this routine was made in the axial computed tomographic examination of the cadaver, which was performed before the freezing. Skin markers were placed before the examination in order to facilitate the accuracy of the level for comparison of the gross anatomic specimen with the cadaver CT. Embalming of the cadaver was avoided to better preserve the natural color of the tissues.

The computed tomographic examinations of the cadaver were supplemented with ones from living individuals at the same anatomic level as accurately as possible. One should take into consideration two important inherent differences. First, the post-mortem changes, particularly of the thorax, include pulmonary edema, and the fact that the examination is performed in expiratory phase. The second factor to be kept in mind is the amount of variation encountered from one patient to another in the location of the different organs, in relationship to the skeleton, particularly to the spine, as well as the relationship of the intra-abdominal viscera. This is particularly true for the position of the stomach, the transverse colon, the sigmoid, and the gallbladder. This difficulty was minimized by carefully selecting the images for the examinations of the living to correspond closely to the anatomic preparations.

It is well understood that the anatomic sections demonstrate the tissues present on the surface of the specimen, whereas the imaging techniques CT and MR demonstrate the structures contained in the three-dimensional slice, which in turn is a function of the thickness of the section. This important difference between the photography and the three-dimensional rendition of the CT and MR imaging explains the apparent discrepancy in the demonstration of small structures shown by imaging but not shown in the photographed specimen.

The photography of the gross section was performed by placing the specimen on a Belcher copy stand totally enclosed with black backdrop paper to block out any extraneous color. The specimens were photographed with a Hasselblad 500 c/m and its complementary 80-mm f2.8 lens.

The computed tomography prints were made from 14 × 17 diagnostic films (hard copies). The image area not pertaining to the anatomy was blocked out with red opaque. This allowed for more contrast control, resulting in an improved print.

The magnetic resonance images were printed by using the Matrix Imager, reversing the desired section from negative to positive, and increasing contrast (width and level) to its maximum without affecting pathology or anatomy. The diagnostic film was used as a contact negative, which allowed more tonal range and more detail, while eliminating the copy negative altogether.

The technical factors used in relation to computed tomography and magnetic resonance imaging are discussed in the Introduction of the individual sections. In order to facilitate and render the study more lucid, seventy-six anatomic sections were made in color. All the axial planes are illustrated with anatomic sections that are accompanied by corresponding CT and magnetic resonance images. Coronal sections of the head are correlated with both computed tomographic and magnetic resonance images, whereas the sagittal anatomic plates of the head are correlated only with magnetic resonance imaging. All the anatomic coronal and sagittal sections of the body and of the extremities are illustrated and correlated with magnetic resonance images. Computed tomographic reconstruction of the coronal and sagittal sections of the body were not considered pertinent for the objectives of this atlas. In each chapter, a number of illustrations that were considered important clinically were added in order to render the study of this atlas more useful for the clinician. The significance of using the appropriate techniques for diagnostic purposes, particularly with magnetic resonance, has been indicated in some of these illustrations.

A. John Christoforidis

Bibliography

Cormack AM: Representation of a function by its line integrals with some radiological applications. *J Appl Physiol* 34:2722, 1963.

Damadian R: Tumor detection by nuclear magnetic resonance. *Science* 171:1151–1153, 1971.

Damadian R et al: Nuclear magnetic resonance as a new tool in cancer research: Human tumor by NMR. *Ann NY Acad Sci* 222:1048–1076, 1973.

Eycleshymer AC, Schoemaker D: A Cross-Section Anatomy. New York and London, Appleton-Century-Crofts, 1911 (Copyright 1970).

Hounsfield GN: A method of an apparatus for examination of the body by radiation such as X or gamma radiation. British Patent No. 1283915, 1972.

Hounsfield GN: Computerized transverse axial scanning (tomography). Part 1. Description of the system. *Br J Radiol* 46:1016–1022, 1973.

Kieffer S A, Heitzman ER: *Atlas of Cross-Sectional Anatomy: Computed Tomography, Ultrasound, Radiography, Gross Anatomy.* New York, Harper and Row Publishers, 1979.

Koritké JG, Sick H: *Atlas of Sectional Human Anatomy: Frontal, Sagittal and Horizontal Planes.* Baltimore, Urban and Schwarzenberg, 1983.

Lauterbur PC: Image formation by induced local interactions: Examples employing nuclear magnetic resonance. *Nature* (Lond.) 242:190–191, 1973.

Pernkopf E: *Atlas of Topographical and Applied Human Anatomy.* 2nd ed. Philadelphia, W.B. Saunders Co., 1964.

Contents

Atlas of Axial, Sagittal, and Coronal

Anatomy

with CT and MRI

Head and Neck

Donald W. Chakeres, M.D.
Delmas J. Allen, Ph.D.
A. John Christoforidis, M.D., Ph.D.

This chapter illustrates the CT and MRI anatomy of the complex head and neck region in the axial, coronal, and sagittal planes. A cross-sectional anatomic plate begins each series of illustrations; it is followed by images made at approximately the same plane and level. The axial sections begin superiorly and end inferiorly and the order is based on the central brain level. The coronal sections begin anteriorly and end posteriorly. The sagittal sections begin in the midline and end laterally. In regions in which the anatomy is complex, the illustrations are expanded to demonstrate the structures to best advantage.

The CT and MR images may not exactly correspond to the anatomic section because of normal patient variations and standard variations in the CT and MRI section planes. The coronal and axial imaging planes are all described in relation to the zero-degree anthropologic baseline, which is defined as the plane that intersects the inferior orbital rim and the superior portion of the external auditory canal. The choice of section plane varies not only because of limitations of patient positioning but also because special planes are required to image specific anatomic structures in an ideal fashion. For example, most coronal MR images are approximately 90 degrees to the anthropologic baseline because this plane parallels the MRI table when the patient is in a neutral supine position. With CT it is very difficult for the patient to be positioned in the gantry for this angle, so most coronal CTs are made at approximately 105 to 120 degrees to the anthropologic baseline. There is no universally accepted ideal axial section plane, so we chose to illustrate two separate interlaced planes. One is made at +20 degrees (similar to routine CT sections), and the second plane is at −10 degrees (used clinically to study the orbits because it parallels the optic canals and nerves). This angle section plane of −10 degrees is similar to the routine MRI axial plane of 0 degrees.

Many different variations of these two imaging modalities are presented to demonstrate the most accurate and sensitive techniques available. One single technique will not display all of the anatomy to best advantage, and special imaging parameters are tailored to the anatomic structure. Different slice thicknesses, contrast agents, section planes, window displays, and raw data algorithms must be used to best advantage with the CT studies. With MRI, variable pulse sequence designs and coils for receiving the signal are used. This diversity is not meant to be confusing but rather to demonstrate that there are many different CT and MRI techniques, which produce widely different appearances of the same structures. Similar techniques are used in routine clinical settings to evaluate specific abnormalities.

Many different pulse sequence designs were used for the same MRI section to illustrate the variable anatomic and physiologic information that can be displayed. The image contrast of a structure (high or low signal intensity) is not constant. The intrinsic tissue parameters (proton density, T1 and T2 relaxation times, movement) can be displayed in many different ways by varying the pulse sequence design. For example, the spinal fluid can be either high or low intensity. When the spinal fluid is of low intensity, we can obtain very high resolution images in a short data acquisition time. When the spinal fluid is of high intensity, the images are usually of lower resolution and may take a long time to acquire. By changing the pulse sequence design, better tissue specificity is possible because many different parameters can be studied. The anatomic structures are also demonstrated using surface coil receivers to enhance the image resolution.

A number of different techniques were utilized to provide optimal CT image quality. Intravenous contrast–enhanced studies were used to make the blood vessels more clearly visible. Intrathecal contrast was used to better demonstrate the outer contours of the brain and spinal cord, which are not visible on routine imaging. Thin-section high-resolution bone techniques were used to demonstrate the bony structures.

Bibliography

Blumenfeld SM, Glover G: Spatial resolution in computed tomography. In Newton TH, Potts DG (eds): *Radiology of the Skull: Technical Aspects of Computed Tomography.* St. Louis, C.V. Mosby Company, 1974.

Bradley WG Jr, Waluch V: Blood flow: magnetic resonance imaging. *Radiology* 154:443–450, 1985.

Bradley WG Jr, Waluch V, Lai K.-S: Appearance of rapidly flowing blood on magnetic resonance images. *AJR* 143:1167–1174, 1984.

Cameron IL, Ord VA, Fullerton GD: Characterization of proton NMR relaxation times in normal and pathological tissues by correlation with other tissue parameters. *Magnetic Resonance Imaging* 2:97–106; 1984.

Carter BL, Morehead J, Hammerschlag SB, Griffiths HJ, Kahn PC: *Cross-Sectional Anatomy.* East Norwalk, Connecticut, Appleton-Century-Crofts; 1977.

Chakeres DW, Kapila A: Brainstem and related structures: Normal CT anatomy using direct longitudinal scanning with metrizamide cisternography. *Radiology* 149:709–715, 1983.

Chakeres DW, Spiegel PK: A systematic technique for comprehensive evaluation of the temporal bone by computed tomography. *Radiology* 146:97–106, 1983.

Chakeres DW, Kapila A: Computed tomography of the temporal bone. *Medical Radiography and Photography* 60:2–29, 1984.

Chakeres, DW: CT of ear structures: A tailored approach. *Radiologic Clinics of North America* 22:3–14, 1984.

Chakeres DW, Kapila A: Radiology of the ambient cistern: Part I. Normal. *Neuroradiology* 27:383–389, 1985.

Chakeres, DW: Clinical significance of partial volume averaging of the temporal bone. *AJNR* 5:297–302, 1984.

Chakeres DW, Kapila A: Normal and pathologic radiographic anatomy of the motor innervation of the face. *AJNR* 5:591–597, 1984.

Chakeres DW, Weider DJ: Computed tomography of the ossicles. *Neuroradiology* 27:99–107, 1985.

Daniels DL, Herkins R, Gager WE, Meyer GA, Koehler PR, Williams AL, Haughton VM: Magnetic resonance imaging of optic nerve and chiasm. *Radiology* 152:79–83, 1984.

Daniels DL, Peck P, Mark L: MRI of the cavernous sinus. *AJNR* 6:187–190, 1985.

Daniels DL, Herfkins R, Koehler PR, Millen SJ, Shaffer KA, Williams AL, Haughton VM: Magnetic resonance imaging of internal auditory canal. *Radiology* 151:105–108, 1984.

Daniels DL, Schneck JR, Foster T, Hart H Jr, Millen SJ, Meyer GA, Pech P, Haughton VM: Magnetic resonance imaging of jugular foramen. *AJNR* 6:699–703, 1985.

Daniels DL, Williams AL, Haughton VM: Computed tomography of the medulla. *Radiology* 145:63–69, 1982.

Daniels DL, Rauschning W, Lovus J, Williams AL, Haughton VM: Pterygopalatine fossa: Computed tomographic studies. *Radiology* 149:511–516, 1983.

Daniels DL, Williams AL, Haughton VM: Computed tomography of the articulations and ligaments at the occipito-atlantoaxial region. *Radiology* 146:709–711, 1983.

Daniels DL, Haughton VM, Williams AL: Flocculus in computed tomography. *AJNR* 2:227–229, 1981.

Flannigan BD, Bradley WG, Mazziotta JC, Rauschning W, Bentson JR, Lufkin RB, Hieshima GB: Magnetic resonance imaging of the brainstem: Normal structure and basic functional anatomy. *Radiology* 154:375–383, 1985.

Kapila A, Chakeres DW: Computed tomography of Meckel's cave. Part I: Normal Anatomy; Part II: Pathology. *Radiology* 152:425–433, 1984.

LaMasters DL, Watanabe TJ, Chambers EF, Norman D, Newton TH: Multiplanar metrizamide-enhanced CT imaging of the foramen magnum. *AJNR* 3:485–494, 1982.

Latchaw RE, Hirsch WL Jr, Horton JA, Bissonette D, Shaw DD: Iodhexol vs metrizamide: Study of efficacy and morbidity in cervical myelography. *AJR* 6:931–934, 1985.

Mafee MF, Schild JA, Valvassori GE, Capek M: Computed tomography of the larynx: Correlation with anatomic and pathologic studies in cases of laryngeal carcinoma. *Radiology* 147:123–128, 1983.

Mancuso AA, Hanafee WN: *Computed Tomography of the Head and Neck.* Baltimore, Williams & Wilkins, 1982.

Mark L, Pech MP, Daniels D, Charles C, Williams AL, Haughton VM: Pituitary fossa: A correlative anatomic and magnetic resonance study. *Radiology* 153:453–457, 1984.

Mawad ME, Silver AJ, Hilal SK, Conti SR: Computed tomography of the brainstem with intrathecal metrizamide. Part I. The normal brainstem. *AJNR* 4:1–11, 1983.

Naidich TP, Leeds NE, Kricheff II, Pudlowski RM, Naidich JB, Zimmerman RD: The tentorium in axial section. Normal CT appearance and non-neoplastic pathology. *Radiology* 123:631–638, 1977.

Osborn AG, McIff EB: Computed tomography of the nose. *Head and Neck Surg* 4:182–199, 1982.

Perman WH, Hilal SK, Simon HE, Maudsley AA: Contrast manipulation in NMR imaging. *Magnetic Resonance Imaging* 2:23–32, 1984.

Pernkopf E: *Atlas of Topographical and Applied Human Anatomy.* Philadelphia, W.B. Saunders Company, 1963.

Pinto RS, Kricheff II, Bergeron RT: Small acoustic neuromas: Detection by high resolution gas CT cisternography. *AJNR* 3:283–286, 1982.

Reede DL, Whelan MA, Bergeron RT: Computed tomography of the infrahyoid neck. *Radiology* 145:397–402, 1982.

Schnitzlein HN, Murtagh FR: *Imaging Anatomy of the Head and Spine.* Baltimore, Urban and Schwarzenberg, 1985.

Scholten ET, Hekster REM: Visualization of the craniocervical subarachnoid spaces. *Neuroradiology* 14:139–141, 1977.

Silver AJ, Mawad ME, Hilal SK, Sane P, Ganti SR: Computed tomography of the nasopharynx and related spaces. *Radiology* 147:725–731, 1983.

Steele JR, Hoffman JC: Brainstem evaluation with CT cisternography. *AJR* 135:287–292, 1981.

Valvassori GE, Potter GD, Hanafee WN: *Radiology of the Ear, Nose and Throat.* Philadelphia, W.B. Saunders Co., 1982.

Wehrli FW, MacFall JR, Glover GH, Grigsby N, Haughton, Johanson J: The dependence of nuclear magnetic resonance (NMR) image contrast on intrinsic and pulse sequence timing parameters. *Magnetic Resonance Imaging* 2:3–16, 1984.

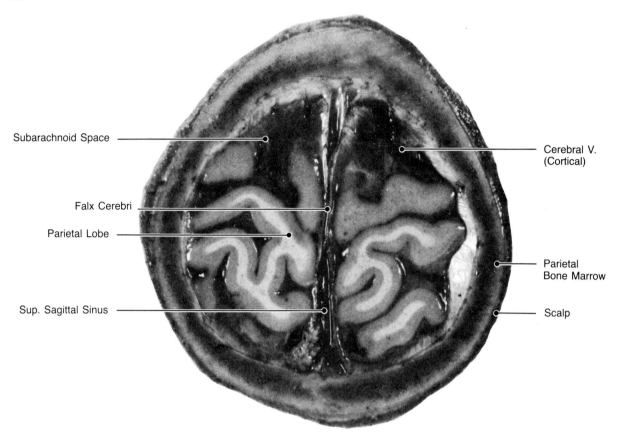

Subarachnoid Space

Cerebral V. (Cortical)

Falx Cerebri

Parietal Lobe

Parietal Bone Marrow

Sup. Sagittal Sinus

Scalp

1A Axial/Anatomic Plate/Vertex. This anatomic axial (+20 degrees) plate is made near the skull vertex. The superior sagittal sinus is formed by two layers of dura with the intervening channel of the dural venous sinus. The cortical draining veins cross from the subarachnoid space through the subdural space into the large venous channel.

1B Axial/CT/Vertex. The falx cerebri and superior sagittal sinus are high density regions on this axial (+20 degrees) intravenous contrast–enhanced vertex section. Because the superior sagittal sinus is a large vascular structure, it should enhance densely with intravenous contrast; however, the sinus commonly appears slightly denser than the brain on noncontrast studies as well owing to dense blood.

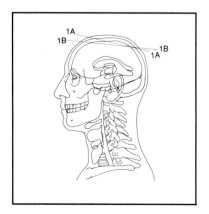

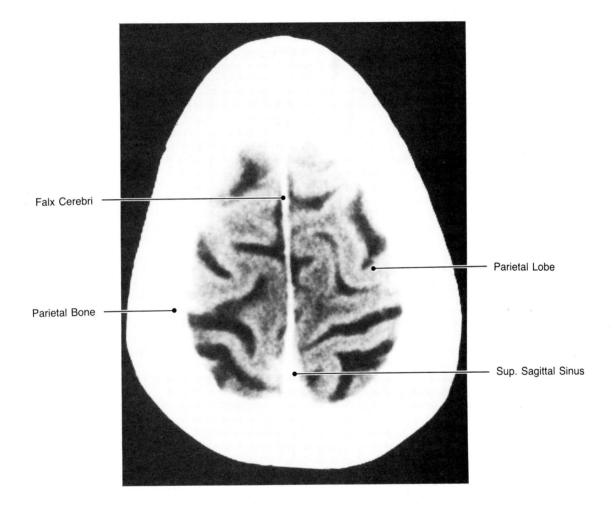

Falx Cerebri

Parietal Lobe

Parietal Bone

Sup. Sagittal Sinus

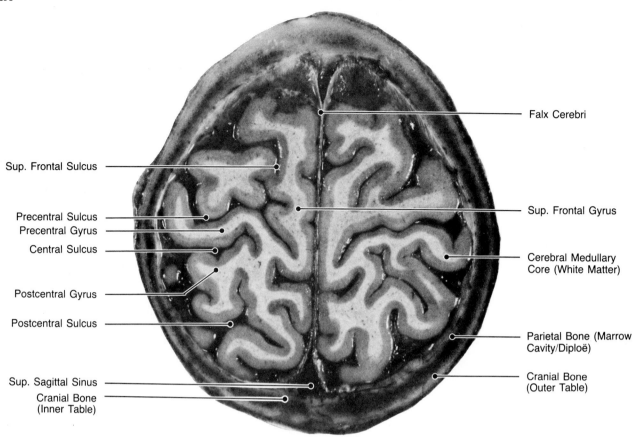

Falx Cerebri

Sup. Frontal Sulcus

Precentral Sulcus
Precentral Gyrus
Central Sulcus

Postcentral Gyrus

Postcentral Sulcus

Sup. Sagittal Sinus
Cranial Bone
(Inner Table)

Sup. Frontal Gyrus

Cerebral Medullary
Core (White Matter)

Parietal Bone (Marrow
Cavity/Diploë)

Cranial Bone
(Outer Table)

2A Axial/Anatomic Plate/Central Sulcus. This anatomic plate is made in the axial plane (+20 degrees) slightly inferior to the vertex. Depending on the section plane, the gyri in this region will be displayed differently. It is frequently difficult to recognize the location of the central sulcus on the axial plane. The sharp distinction between the gray and white matter tracts of the frontal and the parietal lobes are well demonstrated.

2B Axial/CT/Central Sulcus. This intravenous contrast–enhanced axial (+20 degrees) CT scan is made through the superior frontal parietal region. Recognition of the central sulcus on all patients is not possible owing to variations in the configuration of the gyri. One landmark that can be used is the L shaped configuration of the intersection of the superior frontal sulcus and the precentral sulcus.

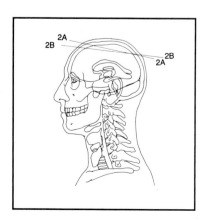

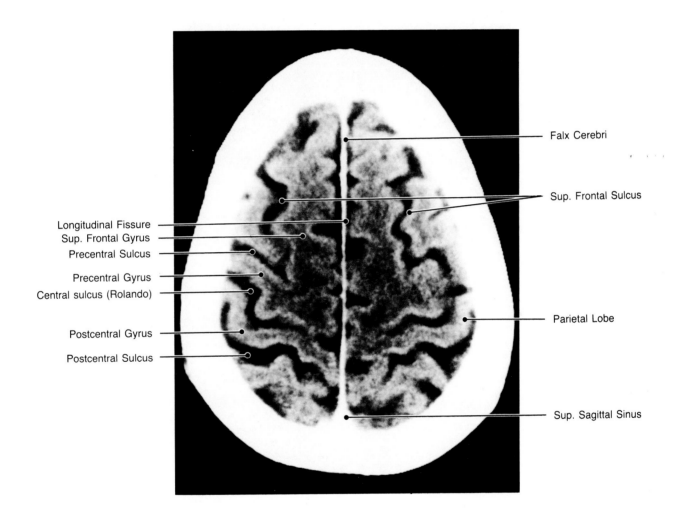

Falx Cerebri

Sup. Frontal Sulcus

Longitudinal Fissure

Sup. Frontal Gyrus

Precentral Sulcus

Precentral Gyrus

Central sulcus (Rolando)

Postcentral Gyrus

Postcentral Sulcus

Parietal Lobe

Sup. Sagittal Sinus

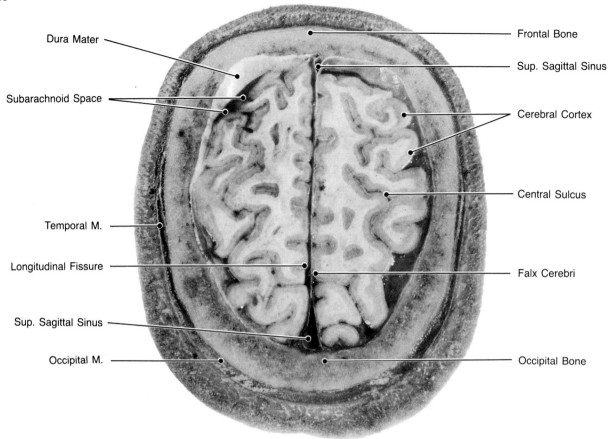

Dura Mater — Frontal Bone

Sup. Sagittal Sinus

Subarachnoid Space — Cerebral Cortex

Central Sulcus

Temporal M.

Longitudinal Fissure — Falx Cerebri

Sup. Sagittal Sinus

Occipital M. — Occipital Bone

3A Axial/Anatomic Plate/ Mid-Superior Parietal. This axial anatomic (−10 degrees) plate demonstrates the thin subarachnoid space of the longitudinal fissure that parallels the falx cerebri. The subarachnoid space is continuous with the remaining cisternal spaces over the convexities.

3B Axial/CT/Mid-Superior Parietal. The spinal fluid in the longitudinal fissure is seen as a low density region on this intravenous contrast–enhanced axial (+20 degrees) CT scan. The white matter tracts have a slightly lower density, while the gray matter has a higher density because it is the densest soft tissue component of the brain. The differences in the attenuation of the x-ray beam account for the contrast on the image.

3C Axial/MRI/Mid-Superior Parietal. This axial (−10 degrees) MRI scan is made with a spin echo technique of TR 2500 and TE 25 msec through the superior frontoparietal region. Note the signal void from the flowing blood in the superior sagittal sinus and the cortical veins extending from the falx, particularly in the posterior parietal region. With this particular pulse sequence technique the white matter structures are slightly lower in intensity than the adjacent gray matter gyri, which appear as serpentine stripes.

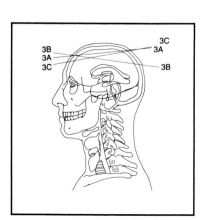

3B

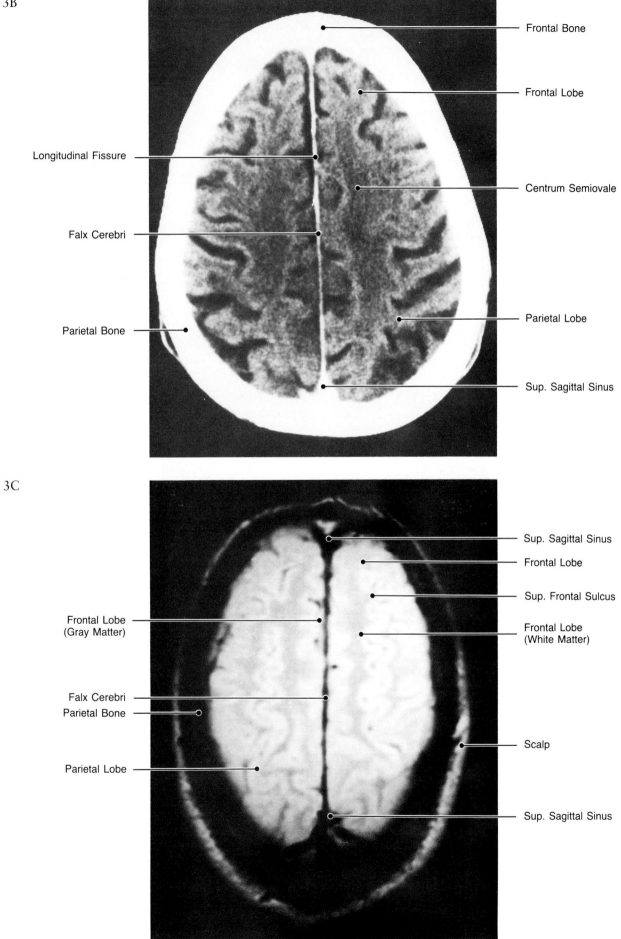

Frontal Bone

Frontal Lobe

Longitudinal Fissure

Centrum Semiovale

Falx Cerebri

Parietal Lobe

Parietal Bone

Sup. Sagittal Sinus

3C

Sup. Sagittal Sinus

Frontal Lobe

Sup. Frontal Sulcus

Frontal Lobe
(Gray Matter)

Frontal Lobe
(White Matter)

Falx Cerebri

Parietal Bone

Scalp

Parietal Lobe

Sup. Sagittal Sinus

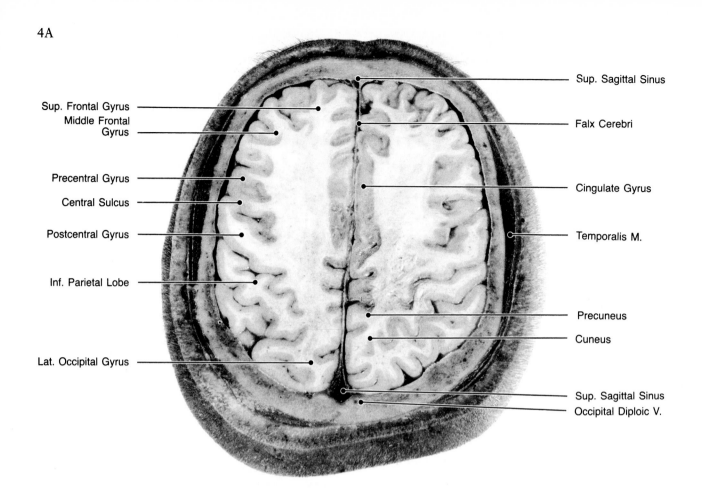

Sup. Sagittal Sinus

Sup. Frontal Gyrus

Middle Frontal Gyrus

Falx Cerebri

Precentral Gyrus

Central Sulcus

Cingulate Gyrus

Postcentral Gyrus

Temporalis M.

Inf. Parietal Lobe

Precuneus

Cuneus

Lat. Occipital Gyrus

Sup. Sagittal Sinus

Occipital Diploic V.

4A Axial/Anatomic Plate/Centrum Semiovale. This axial (−10 degrees) anatomic plate is made just superior to the lateral ventricles through the region of the large white matter tracts. Discoloration of the white matter tracts involving the left posterior parietal region is related to a prior brain infarction.

4B Axial/CT/Centrum Semiovale. The white matter tracts lying lateral and superior to the lateral ventricles are described as the centrum semiovale and are well seen on this axial (+20 degrees) intravenous contrast–enhanced CT scan. The small circular high densities seen just anterior to the corpus callosum are the callosal branches of the anterior cerebral arteries.

4C Axial/MRI/Centrum Semiovale. This axial MRI (−10 degrees) is made through the centrum semiovale in the fronto-parietal regions above the lateral ventricles using a partial saturation technique of TR 800 and TE 25 msec. With this particular imaging technique the white matter tracts are seen as high signal intensity regions outlined by slightly lower signal intensity gray matter gyri. Blood flowing within the vessels, such as the anterior cerebral artery is seen as serpentine low signal voids.

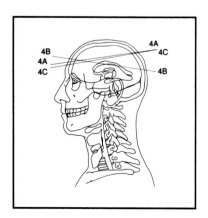

4B

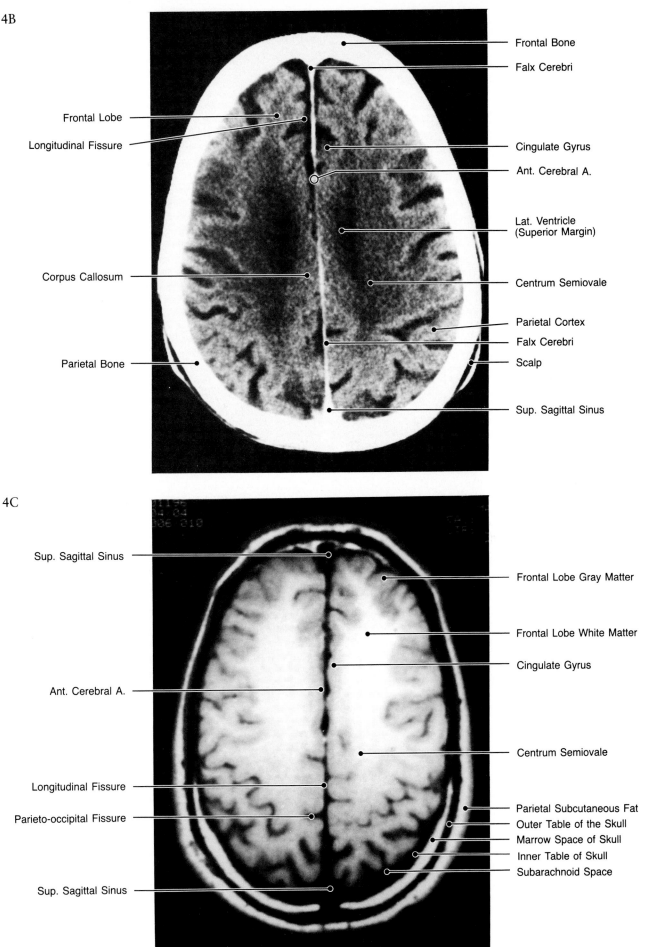

Frontal Bone

Falx Cerebri

Frontal Lobe

Longitudinal Fissure

Cingulate Gyrus

Ant. Cerebral A.

Lat. Ventricle
(Superior Margin)

Corpus Callosum

Centrum Semiovale

Parietal Cortex

Falx Cerebri

Parietal Bone

Scalp

Sup. Sagittal Sinus

4C

Sup. Sagittal Sinus

Frontal Lobe Gray Matter

Frontal Lobe White Matter

Cingulate Gyrus

Ant. Cerebral A.

Centrum Semiovale

Longitudinal Fissure

Parietal Subcutaneous Fat

Parieto-occipital Fissure

Outer Table of the Skull

Marrow Space of Skull

Inner Table of Skull

Subarachnoid Space

Sup. Sagittal Sinus

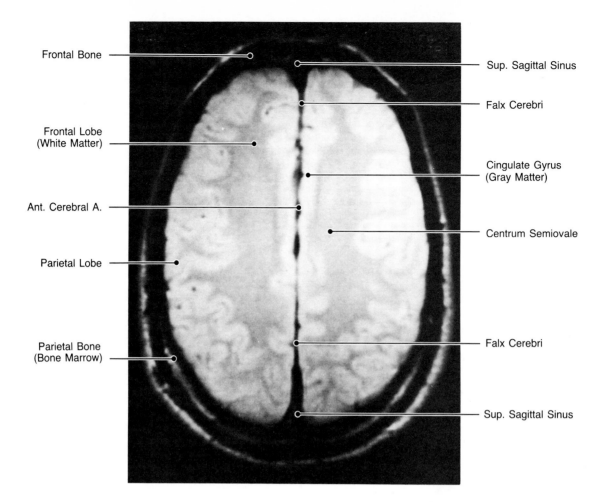

4D

Frontal Bone

Frontal Lobe
(White Matter)

Ant. Cerebral A.

Parietal Lobe

Parietal Bone
(Bone Marrow)

Sup. Sagittal Sinus

Falx Cerebri

Cingulate Gyrus
(Gray Matter)

Centrum Semiovale

Falx Cerebri

Sup. Sagittal Sinus

4D Axial/MRI/Deep White Matter Cerebral Hemisphere. On this spin echo TR 2500 and TE 25 msec axial (−10 degrees) MRI the white matter tracts are seen as low signal intensity regions that are "butterfly wing" shaped in configuration. The gray matter is slightly higher in signal intensity.

4E Axial/MRI/Corona Radiata. This axial (−10 degrees) MRI through the white matter tracts above the lateral ventricles is made with a spin echo technique of TR 2500 and TE 75 msec. Utilizing a long repetition time and an extended echo delay time, the spinal fluid spaces demonstrate the highest signal and are visible as serpentine regions between the gyri of the brain. Note that the distinction between the spinal fluid spaces and the other structures is distinctly more evident than on many of the other pulse sequence techniques.

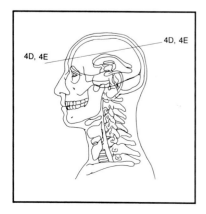

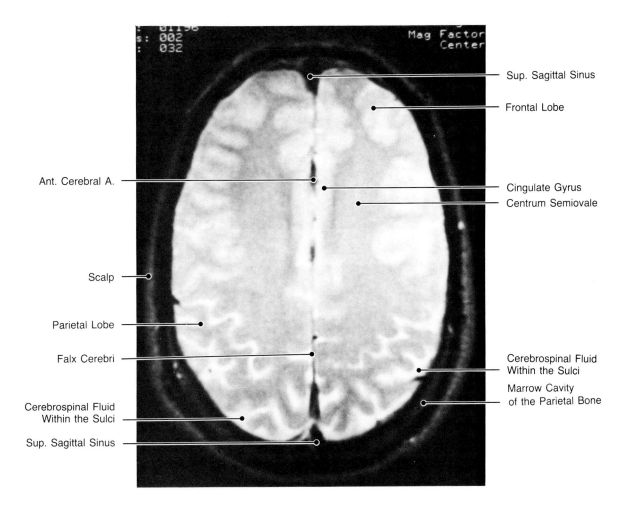

Sup. Sagittal Sinus

Frontal Lobe

Ant. Cerebral A.

Cingulate Gyrus

Centrum Semiovale

Scalp

Parietal Lobe

Falx Cerebri

Cerebrospinal Fluid
Within the Sulci

Marrow Cavity
of the Parietal Bone

Cerebrospinal Fluid
Within the Sulci

Sup. Sagittal Sinus

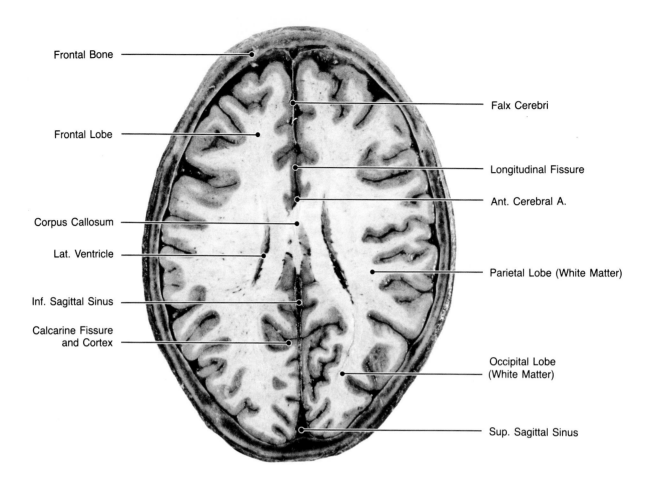

Frontal Bone

Falx Cerebri

Frontal Lobe

Longitudinal Fissure

Ant. Cerebral A.

Corpus Callosum

Lat. Ventricle

Parietal Lobe (White Matter)

Inf. Sagittal Sinus

Calcarine Fissure
and Cortex

Occipital Lobe
(White Matter)

Sup. Sagittal Sinus

5A Axial/Anatomic Plate/Corpus Callosum. This axial anatomic plate (+20 degrees) is made through the superior margins of the lateral ventricles and the corpus callosum. The corpus callosum is seen as a "butterfly wing"–shaped white matter structure with transverse fiber tracts interposed between the two arc-shaped lateral ventricles. Note that posteriorly the inferior sagittal sinus is nearly in direct continuity with the undersurface of the corpus callosum and the falx cerebri extends nearly through the complete longitudinal fissure, while anteriorly the falx cerebri does not extend as deeply.

5B Axial/CT/Superior Corpus Callosum. This intravenous contrast–enhanced axial (+20 degrees) CT demonstrates the lateral ventricles as "butterfly"-shaped spinal fluid structures divided in the midline by the septum pellucidum. The anterior and posterior portions of the corpus callosum are seen as V-shaped white matter bands at the anterior and posterior lateral margins of the lateral ventricles.

5C Axial/MRI/Superior Corpus Callosum. This axial (−10 degrees) partial saturation TR 800 and TE 25 msec image demonstrates the multiple layer appearance of the scalp and skull structures. With this pulse sequence technique, the fat-containing structures are all intense. The scalp is the most superficial layer, and the cortical bone of the outer table of the skull is seen as a low intensity region. The next layer is a high signal intensity region related to the fatty marrow within the diploic space of the skull. Finally, there is another low signal region related to the cortical inner table of the skull as well as the underlying cerebral spinal fluid.

5D Axial/MRI/Corpus Callosum. This axial (−10 degrees) spin echo TR 2500, TE 25 msec image does not demonstrate the ventricular structures as clearly as the corresponding partial saturation image made at the same level. The advantage of this pulse sequence technique, however, is enhancement of the differentiation of the gray and white matter structures.

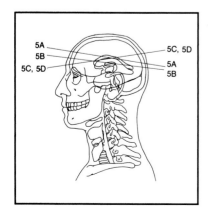

5A
5B
5C, 5D

5C, 5D
5A
5B

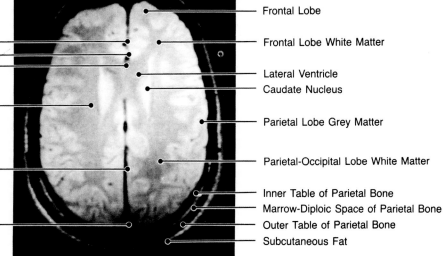

5B

Frontal Bone

Frontal Lobe

Septum Pellucidum

Centrum Semiovale

Lat. Ventricle (Body)

Parietal Lobe

Splenium of the Corpus Callosum

Falx Cerebri (Anterior)

Longitudinal Fissure

Corpus Callosum (Genu)

Ant. Cerebral A.

Subependymal V.

Parietal Lobe White Matter

Falx Cerebri

Occipital Lobe

Sup. Sagittal Sinus

5C

Cingulate Gyrus

Corpus Callosum (Genu)

Deep White Matter

Parietal Lobe Gray Matter

Occipital Lobe

Falx Cerebri

Sup. Sagittal Sinus

Frontal Bone

Longitudinal Fissure (Ant. Portion)

Frontal Lobe White Matter

Pericallosal A.

Body of the Lat. Ventricle

Inner Table of Parietal Bone

Marrow-Diploic Space of Parietal Bone

Outer Table of Parietal Bone

Subcutaneous Fat

5D

Longitudinal Fissure

Pericallosal A.

Corpus Callosum (Genu)

Parietal Lobe White Matter

Falx Cerebri

Sup. Sagittal Sinus

Frontal Lobe

Frontal Lobe White Matter

Lateral Ventricle

Caudate Nucleus

Parietal Lobe Grey Matter

Parietal-Occipital Lobe White Matter

Inner Table of Parietal Bone

Marrow-Diploic Space of Parietal Bone

Outer Table of Parietal Bone

Subcutaneous Fat

15

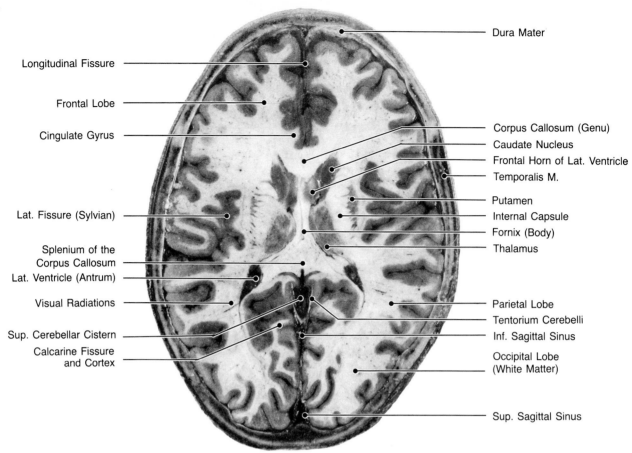

Labels (clockwise/left and right):

Longitudinal Fissure

Frontal Lobe

Cingulate Gyrus

Lat. Fissure (Sylvian)

Splenium of the Corpus Callosum

Lat. Ventricle (Antrum)

Visual Radiations

Sup. Cerebellar Cistern

Calcarine Fissure and Cortex

Dura Mater

Corpus Callosum (Genu)

Caudate Nucleus

Frontal Horn of Lat. Ventricle

Temporalis M.

Putamen

Internal Capsule

Fornix (Body)

Thalamus

Parietal Lobe

Tentorium Cerebelli

Inf. Sagittal Sinus

Occipital Lobe (White Matter)

Sup. Sagittal Sinus

6A Axial/Anatomic Plate/Fornix. This axial (+20 degrees) anatomic plate through the mid lateral ventricles demonstrates the V-shaped configuration of the anterior and posterior margins of the corpus callosum. Note that the bodies of the fornix parallel the septum pellucidum anteriorly but diverge beneath the corpus callosum to parallel the lateral ventricle in the trigone region. The most superior portions of the basal ganglia and thalamus are also seen on this section.

6B Axial/CT/Mid-Corpus Callosum. This contrast-enhanced axial (+20 degrees) CT demonstrates a number of high signal foci related to vascular structures. Note the anterior cerebral arteries just anterior to the corpus callosum in the longitudinal fissure. The subependymal veins are frequently seen paralleling the outer margins of the lateral ventricles. A few linear serpentine branches of the middle cerebral artery are also visible in the sylvian fissures. The choroid plexus within the lateral ventricles stands out intensely because of its high vascularity.

6C Axial/MRI/Corpus Callosum. The signal arising from the frontal bone on this axial (−10 degrees) partial saturation TR 800 and TE 25 msec MRI is quite variable. The more lateral segments of the frontal bone are filled with marrow and have a high signal, while the more central frontal bone is of low signal intensity because of the cortical bone and air of the frontal sinus. Unless fluid or some other soft structure fills the sinus, its exact margins may not be evident on MRI.

6D Axial/MRI/Corona Radiata. This axial (−10 degrees) spin echo TR 2500 and TE 75 msec image highlights the spinal fluid spaces as high signal intensity regions. The separate layers of the cranial contents starting from superficial to deep include the high signal spinal fluid, intermediate signal intensity from the gray matter structures, and finally low signal intensity from the white matter tracts are seen. This type of imaging technique is particularly sensitive for identification of brain pathology that produces increased water content of the brain, such as demyelination, edema, or tumor.

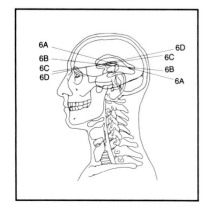

6A
6B
6C
6D
6D
6C
6B
6A

6B

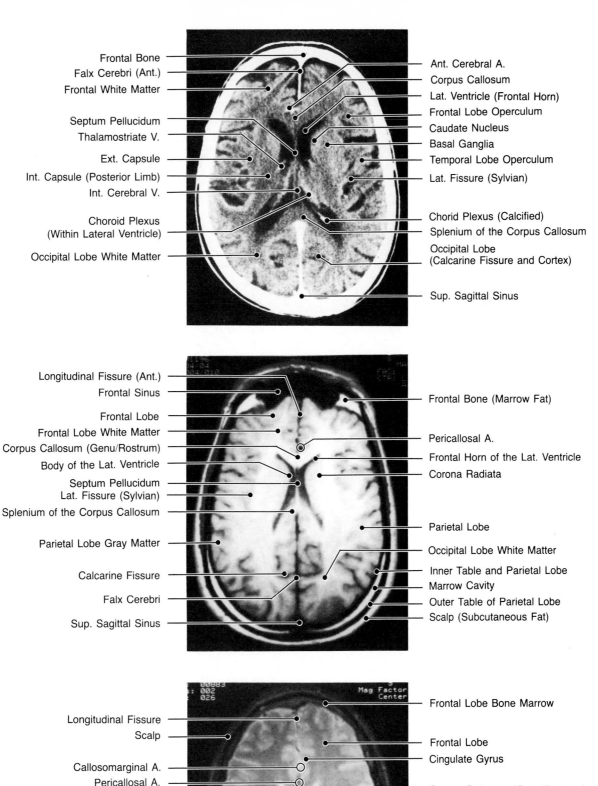

Frontal Bone
Falx Cerebri (Ant.)
Frontal White Matter

Septum Pellucidum
Thalamostriate V.

Ext. Capsule
Int. Capsule (Posterior Limb)
Int. Cerebral V.

Choroid Plexus
(Within Lateral Ventricle)

Occipital Lobe White Matter

Ant. Cerebral A.
Corpus Callosum
Lat. Ventricle (Frontal Horn)
Frontal Lobe Operculum
Caudate Nucleus
Basal Ganglia
Temporal Lobe Operculum
Lat. Fissure (Sylvian)

Chorid Plexus (Calcified)
Splenium of the Corpus Callosum

Occipital Lobe
(Calcarine Fissure and Cortex)

Sup. Sagittal Sinus

6C

Longitudinal Fissure (Ant.)
Frontal Sinus

Frontal Lobe
Frontal Lobe White Matter
Corpus Callosum (Genu/Rostrum)
Body of the Lat. Ventricle
Septum Pellucidum
Lat. Fissure (Sylvian)
Splenium of the Corpus Callosum

Parietal Lobe Gray Matter

Calcarine Fissure

Falx Cerebri

Sup. Sagittal Sinus

Frontal Bone (Marrow Fat)

Pericallosal A.

Frontal Horn of the Lat. Ventricle
Corona Radiata

Parietal Lobe

Occipital Lobe White Matter
Inner Table and Parietal Lobe
Marrow Cavity
Outer Table of Parietal Lobe
Scalp (Subcutaneous Fat)

6D

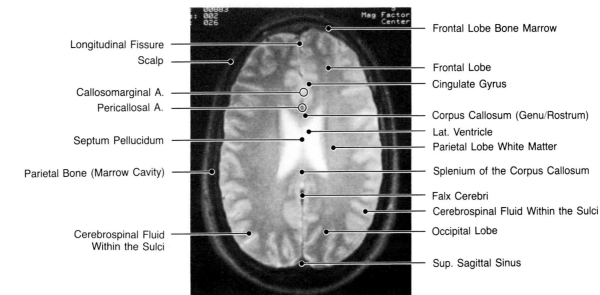

Longitudinal Fissure
Scalp

Callosomarginal A.
Pericallosal A.

Septum Pellucidum

Parietal Bone (Marrow Cavity)

Cerebrospinal Fluid
Within the Sulci

Frontal Lobe Bone Marrow

Frontal Lobe
Cingulate Gyrus

Corpus Callosum (Genu/Rostrum)
Lat. Ventricle
Parietal Lobe White Matter

Splenium of the Corpus Callosum

Falx Cerebri
Cerebrospinal Fluid Within the Sulci

Occipital Lobe

Sup. Sagittal Sinus

17

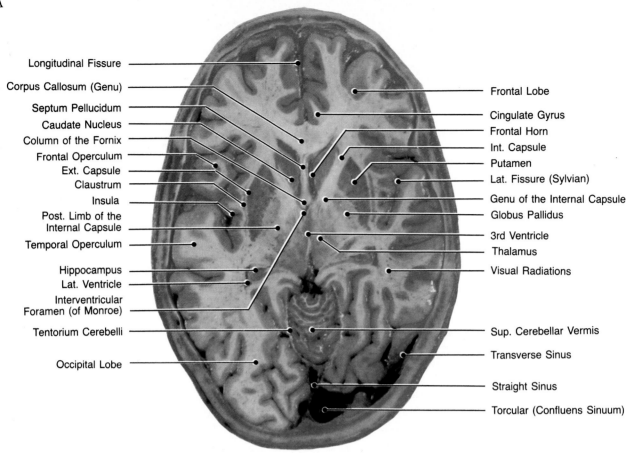

Longitudinal Fissure	Frontal Lobe
Corpus Callosum (Genu)	Cingulate Gyrus
Septum Pellucidum	Frontal Horn
Caudate Nucleus	Int. Capsule
Column of the Fornix	Putamen
Frontal Operculum	Lat. Fissure (Sylvian)
Ext. Capsule	Genu of the Internal Capsule
Claustrum	Globus Pallidus
Insula	3rd Ventricle
Post. Limb of the Internal Capsule	Thalamus
Temporal Operculum	Visual Radiations
Hippocampus	
Lat. Ventricle	
Interventricular Foramen (of Monroe)	
Tentorium Cerebelli	Sup. Cerebellar Vermis
	Transverse Sinus
Occipital Lobe	Straight Sinus
	Torcular (Confluens Sinuum)

7A Axial/Anatomic Plate/Internal Capsule. The multiple layers of the brain are well demonstrated in the central portion of this axial (+20 degrees) anatomic plate. The frontal and temporal opercula cover the gray matter insula of the sylvian fissure. The claustrum and external capsule form the layers just lateral to the basal ganglia. The internal capsule forms a V-shaped white matter tract just lateral to the caudate nucleus and thalamus.

7B Axial/Pixel Highlight CT/Internal Capsule. This axial (0 degrees) CT scan through the region of the internal capsule utilizes a technique in which all of the pixels with Hounsfield units similar to white matter of the anterior corpus callosum are highlighted.

7C Axial/MRI/Internal Capsule. This partial saturation TR 1000 and TE 25 msec axial (0 degrees) MRI is made through the basal ganglia and internal capsule. The white matter tracts are seen as high signal regions whereas the gray matter structures have a lower signal intensity. The multiple layers, including the insula, basal ganglia, and internal capsule, are all well demonstrated.

7D Axial/MRI/Inferior Basal Ganglion. The sylvian fissures are bordered medially by the insula on this partial saturation TR 1000 and TE 25 msec axial (0 degrees) MRI. Laterally, the sylvian fissures are covered by overgrowth of the adjacent frontal and temporal lobes. Because these segments of the brain form as lips, they are described as the operculum.

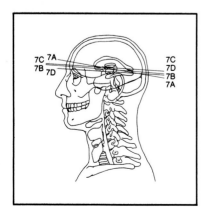

7B

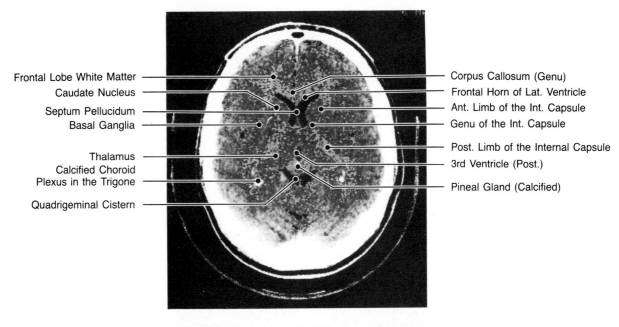

Frontal Lobe White Matter —   — Corpus Callosum (Genu)

Caudate Nucleus —   — Frontal Horn of Lat. Ventricle

Septum Pellucidum —   — Ant. Limb of the Int. Capsule

Basal Ganglia —   — Genu of the Int. Capsule

   — Post. Limb of the Internal Capsule

Thalamus —

Calcified Choroid   — 3rd Ventricle (Post.)

Plexus in the Trigone —

   — Pineal Gland (Calcified)

Quadrigeminal Cistern —

7C

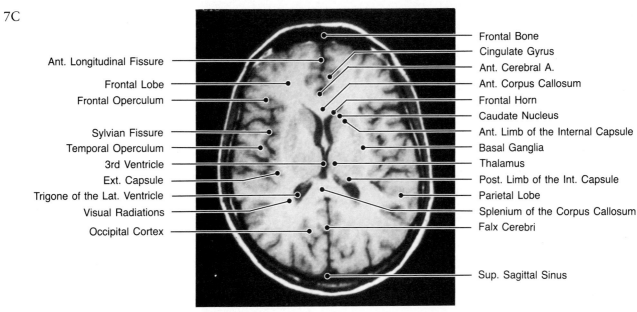

Ant. Longitudinal Fissure —   — Frontal Bone

   — Cingulate Gyrus

   — Ant. Cerebral A.

Frontal Lobe —   — Ant. Corpus Callosum

Frontal Operculum —   — Frontal Horn

   — Caudate Nucleus

Sylvian Fissure —   — Ant. Limb of the Internal Capsule

Temporal Operculum —   — Basal Ganglia

3rd Ventricle —   — Thalamus

Ext. Capsule —   — Post. Limb of the Int. Capsule

Trigone of the Lat. Ventricle —   — Parietal Lobe

Visual Radiations —   — Splenium of the Corpus Callosum

Occipital Cortex —   — Falx Cerebri

   — Sup. Sagittal Sinus

7D

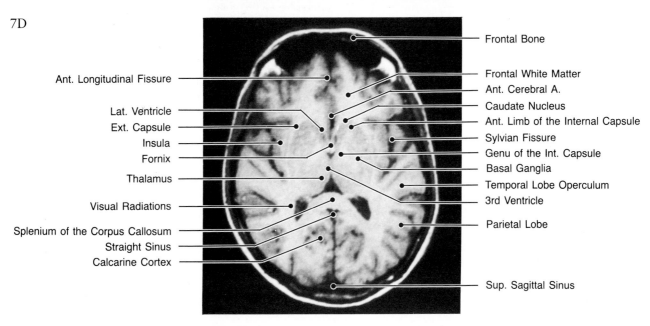

   — Frontal Bone

Ant. Longitudinal Fissure —   — Frontal White Matter

   — Ant. Cerebral A.

Lat. Ventricle —   — Caudate Nucleus

Ext. Capsule —   — Ant. Limb of the Internal Capsule

Insula —   — Sylvian Fissure

Fornix —   — Genu of the Int. Capsule

   — Basal Ganglia

Thalamus —   — Temporal Lobe Operculum

   — 3rd Ventricle

Visual Radiations —

Splenium of the Corpus Callosum —   — Parietal Lobe

Straight Sinus —

Calcarine Cortex —

   — Sup. Sagittal Sinus

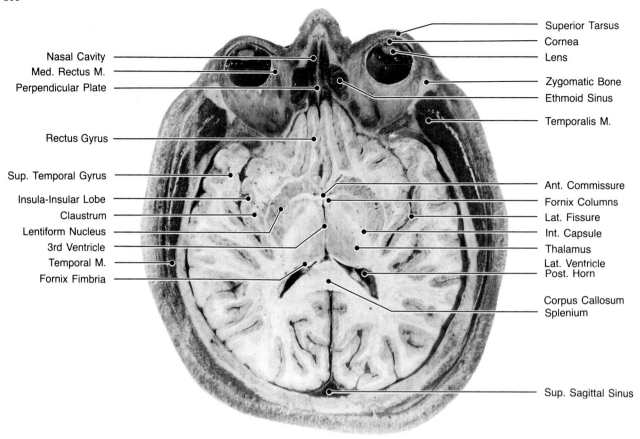

Superior Tarsus
Cornea
Lens
Zygomatic Bone
Ethmoid Sinus
Temporalis M.

Nasal Cavity
Med. Rectus M.
Perpendicular Plate

Rectus Gyrus

Sup. Temporal Gyrus
Insula-Insular Lobe
Claustrum
Lentiform Nucleus
3rd Ventricle
Temporal M.
Fornix Fimbria

Ant. Commissure
Fornix Columns
Lat. Fissure
Int. Capsule
Thalamus
Lat. Ventricle
Post. Horn

Corpus Callosum
Splenium

Sup. Sagittal Sinus

8A Axial/Anatomic Plate/Anterior Commissure. The close relationship of the anterior commissure extending through the basal ganglion region and the ascending columns of the fornix is demonstrated on this axial (−10 degrees) anatomic plate. The columns of the fornix arise in the mamillary body slightly inferiorly and ascend to form the margins of the foramen of Monroe superiorly. The rectus gyri are seen as linear gyri paralleling the midline.

8B Axial/MRI/Inferior Third Ventricle. This partial saturation TR 800 and TE 25 msec axial (−10 degrees) MRI through the globes demonstrates the high signal character of the fat using this pulse sequence technique. The globe and extraocular muscles are outlined by orbital fat.

8C Axial/MRI/Inferior Third Ventricle. This axial (−10 degrees) MRI through the columns of the fornix is made with a spin echo technique of TR 2500 and TE 25 msec. The basal ganglia and white matter tracts have a low signal intensity. Even though the basal ganglia are not all white matter, they are similar in appearance because their higher iron content shortens their T2 relaxation times.

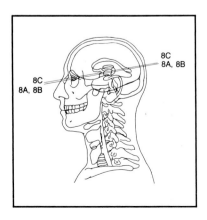

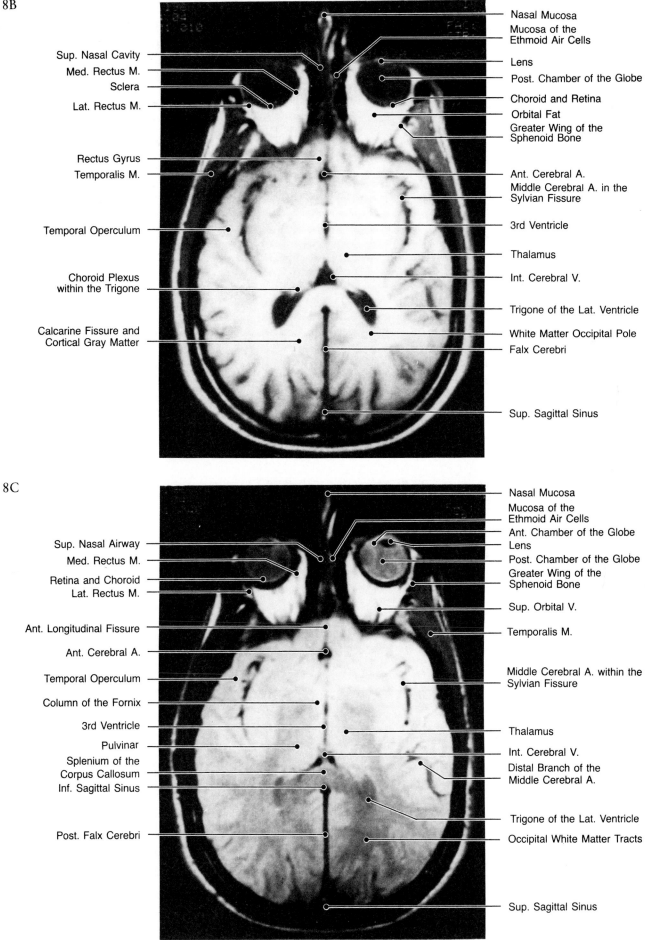

8B

Sup. Nasal Cavity
Med. Rectus M.
Sclera
Lat. Rectus M.

Rectus Gyrus
Temporalis M.

Temporal Operculum

Choroid Plexus
within the Trigone

Calcarine Fissure and
Cortical Gray Matter

Nasal Mucosa
Mucosa of the
Ethmoid Air Cells

Lens
Post. Chamber of the Globe

Choroid and Retina
Orbital Fat
Greater Wing of the
Sphenoid Bone

Ant. Cerebral A.
Middle Cerebral A. in the
Sylvian Fissure

3rd Ventricle

Thalamus

Int. Cerebral V.

Trigone of the Lat. Ventricle

White Matter Occipital Pole
Falx Cerebri

Sup. Sagittal Sinus

8C

Sup. Nasal Airway
Med. Rectus M.

Retina and Choroid
Lat. Rectus M.

Ant. Longitudinal Fissure

Ant. Cerebral A.

Temporal Operculum

Column of the Fornix

3rd Ventricle

Pulvinar
Splenium of the
Corpus Callosum
Inf. Sagittal Sinus

Post. Falx Cerebri

Nasal Mucosa
Mucosa of the
Ethmoid Air Cells
Ant. Chamber of the Globe
Lens
Post. Chamber of the Globe
Greater Wing of the
Sphenoid Bone

Sup. Orbital V.

Temporalis M.

Middle Cerebral A. within the
Sylvian Fissure

Thalamus

Int. Cerebral V.

Distal Branch of the
Middle Cerebral A.

Trigone of the Lat. Ventricle

Occipital White Matter Tracts

Sup. Sagittal Sinus

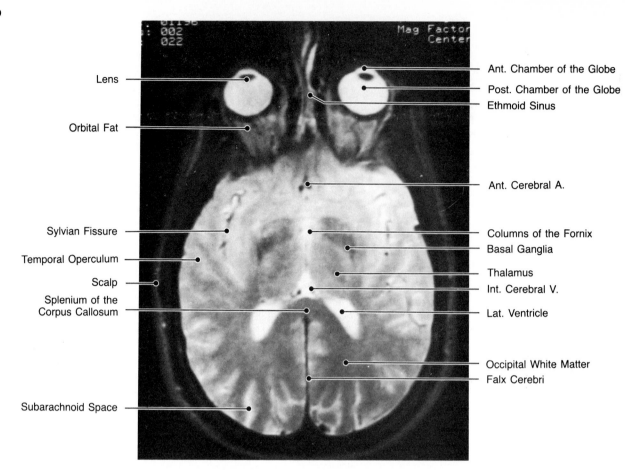

Lens
Orbital Fat
Sylvian Fissure
Temporal Operculum
Scalp
Splenium of the
Corpus Callosum
Subarachnoid Space

Ant. Chamber of the Globe
Post. Chamber of the Globe
Ethmoid Sinus
Ant. Cerebral A.
Columns of the Fornix
Basal Ganglia
Thalamus
Int. Cerebral V.
Lat. Ventricle
Occipital White Matter
Falx Cerebri

8D Axial/MRI/Columns of the Fornix. This axial (−10 degrees) MR image through the orbits is made with a spin echo technique of TR 2500 and TE 75 msec. With this pulse sequence technique all of the fluid spaces are seen to have a high signal character, including the anterior and posterior chambers of the globe and the ventricular structures. Note that many of the vascular structures, including the internal cerebral veins and middle cerebral arteries within the sylvian fissures, are highlighted by the increased signal from the surrounding adjacent tissues.

8E Axial/CT/Superior Orbits. This axial (0 degrees) CT image through the superior orbits is viewed with soft tissue windows. The orbital contents are outlined by low density fat. Lateral to the globe are the lacrimal glands, while medially the superior oblique muscle extends from the medial globe to the trochlea. The superior ophthalmic vein crosses below the superior rectus muscle towards the anterior medial orbit.

8F Axial/CT/Superior Orbits. This axial (0 degrees) high resolution thin section CT displayed at bone windows through the orbits superior to the anatomic plate demonstrates the medial and lateral orbital walls.

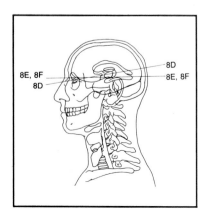

8E

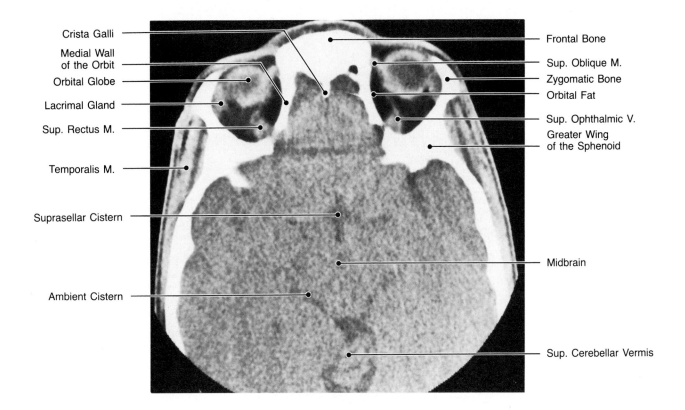

Crista Galli — Frontal Bone

Medial Wall of the Orbit — Sup. Oblique M.

Orbital Globe — Zygomatic Bone

Lacrimal Gland — Orbital Fat

Sup. Rectus M. — Sup. Ophthalmic V.

Greater Wing of the Sphenoid

Temporalis M.

Suprasellar Cistern

Midbrain

Ambient Cistern

Sup. Cerebellar Vermis

8F

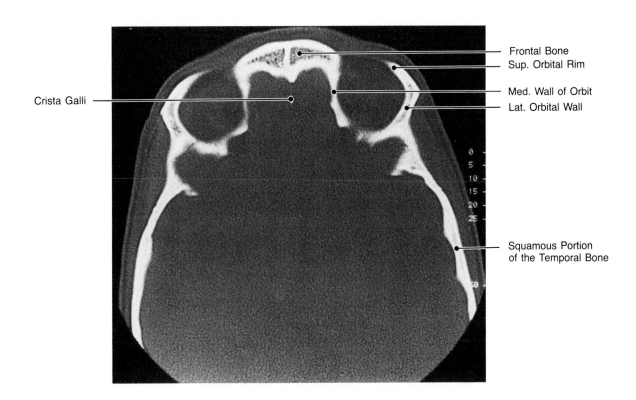

Crista Galli

Frontal Bone
Sup. Orbital Rim

Med. Wall of Orbit
Lat. Orbital Wall

0
5
10
15
20
25

Squamous Portion of the Temporal Bone

50

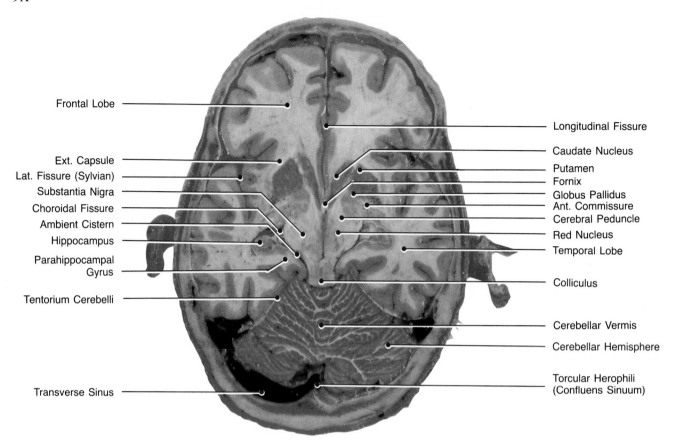

Frontal Lobe

Ext. Capsule
Lat. Fissure (Sylvian)
Substantia Nigra
Choroidal Fissure
Ambient Cistern
Hippocampus
Parahippocampal Gyrus

Tentorium Cerebelli

Transverse Sinus

Longitudinal Fissure
Caudate Nucleus
Putamen
Fornix
Globus Pallidus
Ant. Commissure
Cerebral Peduncle
Red Nucleus
Temporal Lobe

Colliculus

Cerebellar Vermis
Cerebellar Hemisphere

Torcular Herophili (Confluens Sinuum)

9A Axial/Anatomic Plate/Midbrain. There is a component of transtentorial herniation of the parahippocampal gyrus through the tentorial notch on this axial (+20 degrees) anatomic section made at the level of the midbrain secondary to post-mortem brain edema. There is some distortion of the colliculi as well. The confluence of the transverse sinus and the superior sagittal sinus at the level of the torcula is demonstrated. Frequently there is asymmetry between the size of the transverse sinuses, with the right usually being larger than the left.

9B Axial/CT/Superior Midbrain. This intravenous contrast–enhanced axial (+10 degrees) CT scan is made through the superior midbrain region. There are a number of high density structures identified on this image, including the anterior falx cerebri. Occasionally, the falx cerebri is extensively calci-

fied, particularly in older age. Bilateral calcifications within the choroid plexus in the trigones of the lateral ventricles are commonly identified as well. A small focus of high density is seen in the anterior longitudinal fissure related to the anterior cerebral artery.

9C Axial/MRI/Inferior Third Ventricle This axial (+0 degrees) partial saturation TR 1000 and TE 25 msec MRI demonstrates a number of the midline structures well. This section is made at a slightly different angle than the anatomic plate. Anteriorly the anterior longitudinal fissure is seen outlined by adjacent gray matter structures. Separation of the anterior third ventricle from the longitudinal fissure is difficult, since the third ventricle is only marginated by the thin lamina terminalis anteriorly. Just posterior to the third ventricle is a small soft tissue density in the superior cerebellar cistern related to the pineal gland.

9D Axial/MRI/Midbrain. This axial (0 degrees) partial saturation TR 1000 and TE 25 msec MRI is slightly inferior to the anatomic plate and is made through the upper midbrain. The optic tracts of the suprasellar cistern are seen as V-shaped linear densities extending laterally and posteriorly. The optic tracts terminate in the geniculate bodies, which form small protuberances on the lateral posterior portion of the midbrain abutting the ambient cistern. The more midline protuberances of the brainstem are related to the colliculi.

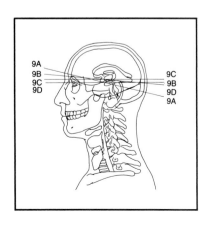

9A
9B
9C
9D

9C
9B
9D
9A

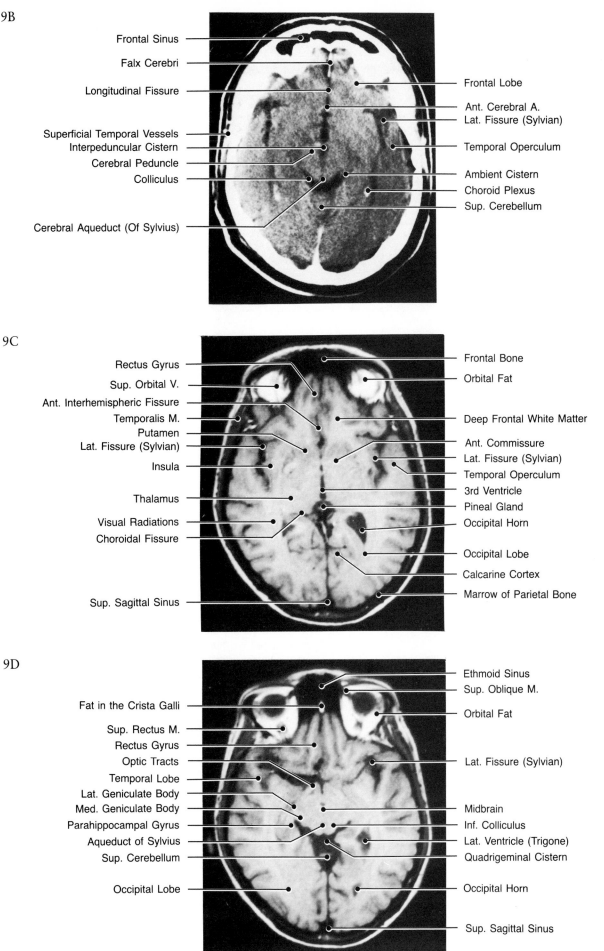

9B

Frontal Sinus
Falx Cerebri
Longitudinal Fissure
Superficial Temporal Vessels
Interpeduncular Cistern
Cerebral Peduncle
Colliculus
Cerebral Aqueduct (Of Sylvius)

Frontal Lobe
Ant. Cerebral A.
Lat. Fissure (Sylvian)
Temporal Operculum
Ambient Cistern
Choroid Plexus
Sup. Cerebellum

9C

Rectus Gyrus
Sup. Orbital V.
Ant. Interhemispheric Fissure
Temporalis M.
Putamen
Lat. Fissure (Sylvian)
Insula
Thalamus
Visual Radiations
Choroidal Fissure
Sup. Sagittal Sinus

Frontal Bone
Orbital Fat
Deep Frontal White Matter
Ant. Commissure
Lat. Fissure (Sylvian)
Temporal Operculum
3rd Ventricle
Pineal Gland
Occipital Horn
Occipital Lobe
Calcarine Cortex
Marrow of Parietal Bone

9D

Fat in the Crista Galli
Sup. Rectus M.
Rectus Gyrus
Optic Tracts
Temporal Lobe
Lat. Geniculate Body
Med. Geniculate Body
Parahippocampal Gyrus
Aqueduct of Sylvius
Sup. Cerebellum
Occipital Lobe

Ethmoid Sinus
Sup. Oblique M.
Orbital Fat
Lat. Fissure (Sylvian)
Midbrain
Inf. Colliculus
Lat. Ventricle (Trigone)
Quadrigeminal Cistern
Occipital Horn
Sup. Sagittal Sinus

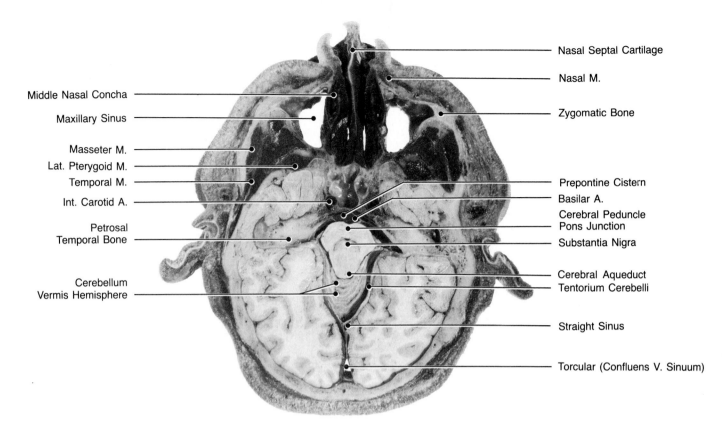

Nasal Septal Cartilage

Nasal M.

Zygomatic Bone

Middle Nasal Concha

Maxillary Sinus

Masseter M.
Lat. Pterygoid M.
Temporal M.
Int. Carotid A.

Petrosal
Temporal Bone

Cerebellum
Vermis Hemisphere

Prepontine Cistern

Basilar A.

Cerebral Peduncle
Pons Junction

Substantia Nigra

Cerebral Aqueduct
Tentorium Cerebelli

Straight Sinus

Torcular (Confluens V. Sinuum)

10A Axial/Anatomic Plate/Midbrain. The aqueduct of Sylvius is seen as a small punctate cavity in the posterior portion of the upper brainstem on this axial (-10 degrees) section. The substantia nigra is also visible as a dark region because of its high mineral content. Note the thin crescent of bone that separates the carotid arteries from the sphenoid sinus.

10B Axial/MRI/Midbrain. This partial saturation TR 800 and TE 25 msec axial (−10 degrees) MRI through the midbrain demonstrates the interpeduncular fossa, which lies between the two cerebral peduncles. The small oval-shaped lower signal regions in the mid portion of the midbrain are related to the red nuclei. The parahippocampal gyrus parallels the lateral margin of the brainstem at this level.

10C Axial/MRI/Midbrain. This spin echo TR 2500 and TE 25 msec axial (−10 degrees) MRI shows the pituitary gland lying between the cavernous carotid arteries. The posterior portion of the sella turcica contains a fat pad, which should not be confused with pathology. The mucoperiosteal coverings of the structures within the nasal cavity are well seen, including the nasolacrimal ducts.

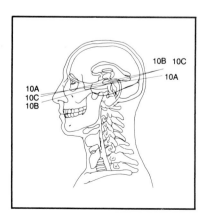

10B

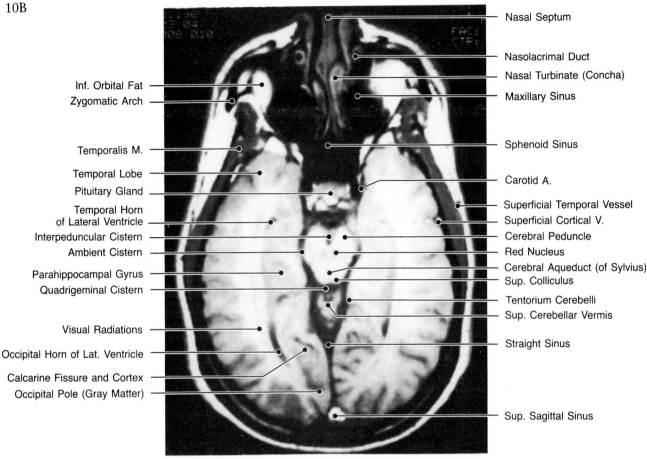

Nasal Septum

Nasolacrimal Duct

Nasal Turbinate (Concha)

Maxillary Sinus

Sphenoid Sinus

Carotid A.

Superficial Temporal Vessel

Superficial Cortical V.

Cerebral Peduncle

Red Nucleus

Cerebral Aqueduct (of Sylvius)

Sup. Colliculus

Tentorium Cerebelli

Sup. Cerebellar Vermis

Straight Sinus

Sup. Sagittal Sinus

Inf. Orbital Fat

Zygomatic Arch

Temporalis M.

Temporal Lobe

Pituitary Gland

Temporal Horn
of Lateral Ventricle

Interpeduncular Cistern

Ambient Cistern

Parahippocampal Gyrus

Quadrigeminal Cistern

Visual Radiations

Occipital Horn of Lat. Ventricle

Calcarine Fissure and Cortex

Occipital Pole (Gray Matter)

10C

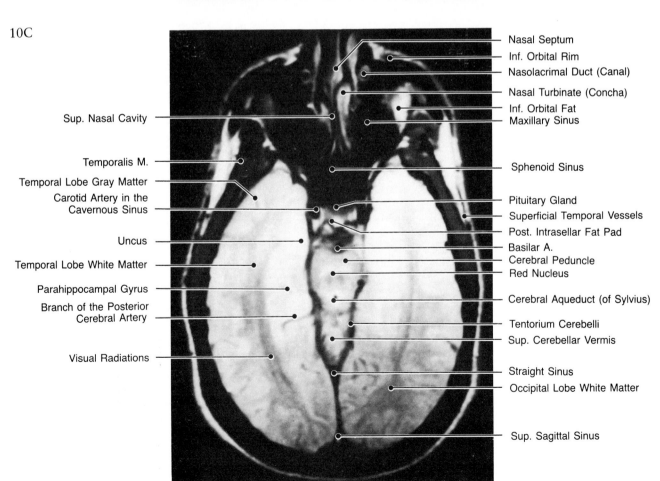

Nasal Septum

Inf. Orbital Rim

Nasolacrimal Duct (Canal)

Nasal Turbinate (Concha)

Inf. Orbital Fat

Maxillary Sinus

Sphenoid Sinus

Pituitary Gland

Superficial Temporal Vessels

Post. Intrasellar Fat Pad

Basilar A.

Cerebral Peduncle

Red Nucleus

Cerebral Aqueduct (of Sylvius)

Tentorium Cerebelli

Sup. Cerebellar Vermis

Straight Sinus

Occipital Lobe White Matter

Sup. Sagittal Sinus

Sup. Nasal Cavity

Temporalis M.

Temporal Lobe Gray Matter

Carotid Artery in the
Cavernous Sinus

Uncus

Temporal Lobe White Matter

Parahippocampal Gyrus

Branch of the Posterior
Cerebral Artery

Visual Radiations

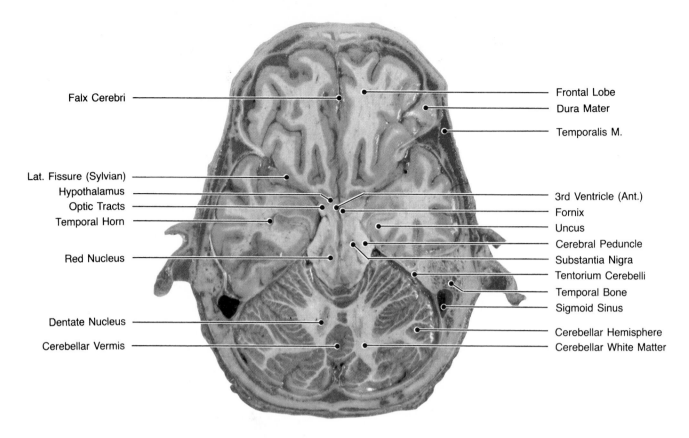

Falx Cerebri

Frontal Lobe
Dura Mater
Temporalis M.

Lat. Fissure (Sylvian)
Hypothalamus
Optic Tracts
Temporal Horn

Red Nucleus

3rd Ventricle (Ant.)
Fornix
Uncus
Cerebral Peduncle
Substantia Nigra
Tentorium Cerebelli
Temporal Bone
Sigmoid Sinus

Dentate Nucleus
Cerebellar Vermis

Cerebellar Hemisphere
Cerebellar White Matter

11A Axial Anatomic plate/Midbrain. Because of a component of postmortem cerebral edema, a number of the structures about the midbrain are distorted on this axial (+20 degrees) section with compression of the normal subarachnoid spaces. The red nuclei are seen as two oval regions in the paramidline location outlined laterally by the darker substantia nigra. The ascending columns of the fornix are seen adjacent to the anterior inferior third ventricle, which is itself outlined by the adjacent hypothalamus and optic tracts.

11B Axial/CT/Midbrain. This intravenous contrast–enhanced axial (+10 degrees) CT demonstrates a number of the vascular structures of the circle of Willis, including the anterior cerebral, supraclinoid carotid, basilar, and posterior cerebral arteries. A number of other vascular structures are also visible, including the basal vein of Rosenthal running in the posterior portion of the ambient cistern, the precentral cerebellar vein in the superior cerebellar cistern, and the choroid plexus in the temporal horn.

11C Axial/MRI/Mamillary Bodies. The mamillary bodies, optic tracts, and portions of the circle of Willis are seen surrounded by high signal intensity spinal fluid on this spin echo TR 2500, TE 75 msec axial (0 degrees) MRI. The posterior cerebral arteries are seen as low signal regions, which parallel the cerebral peduncles in the ambient cisterns.

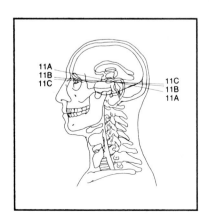

11A
11B
11C

11C
11B
11A

11B

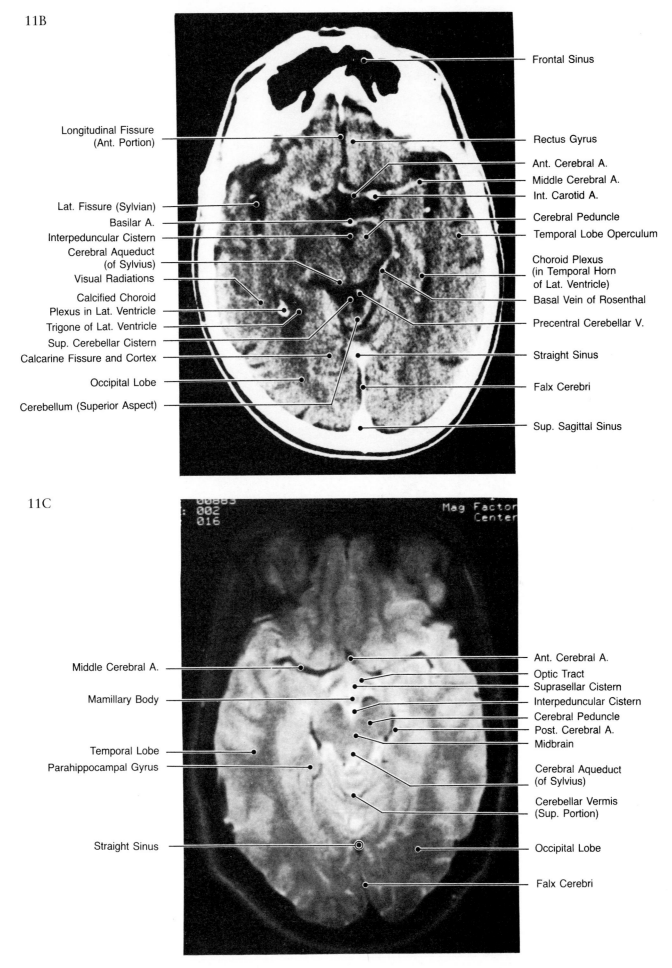

Frontal Sinus

Longitudinal Fissure (Ant. Portion)

Rectus Gyrus

Ant. Cerebral A.

Middle Cerebral A.

Int. Carotid A.

Lat. Fissure (Sylvian)

Basilar A.

Interpeduncular Cistern

Cerebral Peduncle

Temporal Lobe Operculum

Cerebral Aqueduct (of Sylvius)

Visual Radiations

Choroid Plexus (in Temporal Horn of Lat. Ventricle)

Calcified Choroid Plexus in Lat. Ventricle

Basal Vein of Rosenthal

Trigone of Lat. Ventricle

Precentral Cerebellar V.

Sup. Cerebellar Cistern

Calcarine Fissure and Cortex

Straight Sinus

Occipital Lobe

Falx Cerebri

Cerebellum (Superior Aspect)

Sup. Sagittal Sinus

11C

00883
: 002
016

Mag Factor
Center

Middle Cerebral A.

Ant. Cerebral A.

Optic Tract

Suprasellar Cistern

Mamillary Body

Interpeduncular Cistern

Cerebral Peduncle

Post. Cerebral A.

Midbrain

Temporal Lobe

Parahippocampal Gyrus

Cerebral Aqueduct (of Sylvius)

Cerebellar Vermis (Sup. Portion)

Straight Sinus

Occipital Lobe

Falx Cerebri

29

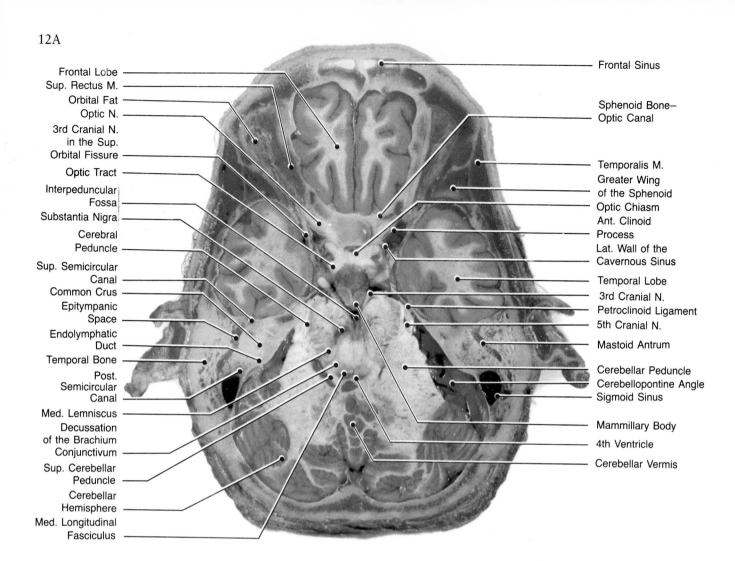

Frontal Lobe
Sup. Rectus M.
Orbital Fat
Optic N.
3rd Cranial N. in the Sup. Orbital Fissure
Optic Tract
Interpeduncular Fossa
Substantia Nigra
Cerebral Peduncle
Sup. Semicircular Canal
Common Crus
Epitympanic Space
Endolymphatic Duct
Temporal Bone
Post. Semicircular Canal
Med. Lemniscus
Decussation of the Brachium Conjunctivum
Sup. Cerebellar Peduncle
Cerebellar Hemisphere
Med. Longitudinal Fasciculus

Frontal Sinus
Sphenoid Bone— Optic Canal
Temporalis M.
Greater Wing of the Sphenoid
Optic Chiasm
Ant. Clinoid Process
Lat. Wall of the Cavernous Sinus
Temporal Lobe
3rd Cranial N.
Petroclinoid Ligament
5th Cranial N.
Mastoid Antrum
Cerebellar Peduncle
Cerebellopontine Angle
Sigmoid Sinus
Mammillary Body
4th Ventricle
Cerebellar Vermis

12A Axial/Anatomic Plates/Optic Chiasm. The optic nerves are seen within the optic canals on this axial (+20 degrees) section. The anterior clinoid processes are seen posterolaterally and contain marrow. The remaining walls of the canal area are formed by cortical sphenoid bone. The posterior optic nerves, optic chiasm, and anterior optic tracts form an ×-shaped structure in the suprasellar cistern. In the interpeduncular cistern the mammillary bodies are seen and a branch of the third cranial nerve projects anteriorly towards the cavernous sinus. Because of postmortem brain swelling the fourth ventricle is slit-like, but the cerebellopontine cisterns adjacent to the temporal bones are still intact.

12B Axial/CT/Fifth Cranial Nerves. This axial (+10 degrees) CT was made following injection of intrathecal metrizamide contrast. The fifth cranial nerves are seen outlined by contrast, extending from the mid portion of the pons anteriorly towards Meckel's cave and bridging the cerebellopontine angle cistern. The undulating contour of the midline inferior vermis is also seen as contrast fills the small CSF spaces.

12C Axial/CT/Pons. The inverted pear–shaped pons is outlined by metrizamide contrast which was injected into the lumbar subarachnoid space on this axial (+0 degrees) CT. A small amount of contrast has refluxed into the fourth ventricle. Note that the carotid arteries are calcified bilaterally within the cavernous sinuses. Calcification of the arterial structures, such as superficial temporal arteries in this case, is commonly seen in patients with arteriosclerotic disease. (From Chakeres DW, Kapila A: Radiology of the ambient cistern. Part I. Normal. 27:383–389, 1985.)

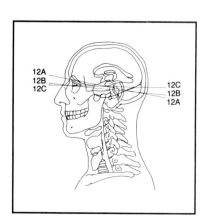

12B

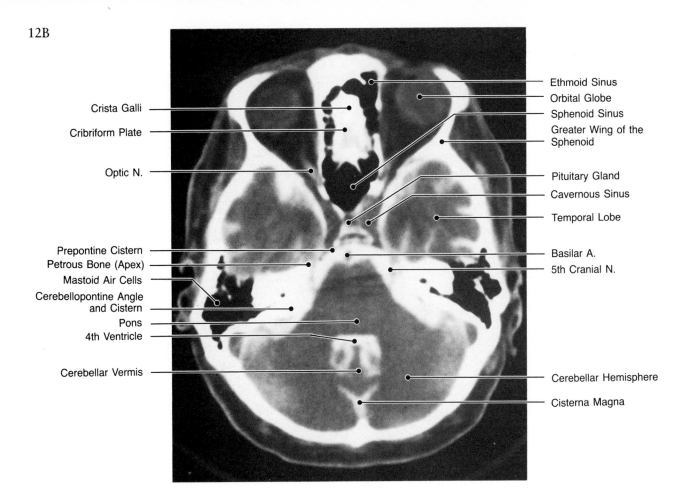

Crista Galli

Cribriform Plate

Optic N.

Prepontine Cistern
Petrous Bone (Apex)
Mastoid Air Cells
Cerebellopontine Angle
and Cistern
Pons
4th Ventricle

Cerebellar Vermis

Ethmoid Sinus
Orbital Globe
Sphenoid Sinus
Greater Wing of the
Sphenoid

Pituitary Gland
Cavernous Sinus
Temporal Lobe

Basilar A.
5th Cranial N.

Cerebellar Hemisphere

Cisterna Magna

12C

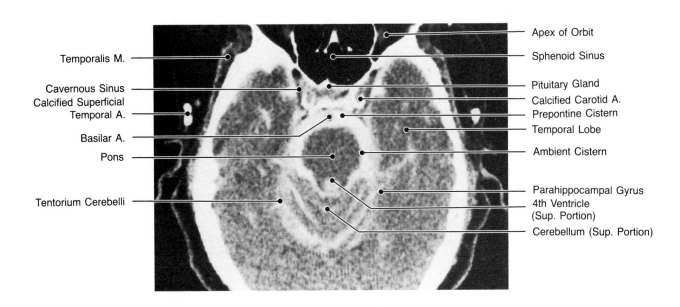

Temporalis M.

Cavernous Sinus
Calcified Superficial
Temporal A.

Basilar A.

Pons

Tentorium Cerebelli

Apex of Orbit

Sphenoid Sinus

Pituitary Gland
Calcified Carotid A.
Prepontine Cistern
Temporal Lobe

Ambient Cistern

Parahippocampal Gyrus
4th Ventricle
(Sup. Portion)
Cerebellum (Sup. Portion)

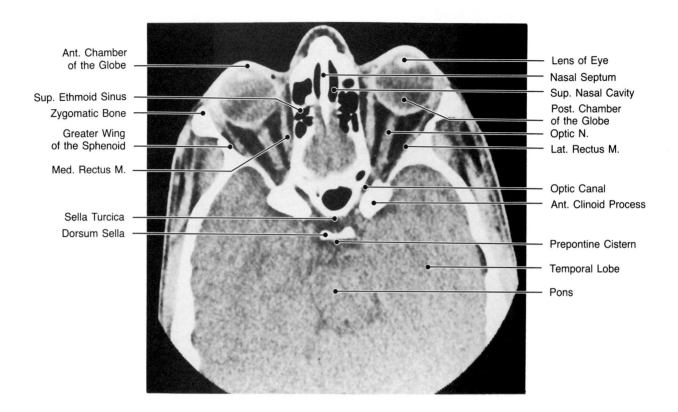

Ant. Chamber of the Globe

Sup. Ethmoid Sinus

Zygomatic Bone

Greater Wing of the Sphenoid

Med. Rectus M.

Sella Turcica

Dorsum Sella

Lens of Eye

Nasal Septum

Sup. Nasal Cavity

Post. Chamber of the Globe

Optic N.

Lat. Rectus M.

Optic Canal

Ant. Clinoid Process

Prepontine Cistern

Temporal Lobe

Pons

12D Axial/CT/Optic Canals. This axial (0 degrees) CT displayed at soft tissue windows is made through the optic canals and optic nerves at a slightly different angle than the anatomic plate. The anterior and posterior chambers of the globe are separated by the higher density lens. The insertion of the optic nerve on the posterior margin of the globe is evident and is outlined by orbital fat within the muscle cone of the extraocular muscles.

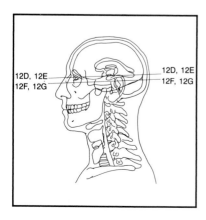

12E Axial/CT/Optic Canals. This axial (0 degrees) CT through the optic canals is displayed with bone windows. The optic canals diverge from the region of the sella turcica at approximately a 90-degree angle to one another. The medial margin is outlined by the sphenoid sinus, while the lateral margin is the anterior clinoid process.

12F Axial/CT/Optic Nerves. This axial (0 degrees) CT is made through the superior ethmoid sinus and is displayed with soft tissue windows. The lamina papyracea is the thin bony lateral margin of the ethmoid sinus. A small amount of periorbital fat is seen interposed between the lamina papyracea and the medial rectus muscle. Small apparent defects are occasionally seen within the thin bony walls of the ethmoid sinus. They may be artifactual owing to partial volume averaging.

12G Axial/CT/Orbital Optic Nerve. The ethmoid sinuses are divided into a series of individual air cells by fine septations on this axial (0 degrees) CT. The perpendicular plate of the ethmoid bone separates the superior nasal cavity and lies just inferior to the cribriform plates. The sphenoid sinus is seen farther posterior and usually has only a few septations. There are a number of lucencies within the osseous structures related to sutures which may appear as fractures. For example, the spheno-squamosal suture is visible, as is the suture between the greater wing of the sphenoid and the zygomatic bone, which forms the lateral wall of the orbit.

12E

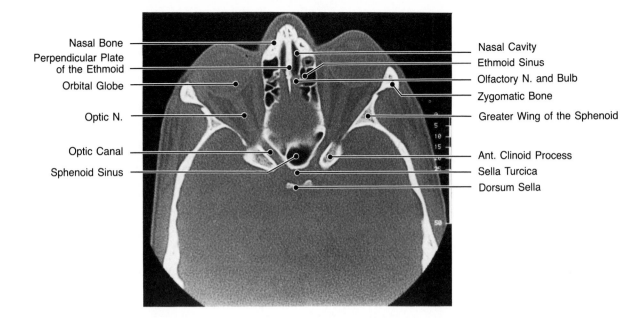

Nasal Bone
Perpendicular Plate of the Ethmoid
Orbital Globe

Optic N.

Optic Canal
Sphenoid Sinus

Nasal Cavity
Ethmoid Sinus
Olfactory N. and Bulb
Zygomatic Bone
Greater Wing of the Sphenoid

Ant. Clinoid Process
Sella Turcica
Dorsum Sella

12F

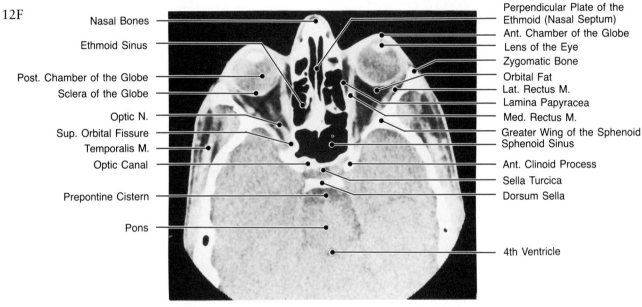

Nasal Bones

Ethmoid Sinus

Post. Chamber of the Globe
Sclera of the Globe

Optic N.
Sup. Orbital Fissure
Temporalis M.

Optic Canal

Prepontine Cistern

Pons

Perpendicular Plate of the Ethmoid (Nasal Septum)
Ant. Chamber of the Globe
Lens of the Eye
Zygomatic Bone
Orbital Fat
Lat. Rectus M.
Lamina Papyracea
Med. Rectus M.
Greater Wing of the Sphenoid
Sphenoid Sinus
Ant. Clinoid Process
Sella Turcica
Dorsum Sella

4th Ventricle

12G

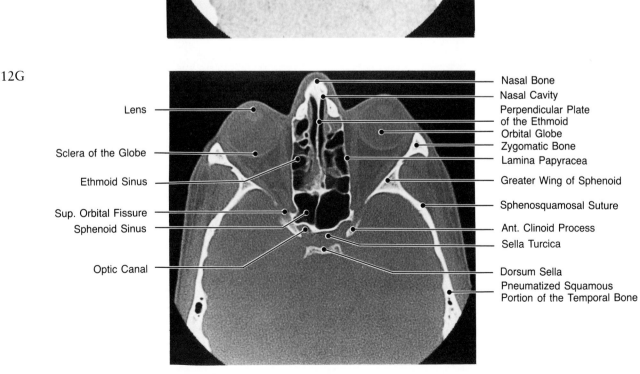

Lens

Sclera of the Globe

Ethmoid Sinus

Sup. Orbital Fissure
Sphenoid Sinus

Optic Canal

Nasal Bone
Nasal Cavity
Perpendicular Plate of the Ethmoid
Orbital Globe
Zygomatic Bone
Lamina Papyracea
Greater Wing of Sphenoid
Sphenosquamosal Suture
Ant. Clinoid Process
Sella Turcica
Dorsum Sella
Pneumatized Squamous Portion of the Temporal Bone

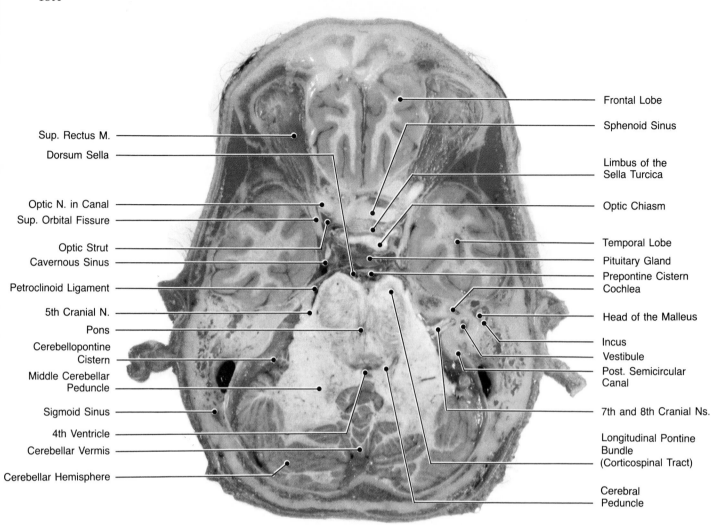

Sup. Rectus M.
Dorsum Sella
Optic N. in Canal
Sup. Orbital Fissure
Optic Strut
Cavernous Sinus
Petroclinoid Ligament
5th Cranial N.
Pons
Cerebellopontine Cistern
Middle Cerebellar Peduncle
Sigmoid Sinus
4th Ventricle
Cerebellar Vermis
Cerebellar Hemisphere

Frontal Lobe
Sphenoid Sinus
Limbus of the Sella Turcica
Optic Chiasm
Temporal Lobe
Pituitary Gland
Prepontine Cistern
Cochlea
Head of the Malleus
Incus
Vestibule
Post. Semicircular Canal
7th and 8th Cranial Ns.
Longitudinal Pontine Bundle (Corticospinal Tract)
Cerebral Peduncle

13A Axial/Anatomic Plate/Pituitary Gland. The seventh and eighth cranial nerves are seen bridging the cerebellopontine cistern on this axial (+ 20 degree) section. Four separate nerves enter the internal auditory canal, including the facial nerve, cochlear nerve, and two divisions of the vestibular nerve. The cranial nerves traversing the cavernous sinus and entering the superior orbital fissures are also well seen. The fifth cranial nerve extends anteriorly from the pons towards the region of Meckel's cave.

13B Axial/CT/Cerebellar Peduncle. This axial (+10 degrees) intravenous contrast–enhanced CT scan is made at the level of the cerebellar peduncles. The cisternal spaces about the cerebellum are quite prominent secondary to cerebellar atrophy in this case. The fourth ventricle is enlarged, as are the cerebellopontine angle cisterns and folia about the cerebellum.

13C Axial/MRI/Internal Auditory Canal. This TR 2500 and TE 75 msec axial (−10 degrees) MR spin echo image is made at the level of the internal auditory canals. The internal auditory canal, otic capsule structures, and spinal fluid spaces filling the cerebellopontine angles are seen as high signal regions. This type of technique allows for a detailed evaluation of this region without the need to inject intra-thecal contrast agents.

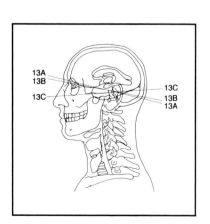

13B

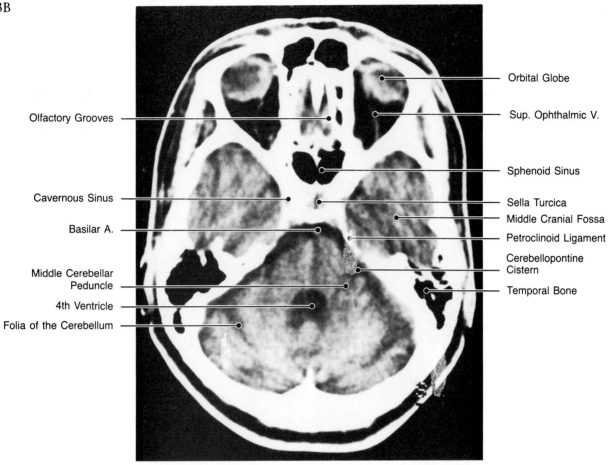

Olfactory Grooves

Cavernous Sinus

Basilar A.

Middle Cerebellar Peduncle

4th Ventricle

Folia of the Cerebellum

Orbital Globe

Sup. Ophthalmic V.

Sphenoid Sinus

Sella Turcica

Middle Cranial Fossa

Petroclinoid Ligament

Cerebellopontine Cistern

Temporal Bone

13C

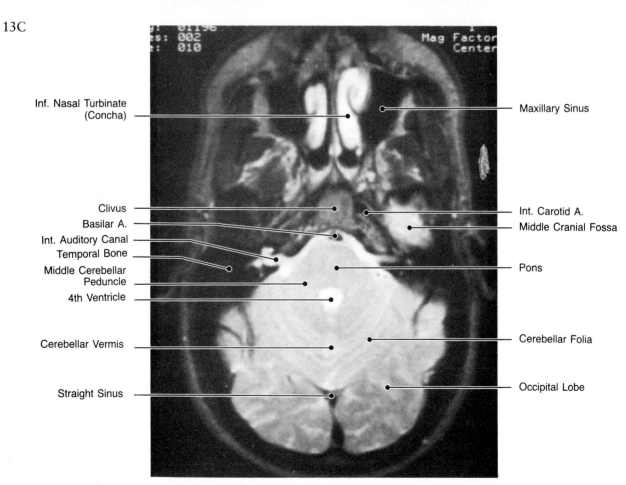

Inf. Nasal Turbinate (Concha)

Clivus

Basilar A.

Int. Auditory Canal

Temporal Bone

Middle Cerebellar Peduncle

4th Ventricle

Cerebellar Vermis

Straight Sinus

Maxillary Sinus

Int. Carotid A.

Middle Cranial Fossa

Pons

Cerebellar Folia

Occipital Lobe

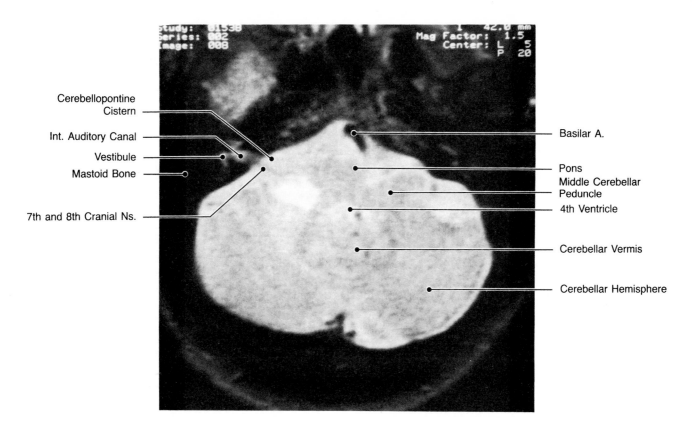

Cerebellopontine Cistern

Int. Auditory Canal

Vestibule

Mastoid Bone

7th and 8th Cranial Ns.

Basilar A.

Pons

Middle Cerebellar Peduncle

4th Ventricle

Cerebellar Vermis

Cerebellar Hemisphere

13D Axial/MRI/Internal Auditory Canal. On this TR 2500 and TE 75 msec spin echo axial (0 degrees) MRI at the ponto-medullary junction the seventh and eighth cranial nerves are seen on the right crossing the cerebellopontine angle cistern to enter the internal auditory canal. The spinal fluid is of high signal intensity and outlines the lower signal linear nerve. The high signal lesion in the right middle cerebellar peduncle and pons is related to a brain infarction.

13E Axial/MRI/Middle Ear Cavity Filled with Fluid. This spin echo TR 2500 and TE 75 msec MRI is made in the axial (0 degree) plane through the internal auditory canals. A high signal is seen arising from the middle ear cavities (which are usually signal voids) because of the presence of abnormal fluid. The hourglass-shaped air space is visible only because of the presence of pathology. A small defect related to the ossicles is seen anteriorly in the left epitympanic space.

13F Axial/MRI/Internal Auditory Canal Blowup. This TR 2500 and TE 75 msec spin echo MRI image is made axially (0 degrees) through the internal auditory canal region. Signals arising from the otic capsule structures, including the internal auditory canal, cochlea, vestibule, and lateral semicircular canal are seen. The remaining temporal bone structures, including the air spaces, are black. The spinal fluid within the trigeminal cistern of Meckel's cave is outlined laterally by a layer of dura.

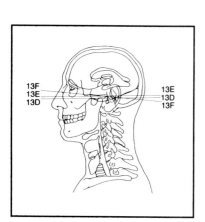

13E

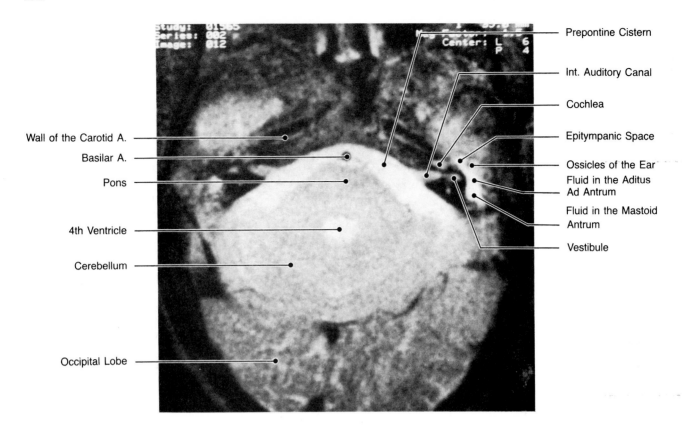

Prepontine Cistern

Int. Auditory Canal

Cochlea

Epitympanic Space

Ossicles of the Ear

Fluid in the Aditus Ad Antrum

Fluid in the Mastoid Antrum

Vestibule

Wall of the Carotid A.

Basilar A.

Pons

4th Ventricle

Cerebellum

Occipital Lobe

13F

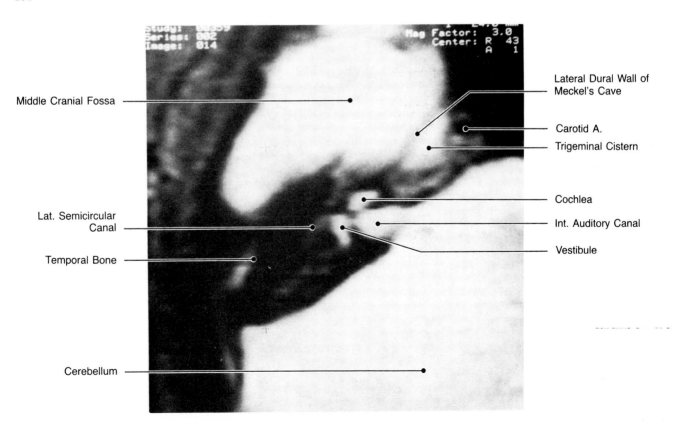

Lateral Dural Wall of Meckel's Cave

Carotid A.

Trigeminal Cistern

Cochlea

Int. Auditory Canal

Vestibule

Middle Cranial Fossa

Lat. Semicircular Canal

Temporal Bone

Cerebellum

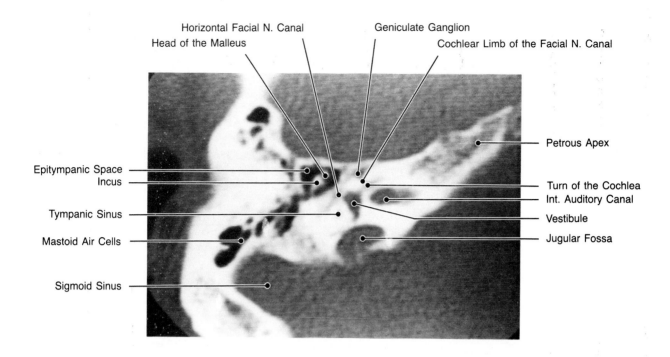

Horizontal Facial N. Canal
Head of the Malleus
Geniculate Ganglion
Cochlear Limb of the Facial N. Canal
Epitympanic Space
Incus
Tympanic Sinus
Mastoid Air Cells
Sigmoid Sinus
Petrous Apex
Turn of the Cochlea
Int. Auditory Canal
Vestibule
Jugular Fossa

13G Axial/CT/Facial Nerve Canal. This axial (+30 degrees) CT section is made with high resolution thin section bone algorithm technique. This section parallels the course of the facial nerve from the internal auditory canal through the geniculate ganglion and horizontal segment. The geniculate ganglion segment is adjacent to the cochlea, while the horizontal segment lies between the vestibule and the epitympanic space. (From Chakeres DW, Kapila A: Normal and pathologic radiologic anatomy of the motor innervation of the face. AJNR 5:591–597, 1984.)

13H Axial/CT/Lateral Semicircular Canal. This high resolution thin section bone algorithm axial (+30 degree) CT demonstrates the lateral semicircular canal and vestibule. The air spaces just lateral to the promontory of the lateral semicircular canal, forming an hourglass configuration, are the epitympanic space anteriorly and the mastoid antrum posteriorly, with the aditus ad antrum interposed.

13I Axial/CT/Air Cerebellopontine Cisternogram. To most accurately evaluate the cerebellopontine angle with axial (0 degrees) CT, air is injected into the cerebrospinal fluid and then directed into the region. In this case, we are able to see the seventh and eighth cranial nerves outlined by air. The small looping density adjacent to the cranial nerves is related to vascular structures such as the anterior inferior cerebellar artery.

13J Axial/CT/Stapes. This axial (+30 degrees) high resolution thin section bone algorithm CT of the right temporal bone is made through the vestibule. The Y shaped stapes is visible interposed between the oval window of the vestibule medially and the long process of the incus laterally. The neck of the malleus is seen just anteriorly in the middle ear cavity.

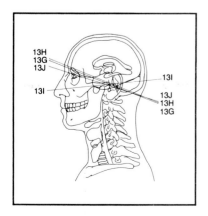

13H

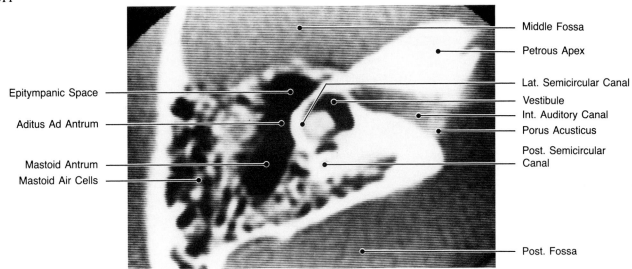

Epitympanic Space

Aditus Ad Antrum

Mastoid Antrum
Mastoid Air Cells

Middle Fossa
Petrous Apex
Lat. Semicircular Canal
Vestibule
Int. Auditory Canal
Porus Acusticus
Post. Semicircular
Canal

Post. Fossa

13I

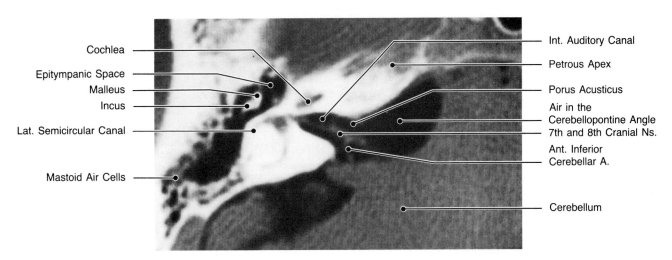

Cochlea
Epitympanic Space
Malleus
Incus
Lat. Semicircular Canal

Mastoid Air Cells

Int. Auditory Canal
Petrous Apex
Porus Acusticus
Air in the
Cerebellopontine Angle
7th and 8th Cranial Ns.
Ant. Inferior
Cerebellar A.

Cerebellum

13J

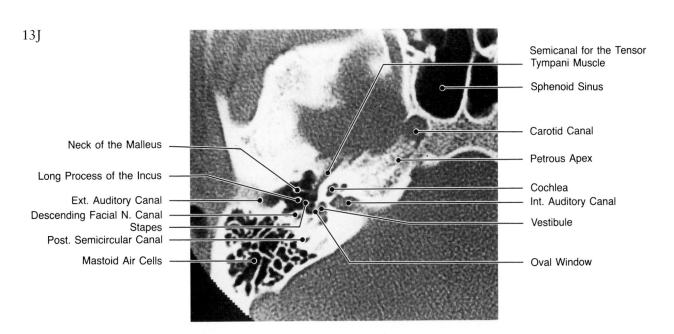

Neck of the Malleus

Long Process of the Incus

Ext. Auditory Canal
Descending Facial N. Canal
Stapes
Post. Semicircular Canal
Mastoid Air Cells

Semicanal for the Tensor
Tympani Muscle
Sphenoid Sinus

Carotid Canal

Petrous Apex

Cochlea
Int. Auditory Canal

Vestibule

Oval Window

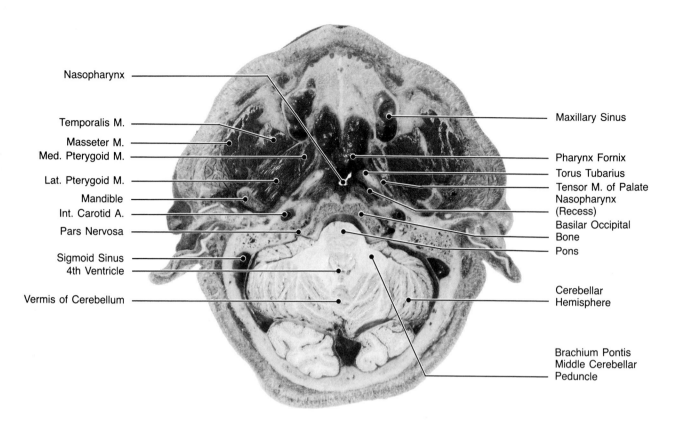

Nasopharynx

Temporalis M.

Masseter M.
Med. Pterygoid M.

Lat. Pterygoid M.

Mandible
Int. Carotid A.

Pars Nervosa

Sigmoid Sinus
4th Ventricle

Vermis of Cerebellum

Maxillary Sinus

Pharynx Fornix
Torus Tubarius
Tensor M. of Palate
Nasopharynx
(Recess)
Basilar Occipital
Bone
Pons

Cerebellar
Hemisphere

Brachium Pontis
Middle Cerebellar
Peduncle

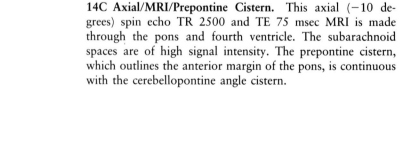

14A Axial/Anatomic Plate/Inferior Pons. The structures of the nasopharynx are well seen on this axial (−20 degrees) section. The cylindrical cartilaginous end of the eustachian tube, the torus tubarius, is seen projecting into the superior lateral nasopharynx anterior to the nasopharyngeal recess. This recess is also called the fossa of Rosenmüller. The muscles of mastication filling the subtemporal fossa at the skull base are also well seen. The lower cranial nerves are visible entering the neural portion of the jugular fossa at the pars nervosa. This space is particularly well seen on magnetic resonance images.

14B Axial/MRI/Pons. A number of the bony structures of the skull base are outlined by fat spaces on this partial saturation TR 800 and TE 25 msec axial (−10 degrees) MRI. The section plane is slightly different than the anatomic plate. The clivus and posterior portions of the petrous apices both contain marrow. Fat is seen surrounding the pterygoid plates farther anteriorly. The presence and distribution of marrow spaces within the skull base structures are widely variable, making recognition of subtle bony abnormalities of the skull base somewhat difficult to recognize on MRI compared with CT.

14C Axial/MRI/Prepontine Cistern. This axial (−10 degrees) spin echo TR 2500 and TE 75 msec MRI is made through the pons and fourth ventricle. The subarachnoid spaces are of high signal intensity. The prepontine cistern, which outlines the anterior margin of the pons, is continuous with the cerebellopontine angle cistern.

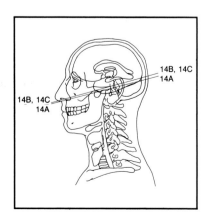

14B

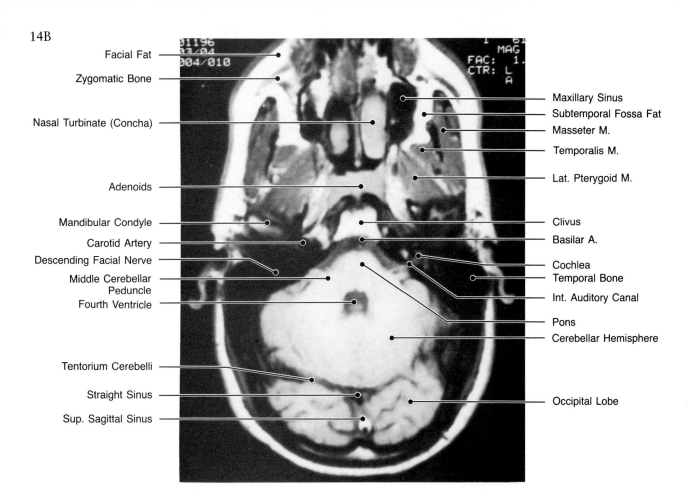

Facial Fat
Zygomatic Bone
Nasal Turbinate (Concha)
Adenoids
Mandibular Condyle
Carotid Artery
Descending Facial Nerve
Middle Cerebellar Peduncle
Fourth Ventricle
Tentorium Cerebelli
Straight Sinus
Sup. Sagittal Sinus

Maxillary Sinus
Subtemporal Fossa Fat
Masseter M.
Temporalis M.
Lat. Pterygoid M.
Clivus
Basilar A.
Cochlea
Temporal Bone
Int. Auditory Canal
Pons
Cerebellar Hemisphere
Occipital Lobe

14C

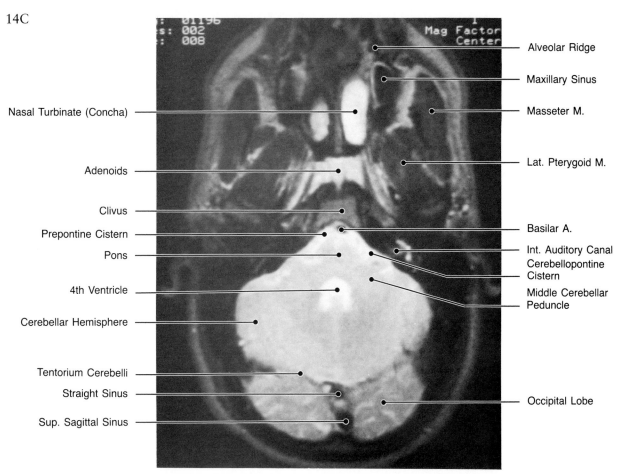

Nasal Turbinate (Concha)
Adenoids
Clivus
Prepontine Cistern
Pons
4th Ventricle
Cerebellar Hemisphere
Tentorium Cerebelli
Straight Sinus
Sup. Sagittal Sinus

Alveolar Ridge
Maxillary Sinus
Masseter M.
Lat. Pterygoid M.
Basilar A.
Int. Auditory Canal
Cerebellopontine Cistern
Middle Cerebellar Peduncle
Occipital Lobe

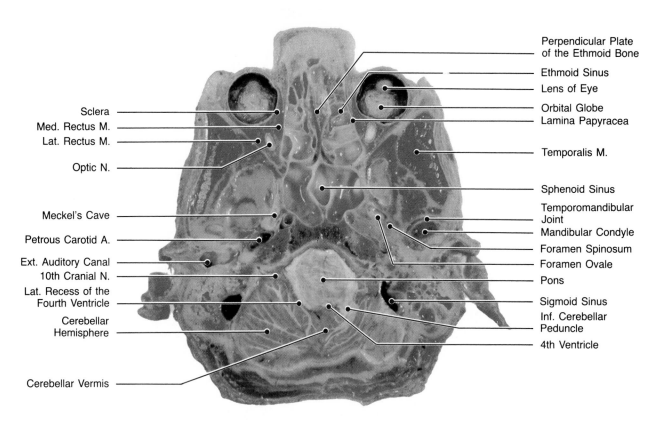

Perpendicular Plate of the Ethmoid Bone
Ethmoid Sinus
Lens of Eye
Orbital Globe
Lamina Papyracea
Temporalis M.
Sphenoid Sinus
Temporomandibular Joint
Mandibular Condyle
Foramen Spinosum
Foramen Ovale
Pons
Sigmoid Sinus
Inf. Cerebellar Peduncle
4th Ventricle

Sclera
Med. Rectus M.
Lat. Rectus M.
Optic N.
Meckel's Cave
Petrous Carotid A.
Ext. Auditory Canal
10th Cranial N.
Lat. Recess of the Fourth Ventricle
Cerebellar Hemisphere
Cerebellar Vermis

15A Axial/Anatomic Plate/Inferior Pons. This axial (+20 degrees) anatomic plate is made through the lower pons and mid orbit region. Meckel's cave contains the gasserian ganglion of the fifth cranial nerve. The ganglion lies just anterior and lateral to the petrous apex in the middle fossa. Its lateral margin faces the middle fossa, while the medial margin abuts the sphenoid sinus.

15B Axial/CT/Lateral Recesses of the Fourth Ventricle. This axial (+10 degrees) CT was made following instillation of intrathecal metrizamide contrast. The convoluted contours of the pontomedullary junction are evident. The cerebellopontine angle cisterns are seen laterally as high density regions. The seventh and eighth cranial nerves are visible coursing towards the internal auditory canal on the right.

15C Axial/CT/Inferior Orbits. This axial (0 degrees) CT is made at the level of the inferior orbit and is displayed with soft tissue windows. The inferior rectus muscle is seen as a linear density within the orbital fat. The inferior orbital fissure is seen interposed between the greater wing of the sphenoid laterally and the superior maxillary sinus medially.

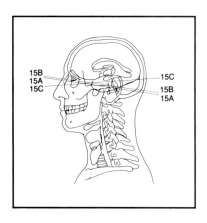

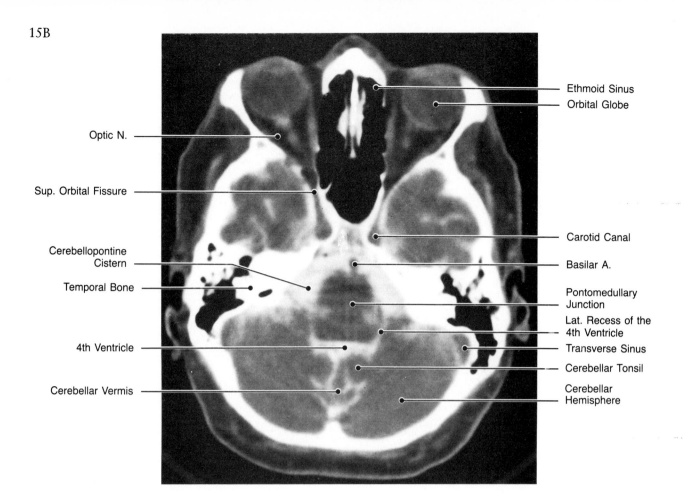

Ethmoid Sinus

Orbital Globe

Optic N.

Sup. Orbital Fissure

Carotid Canal

Basilar A.

Cerebellopontine Cistern

Pontomedullary Junction

Temporal Bone

Lat. Recess of the 4th Ventricle

4th Ventricle

Transverse Sinus

Cerebellar Tonsil

Cerebellar Vermis

Cerebellar Hemisphere

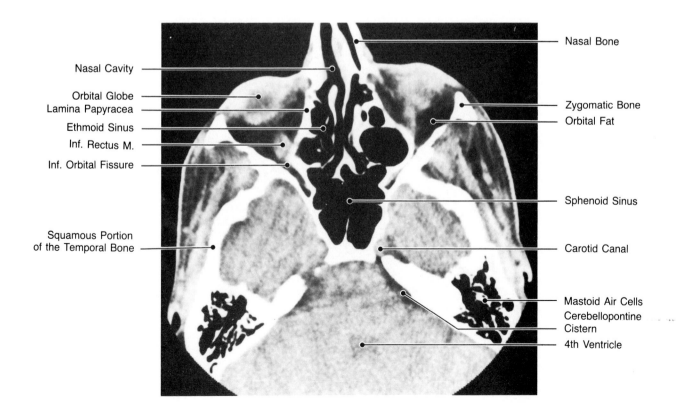

Nasal Bone

Nasal Cavity

Orbital Globe
Lamina Papyracea

Zygomatic Bone

Ethmoid Sinus

Orbital Fat

Inf. Rectus M.

Inf. Orbital Fissure

Sphenoid Sinus

Squamous Portion of the Temporal Bone

Carotid Canal

Mastoid Air Cells

Cerebellopontine Cistern

4th Ventricle

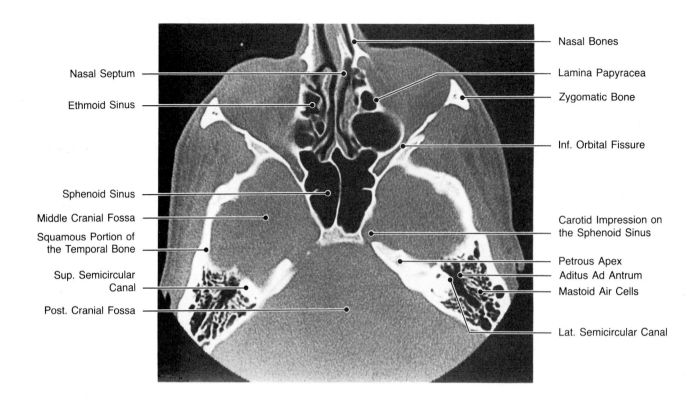

Nasal Septum

Ethmoid Sinus

Sphenoid Sinus

Middle Cranial Fossa

Squamous Portion of the Temporal Bone

Sup. Semicircular Canal

Post. Cranial Fossa

Nasal Bones

Lamina Papyracea

Zygomatic Bone

Inf. Orbital Fissure

Carotid Impression on the Sphenoid Sinus

Petrous Apex
Aditus Ad Antrum
Mastoid Air Cells

Lat. Semicircular Canal

15D Axial/CT/Inferior Orbit. This axial (0 degrees) CT image is made through the inferior orbits and is displayed at bone windows. The bony crevice between the superior posterior margin of the maxillary sinus and the greater wing of the sphenoid is the inferior orbital fissure. This is in direct continuity with the pterygopalatine fossa slightly farther inferiorly.

15E Axial/CT/Foramen Rotundum. This axial (0 degrees) high resolution thin section CT is displayed at bone windows. A small canal is seen connecting the pterygopalatine fossa anteriorly with the middle fossa posteriorly on the left. This is related to the foramen rotundum, through which the second division of the fifth cranial nerve courses. This is in close association with the pterygoid canal, which lies slightly farther medially and inferiorly.

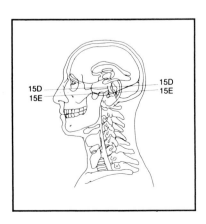

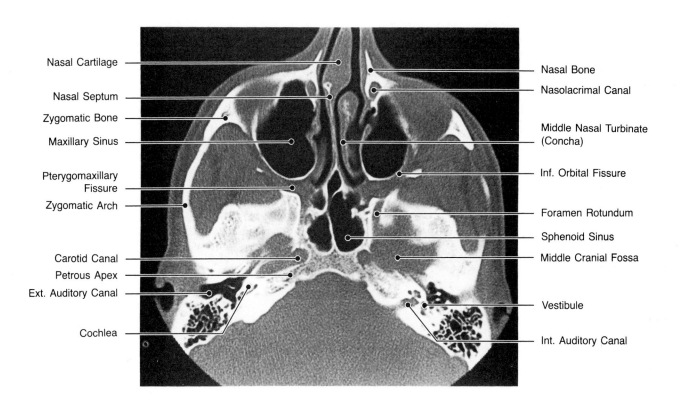

Nasal Cartilage

Nasal Septum

Zygomatic Bone

Maxillary Sinus

Pterygomaxillary Fissure

Zygomatic Arch

Carotid Canal

Petrous Apex

Ext. Auditory Canal

Cochlea

Nasal Bone

Nasolacrimal Canal

Middle Nasal Turbinate (Concha)

Inf. Orbital Fissure

Foramen Rotundum

Sphenoid Sinus

Middle Cranial Fossa

Vestibule

Int. Auditory Canal

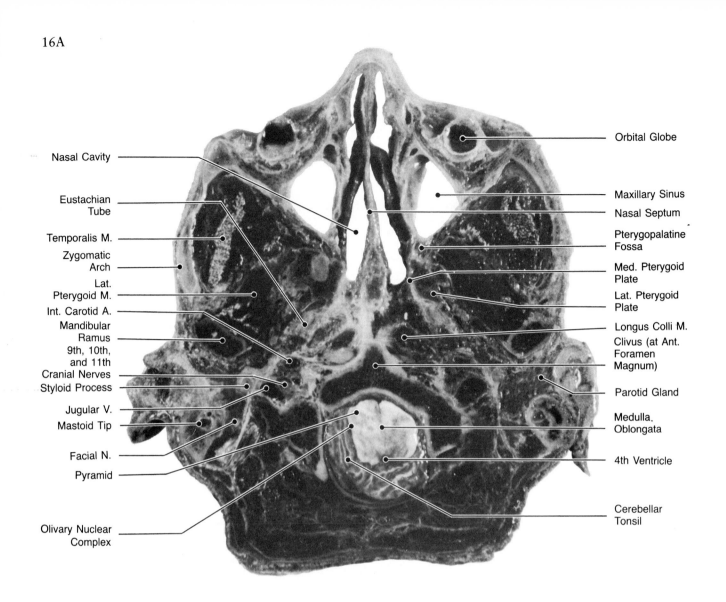

Nasal Cavity

Eustachian Tube

Temporalis M.

Zygomatic Arch

Lat. Pterygoid M.

Int. Carotid A.

Mandibular Ramus

9th, 10th, and 11th Cranial Nerves

Styloid Process

Jugular V.

Mastoid Tip

Facial N.

Pyramid

Olivary Nuclear Complex

Orbital Globe

Maxillary Sinus

Nasal Septum

Pterygopalatine Fossa

Med. Pterygoid Plate

Lat. Pterygoid Plate

Longus Colli M.

Clivus (at Ant. Foramen Magnum)

Parotid Gland

Medulla. Oblongata

4th Ventricle

Cerebellar Tonsil

16A Axial/Anatomic Plates/Foramen Magnum. The cerebellar tonsils are herniated into the foramen magnum as a postmortem artifact on this axial (+20 degrees) section. The medulla oblongata has lost its normal "cloverleaf" shape related to the herniation. The facial nerve is just below the stylomastoid foramen in a small fatty space bordered by the occipital bone medially, the styloid process anteriorly, and the mastoid tip laterally.

16B Axial/CT/Lateral Recesses of Fourth Ventricle. This axial (+10 degrees) CT scan was made following intrathecal injection of metrizamide contrast. The medulla has a cloverleaf configuration related to the pyramids anteriorly. The fourth ventricle is somewhat T-shaped in configuration with the foramen Magendie extending inferiorly between the two cerebellar tonsils. The lateral recesses of the fourth ventricle extend anteriorly and laterally into the cerebellopontine angle region. This space is traversed by a number of vascular structures, such as the posterior inferior cerebellar artery.

16C Axial/MRI/Medulla. The cloverleaf-shaped medulla oblongata is well outlined by the lower signal intensity spinal fluid spaces of the lateral recesses of the fourth ventricle on this partial saturation TR 800 and TE 25 msec axial (−10 degrees) MRI. This section is made at a slightly different angle than the anatomic section. A number of the muscles about the mandible, including the masseter, temporalis, and lateral pterygoid, are well outlined by facial fat. A small subarachnoid space, which is related to the 9th, 10th, and 11th cranial nerves entering the pars nervosa of the jugular fossa, is seen projecting into the medial portion of the temporal bone.

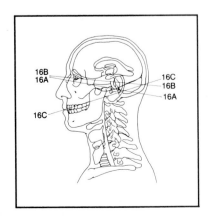

16B
16A

16C
16B

16A

16C

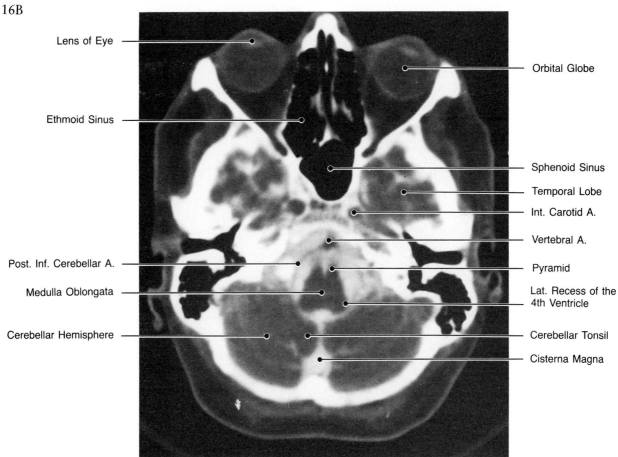

Lens of Eye

Orbital Globe

Ethmoid Sinus

Sphenoid Sinus

Temporal Lobe

Int. Carotid A.

Vertebral A.

Post. Inf. Cerebellar A.

Pyramid

Medulla Oblongata

Lat. Recess of the 4th Ventricle

Cerebellar Hemisphere

Cerebellar Tonsil

Cisterna Magna

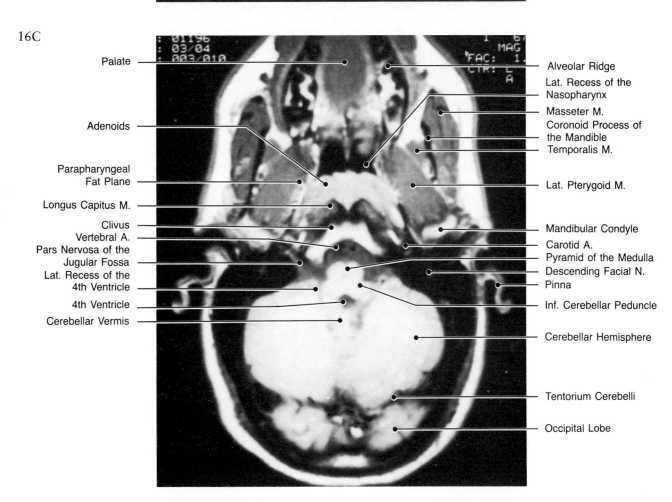

Palate

Alveolar Ridge

Lat. Recess of the Nasopharynx

Masseter M.

Coronoid Process of the Mandible

Temporalis M.

Adenoids

Parapharyngeal Fat Plane

Lat. Pterygoid M.

Longus Capitus M.

Clivus

Mandibular Condyle

Vertebral A.

Carotid A.

Pars Nervosa of the Jugular Fossa

Pyramid of the Medulla

Lat. Recess of the 4th Ventricle

Descending Facial N.

Pinna

4th Ventricle

Inf. Cerebellar Peduncle

Cerebellar Vermis

Cerebellar Hemisphere

Tentorium Cerebelli

Occipital Lobe

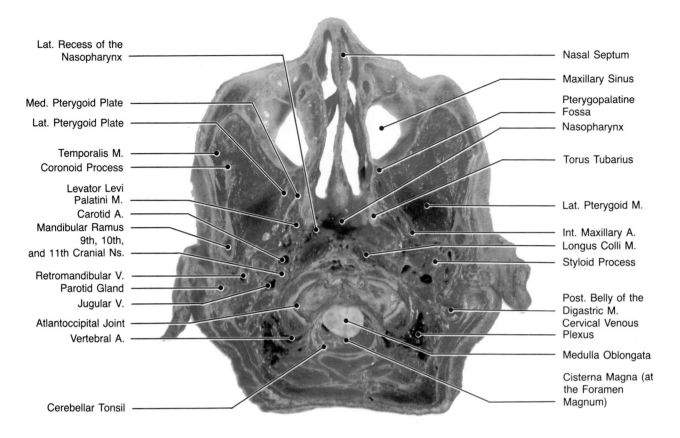

Lat. Recess of the Nasopharynx

Med. Pterygoid Plate

Lat. Pterygoid Plate

Temporalis M.

Coronoid Process

Levator Levi Palatini M.

Carotid A.

Mandibular Ramus

9th, 10th, and 11th Cranial Ns.

Retromandibular V.

Parotid Gland

Jugular V.

Atlantoccipital Joint

Vertebral A.

Cerebellar Tonsil

Nasal Septum

Maxillary Sinus

Pterygopalatine Fossa

Nasopharynx

Torus Tubarius

Lat. Pterygoid M.

Int. Maxillary A.

Longus Colli M.

Styloid Process

Post. Belly of the Digastric M.

Cervical Venous Plexus

Medulla Oblongata

Cisterna Magna (at the Foramen Magnum)

17A Axial/Anatomic Plate/Occipital Condyles. The structures of the upper carotid sheath are well demonstrated on this axial (+20 degrees) anatomic plate. The carotid artery lies anteriorly, and the jugular vein lies posteriorly. The 9th, 10th, and 11th cranial nerves are seen interposed in the carotid sheath. The partoid gland is seen as a fatty structure surrounding the posterior margin of the mandible. There is a large venous channel related to the retromandibular vein within the gland that can be used as a landmark to approximate the course of the facial nerve, which lies laterally.

17B Axial/CT/Pterygoid Plates. The pterygopalatine fossa lies between the posterior margin of the maxillary sinus and the pterygoid plates of the sphenoid bone on this axial (0 degrees) CT. A number of neurovascular structures, including the ciliary ganglion and branches of the internal maxillary artery, are in this region. This is an important location in the extension of tumors of the skull base. Note the close relationship of the carotid artery and jugular fossa in the temporal bone, forming a snowman configuration.

17C Axial/CT/Pterygoid Canals. This axial (0 degrees) CT is displayed with high resolution bone window settings. The pterygopalatine fossae are seen between the posterior margin of the maxillary sinuses and anterior margin of the pterygoid plates.

17D Axial/CT/Nasopharynx. The nasopharyngeal soft tissues are well demonstrated on this axial (0 degrees) CT image displayed at soft tissue windows. The soft tissue protuberance in the lateral portion of the nasopharynx is related to the torus tubarius, which is the cartilaginous portion of the eustachian tube. The air space posterior to the torus tubarius is the lateral recess of the nasopharynx. Note the fat planes about the medial margins of the mandibular condyles, which extend anteriorly towards the lateral pterygoid plates. The distortion of these fat planes is important in the recognition of potential tumors in this region.

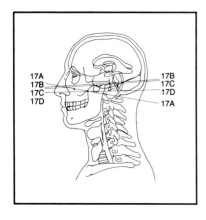

17A
17B
17C
17D

17B
17C
17D

17A

17B

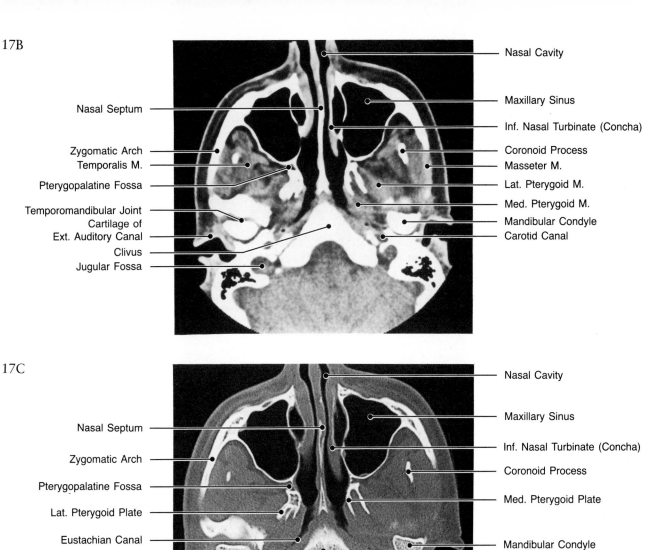

Nasal Septum

Zygomatic Arch
Temporalis M.

Pterygopalatine Fossa

Temporomandibular Joint
Cartilage of
Ext. Auditory Canal
Clivus
Jugular Fossa

Nasal Cavity

Maxillary Sinus

Inf. Nasal Turbinate (Concha)

Coronoid Process
Masseter M.

Lat. Pterygoid M.

Med. Pterygoid M.

Mandibular Condyle

Carotid Canal

17C

Nasal Septum

Zygomatic Arch

Pterygopalatine Fossa

Lat. Pterygoid Plate

Eustachian Canal
Temporomandibular Joint
Carotid Canal

Jugular Fossa

Nasal Cavity

Maxillary Sinus

Inf. Nasal Turbinate (Concha)

Coronoid Process

Med. Pterygoid Plate

Mandibular Condyle
Ext. Auditory Canal

Clivus

17D

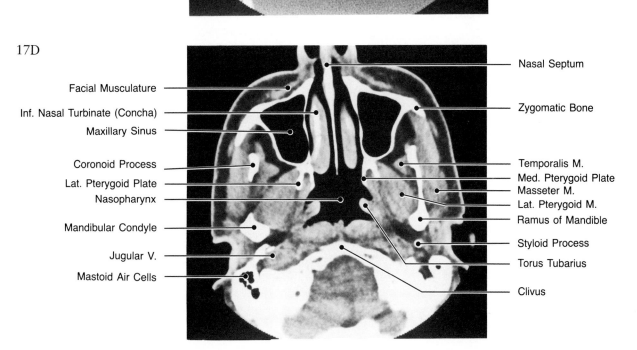

Facial Musculature

Inf. Nasal Turbinate (Concha)

Maxillary Sinus

Coronoid Process

Lat. Pterygoid Plate

Nasopharynx

Mandibular Condyle

Jugular V.

Mastoid Air Cells

Nasal Septum

Zygomatic Bone

Temporalis M.
Med. Pterygoid Plate
Masseter M.
Lat. Pterygoid M.
Ramus of Mandible

Styloid Process

Torus Tubarius

Clivus

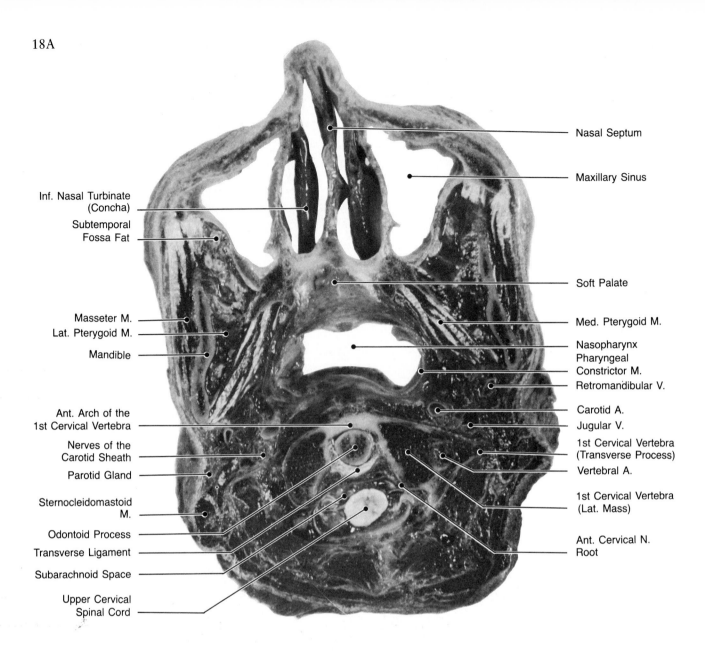

Nasal Septum

Maxillary Sinus

Inf. Nasal Turbinate (Concha)

Subtemporal Fossa Fat

Soft Palate

Masseter M.

Lat. Pterygoid M.

Med. Pterygoid M.

Mandible

Nasopharynx Pharyngeal Constrictor M.

Retromandibular V.

Ant. Arch of the 1st Cervical Vertebra

Carotid A.

Jugular V.

Nerves of the Carotid Sheath

1st Cervical Vertebra (Transverse Process)

Parotid Gland

Vertebral A.

Sternocleidomastoid M.

1st Cervical Vertebra (Lat. Mass)

Odontoid Process

Transverse Ligament

Ant. Cervical N. Root

Subarachnoid Space

Upper Cervical Spinal Cord

18A Axial/Anatomic Plate/Odontoid. The articulation between the anterior arch of C1 and the odontoid is well demonstrated on this axial (+20 degrees) anatomic plate. The stability of this joint is also defined by the transverse ligament, which extends about the posterior margin of the odontoid and inserts on the lateral masses. The canal for the vertebral artery is visible in the transverse process of C1.

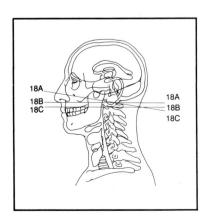

18B Axial/CT/Alveolar Ridge. This axial (0 degrees) CT image is displayed with soft tissue windows at the level of the anterior arch of C1 and the superior odontoid. Within the subarachnoid space of the high cervical region the cervical cord is seen centrally, with two smaller densities laterally that are related to the vertebral arteries. The carotid sheath structures lie medial to the styloid processes adjacent to the upper cervical spine. The carotid artery lies farther anteriorly; the jugular vein is larger and lies directly posterior to the styloid process.

18C Axial/CT/Alveolar Ridge. This axial (0 degrees) CT image is displayed at bone windows and is made at the level of the atlantoaxial joint. The anterior face is sectioned at a slightly different angle than the anatomic plate. The odontoid is interposed between the transverse processes of the first cervical vertebra and just posterior to the anterior arch. The small bony prominence from the lateral mass of C1 is related to the insertion site of the transverse ligament. Disruption of this joint can produce significant compression of the adjacent neural structures.

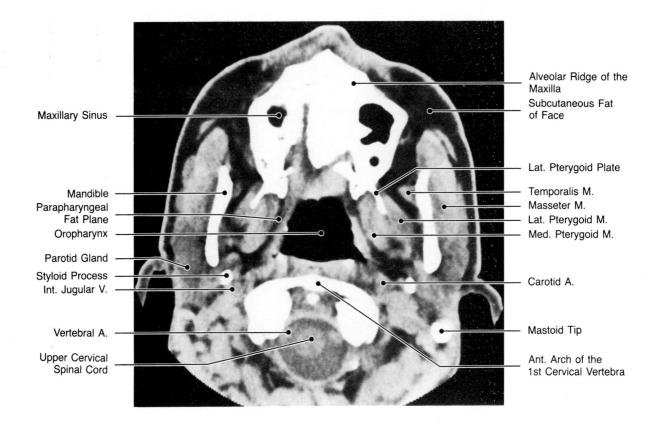

Maxillary Sinus

Mandible
Parapharyngeal
Fat Plane
Oropharynx

Parotid Gland
Styloid Process
Int. Jugular V.

Vertebral A.

Upper Cervical
Spinal Cord

Alveolar Ridge of the
Maxilla
Subcutaneous Fat
of Face

Lat. Pterygoid Plate
Temporalis M.
Masseter M.
Lat. Pterygoid M.
Med. Pterygoid M.

Carotid A.

Mastoid Tip

Ant. Arch of the
1st Cervical Vertebra

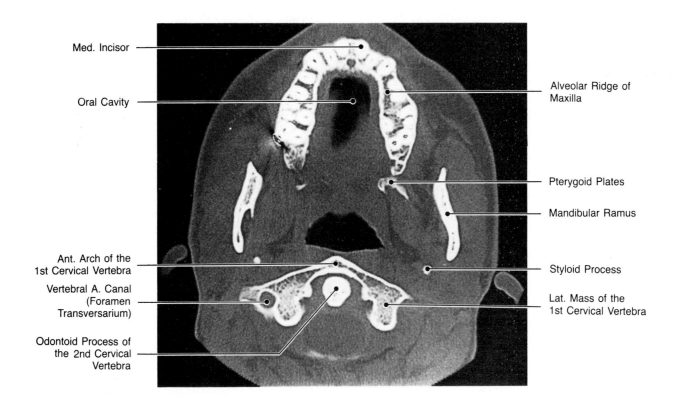

Med. Incisor

Oral Cavity

Ant. Arch of the
1st Cervical Vertebra

Vertebral A. Canal
(Foramen
Transversarium)

Odontoid Process of
the 2nd Cervical
Vertebra

Alveolar Ridge of
Maxilla

Pterygoid Plates

Mandibular Ramus

Styloid Process

Lat. Mass of the
1st Cervical Vertebra

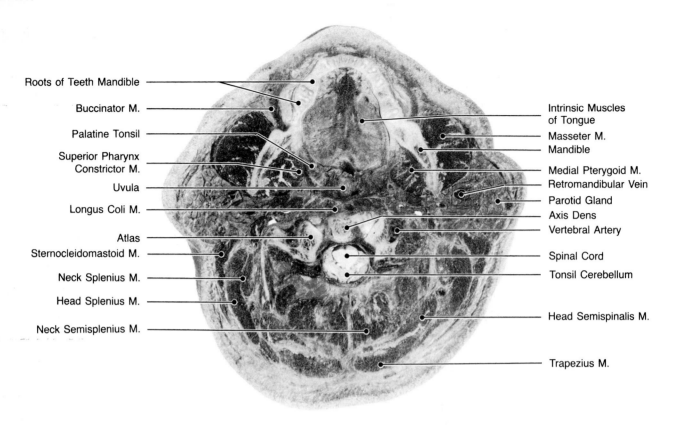

Roots of Teeth Mandible
Buccinator M.
Palatine Tonsil
Superior Pharynx Constrictor M.
Uvula
Longus Coli M.
Atlas
Sternocleidomastoid M.
Neck Splenius M.
Head Splenius M.
Neck Semisplenius M.

Intrinsic Muscles of Tongue
Masseter M.
Mandible
Medial Pterygoid M.
Retromandibular Vein
Parotid Gland
Axis Dens
Vertebral Artery
Spinal Cord
Tonsil Cerebellum
Head Semispinalis M.
Trapezius M.

19A Axial/Anatomic Plate/Occipital Condyles. This axial (−10 degrees) anatomic section is made at the alveolar ridge of the mandible. The masseter muscles lie lateral to the mandible, while the medial pterygoid muscles lie on the medial margin. The parotid gland, through which the retromandibular vein courses, is seen posterior to the angle of the mandible. The sternocleidomastoid muscle is an arc-shaped muscular band in the more superficial lateral neck. Underlying this muscle is the posterior triangle, which is important in evaluation of possible metastatic lymph node spread.

19B Axial/CT/Mandibular Alveolar Ridge. The sternocleidomastoid muscle lies just posterior to the angle of the mandible and forms an arc-shaped muscle band on this axial (0 degrees) CT displayed at soft tissue windows. Laterally, the parotid gland is seen as a structure of fatty density posterior and lateral to the angle of the mandible. Medial to the sternocleidomastoid muscle is the posterior triangle fat space. Occasionally, a few small lymph nodes and neurovascular structures are seen in this space.

19C Axial/MRI/Odontoid. This axial (−10 degrees) partial saturation TR 800 and TE 25 msec MRI is made at the level of the odontoid. A number of the muscle bundles of the tongue are evident, including the medial genioglossus and more lateral hypoglossus muscles. The lateral margins of the oropharynx are bordered by the palatine tonsils. Note that the partoid gland has a significant amount of fat within it, producing a rather high signal on this T1 weighted image.

19D Axial/MRI/Second Cervical Vertebra. This axial (−10 degrees) MR partial saturation TR 800 and TE 25 msec image is made through the base of the odontoid. The jugular vein on the right demonstrates a high signal character, probably related to laminar venous flow. The adjacent carotid artery has a low signal. Just lateral to the carotid sheath structures lies the posterior belly of the digastric muscle. The muscles of the tongue are quite well demonstrated on this image and are outlined by fat. The submandibular glands are interposed between the hypoglossus and mylohyoid muscles.

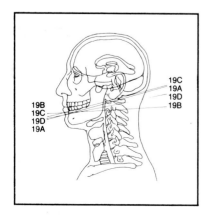

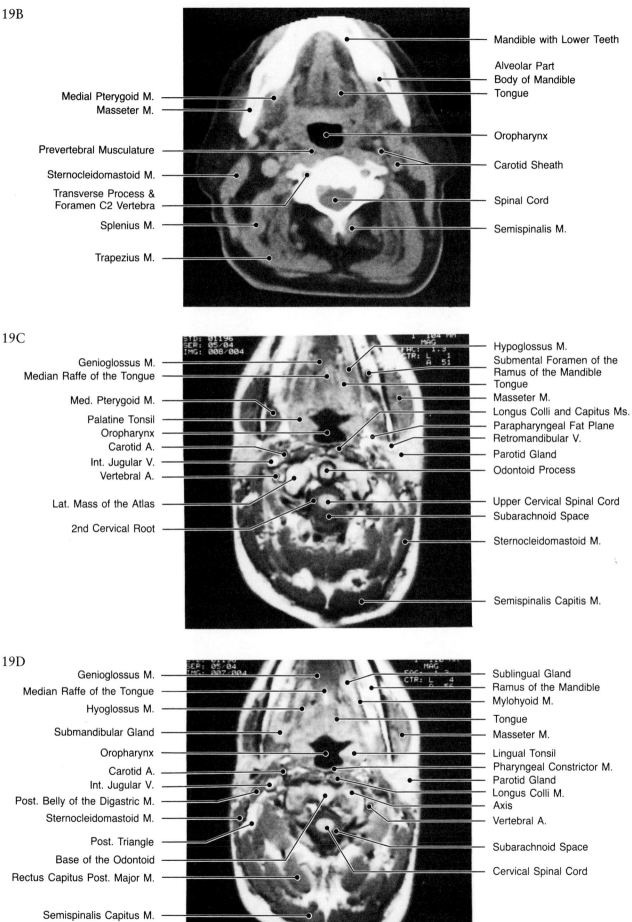

19B

Medial Pterygoid M.
Masseter M.

Prevertebral Musculature

Sternocleidomastoid M.

Transverse Process & Foramen C2 Vertebra

Splenius M.

Trapezius M.

Mandible with Lower Teeth

Alveolar Part
Body of Mandible
Tongue

Oropharynx

Carotid Sheath

Spinal Cord

Semispinalis M.

19C

Genioglossus M.
Median Raffe of the Tongue

Med. Pterygoid M.

Palatine Tonsil
Oropharynx
Carotid A.
Int. Jugular V.
Vertebral A.

Lat. Mass of the Atlas

2nd Cervical Root

Hypoglossus M.
Submental Foramen of the Ramus of the Mandible
Tongue
Masseter M.
Longus Colli and Capitus Ms.
Parapharyngeal Fat Plane
Retromandibular V.
Parotid Gland
Odontoid Process

Upper Cervical Spinal Cord
Subarachnoid Space
Sternocleidomastoid M.

Semispinalis Capitis M.

19D

Genioglossus M.
Median Raffe of the Tongue
Hyoglossus M.
Submandibular Gland
Oropharynx
Carotid A.
Int. Jugular V.
Post. Belly of the Digastric M.
Sternocleidomastoid M.
Post. Triangle
Base of the Odontoid
Rectus Capitus Post. Major M.

Semispinalis Capitus M.

Sublingual Gland
Ramus of the Mandible
Mylohyoid M.
Tongue
Masseter M.
Lingual Tonsil
Pharyngeal Constrictor M.
Parotid Gland
Longus Colli M.
Axis
Vertebral A.
Subarachnoid Space
Cervical Spinal Cord

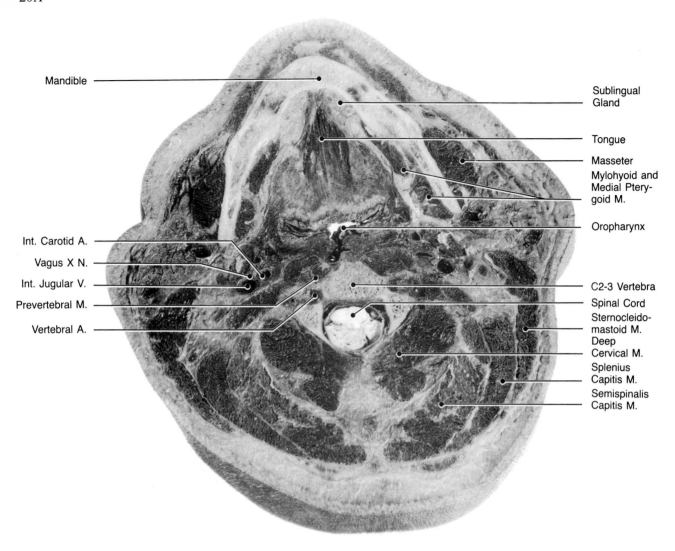

Mandible

Sublingual Gland

Tongue

Masseter

Mylohyoid and Medial Pterygoid M.

Oropharynx

Int. Carotid A.

Vagus X N.

Int. Jugular V.

Prevertebral M.

Vertebral A.

C2-3 Vertebra

Spinal Cord

Sternocleido-mastoid M.

Deep Cervical M.

Splenius Capitis M.

Semispinalis Capitis M.

20A Axial/Anatomic Plate/Mandible. A number of the salivary glands are well seen on this axial (−10 degrees) anatomic plate, including the parotid glands, which lie lateral to the mandible, as well as the sublingual glands, which lie medial to the anterior mandible adjacent to the tongue muscles. This section also demonstrates the upper cervical vertebral body, with the pedicle and the facet forming the anteriolateral margin of the cervical spinal canal.

20B Axial/CT/Mandible. This axial (0 degrees) CT is made at the level of the superior hyoid bone. The epiglottis is seen between the air spaces of the vallecula of the tongue anteriorly and the oropharynx posteriorly. The carotid sheath structures, including the carotid artery and internal jugular vein, are seen at the anterior margin of the posterior triangle of the neck.

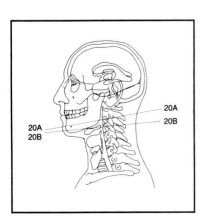

20A
20B

20A
20B

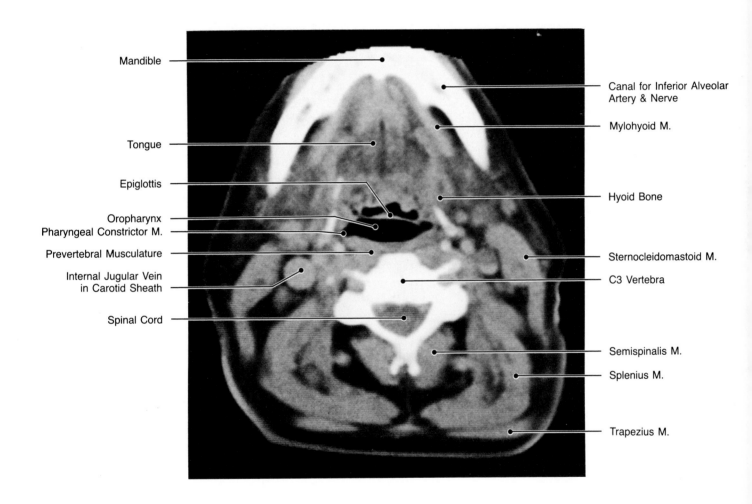

Mandible

Canal for Inferior Alveolar
Artery & Nerve

Mylohyoid M.

Tongue

Epiglottis

Hyoid Bone

Oropharynx
Pharyngeal Constrictor M.

Prevertebral Musculature

Sternocleidomastoid M.

Internal Jugular Vein
in Carotid Sheath

C3 Vertebra

Spinal Cord

Semispinalis M.

Splenius M.

Trapezius M.

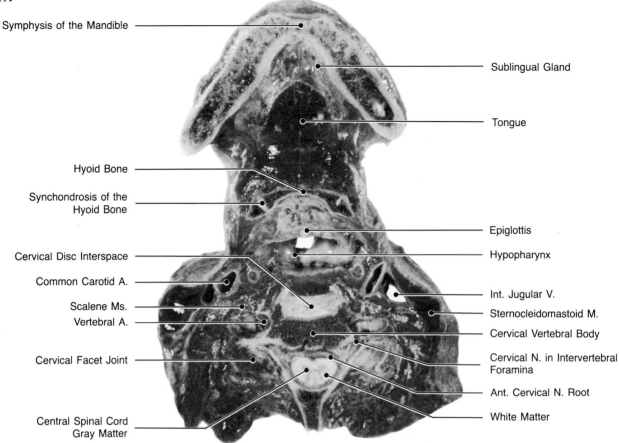

Symphysis of the Mandible

Sublingual Gland

Tongue

Hyoid Bone

Synchondrosis of the Hyoid Bone

Epiglottis

Cervical Disc Interspace

Hypopharynx

Common Carotid A.

Int. Jugular V.

Scalene Ms.

Sternocleidomastoid M.

Vertebral A.

Cervical Vertebral Body

Cervical Facet Joint

Cervical N. in Intervertebral Foramina

Ant. Cervical N. Root

Central Spinal Cord Gray Matter

White Matter

21A Axial/Anatomic Plate/Hyoid Bone. This axial (−10 degrees) anatomic plate is made at the level of the hyoid bone. Just posterior to the hyoid is the epiglottic fat space. The hyoid bone is divided into three segments. This section is also through a neural foramen of the upper cervical spine. The lateral posterior margins of the vertebral body form the anterior margin of the neural foramina, while the facet forms the posterior margin. Note the "butterfly" configuration of the gray matter tracts within the spinal cord.

21B Axial/MRI/Cervical Interspace. The upper cervical spinal cord is oval in configuration on this TR 800, TE 25 msec partial saturation axial (−10 degrees) MRI. The small nerve roots are seen arising from the lateral margin of the cord and are directed anteriorly and laterally towards the neural foramen. The neural foramen frequently contains epidural fat and veins. With this pulse sequence technique, these structures usually have a high signal intensity. The marrow-containing spaces of the facets and vertebral body outline the cortical bony portions of the neural foramen.

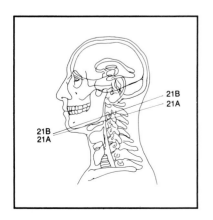

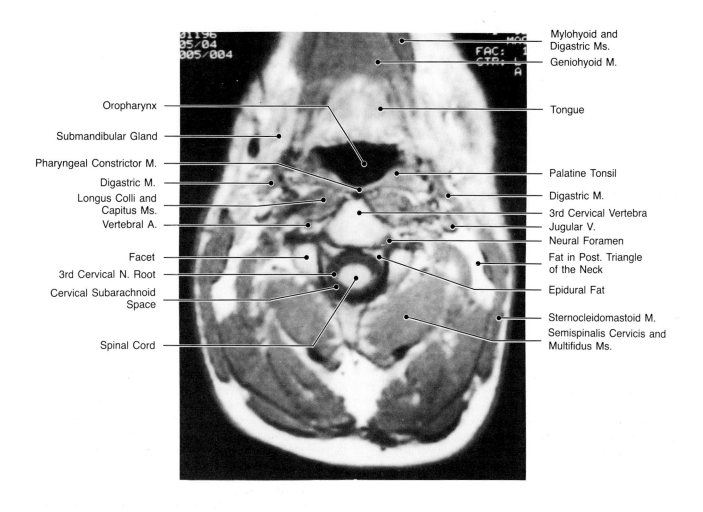

Mylohyoid and Digastric Ms.

Geniohyoid M.

Oropharynx

Submandibular Gland

Pharyngeal Constrictor M.

Digastric M.

Longus Colli and Capitus Ms.

Vertebral A.

Facet

3rd Cervical N. Root

Cervical Subarachnoid Space

Spinal Cord

Tongue

Palatine Tonsil

Digastric M.

3rd Cervical Vertebra

Jugular V.

Neural Foramen

Fat in Post. Triangle of the Neck

Epidural Fat

Sternocleidomastoid M.

Semispinalis Cervicis and Multifidus Ms.

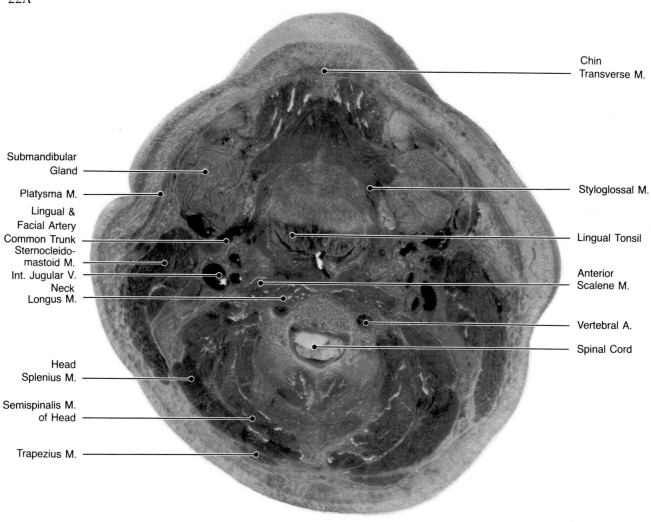

Chin
Transverse M.

Submandibular
Gland

Styloglossal M.

Platysma M.

Lingual &
Facial Artery
Common Trunk

Lingual Tonsil

Sternocleido-
mastoid M.

Int. Jugular V.

Anterior
Scalene M.

Neck
Longus M.

Vertebral A.

Spinal Cord

Head
Splenius M.

Semispinalis M.
of Head

Trapezius M.

22A Axial/Anatomic Plate/Submandibular Glands. The submandibular glands are circularly shaped salivary structures that lie lateral and inferior to the tongue and angle of the mandible on this axial (−10 degrees) section. Note the large lingual tonsil seen at the base of the tongue. This frequently protrudes into the hypopharynx and has a multinodular appearance.

22B Axial/CT/Hyoid Bone. This axial (0 degrees) CT scan is displayed at soft tissue windows. The submandibular glands are seen as oval-shaped soft tissue densities just lateral to the tongue and hyoid bone. The hyoid bone is composed of three separate segments. The epiglottis is seen interposed between the vallecula of the tongue base anteriorly and the laryngeal airway posteriorly.

22C Axial/MRI/Vallecula. The small air spaces of the vallecula just anterior to the epiglottis and at the base of the tongue are well seen on this partial saturation TR 800 and TE25 msec MRI. The musculature about the hypopharynx is well defined. Just anterior and lateral to the cervical spine lie the scalene and longus colli muscles.

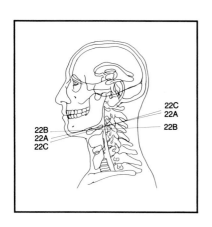

22B

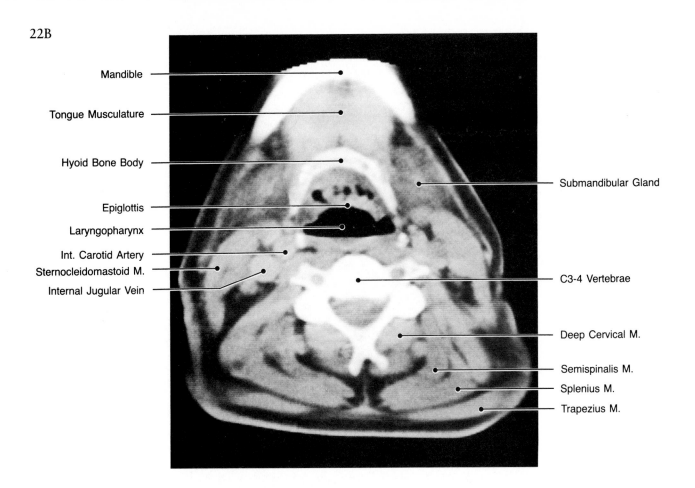

Mandible

Tongue Musculature

Hyoid Bone Body

Epiglottis

Laryngopharynx

Int. Carotid Artery

Sternocleidomastoid M.

Internal Jugular Vein

Submandibular Gland

C3-4 Vertebrae

Deep Cervical M.

Semispinalis M.

Splenius M.

Trapezius M.

22C

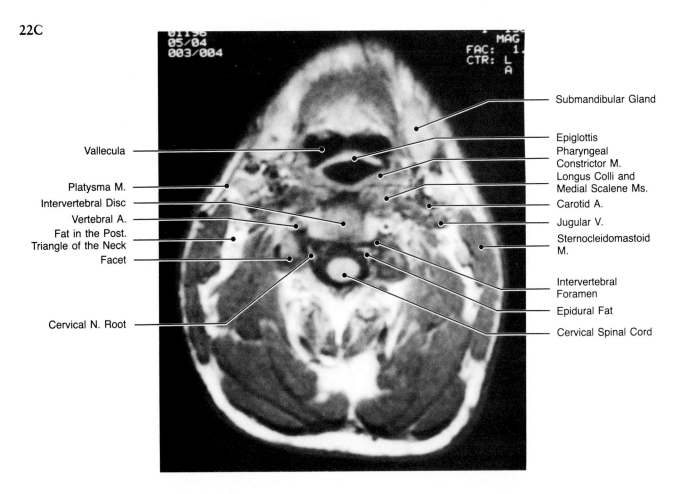

Vallecula

Platysma M.

Intervertebral Disc

Vertebral A.

Fat in the Post.
Triangle of the Neck

Facet

Cervical N. Root

Submandibular Gland

Epiglottis

Pharyngeal
Constrictor M.

Longus Colli and
Medial Scalene Ms.

Carotid A.

Jugular V.

Sternocleidomastoid
M.

Intervertebral
Foramen

Epidural Fat

Cervical Spinal Cord

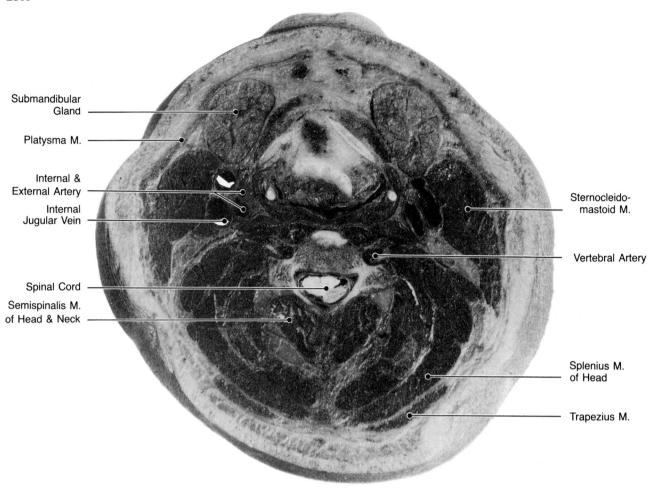

Submandibular Gland

Platysma M.

Internal & External Artery

Internal Jugular Vein

Spinal Cord

Semispinalis M. of Head & Neck

Sternocleido-mastoid M.

Vertebral Artery

Splenius M. of Head

Trapezius M.

23A Axial/Anatomic Plate/Superior Thyroid Cartilage. The epiglottis is seen as a crescent-shaped cartilaginous structure on this axial (−10 degrees) anatomic plate. Anterior to the epiglottis is a normal fat space that is directly posterior to the hyoid bone and superior margin of the thyroid cartilage. The oropharynx broadens into a crescent-shaped cavity, with the two lateral margins forming the superior portions of the pyriform recesses.

23B Axial/CT/Epiglottis. This axial (0 degrees) CT displayed at soft tissue windows is made at the level of the hyoid bone and superior horn of the thyroid cartilage. The hypopharyngeal airway is divided into three cavities; the central is the laryngeal pharynx vestibule. Small tissue bands are seen at the lateral margins of this, related to the aryepiglottic folds dividing the pyriform sinuses laterally.

23C Axial/MRI/Aryepiglottic Folds. This axial (O degrees) partial saturation TR 1000 and TE 25 msec MRI demonstrates the anterior bellies of the diagastric muscles inserting on the mandible and extending posteriorly and laterally. The sternohyoid and omohyoid muscles overlie the anterior margin of the superior thyroid cartilage, which contains high signal intensity marrow. The aryepiglottic folds separate the pyriform recesses from the laryngeal airway.

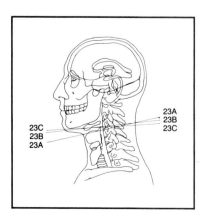

23B

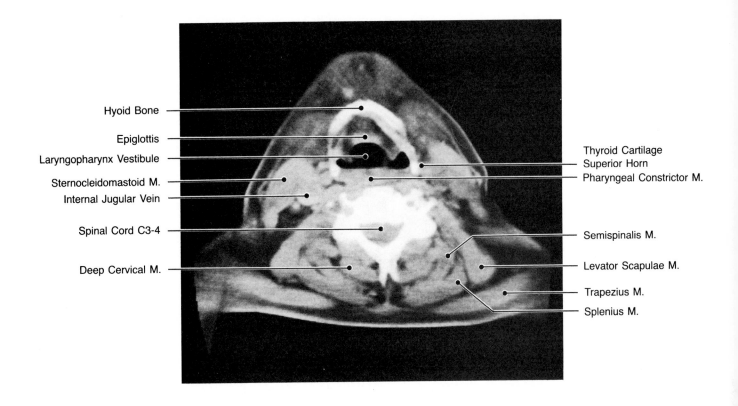

Hyoid Bone

Epiglottis

Laryngopharynx Vestibule

Sternocleidomastoid M.

Internal Jugular Vein

Spinal Cord C3-4

Deep Cervical M.

Thyroid Cartilage
Superior Horn

Pharyngeal Constrictor M.

Semispinalis M.

Levator Scapulae M.

Trapezius M.

Splenius M.

23C

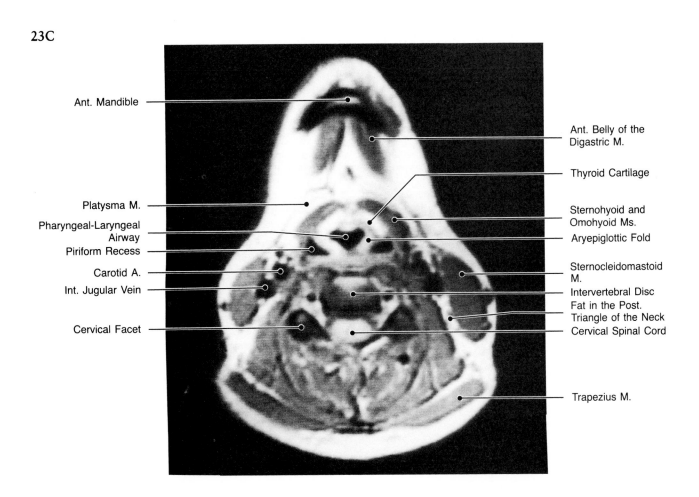

Ant. Mandible

Platysma M.

Pharyngeal-Laryngeal
Airway

Piriform Recess

Carotid A.

Int. Jugular Vein

Cervical Facet

Ant. Belly of the
Digastric M.

Thyroid Cartilage

Sternohyoid and
Omohyoid Ms.

Aryepiglottic Fold

Sternocleidomastoid
M.

Intervertebral Disc

Fat in the Post.
Triangle of the Neck

Cervical Spinal Cord

Trapezius M.

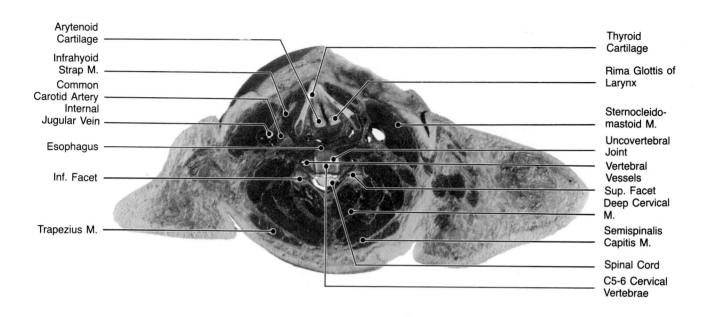

Arytenoid Cartilage

Infrahyoid Strap M.

Common Carotid Artery

Internal Jugular Vein

Esophagus

Inf. Facet

Trapezius M.

Thyroid Cartilage

Rima Glottis of Larynx

Sternocleido-mastoid M.

Uncovertebral Joint

Vertebral Vessels

Sup. Facet

Deep Cervical M.

Semispinalis Capitis M.

Spinal Cord

C5-6 Cervical Vertebrae

24A Axial/Anatomic Plate/Arytenoid Cartilages. This axial (0 degrees) anatomic plate demonstrates the uncovertebral joints of the cervical vertebrae. The vertebral end plates form a cup and saucer configuration with the inferior end plate representing the cup. The saucer lips of the superior vertebral endplate project superiorly and laterally on axial sections, with joint spaces seen at the lateral margins of the vertebral bodies.

24B Axial/CT/False Cords. This axial (0 degrees) CT image is displayed with soft tissue windows at the level of the false cords. The triangular thyroid cartilage is seen with the superior notch evident as a defect in the bridging thyroid cartilage. The arytenoid cartilages are partially calcified and seen at the posterior margin of the vestibule. The base of the pyriform recesses is also visible just lateral to the arytenoids.

24C Axial/CT/True Cords. The superior posterior ring of the cricoid cartilage is seen on this axial CT image displayed at soft tissue windows. The section is made at the level of the true cords. Note that there is only a minimal amount of soft tissue at the anterior portion of the airway related to the thin anterior commissure.

24D Axial/MRI/False Cords. The carotid artery and internal jugular veins are seen as circular low signal regions underlying the sternocleidomastoid muscle on this TR 1000 and TE 25 msec axial (0 degrees) partial saturation MRI. Posterior to the carotid sheath structures lies the fatty space of the posterior triangle. The cervical vertebral body is well seen, with high signal intensity related to the marrow spaces. Note that the vertebral artery in the transverse process demonstrates low signal intensity.

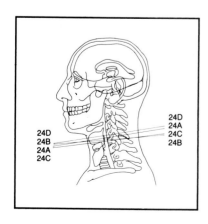

24B

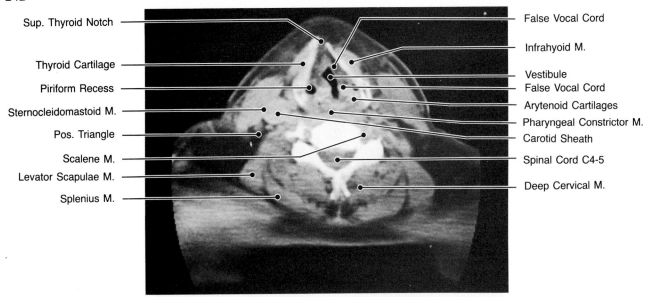

Sup. Thyroid Notch — False Vocal Cord

Thyroid Cartilage — Infrahyoid M.

Piriform Recess — Vestibule
False Vocal Cord

Sternocleidomastoid M. — Arytenoid Cartilages
Pharyngeal Constrictor M.

Pos. Triangle — Carotid Sheath

Scalene M. — Spinal Cord C4-5

Levator Scapulae M. — Deep Cervical M.

Splenius M.

24C

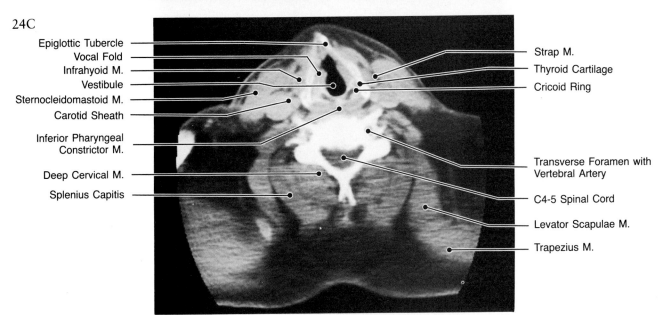

Epiglottic Tubercle — Strap M.
Vocal Fold — Thyroid Cartilage
Infrahyoid M. — Cricoid Ring
Vestibule
Sternocleidomastoid M.
Carotid Sheath

Inferior Pharyngeal
Constrictor M.

Deep Cervical M. — Transverse Foramen with
Vertebral Artery

Splenius Capitis — C4-5 Spinal Cord

Levator Scapulae M.

Trapezius M.

24D

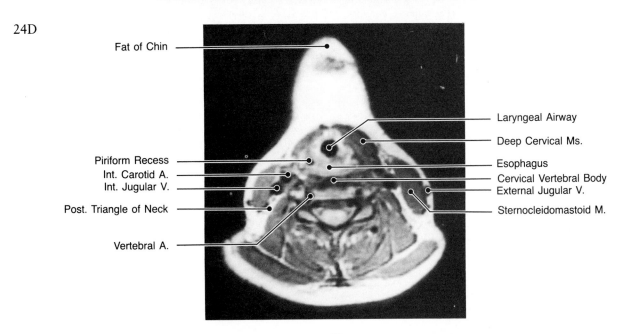

Fat of Chin — Laryngeal Airway

Deep Cervical Ms.

Piriform Recess — Esophagus
Int. Carotid A. — Cervical Vertebral Body
Int. Jugular V. — External Jugular V.

Post. Triangle of Neck — Sternocleidomastoid M.

Vertebral A.

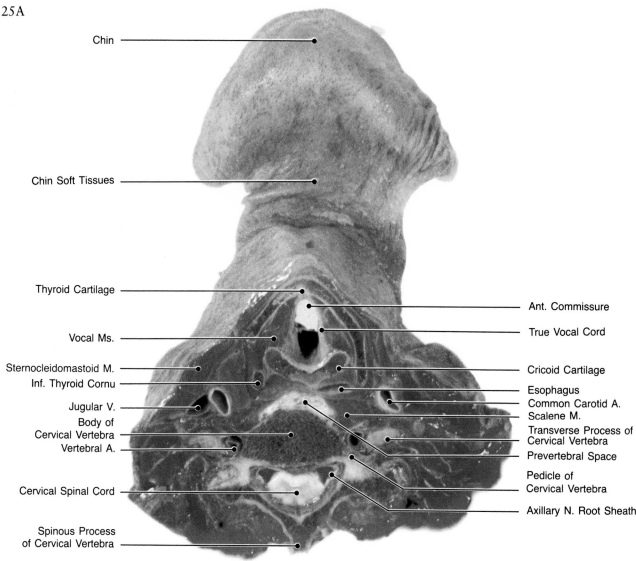

Chin

Chin Soft Tissues

Thyroid Cartilage

Vocal Ms.

Sternocleidomastoid M.
Inf. Thyroid Cornu

Jugular V.
Body of
Cervical Vertebra
Vertebral A.

Cervical Spinal Cord

Spinous Process
of Cervical Vertebra

Ant. Commissure

True Vocal Cord

Cricoid Cartilage

Esophagus
Common Carotid A.
Scalene M.

Transverse Process of
Cervical Vertebra

Prevertebral Space

Pedicle of
Cervical Vertebra

Axillary N. Root Sheath

25A Axial/Anatomic Plate/True Vocal Cords. The true vocal cords are seen on this axial (0 degrees) anatomic plate extending from the region of the superior posterior cricoid ring anteriorly to the thyroid cartilage. The anterior commissure is a thin structure separating the two true vocal cords. The scalene muscles lie just anterior and lateral to the cervical vertebral body and overlie the region of the exiting nerve roots.

25B Axial/CT/Cricoid Ring. This axial (0 degrees) CT is displayed at soft tissue windows. The cricoid ring is seen at the level of the inferior thyroid cornu. Note that there is no visible soft tissue component within the inner ring of the laryngeal structures at this level.

25C Axial/MRI/True Vocal Cords. The airway at the level of the true vocal cords is a sharply pointed triangular space on this TR 1000 and TE 25 msec axial (0 degrees) partial saturation MRI. The anterior margin of the airway is defined by the thyroid cartilage, while the lateral margins are the true cords themselves. Posteriorly the arc-shaped ring of the cricoid is visible. The esophageal airway lies posterior to the cricoid ring and is usually not visible since it is collapsed.

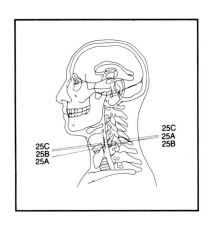

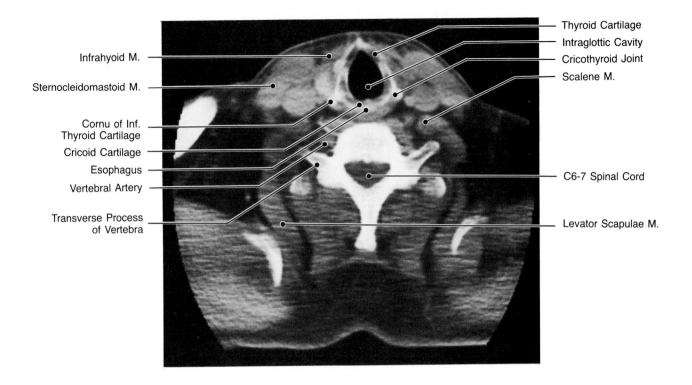

Infrahyoid M.

Sternocleidomastoid M.

Cornu of Inf.
Thyroid Cartilage

Cricoid Cartilage

Esophagus

Vertebral Artery

Transverse Process
of Vertebra

Thyroid Cartilage

Intraglottic Cavity

Cricothyroid Joint

Scalene M.

C6-7 Spinal Cord

Levator Scapulae M.

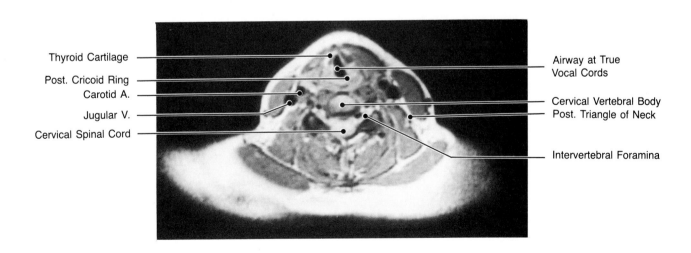

Thyroid Cartilage

Post. Cricoid Ring

Carotid A.

Jugular V.

Cervical Spinal Cord

Airway at True
Vocal Cords

Cervical Vertebral Body

Post. Triangle of Neck

Intervertebral Foramina

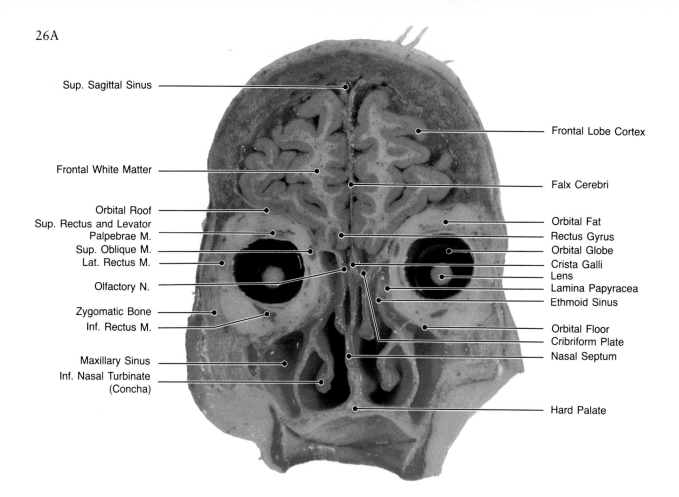

Sup. Sagittal Sinus

Frontal White Matter

Orbital Roof
Sup. Rectus and Levator
Palpebrae M.
Sup. Oblique M.
Lat. Rectus M.

Olfactory N.

Zygomatic Bone
Inf. Rectus M.

Maxillary Sinus
Inf. Nasal Turbinate
(Concha)

Frontal Lobe Cortex

Falx Cerebri

Orbital Fat
Rectus Gyrus
Orbital Globe
Crista Galli
Lens
Lamina Papyracea
Ethmoid Sinus

Orbital Floor
Cribriform Plate
Nasal Septum

Hard Palate

26A Coronal/Anatomic Plate/Anterior Globe. This coronal (+90 degrees) anatomic plate is made through the anterior frontal lobes and orbits. The lenses of the globes have collapsed into the globes as an artifact of liquefication of the vitreous. The close relationship of the frontal lobes, orbital contents, and sinus cavities is depicted. The thin cribriform plates separate the olfactory tracts from the superior nasal cavity. The peduncular plate of the ethmoid bone projects inferiorly in continuity with the nasal septum. The ethmoid sinus forms the medial wall of the orbit.

26B Coronal/CT/Anterior Face. The anterior superficial orbital structures are well delineated on this soft tissue window coronal (+100 degrees) CT with visualization of the dense lens to the vitreous. Cataracts are seen as calcifications. The medial and lateral canthi of the eyes are seen as thin air-filled strips outlined by the soft tissue of the lids. The cartilaginous and bony components of the anterior nasal cavity are seen.

26C Coronal/MRI/Anterior Face. This partial saturation TR 800, TE 25 msec (+90 degrees) MRI demonstrates the fatty portions of the face, scalp, and orbit to have high signal character. The orbital fat outlines the anterior globe. The bone and air-containing spaces of the face and sinuses are seen as low signal areas. Note that the soft tissues of the nasal cavity are well delineated by virtue of visualization of their soft tissue components rather than their bony members.

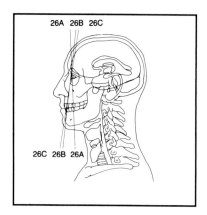

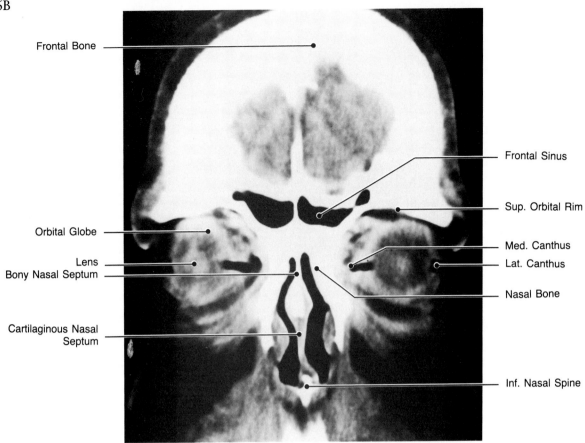

Frontal Bone

Frontal Sinus

Sup. Orbital Rim

Orbital Globe

Med. Canthus

Lens

Lat. Canthus

Bony Nasal Septum

Nasal Bone

Cartilaginous Nasal Septum

Inf. Nasal Spine

26C

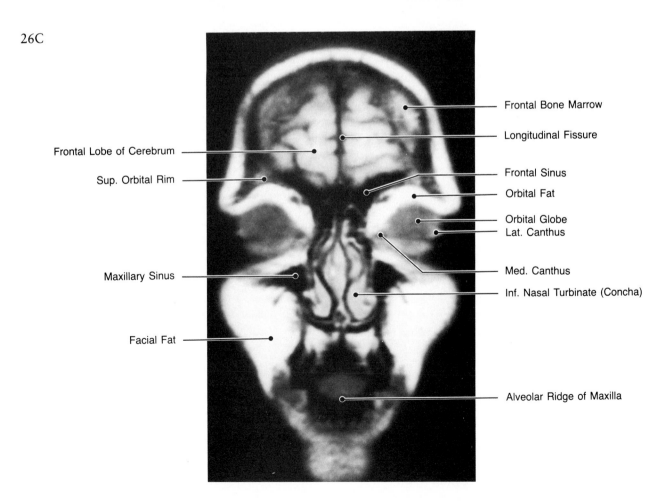

Frontal Bone Marrow

Longitudinal Fissure

Frontal Lobe of Cerebrum

Sup. Orbital Rim

Frontal Sinus

Orbital Fat

Orbital Globe

Lat. Canthus

Maxillary Sinus

Med. Canthus

Inf. Nasal Turbinate (Concha)

Facial Fat

Alveolar Ridge of Maxilla

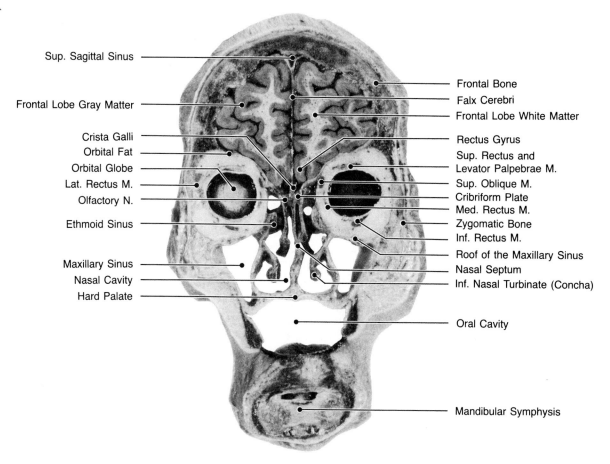

Sup. Sagittal Sinus

Frontal Lobe Gray Matter

Crista Galli
Orbital Fat
Orbital Globe
Lat. Rectus M.
Olfactory N.

Ethmoid Sinus

Maxillary Sinus
Nasal Cavity
Hard Palate

Frontal Bone
Falx Cerebri
Frontal Lobe White Matter

Rectus Gyrus
Sup. Rectus and
Levator Palpebrae M.
Sup. Oblique M.
Cribriform Plate
Med. Rectus M.
Zygomatic Bone
Inf. Rectus M.
Roof of the Maxillary Sinus
Nasal Septum
Inf. Nasal Turbinate (Concha)

Oral Cavity

Mandibular Symphysis

27A Coronal/Anatomic Plate/Midglobe. This coronal (+90 degrees) section is made through the crista galli, which is seen as a bony spur projecting superiorly from the midline ethmoid bone. The falx cerebri inserts from above, while the perpendicular plate of the superior nasal cavity projects inferiorly. The adjacent cribriform plates and olfactory nerves are evident. The external ocular muscles form thin bands near their insertion to the globe. Note the extensive fatty spaces of the face.

27B Coronal/CT Soft Tissue Windows/Midglobe. This coronal (+90 degrees) CT demonstrates a number of the extraocular muscles, which are seen as water density structures surrounded by lower density orbital fat. Anteriorly the muscles broaden into a band and conform to the contour of the globe. The lamina papyracea is the medial wall of the orbit and the lateral wall of the ethmoid sinus. The midline crista galli of the ethmoid bone is evident, with dense falx cerebri extending superiorly from its insertion. The relationship of the nasal turbinates to the nasal septum is well depicted.

27C Coronal/MRI/Globe. This coronal (+90 degrees) partial saturation TR 400, TE 25 msec MRI was made with a surface coil of the single left orbit. A number of the superficial structures of the face are well seen, including a small vein running in the fat of the face. Note that the sclera is seen as a signal void, which is related to the fact that the hydrogen is bound in a solid chemical-like environment. The choroid and retina are seen as a thin circular strip of tissue bordering the vitreous of the posterior chamber.

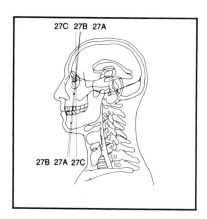

27C 27B 27A

27B 27A 27C

27B

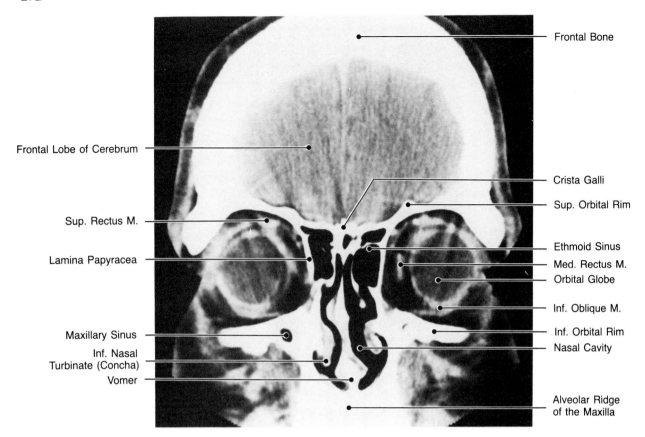

Frontal Bone

Frontal Lobe of Cerebrum

Crista Galli

Sup. Orbital Rim

Sup. Rectus M.

Ethmoid Sinus

Lamina Papyracea

Med. Rectus M.

Orbital Globe

Inf. Oblique M.

Maxillary Sinus

Inf. Orbital Rim

Inf. Nasal
Turbinate (Concha)

Nasal Cavity

Vomer

Alveolar Ridge
of the Maxilla

27C

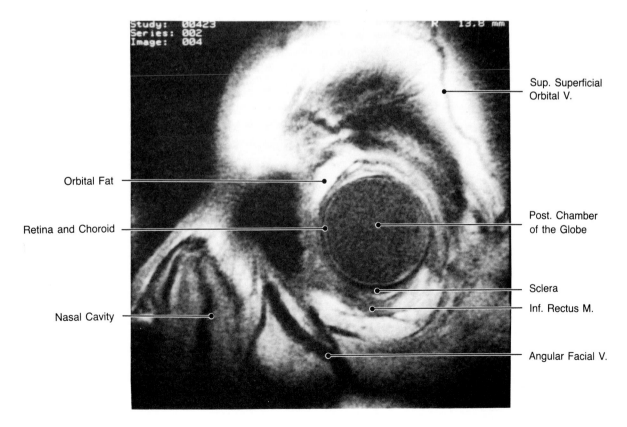

Sup. Superficial
Orbital V.

Orbital Fat

Post. Chamber
of the Globe

Retina and Choroid

Sclera

Inf. Rectus M.

Nasal Cavity

Angular Facial V.

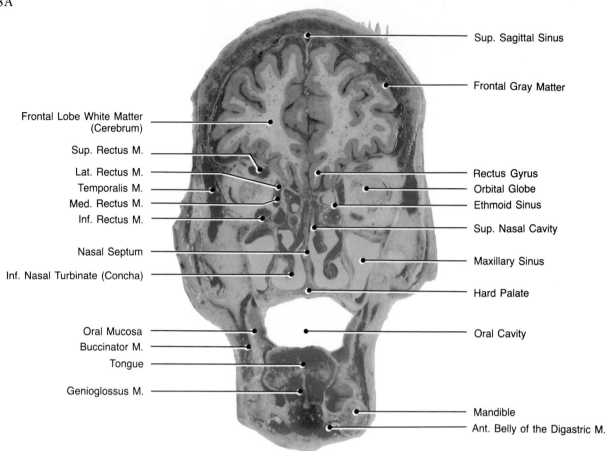

Sup. Sagittal Sinus

Frontal Gray Matter

Frontal Lobe White Matter (Cerebrum)

Sup. Rectus M.

Lat. Rectus M.

Temporalis M.

Med. Rectus M.

Inf. Rectus M.

Nasal Septum

Inf. Nasal Turbinate (Concha)

Oral Mucosa

Buccinator M.

Tongue

Genioglossus M.

Rectus Gyrus

Orbital Globe

Ethmoid Sinus

Sup. Nasal Cavity

Maxillary Sinus

Hard Palate

Oral Cavity

Mandible

Ant. Belly of the Digastric M.

28A Coronal/Anatomic Plate/Posterior Orbit. The rectus gyrus is seen on this coronal anatomic (+90 degrees) plate as the most medial gyrus of the inferior frontal lobe directly above the ethmoid sinus. The olfactory tracts parallel the undersurface of the gyrus and end in the cribriform plates. The extraocular muscles are nearly rectangular in configuration in the mid orbits.

28B Coronal/MRI/Midglobe. This partial saturation TR 800, TE 25 msec coronal (+90 degrees) MRI demonstrates the extraocular muscles as lower signal structures within the high signal orbital fat. The globes appear as low signal structures due to the long relaxation times of the vitreous. The mucoperiosteum covering the ethmoid sinus structures and the soft tissue coverings of the nasal cavity are evident. The orbital roof is between the frontal lobe and the orbital fat. The floor of the orbit, however, is not seen because no signal arises from either the floor of the orbit or the air within it. Note the extensive fat of the scalp and face.

28C Coronal/CT Soft Tissue Windows/Posterior Orbit. The medial wall of the orbit (lamina papyracea) and the floor of the orbit (roof of the maxillary sinus) form an obtuse angle on this coronal (+90 degrees) CT. The superior and inferior nasal turbinates are seen in the nasal cavity and lie between the maxillary sinuses. The temporalis muscle is interposed between the lateral frontal bone and the zygomatic bone.

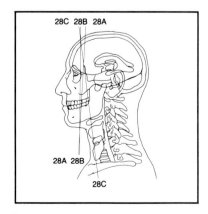

28B

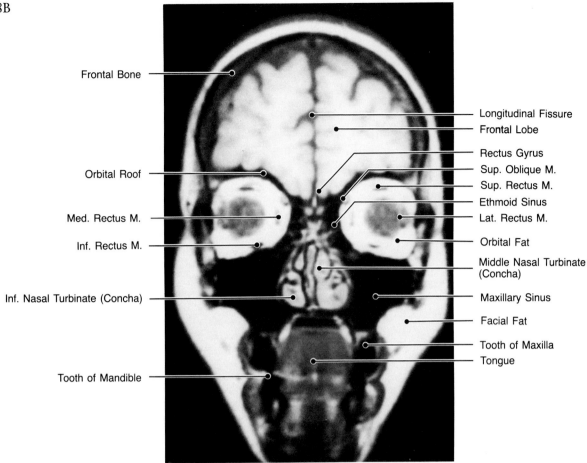

Frontal Bone

Longitudinal Fissure

Frontal Lobe

Rectus Gyrus

Sup. Oblique M.

Orbital Roof

Sup. Rectus M.

Ethmoid Sinus

Lat. Rectus M.

Med. Rectus M.

Orbital Fat

Inf. Rectus M.

Middle Nasal Turbinate
(Concha)

Maxillary Sinus

Inf. Nasal Turbinate (Concha)

Facial Fat

Tooth of Maxilla

Tongue

Tooth of Mandible

28C

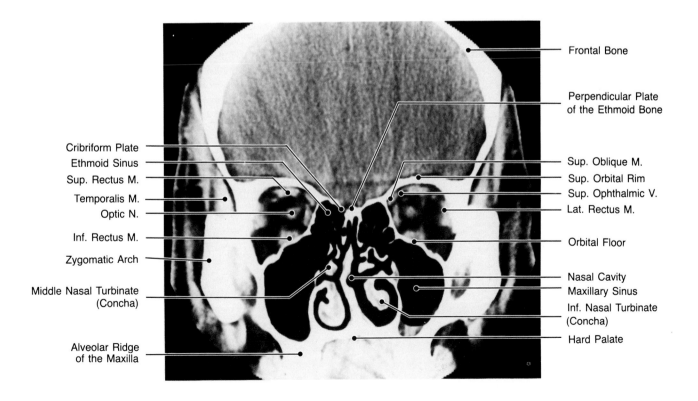

Frontal Bone

Perpendicular Plate
of the Ethmoid Bone

Cribriform Plate

Ethmoid Sinus

Sup. Oblique M.

Sup. Rectus M.

Sup. Orbital Rim

Temporalis M.

Sup. Ophthalmic V.

Optic N.

Lat. Rectus M.

Inf. Rectus M.

Orbital Floor

Zygomatic Arch

Nasal Cavity

Middle Nasal Turbinate
(Concha)

Maxillary Sinus

Inf. Nasal Turbinate
(Concha)

Alveolar Ridge
of the Maxilla

Hard Palate

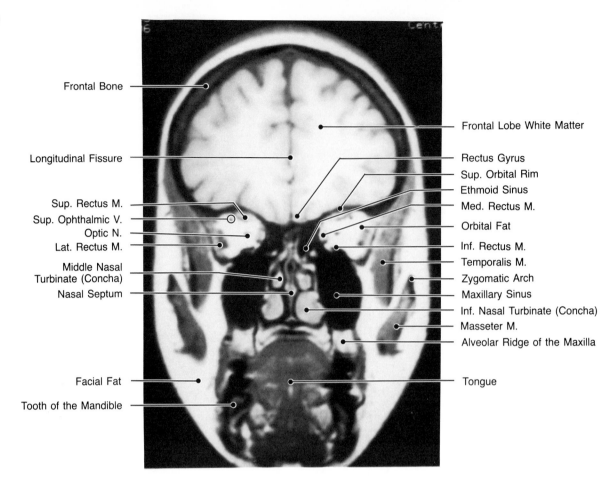

28D Coronal/MRI/Posterior Orbit. This partial saturation TR 800, TE 25 msec coronal (+90 degrees) MRI demonstrates the temporalis muscle interposed between the lateral wall of the orbit (innominate line) and the zygomatic arch. The masseter muscle arises from the zygomatic arch and is outlined by adjacent fat of the face. The optic nerve and extraocular muscles are lower signal regions than the surrounding orbital fat.

28E Coronal/MRI/Optic Nerve. This partial saturation coronal (+90 degrees) TR 400, TE 25 msec MRI was made with a surface coil of the single left globe. Utilization of surface coils allows for exquisite detail of the anatomic structures. The optic nerve is seen surrounded by a low signal halo related to the spinal fluid of the optic nerve sheath. The superior ophthalmic vein turns medially beneath the superior rectus muscle and is also seen as a low signal region.

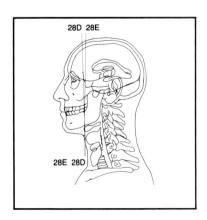

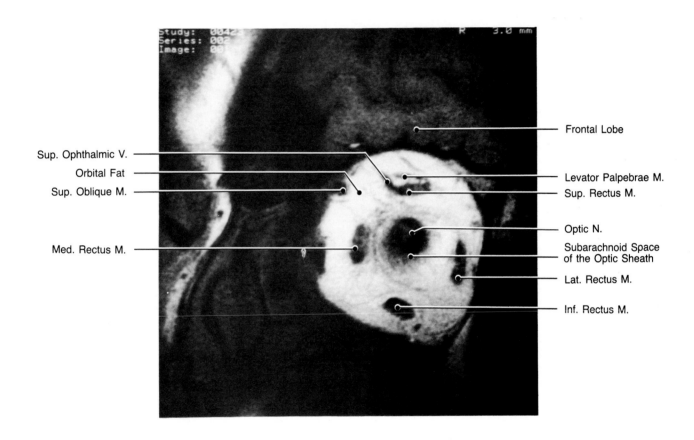

Sup. Ophthalmic V.

Orbital Fat

Sup. Oblique M.

Med. Rectus M.

Frontal Lobe

Levator Palpebrae M.

Sup. Rectus M.

Optic N.

Subarachnoid Space
of the Optic Sheath

Lat. Rectus M.

Inf. Rectus M.

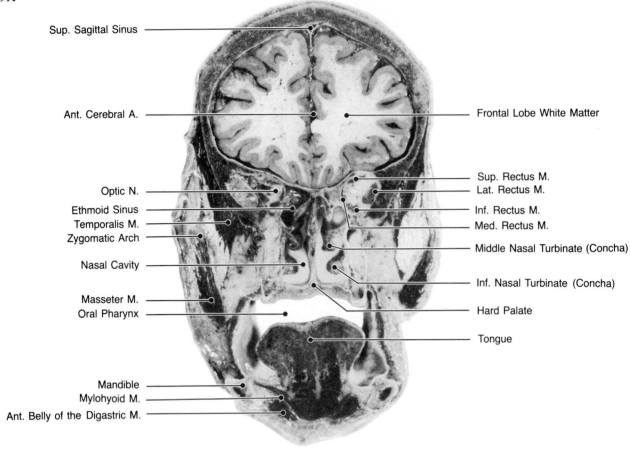

Sup. Sagittal Sinus

Ant. Cerebral A.

Frontal Lobe White Matter

Optic N.
Ethmoid Sinus
Temporalis M.
Zygomatic Arch

Sup. Rectus M.
Lat. Rectus M.
Inf. Rectus M.
Med. Rectus M.

Nasal Cavity

Middle Nasal Turbinate (Concha)

Masseter M.
Oral Pharynx

Inf. Nasal Turbinate (Concha)

Hard Palate

Tongue

Mandible
Mylohyoid M.
Ant. Belly of the Digastric M.

29A Coronal/Anatomic Plate/Optic Nerve. The extraocular muscles converge on the central optic nerve at the orbital apex on this coronal (+90 degrees) anatomic section. The muscles lose their rectangular configuration as they form tendinous insertions. The optic nerve is surrounded by a thin subarachnoid space within the dural optic nerve sheath. The middle and inferior nasal turbinates are seen projecting into the nasal cavity. The temporalis muscles are interposed between the lateral margin of the frontal bone and the zygomatic arch. The masseter muscle is seen extending inferiorly from the zygomatic arch. The tongue is suspended by a sling formed by the mylohyoid muscle. Directly beneath this muscle lies the anterior belly of the digastric muscle.

29B Coronal/MRI/Anterior Cerebral Arteries. This coronal (+90 degrees) spin echo MRI was made with TR 4000, TE 25 msec. The close association of the anterior corpus callosum to the pericallosal arteries is evident. The white matter tracts of the frontal lobes stand out as low signal regions because their proton density (protons per volume) is lower than that of the adjacent gray matter structures.

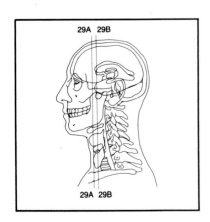

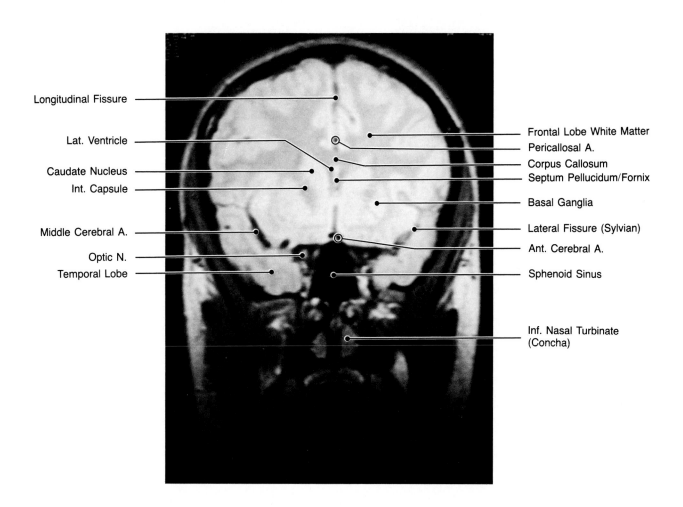

Longitudinal Fissure

Lat. Ventricle

Caudate Nucleus

Int. Capsule

Middle Cerebral A.

Optic N.

Temporal Lobe

Frontal Lobe White Matter

Pericallosal A.

Corpus Callosum

Septum Pellucidum/Fornix

Basal Ganglia

Lateral Fissure (Sylvian)

Ant. Cerebral A.

Sphenoid Sinus

Inf. Nasal Turbinate
(Concha)

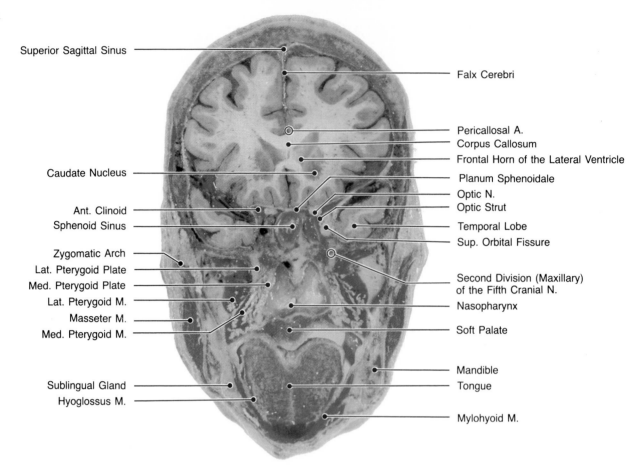

Superior Sagittal Sinus

Caudate Nucleus

Ant. Clinoid
Sphenoid Sinus

Zygomatic Arch
Lat. Pterygoid Plate
Med. Pterygoid Plate
Lat. Pterygoid M.
Masseter M.
Med. Pterygoid M.

Sublingual Gland
Hyoglossus M.

Falx Cerebri

Pericallosal A.
Corpus Callosum
Frontal Horn of the Lateral Ventricle
Planum Sphenoidale
Optic N.
Optic Strut
Temporal Lobe
Sup. Orbital Fissure

Second Division (Maxillary)
of the Fifth Cranial N.
Nasopharynx
Soft Palate

Mandible
Tongue

Mylohyoid M.

30A Coronal/Anatomic Plate/Optic Canals. The pericallosal artery is seen in the longitudinal fissure directly above the corpus callosum (which bridges the two hemispheres) on this coronal (+90 degrees) anatomic section. The optic nerves are within the optic canals. The anterior clinoid lies laterally and superiorly, while the optic strut forms the lateral inferior margin of the optic canal. The medial inferior margin is formed by the sphenoid sinus. The medial and lateral pterygoid plates are interposed between the nasopharynx and the adjacent medial and lateral pterygoid muscles. The mylohyoid muscle is suspended between the bodies of the mandible. The hyoglossus muscle is just medial to the mylohyoid, while the sublingual gland lies just superficially.

30B Coronal/MRI/Optic Canals. This coronal (+90 degrees) partial saturation MRI TR 800, TE 25 msec demonstrates the optic nerves surrounded by a signal void related to

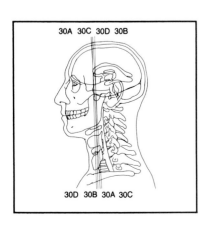

30A 30C 30D 30B

30D 30B 30A 30C

the bony segments of the optic canals. The superior orbital fissures form thin inverted triangular regions of high signal related to their fat content. The pterygoid plates are seen as low signal arcs outlined by the adjacent intermediate signal muscles and high signal fat spaces. The small low signal foci within the fat spaces of the deep face are primarily related to blood vessels, such as the internal maxillary artery branches. The soft palate is a high signal area related to its fat content.

30C Coronal/MRI/Optic Canals. This high resolution coronal (+90 degrees) spin echo TR 2500, TE 75 msec MRI demonstrates less contrast between the fat and brain tissues than the shorter repetition time images. The bony structures are more accurately outlined because of the increased surrounding signal of all of the adjacent tissues. The optic nerves are seen as ovals surrounded by a ring of absent signal related to the anterior clinoid, optic strut, and sphenoid sinus. Directly inferior to the optic canal lies the superior orbital fissure. Small signal voids can be identified in this region caused by the cranial nerves. The middle nasal turbinate demonstrates a high signal because of the high water content of the mucosa.

30D Coronal/CT Bone Window/Optic Canals. This is a coronal (+110 degrees) CT obtained with high resolution bone window settings; it demonstrates the optic canals. Laterally the optic canal is bordered by the anterior clinoid. Laterally and inferiorly the canal is formed by the optic strut, and farther medially lies the sphenoid sinus. Septations within the sphenoid sinus are commonly seen. At the base of the pterygoid plates lie two canals. The foramen rotundum lies at the inferior margin of the superior orbital fissure. The pterygoid canal, which contains the vidian nerve, is seen in the base of the pterygoid plates.

30B

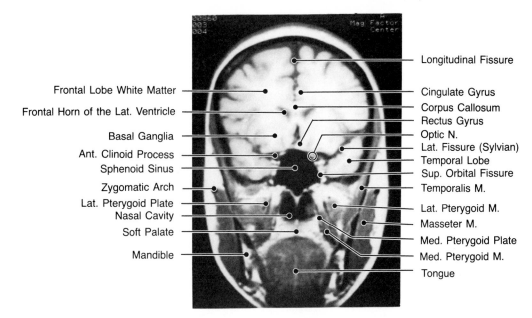

Frontal Lobe White Matter —
Frontal Horn of the Lat. Ventricle —
Basal Ganglia —
Ant. Clinoid Process —
Sphenoid Sinus —
Zygomatic Arch —
Lat. Pterygoid Plate —
Nasal Cavity —
Soft Palate —
Mandible —

— Longitudinal Fissure
— Cingulate Gyrus
— Corpus Callosum
— Rectus Gyrus
— Optic N.
— Lat. Fissure (Sylvian)
— Temporal Lobe
— Sup. Orbital Fissure
— Temporalis M.
— Lat. Pterygoid M.
— Masseter M.
— Med. Pterygoid Plate
— Med. Pterygoid M.
— Tongue

30C

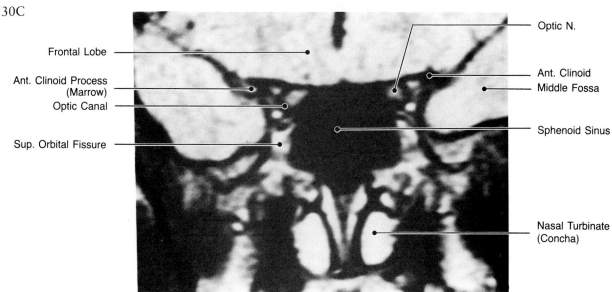

Frontal Lobe —
Ant. Clinoid Process (Marrow) —
Optic Canal —
Sup. Orbital Fissure —

— Optic N.
— Ant. Clinoid
— Middle Fossa
— Sphenoid Sinus
— Nasal Turbinate (Concha)

30D

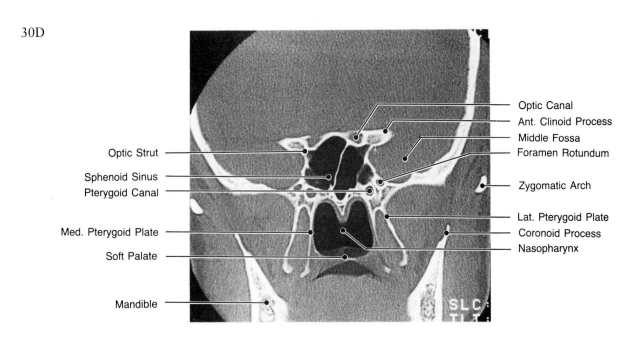

Optic Strut —
Sphenoid Sinus —
Pterygoid Canal —
Med. Pterygoid Plate —
Soft Palate —
Mandible —

— Optic Canal
— Ant. Clinoid Process
— Middle Fossa
— Foramen Rotundum
— Zygomatic Arch
— Lat. Pterygoid Plate
— Coronoid Process
— Nasopharynx

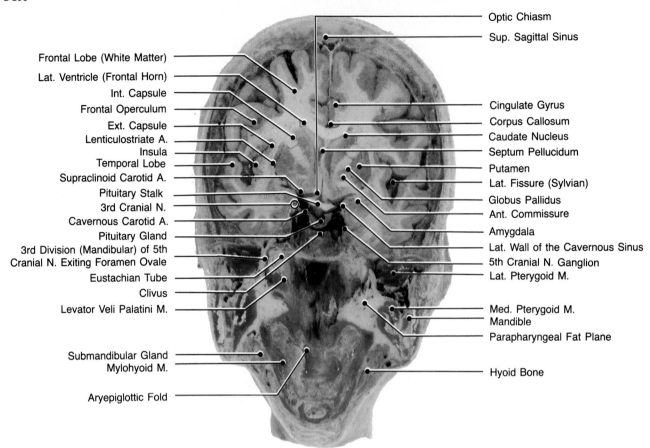

Optic Chiasm
Sup. Sagittal Sinus

Frontal Lobe (White Matter)
Lat. Ventricle (Frontal Horn)
Int. Capsule
Frontal Operculum
Ext. Capsule
Lenticulostriate A.
Insula
Temporal Lobe
Supraclinoid Carotid A.
Pituitary Stalk
3rd Cranial N.
Cavernous Carotid A.
Pituitary Gland
3rd Division (Mandibular) of 5th
Cranial N. Exiting Foramen Ovale
Eustachian Tube
Clivus
Levator Veli Palatini M.

Cingulate Gyrus
Corpus Callosum
Caudate Nucleus
Septum Pellucidum
Putamen
Lat. Fissure (Sylvian)
Globus Pallidus
Ant. Commissure
Amygdala
Lat. Wall of the Cavernous Sinus
5th Cranial N. Ganglion
Lat. Pterygoid M.

Med. Pterygoid M.
Mandible
Parapharyngeal Fat Plane

Submandibular Gland
Mylohyoid M.

Hyoid Bone

Aryepiglottic Fold

31A Coronal/Anatomic Plate/Optic Chiasm. The corpus callosum forms the roof of the lateral ventricles on this coronal (+90 degrees) section. The ventricles are separated by the septum pellucidum. The internal capsule is interposed between the caudate nucleus and the globus pallidus. The amygdala is a grey matter structure in the medial portion of the temporal lobe just anterior to the temporal horn. This coronal image well depicts the close relationship of the optic chiasm and the pituitary stalk, which courses towards the pituitary gland. The carotid arteries are seen within the cavernous sinus. A number of the cranial nerves are visible within the lateral margin of the cavernous sinus. Inferiorly the gasserian ganglion of the fifth cranial nerve is seen in Meckel's cave. The eustachian canal lies at the superior lateral aspect of the nasopharynx. Directly inferiorly is the levator veli palatini muscle, while superiorly and laterally is the third division of the fifth cranial nerve exiting the foramen ovale.

31B Coronal/MRI/Pituitary Infundibulum. This coronal (+90 degrees) partial saturation TR 800, TE 25 msec MRI accurately delineates the relationship of the anterior inferior third ventricle, optic chiasm, pituitary stalk, and pituitary gland. The spinal fluid spaces are all of low signal intensity because of their long relaxation times. A thin vessel is seen running through the basal ganglia, which is related to a lenticular striate artery. The region of Meckel's cave is seen as an intermediate signal intensity lateral to the sphenoid sinus and medial to the temporal lobes. The third division of the fifth cranial nerve at the foramen ovale is seen as a soft tissue structure bridging the middle fossa and the subtemporal space. The parapharyngeal fat plane is seen interposed between the pharyngeal constrictors medially and the pterygoid muscles laterally.

31C Coronal/MRI/Optic Chiasm. This coronal (+90 degrees) spin echo TR 3000, TE 75 msec MRI demonstrates the spinal fluid spaces as high signal regions related to their long relaxation times. Note that the optic chiasm and the anterior cerebral arteries are outlined by the suprasellar cistern. The white matter tracts are seen as low signal regions in relation to the gray matter structures such as the caudate nucleus and basal ganglia. This technique produces information similar to that obtained with a metrizamide cisternogram.

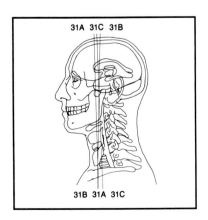

31A 31C 31B

31B 31A 31C

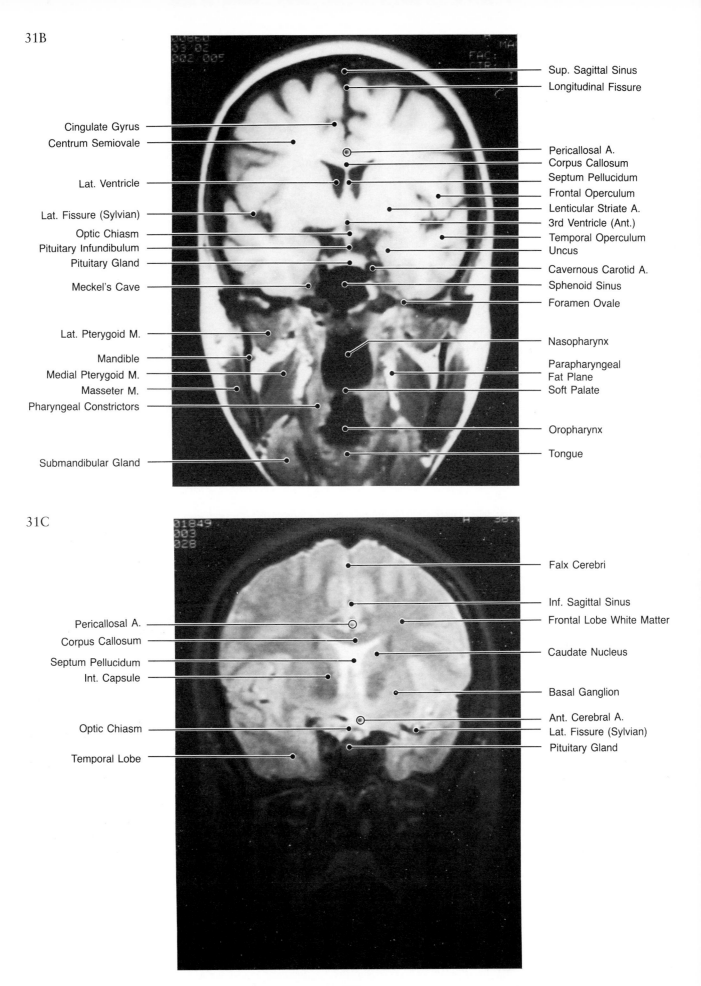

Sup. Sagittal Sinus

Longitudinal Fissure

Cingulate Gyrus

Centrum Semiovale

Pericallosal A.

Corpus Callosum

Lat. Ventricle

Septum Pellucidum

Frontal Operculum

Lenticular Striate A.

Lat. Fissure (Sylvian)

3rd Ventricle (Ant.)

Optic Chiasm

Temporal Operculum

Pituitary Infundibulum

Uncus

Pituitary Gland

Cavernous Carotid A.

Meckel's Cave

Sphenoid Sinus

Foramen Ovale

Lat. Pterygoid M.

Nasopharynx

Mandible

Parapharyngeal
Fat Plane

Medial Pterygoid M.

Masseter M.

Soft Palate

Pharyngeal Constrictors

Oropharynx

Tongue

Submandibular Gland

31C

Falx Cerebri

Inf. Sagittal Sinus

Pericallosal A.

Frontal Lobe White Matter

Corpus Callosum

Septum Pellucidum

Caudate Nucleus

Int. Capsule

Basal Ganglion

Optic Chiasm

Ant. Cerebral A.

Lat. Fissure (Sylvian)

Temporal Lobe

Pituitary Gland

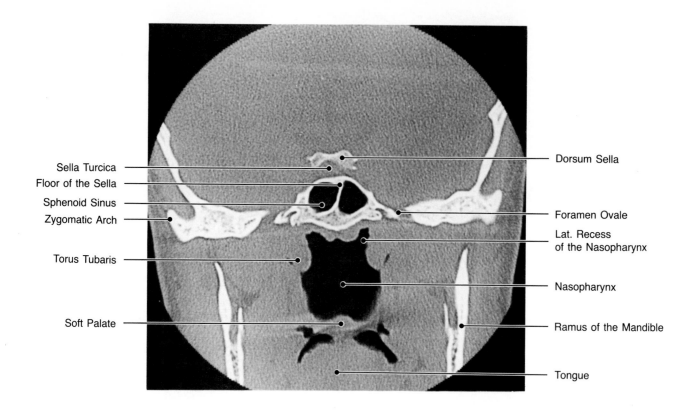

Sella Turcica
Floor of the Sella
Sphenoid Sinus
Zygomatic Arch
Torus Tubaris
Soft Palate

Dorsum Sella
Foramen Ovale
Lat. Recess of the Nasopharynx
Nasopharynx
Ramus of the Mandible
Tongue

31D Coronal/CT Bone Window/Sella Turcica. This high resolution thin section coronal (+110 degrees) CT through the sella turcica demonstrates the floor of the pituitary fossa. The dorsum sella is imaged simultaneously and demonstrates the upper aspect of the pituitary fossa. The foramen adjacent to the lateral margin of the sphenoid sinus is the foramen ovale.

31E Coronal/CT/Optic Chiasm. A coronal (+70 degrees) metrizamide cisternogram was used to accentuate visualization of the subarachnoid spaces at the skull base. The optic chiasm lies between the anterior communicating artery superiorly and the pituitary infundibulum inferiorly. The supraclinoid carotids are seen in the suprasellar cistern. The basilar and vertebral arteries form linear defects within the metrizamide filling the prepontine cistern. This technique is particularly valuable in the evaluation of surface abnormalities of the brain but is of less value in determining subtle abnormalities of the parenchyma that do not deform the outer contours. (From Chakeres DW, Kapila A: Brainstem and related structures: Normal CT anatomy using direct longitudinal scanning with metrizamide cisternography. Radiology 149:709–715, 1983.)

31F Coronal/CT/Sella Turcica. This modified coronal (+70 degrees) CT was made with a section plane parallel to the clivus. The study was completed following lumbar injection of intrathecal metrizamide contrast. This technique highlights the subarachnoid spaces to best advantage. The pituitary gland is seen between the floor of the sella and the suprasellar cistern. (From Chakeres DW, Kapila A: Brainstem and related structures: Normal CT anatomy using direct longitudinal scanning with metrizamide cisternography. Radiology 149:709–715, 1983.)

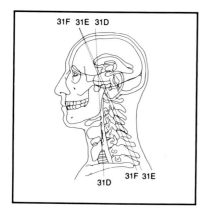

31F 31E 31D

31D

31F 31E

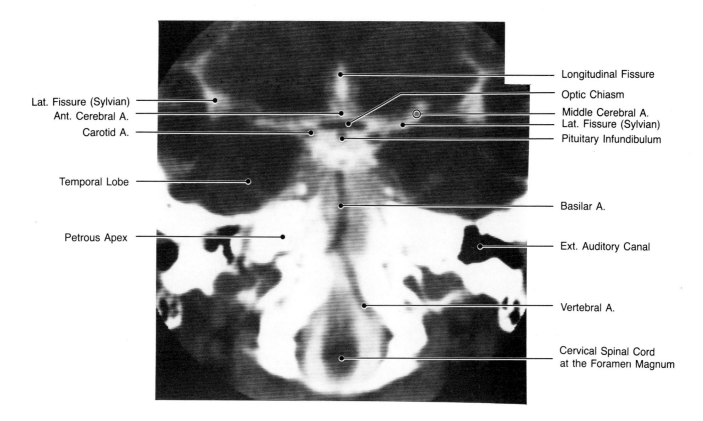

Longitudinal Fissure

Lat. Fissure (Sylvian)
Ant. Cerebral A.
Carotid A.

Optic Chiasm
Middle Cerebral A.
Lat. Fissure (Sylvian)
Pituitary Infundibulum

Temporal Lobe

Basilar A.

Petrous Apex

Ext. Auditory Canal

Vertebral A.

Cervical Spinal Cord
at the Foramen Magnum

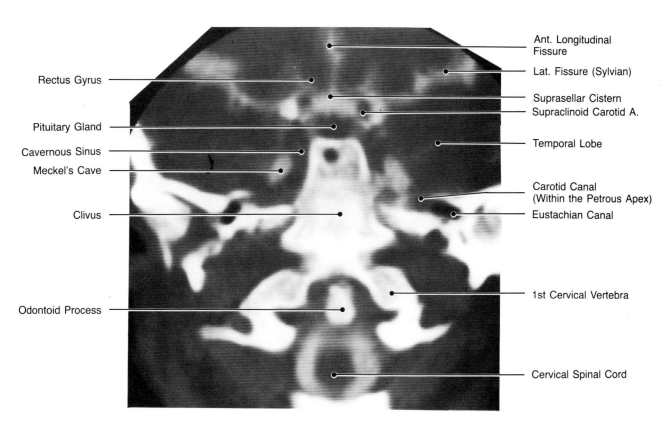

Ant. Longitudinal
Fissure

Rectus Gyrus

Lat. Fissure (Sylvian)

Suprasellar Cistern
Supraclinoid Carotid A.

Pituitary Gland

Temporal Lobe

Cavernous Sinus
Meckel's Cave

Carotid Canal
(Within the Petrous Apex)

Clivus

Eustachian Canal

1st Cervical Vertebra

Odontoid Process

Cervical Spinal Cord

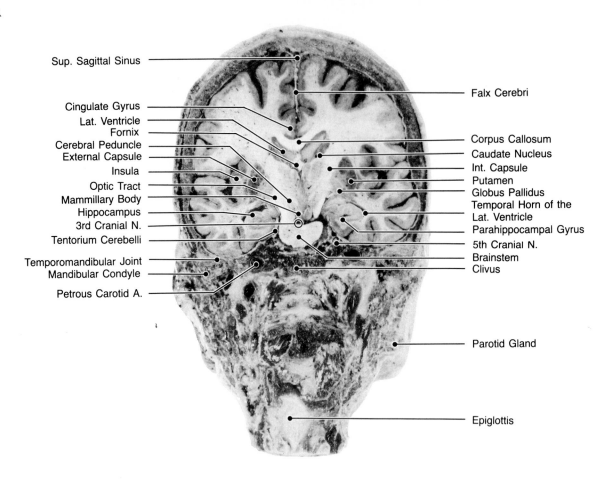

Sup. Sagittal Sinus

Cingulate Gyrus
Lat. Ventricle
Fornix
Cerebral Peduncle
External Capsule
Insula
Optic Tract
Mammillary Body
Hippocampus
3rd Cranial N.
Tentorium Cerebelli

Temporomandibular Joint
Mandibular Condyle

Petrous Carotid A.

Falx Cerebri

Corpus Callosum
Caudate Nucleus
Int. Capsule
Putamen
Globus Pallidus
Temporal Horn of the
Lat. Ventricle
Parahippocampal Gyrus
5th Cranial N.
Brainstem
Clivus

Parotid Gland

Epiglottis

32A Coronal/Anatomic Plate/Caudate Nucleus. This coronal (+90 degrees) anatomic plate is made through the region of the interpeduncular cistern. Because of postmortem brain herniation the interpeduncular cistern is compressed. We are still able to identify the mammillary bodies and third cranial nerves in this region. The hippocampal complex is particularly well seen on this section, with the crescent-shaped temporal horns visible. The hippocampus projects into the lateral ventricle, forming its medial margin.

32B Coronal/MRI/Cervical Carotids. The fifth cranial nerves are seen as soft tissue densities outlined by adjacent low signal spinal fluid on this partial saturation coronal (+90 degrees) TR 800, TE 25 msec MRI. They lie below the tentorium cerebelli and just above the petrous apex. They are in close association with the adjacent carotid artery within the petrous apex. The walls of the cervical carotid arteries are outlined by adjacent soft tissue structures. There is a small intense signal arising over the left temporal lobe related to a cortical vein. This is a common and well described finding seen best on short repetition time images; it is related to laminar flow within venous structures. The occurrence of this high signal is dependent on the relationship of the flow to the gradients, rate of flow, and pulse sequence design.

32C Coronal/MRI/Red Nucleus. This TR 3000, TE 25 msec spin echo coronal (+90 degrees) MRI is made through the red nuclei. Many of the nuclei, including the dentate, globus pallidus, and putamen are seen as low signal regions related to their high iron content. The iron composition may alter the T2 relaxation times to produce a low signal appearance. The low signal is not related to a decrease in the proton density, as is seen in the white matter tracts.

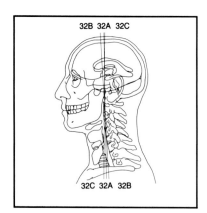

32B 32A 32C

32C 32A 32B

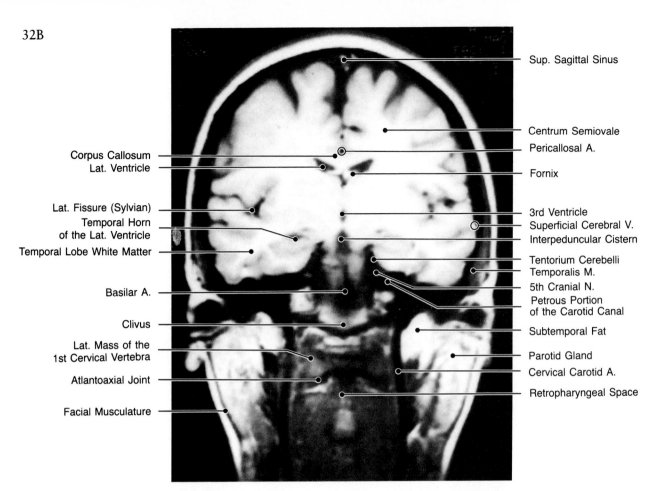

Sup. Sagittal Sinus

Corpus Callosum
Lat. Ventricle

Lat. Fissure (Sylvian)
Temporal Horn
of the Lat. Ventricle
Temporal Lobe White Matter

Basilar A.

Clivus

Lat. Mass of the
1st Cervical Vertebra

Atlantoaxial Joint

Facial Musculature

Centrum Semiovale
Pericallosal A.

Fornix

3rd Ventricle
Superficial Cerebral V.
Interpeduncular Cistern

Tentorium Cerebelli
Temporalis M.
5th Cranial N.
Petrous Portion
of the Carotid Canal

Subtemporal Fat

Parotid Gland

Cervical Carotid A.

Retropharyngeal Space

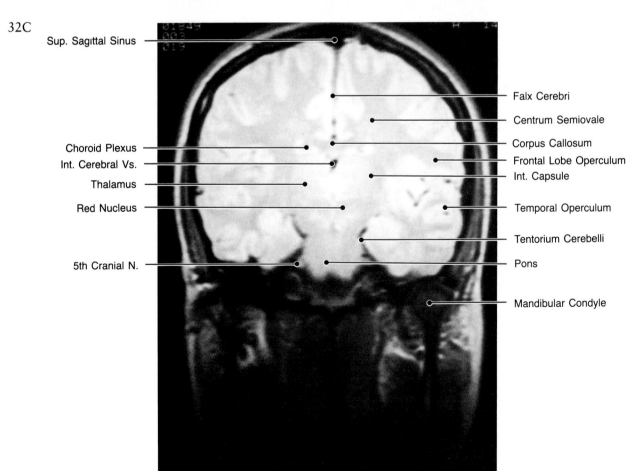

Sup. Sagittal Sinus

Choroid Plexus
Int. Cerebral Vs.

Thalamus

Red Nucleus

5th Cranial N.

Falx Cerebri

Centrum Semiovale

Corpus Callosum

Frontal Lobe Operculum

Int. Capsule

Temporal Operculum

Tentorium Cerebelli

Pons

Mandibular Condyle

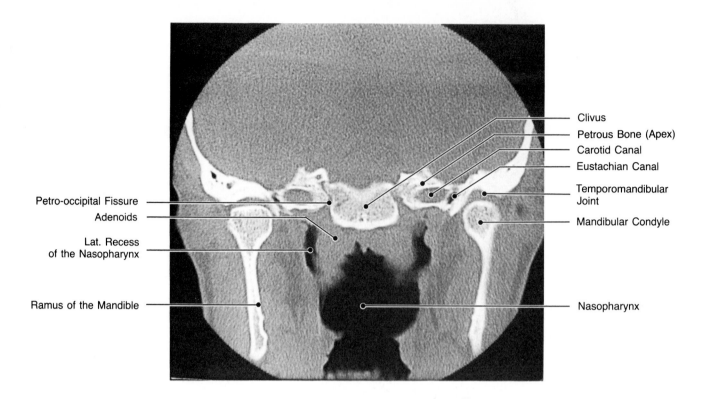

Clivus

Petrous Bone (Apex)

Carotid Canal

Eustachian Canal

Temporomandibular Joint

Petro-occipital Fissure

Adenoids

Lat. Recess of the Nasopharynx

Mandibular Condyle

Ramus of the Mandible

Nasopharynx

32D Coronal/CT/Horizontal Carotid Canals. This coronal (+110 degrees) high resolution thin section bone CT image is made through the horizontal portions of the carotid canals in the petrous apex. Just lateral to the canal lies the air-containing bony eustachian canal. The temporomandibular joints are also well demonstrated, with the mandibular fossa above and the mandibular condyle below. The cleft between the petrous apex and the clivus is the petro-occipital fissure.

32E Coronal/CT/Temporomandibular Joint. This high resolution thin section coronal (+110 degrees) CT of the temporal bones demonstrates the horizontal portions of the carotid canals. The nasopharyngeal soft tissues are seen interposed between the clivus and the nasopharyngeal airway. In younger patients, large adenoids can be seen in this region.

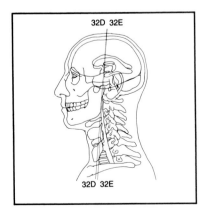

32D 32E

32D 32E

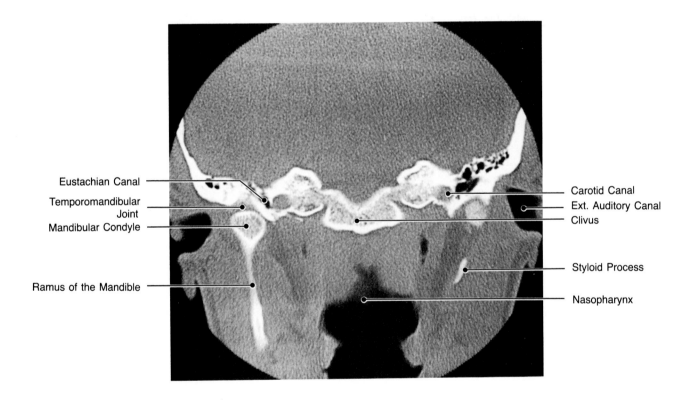

Eustachian Canal

Temporomandibular
Joint

Mandibular Condyle

Ramus of the Mandible

Carotid Canal

Ext. Auditory Canal

Clivus

Styloid Process

Nasopharynx

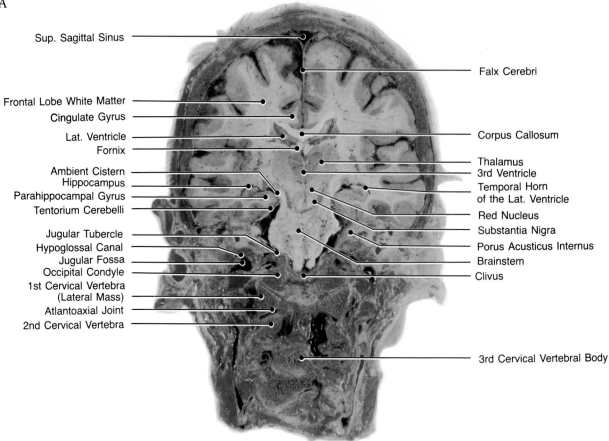

33A

Sup. Sagittal Sinus

Frontal Lobe White Matter
Cingulate Gyrus
Lat. Ventricle
Fornix
Ambient Cistern
Hippocampus
Parahippocampal Gyrus
Tentorium Cerebelli
Jugular Tubercle
Hypoglossal Canal
Jugular Fossa
Occipital Condyle
1st Cervical Vertebra
(Lateral Mass)
Atlantoaxial Joint
2nd Cervical Vertebra

Falx Cerebri

Corpus Callosum

Thalamus
3rd Ventricle
Temporal Horn
of the Lat. Ventricle
Red Nucleus
Substantia Nigra
Porus Acusticus Internus
Brainstem
Clivus

3rd Cervical Vertebral Body

33A Coronal/Anatomic Plate/Midbrain. The brainstem is seen bridging the tentorial incisura from the posterior fossa to the supratentorial compartment on this coronal (+90 degrees) section. The ambient cisterns are arch shaped subarachnoid spaces paralleling the lateral margins of the brainstem. They are continuous with the choroid fissures, which separate the ambient cistern from the temporal horn. The substantia nigra forms a U-shaped structure at the midbrain level. Within the "U" are the red nuclei. The jugular tubercle is seen projecting into the cerebellopontine cistern. The hypoglossal canal is seen coursing between the jugular tubercle superiorly and the occipital condyle inferiorly.

33B Coronal/MRI/Third Ventricle. This partial saturation TR 800, TE 25 msec coronal (+90 degrees) MRI is made at the level of the foramen of Monroe. The columns of the fornix are suspended by the septum pellucidum. The mammillary bodies protrude into the suprasellar cistern. The amygdala and uncus form the lateral margins of the suprasellar cistern. The close association between Meckel's cave (trigeminal cistern) and the horizontal portion of the petrous carotid artery is seen. The fatty spaces about the mandibular ramus are evident. On the medial margin small low signal foci are related to vascular structures such as branches of the internal maxillary artery.

33C Coronal/MRI/Odontoid. The articulation of the upper cervical vertebra is well demonstrated on this coronal (90 degrees) partial saturation TR 800, TE 25 msec MRI. The lateral masses of the first cervical vertebra are seen as a higher signal region centrally because of their marrow contents, with a low signal rim secondary to cortical bone. The odontoid is seen extending from the body of the second cervical vertebra superiorly between the two lateral masses of the first cervical vertebra.

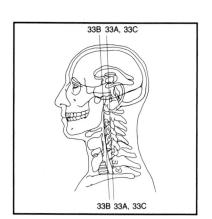

33B 33A, 33C

33B 33A, 33C

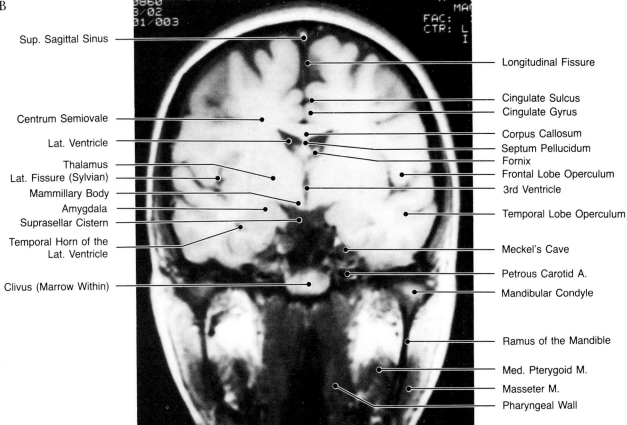

Sup. Sagittal Sinus

Centrum Semiovale

Lat. Ventricle

Thalamus
Lat. Fissure (Sylvian)
Mammillary Body
Amygdala
Suprasellar Cistern

Temporal Horn of the
Lat. Ventricle

Clivus (Marrow Within)

Longitudinal Fissure

Cingulate Sulcus
Cingulate Gyrus

Corpus Callosum
Septum Pellucidum
Fornix
Frontal Lobe Operculum

3rd Ventricle

Temporal Lobe Operculum

Meckel's Cave

Petrous Carotid A.

Mandibular Condyle

Ramus of the Mandible

Med. Pterygoid M.

Masseter M.

Pharyngeal Wall

33C

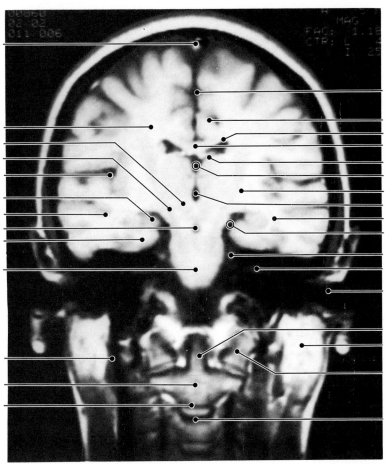

Sup. Sagittal Sinus

Corona Radiata
Red Nucleus
Cerebral Peduncle
Lat. Fissure (Sylvian)

Subiculum
Temporal Lobe
Interpeduncular Cistern
Parahippocampal Gyrus

Pons

Cervical Carotid A.

Body of the
2nd Cervical Vertebra

2nd-3rd Cervical Interspace

Longitudinal Fissure

Cingulate Gyrus
Lat. Ventricle
Corpus Callosum
Fornix
Int. Cerebral Vs.

Int. Capsule
3rd Ventricle
Temporal Lobe
Post. Cerebral A.

5th Cranial N.
Int. Auditory Canal

Ext. Auditory Canal

Odontoid Process
Parotid Gland

Lat. Mass of the
First Cervical Vertebra

3rd Cervical Vertebra

33D

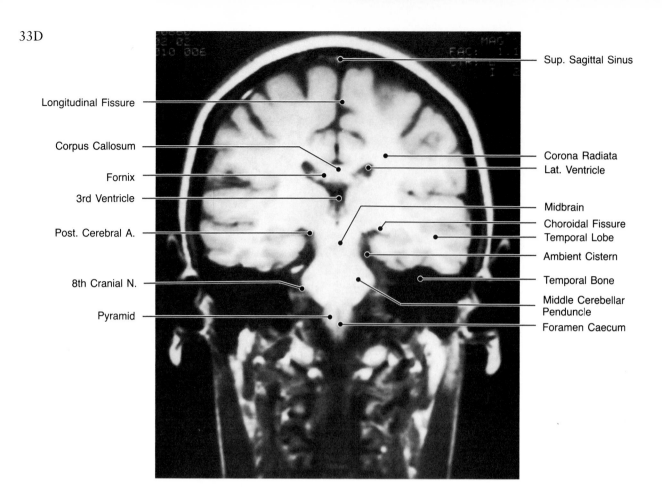

Longitudinal Fissure

Corpus Callosum

Fornix

3rd Ventricle

Post. Cerebral A.

8th Cranial N.

Pyramid

Sup. Sagittal Sinus

Corona Radiata

Lat. Ventricle

Midbrain

Choroidal Fissure

Temporal Lobe

Ambient Cistern

Temporal Bone

Middle Cerebellar Penduncle

Foramen Caecum

33D Coronal/MRI/Brainstem. This coronal (+90 degrees) partial saturation TR 800, TE 25 msec MRI demonstrates the diverging limbs of the bodies of the fornix beneath the corpus callosum. The seventh and eighth cranial nerves arise from the pontomedullary junction and are seen extending laterally towards the internal auditory canals. The ambient cistern lies between the brainstem and the temporal lobes. Superiorly the ambient cistern is bordered by the optic tracts and geniculate bodies.

33E Coronal/CT/Brainstem. The modified coronal (+70 degrees) CT is made parallel to the clivus following intrathecal injection of metrizamide contrast. The pontomedullary junction is evident with a small cisternal space related to the foramen caecum in the midline. The fifth cranial nerves lie just below the tentorial edge and superior to the petrous apex. The suprasellar cistern is seen filled with contrast, with the optic chiasm and optic tracts forming a flat triangle-shaped structure. (From Chakeres DW, Kapila A: Brainstem and related structures: Normal CT anatomy using direct longitudinal scanning with metrizamide cisternography. Radiology 149:709–715, 1983).

33F Coronal/CT/External Auditory Canal. This high resolution thin section coronal (+105 degrees) CT of the temporal bones demonstrates the close association between the turns of the cochlea and the facial nerve canal at the geniculate ganglion. The carotid canals are just inferior to the cochlea. The styloid processes are seen extending inferiorly from the undersurface of the temporal bone.

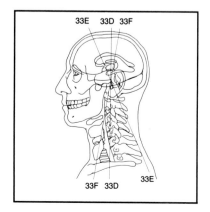

33E 33D 33F

33F 33D

33E

33E

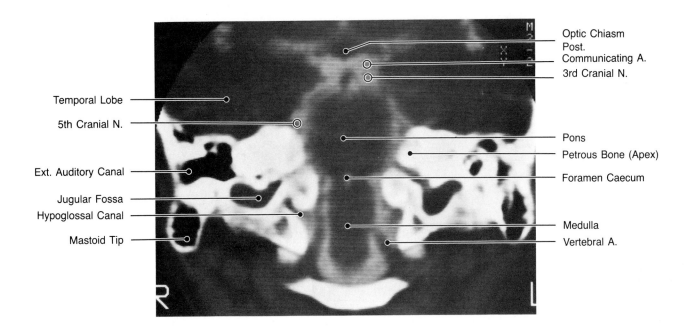

Optic Chiasm
Post. Communicating A.
3rd Cranial N.

Temporal Lobe
5th Cranial N.

Pons
Petrous Bone (Apex)

Ext. Auditory Canal
Foramen Caecum

Jugular Fossa
Hypoglossal Canal

Medulla
Vertebral A.

Mastoid Tip

33F

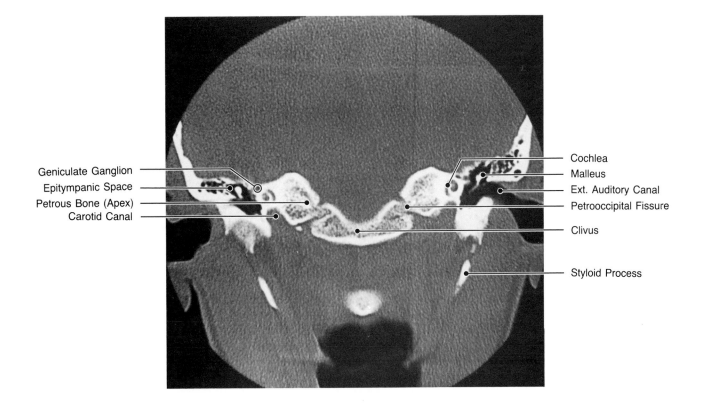

Geniculate Ganglion
Epitympanic Space
Petrous Bone (Apex)
Carotid Canal

Cochlea
Malleus
Ext. Auditory Canal
Petrooccipital Fissure
Clivus

Styloid Process

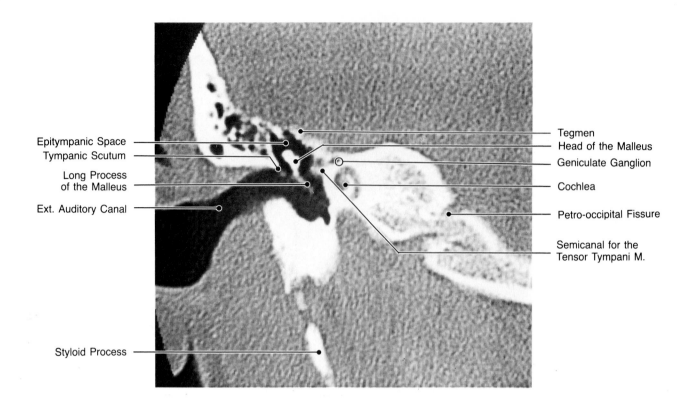

Epitympanic Space
Tympanic Scutum
Long Process of the Malleus
Ext. Auditory Canal

Styloid Process

Tegmen
Head of the Malleus
Geniculate Ganglion
Cochlea
Petro-occipital Fissure
Semicanal for the Tensor Tympani M.

33G Coronal/CT/Malleus. The middle ear cavity is well seen on this coronal (+105 degrees) high resolution thin section CT of the temporal bone. The head, neck, and long process of the malleus are seen. The middle ear cavity lies medial to the scutum, long process of the malleus, and the small bony protuberance from the inferior margin of the hypotympanum called the limbus, where the tympanic membrane inserts. The close association of the limbs of the geniculate ganglion, forming snake-eye lucencies, and the underlying semicanal for the tensor tympani muscle are well seen.

33H Coronal/CT/Incudostapedial Joint. This is a coronal (+105 degrees) CT scan through the temporal bone made with high resolution thin section technique. The falciform crest is a thin spur that divides the anterior lateral internal auditory canal into a superior component containing the facial nerve and an inferior component containing the cochlear branch of the eighth cranial nerve. Loss of this bony structure can suggest the presence of an acoustic neuroma. Coronal imaging best depicts the relationship between the long process of the incus and the stapes at the incudostapedial joint. The incus and head of the malleus lie in the epitympanic space.

33I Coronal/CT/Incudostapedial Joint. This high resolution thin section coronal (+105 degrees) CT of the temporal bone demonstrates the stapes at the oval window. The oval window is seen as a bony defect between the vestible and the middle ear cavity. Visualization of this structure is important in the evaluation of patients with otosclerosis and congenital abnormalities.

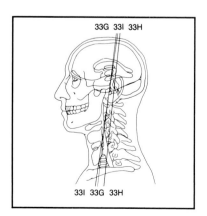

33G 33I 33H

33I 33G 33H

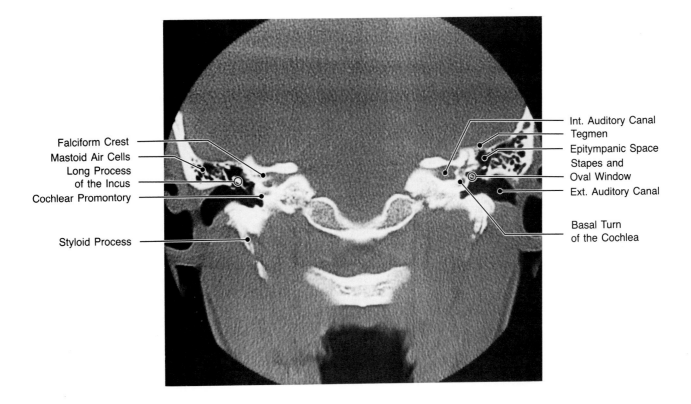

Falciform Crest
Mastoid Air Cells
Long Process
of the Incus
Cochlear Promontory

Styloid Process

Int. Auditory Canal
Tegmen
Epitympanic Space
Stapes and
Oval Window
Ext. Auditory Canal

Basal Turn
of the Cochlea

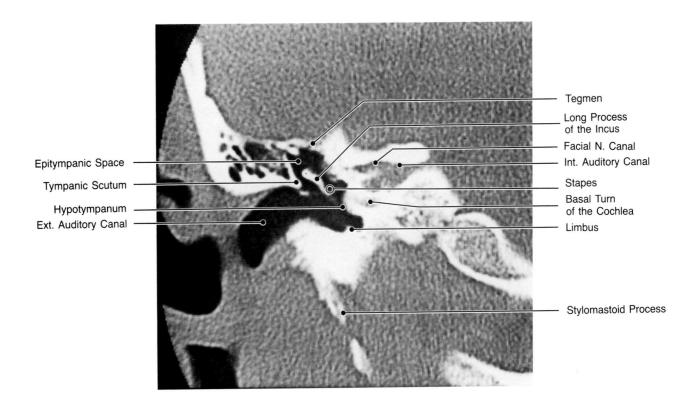

Epitympanic Space
Tympanic Scutum
Hypotympanum
Ext. Auditory Canal

Tegmen
Long Process
of the Incus
Facial N. Canal
Int. Auditory Canal
Stapes
Basal Turn
of the Cochlea
Limbus

Stylomastoid Process

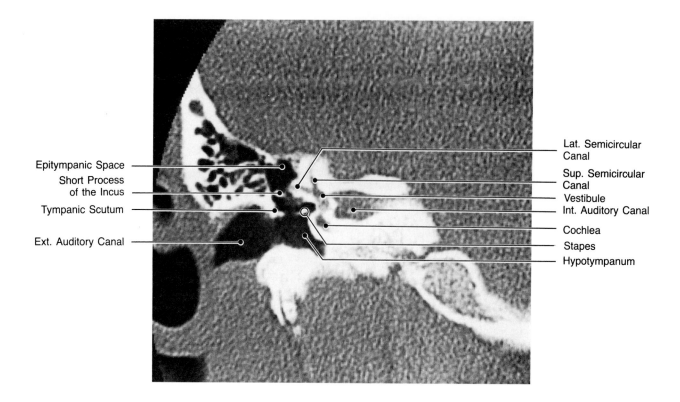

Epitympanic Space —
Short Process of the Incus —
Tympanic Scutum —
Ext. Auditory Canal —

— Lat. Semicircular Canal
— Sup. Semicircular Canal
— Vestibule
— Int. Auditory Canal
— Cochlea
— Stapes
— Hypotympanum

33J Coronal/CT/Oval Window. This high resolution coronal (+105 degrees) CT of the temporal bone demonstrates the fundus of the internal auditory canal ending in the vestibule, superior and lateral semicircular canals, and cochlea. The air space bordering the lateral margin of the cochlear promontory is the hypotympanum. The air space bordering the lateral margin and the lateral semicircular canal is the aditus ad antrum. The small bony structure in the aditus is the short process of the incus. The scutum forms the superior margin of the external auditory canal medially and is the site of insertion of the tympanic membrane.

33K Coronal/CT/Hypoglossal Canal. The jugular tubercle is seen just medial to the jugular fossa on this coronal (+110 degrees) CT. Just inferior are the hypoglossal canal and the occipital condyle. This coronal image parallels the descending portion of the facial nerve canal, which ends at the stylomastoid foramina inferiorly. The tympanic sinus is just lateral to the vestibule and medial to the descending facial nerve canal.

33L Coronal/CT/Facial Nerve Canal. This high resolution thin section coronal (+105 degrees) CT of the temporal bones demonstrates the close relationship of the lateral semicircular canal and the air-filled tympanic sinus. The jugular fossa is seen at the undersurface of the posterior temporal bone. Directly above this is a small linear opening directed medially and inferiorly, which is the cochlear aqueduct. The descending facial nerve canal is seen just lateral to the jugular fossa.

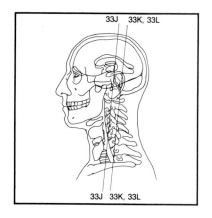

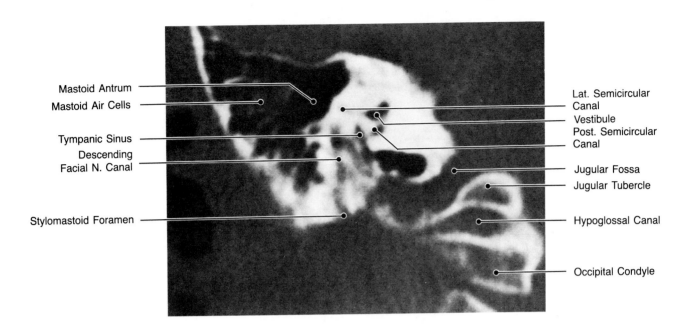

Mastoid Antrum

Mastoid Air Cells

Tympanic Sinus

Descending
Facial N. Canal

Stylomastoid Foramen

Lat. Semicircular
Canal

Vestibule

Post. Semicircular
Canal

Jugular Fossa

Jugular Tubercle

Hypoglossal Canal

Occipital Condyle

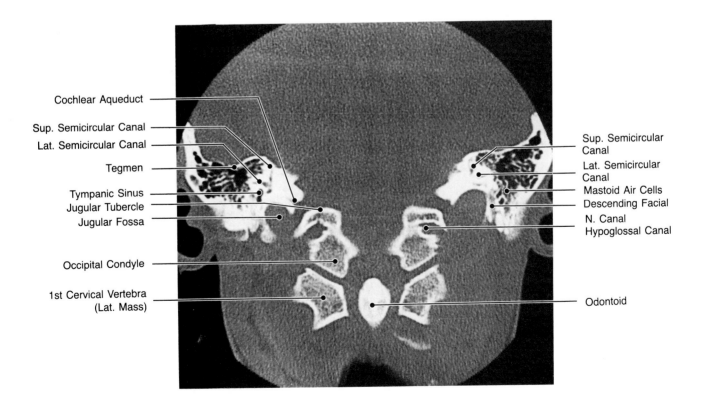

Cochlear Aqueduct

Sup. Semicircular Canal

Lat. Semicircular Canal

Tegmen

Tympanic Sinus

Jugular Tubercle

Jugular Fossa

Occipital Condyle

1st Cervical Vertebra
(Lat. Mass)

Sup. Semicircular
Canal

Lat. Semicircular
Canal

Mastoid Air Cells

Descending Facial

N. Canal
Hypoglossal Canal

Odontoid

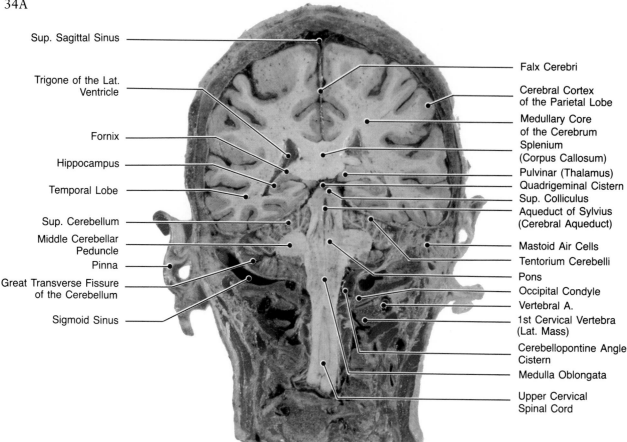

Sup. Sagittal Sinus

Trigone of the Lat. Ventricle

Fornix

Hippocampus

Temporal Lobe

Sup. Cerebellum

Middle Cerebellar Peduncle

Pinna

Great Transverse Fissure of the Cerebellum

Sigmoid Sinus

Falx Cerebri

Cerebral Cortex of the Parietal Lobe

Medullary Core of the Cerebrum

Splenium (Corpus Callosum)

Pulvinar (Thalamus)

Quadrigeminal Cistern

Sup. Colliculus

Aqueduct of Sylvius (Cerebral Aqueduct)

Mastoid Air Cells

Tentorium Cerebelli

Pons

Occipital Condyle

Vertebral A.

1st Cervical Vertebra (Lat. Mass)

Cerebellopontine Angle Cistern

Medulla Oblongata

Upper Cervical Spinal Cord

34A Coronal/Anatomic Plate/Foramen Magnum. The splenium of the corpus callosum is seen as a broad band between the two trigones of the lateral ventricle on this coronal (+90 degrees) section. The body of the fornix is seen looping laterally and inferiorly. It thins to a bandlike white matter structure called the alveus, which surrounds the hippocampus. The brainstem is seen extending from the colliculi region and aqueduct superiorly to the foramen magnum inferiorly. The middle cerebral peduncles are large white matter tracts forming oval-shaped structures extending into the central portion of the cerebellar hemispheres.

34B Coronal/MRI/Colliculi. This coronal (+90 degrees) partial saturation MRI made with a TR 800 and TE 25 msec demonstrates a linear black signal region related to the aqueduct of Sylvius in the upper brainstem. The adjacent colliculi are identified as four circular structures in a square array. The aqueduct has a low signal because of movement of the spinal fluid, similar to the signal void identified in the internal cerebral veins directly above the pineal gland. If signal is seen from the aqueduct, it is suggestive of obstruction because of the lack of normal movement of the spinal fluid. The lateral recesses of the fourth ventricle are seen as spinal fluid spaces interposed among the middle cerebral peduncle, the inferior cerebellum, and the adjacent lower pons and medulla. The small oval structure at the lateral margin of the lateral recess of the fourth ventricle is the flocculus.

34C Coronal/CT/Occipital Condyles. This high resolution thin section coronal (+110 degrees) CT scan of the skull base demonstrates the articulation between the occipital condyle and the superior articular surface of the lateral mass of the first cervical vertebra. The atlantoaxial joint is also evident. The occipital condyles form the lateral margins of the foramen magnum.

34D Coronal/CT/Odontoid. This coronal (+110 degrees) high resolution CT of the skull base and upper cervical spine demonstrates the lateral masses of the first cervical vertebra as truncated triangles between the occipital condyles above and the superior articular surfaces of the body of the second cervical vertebra below. The odontoid is between the lateral masses.

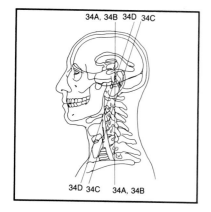

34A, 34B 34D 34C

34D 34C 34A, 34B

34B

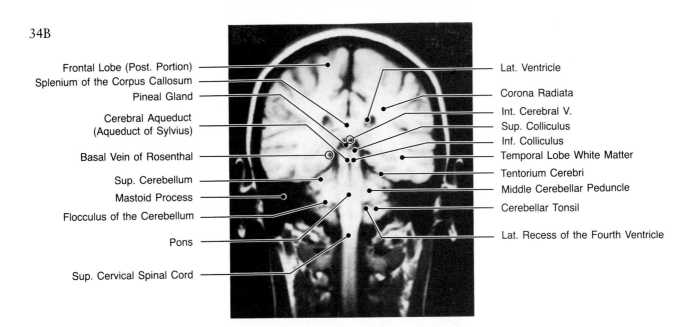

Frontal Lobe (Post. Portion)
Splenium of the Corpus Callosum
Pineal Gland
Cerebral Aqueduct
(Aqueduct of Sylvius)
Basal Vein of Rosenthal
Sup. Cerebellum
Mastoid Process
Flocculus of the Cerebellum
Pons
Sup. Cervical Spinal Cord

Lat. Ventricle
Corona Radiata
Int. Cerebral V.
Sup. Colliculus
Inf. Colliculus
Temporal Lobe White Matter
Tentorium Cerebri
Middle Cerebellar Peduncle
Cerebellar Tonsil
Lat. Recess of the Fourth Ventricle

34C

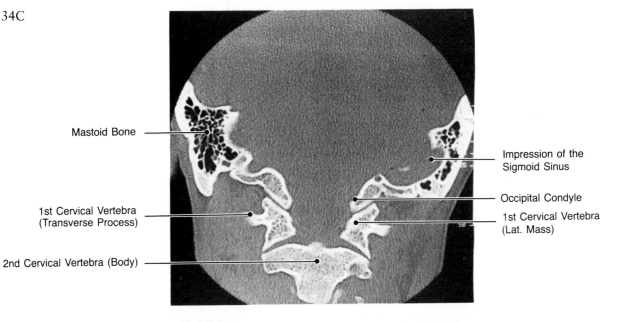

Mastoid Bone

1st Cervical Vertebra
(Transverse Process)

2nd Cervical Vertebra (Body)

Impression of the
Sigmoid Sinus

Occipital Condyle

1st Cervical Vertebra
(Lat. Mass)

34D

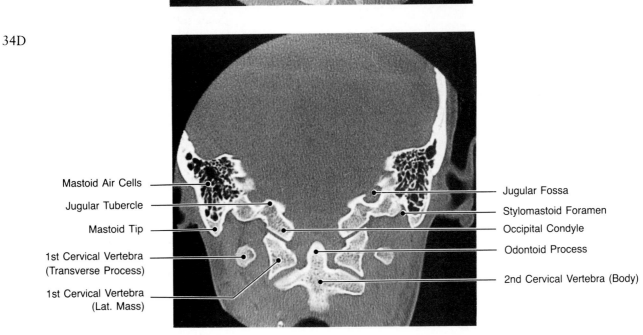

Mastoid Air Cells
Jugular Tubercle
Mastoid Tip
1st Cervical Vertebra
(Transverse Process)
1st Cervical Vertebra
(Lat. Mass)

Jugular Fossa
Stylomastoid Foramen
Occipital Condyle
Odontoid Process
2nd Cervical Vertebra (Body)

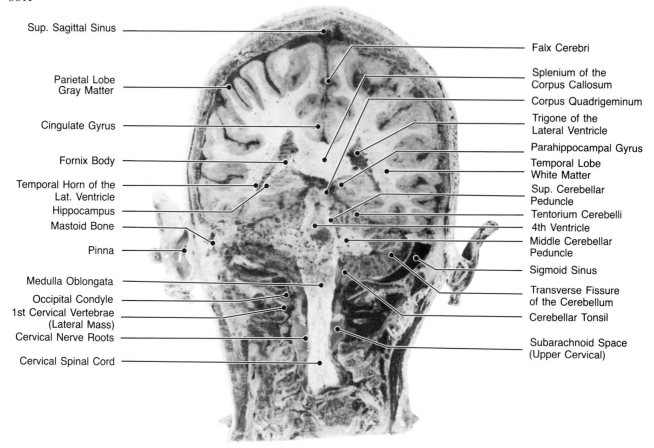

35A

Sup. Sagittal Sinus

Parietal Lobe
Gray Matter

Cingulate Gyrus

Fornix Body

Temporal Horn of the
Lat. Ventricle
Hippocampus
Mastoid Bone

Pinna

Medulla Oblongata

Occipital Condyle
1st Cervical Vertebrae
(Lateral Mass)
Cervical Nerve Roots

Cervical Spinal Cord

Falx Cerebri

Splenium of the
Corpus Callosum

Corpus Quadrigeminum

Trigone of the
Lateral Ventricle

Parahippocampal Gyrus

Temporal Lobe
White Matter

Sup. Cerebellar
Peduncle

Tentorium Cerebelli

4th Ventricle

Middle Cerebellar
Peduncle

Sigmoid Sinus

Transverse Fissure
of the Cerebellum

Cerebellar Tonsil

Subarachnoid Space
(Upper Cervical)

35A Coronal/Anatomic Plate/Fourth Ventricle. The colliculi are seen just superior to the diamond-shaped fourth ventricle on this coronal (+90 degrees) section. The fourth ventricle is outlined by the superior cerebellar peduncles superiorly and the middle cerebellar peduncles laterally. The cerebellar tonsils, partially herniated as a postmortem change, lie adjacent to the brainstem as it exits the foramen magnum. The medulla and upper cervical spinal cord are margined laterally by the occipital condyles and the lateral masses of the first cervical vertebra.

35B Coronal/MRI/Fourth Ventricle. This coronal (+90 degrees) partial saturation TR 800, TE 25 msec MRI demonstrates the diamond-shaped fourth ventricle. The spinal fluid spaces extending from the lateral inferior margins of the fourth ventricle are the lateral recesses. The midline spinal fluid space descending from the fourth ventricle is the foramen Magendie. Above the cerebellar vermis in the superior cerebellar cistern is a linear vascular low signal structure related to the precentral cerebellar vein, which lies posterior to the colliculi and anterior to the cerebellum.

35C Coronal/MRI/Fourth Ventricle. This coronal (+90 degrees) spin echo TR 4000, TE 75 msec MRI demonstrates the spinal fluid spaces of the fourth ventricle, occipital horn, and cisterns as having high signal intensity. This long TR TE image also enhances the difference between the grey and white matter tracts.

35D Coronal/CT/Tentorial Notch. This coronal (+70 degrees) CT displayed at soft tissue windows is made parallel to the clivus following intrathecal injection of contrast material. The cerebellum is outlined by metrizamide contrast, and the transverse fissures are visible. The ambient cistern forms arc-shaped areas of high density outlining the lateral margins of the brainstem at the tentorial incisura. The fourth ventricular lateral and midline foramen are seen. (From Chakeres DW, Kapila A Brainstem and related structures: Normal CT anatomy using direct longitudinal scanning with metrizamide cisternography. Radiology 149:709–715, 1983.)

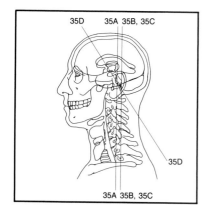

35B

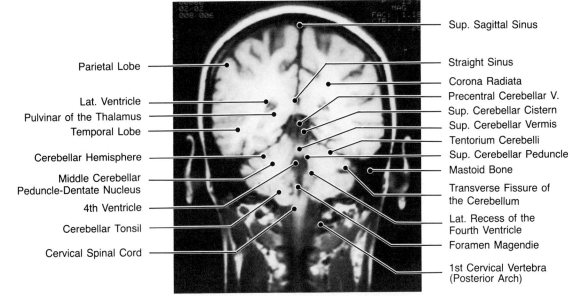

Parietal Lobe

Lat. Ventricle

Pulvinar of the Thalamus

Temporal Lobe

Cerebellar Hemisphere

Middle Cerebellar
Peduncle-Dentate Nucleus

4th Ventricle

Cerebellar Tonsil

Cervical Spinal Cord

Sup. Sagittal Sinus

Straight Sinus

Corona Radiata

Precentral Cerebellar V.

Sup. Cerebellar Cistern

Sup. Cerebellar Vermis

Tentorium Cerebelli

Sup. Cerebellar Peduncle

Mastoid Bone

Transverse Fissure of
the Cerebellum

Lat. Recess of the
Fourth Ventricle

Foramen Magendie

1st Cervical Vertebra
(Posterior Arch)

35C

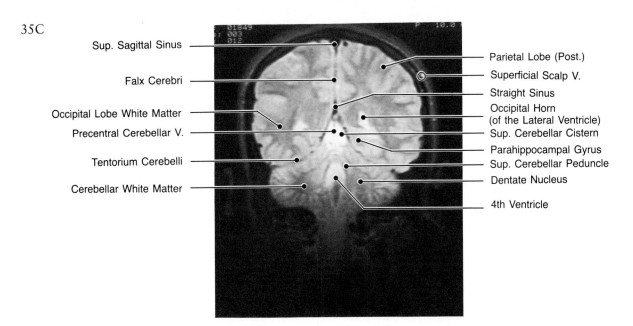

Sup. Sagittal Sinus

Falx Cerebri

Occipital Lobe White Matter

Precentral Cerebellar V.

Tentorium Cerebelli

Cerebellar White Matter

Parietal Lobe (Post.)

Superficial Scalp V.

Straight Sinus

Occipital Horn
(of the Lateral Ventricle)

Sup. Cerebellar Cistern

Parahippocampal Gyrus

Sup. Cerebellar Peduncle

Dentate Nucleus

4th Ventricle

35D

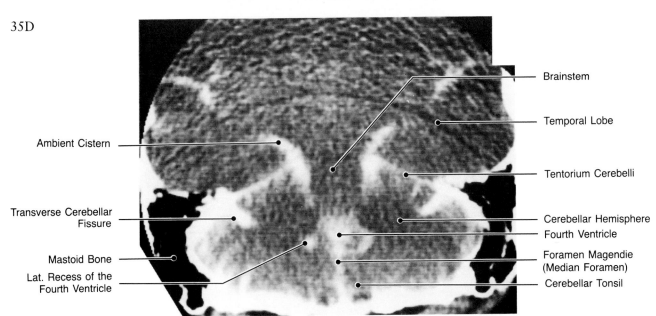

Ambient Cistern

Transverse Cerebellar
Fissure

Mastoid Bone

Lat. Recess of the
Fourth Ventricle

Brainstem

Temporal Lobe

Tentorium Cerebelli

Cerebellar Hemisphere

Fourth Ventricle

Foramen Magendie
(Median Foramen)

Cerebellar Tonsil

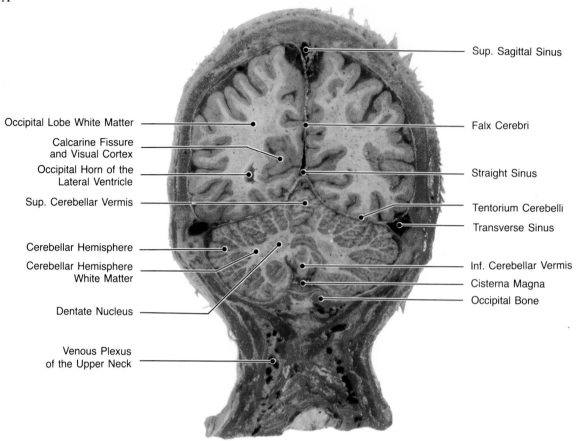

Sup. Sagittal Sinus

Occipital Lobe White Matter

Calcarine Fissure and Visual Cortex

Occipital Horn of the Lateral Ventricle

Sup. Cerebellar Vermis

Cerebellar Hemisphere

Cerebellar Hemisphere White Matter

Dentate Nucleus

Venous Plexus of the Upper Neck

Falx Cerebri

Straight Sinus

Tentorium Cerebelli

Transverse Sinus

Inf. Cerebellar Vermis

Cisterna Magna

Occipital Bone

36A Coronal/Anatomic Plate/Apex Of The Tentorium. The dural venous sinuses are well seen on this coronal (+90 degrees) anatomic plate. They are outlined by two layers of dura, a visceral layer that faces the brain, and a parietal layer that faces the bone. In most areas of the skull these layers are fused. In the regions of the dural sinuses, however, they are separate, with the venous canals suspended between. The superior sagittal and straight sinuses as well as both transverse sinuses are outlined by their dural investments. The dentate nuclei of the cerebellum are seen centrally within the white matter tracts in the mid-cerebellum.

36B Coronal/MRI/Apex Of The Tentorium. This coronal (+90 degrees) MRI uses a partial saturation TR 800, TE 25 msec technique. It is made at the level of the junction of the falx cerebri and the two limbs of the tentorium cerebelli above the superior cerebellar vermis. The fourth ventricle is seen as an inverted arch-shaped low signal region between the cerebellar hemispheres. The inferior vermian structure protruding into the fourth ventricle is the nodulus. The cisterna magna is in the subarachnoid space below the cerebellum adjacent to the cerebellar tonsils. Occasionally small vessels are seen in this region related to the posterior inferior cerebellar artery.

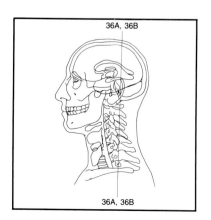

36A, 36B

36A, 36B

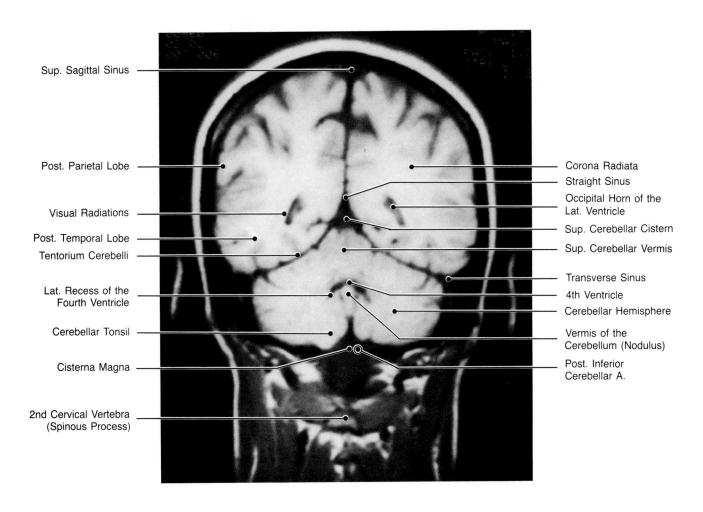

Sup. Sagittal Sinus

Post. Parietal Lobe

Visual Radiations

Post. Temporal Lobe

Tentorium Cerebelli

Lat. Recess of the
Fourth Ventricle

Cerebellar Tonsil

Cisterna Magna

2nd Cervical Vertebra
(Spinous Process)

Corona Radiata

Straight Sinus

Occipital Horn of the
Lat. Ventricle

Sup. Cerebellar Cistern

Sup. Cerebellar Vermis

Transverse Sinus

4th Ventricle

Cerebellar Hemisphere

Vermis of the
Cerebellum (Nodulus)

Post. Inferior
Cerebellar A.

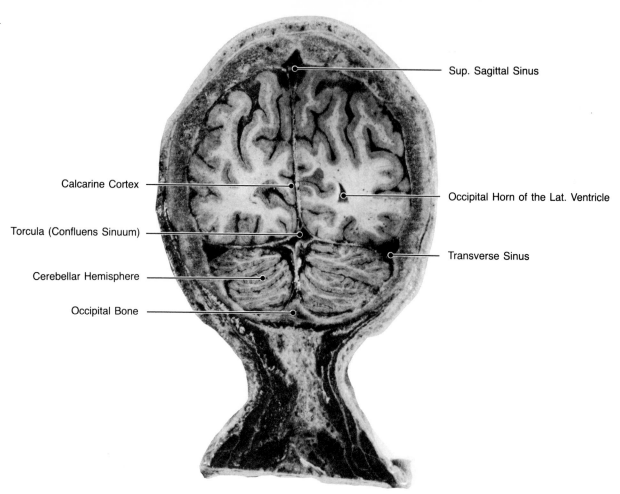

Sup. Sagittal Sinus

Calcarine Cortex

Occipital Horn of the Lat. Ventricle

Torcula (Confluens Sinuum)

Transverse Sinus

Cerebellar Hemisphere

Occipital Bone

37A Coronal/Anatomic Plate/Torcula. The confluence of the transverse and superior sagittal sinuses is seen at the torcula on this coronal (+90 degrees) section. The tentorium cerebelli is seen separating the cerebellar hemispheres inferiorly from the occipital lobes superiorly. The occipital pole of the lateral ventricle is seen within the calcarine cortex.

37B Coronal/MRI/Tentorium. This coronal (+90 degrees) partial saturation TR 800, TE 25 msec MRI demonstrates a low signal region arising from blood flow in the superior sagittal sinus. Note that in the transverse sinuses there is a higher signal. This is related to slow laminar venous blood flow enhancement. The cisterna magna is the subarachnoid space interposed between the cerebellar tonsils and the inferior cerebellar vermis.

37C Coronal/MRI/Dentate Nuclei. This coronal (+90 degrees) spin echo MRI made with TR 4000, TE 25 msec demonstrates a low signal frond-like region of the white matter tracts of the cerebellar hemispheres. This particular pulse sequence accentuates the internal anatomy of the brain, particularly the gray and white matter structures. It is less helpful for evaluation of the surface anatomy of the brain, since the spinal fluid and the gray matter are similar in signal intensity.

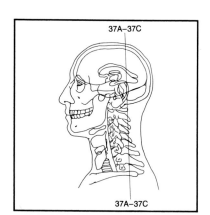

37A–37C

37A–37C

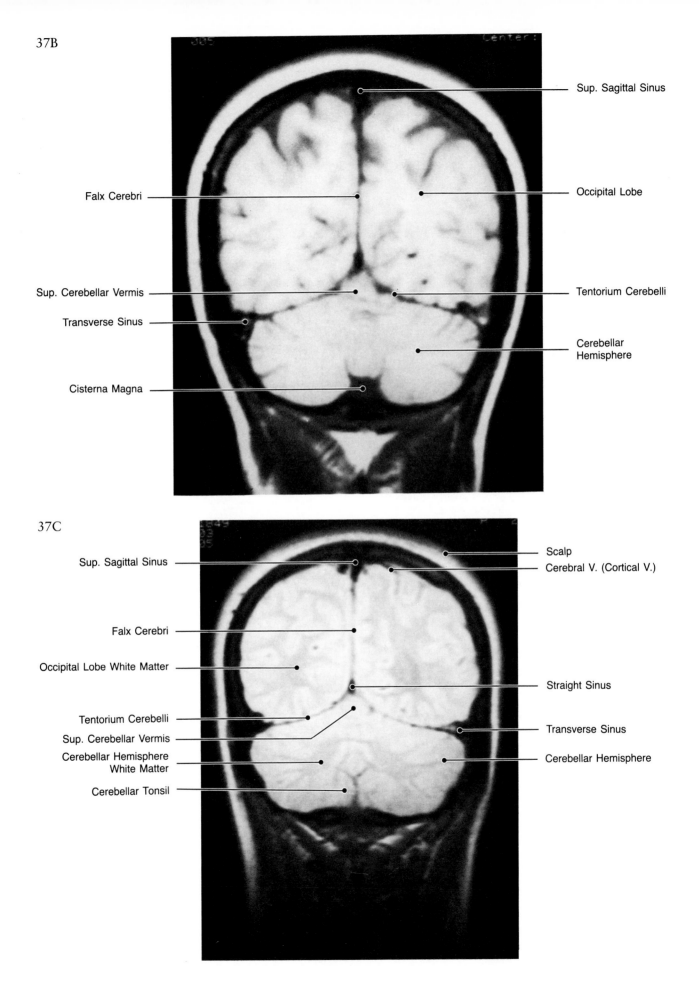

37B

Sup. Sagittal Sinus

Falx Cerebri

Occipital Lobe

Sup. Cerebellar Vermis

Tentorium Cerebelli

Transverse Sinus

Cerebellar Hemisphere

Cisterna Magna

37C

Sup. Sagittal Sinus

Scalp

Cerebral V. (Cortical V.)

Falx Cerebri

Occipital Lobe White Matter

Straight Sinus

Tentorium Cerebelli

Sup. Cerebellar Vermis

Transverse Sinus

Cerebellar Hemisphere White Matter

Cerebellar Hemisphere

Cerebellar Tonsil

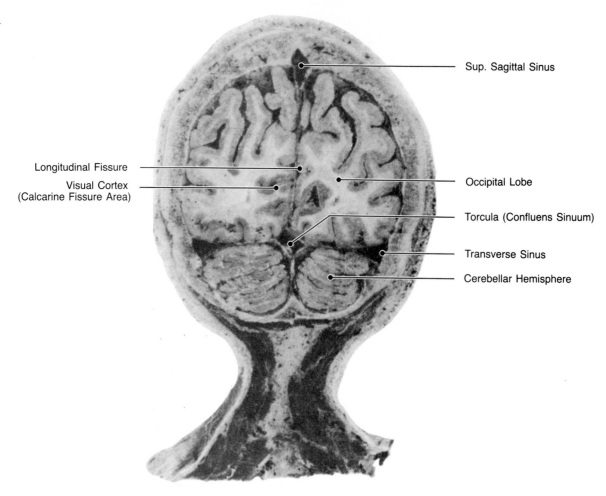

Sup. Sagittal Sinus

Longitudinal Fissure

Visual Cortex
(Calcarine Fissure Area)

Occipital Lobe

Torcula (Confluens Sinuum)

Transverse Sinus

Cerebellar Hemisphere

38A Coronal/Anatomic Plate/Posterior Occiput. The torcula is seen at the confluence of the transverse sinuses and superior sagittal sinus in the posterior occipital region on this coronal (+90 degrees) section.

38B Coronal/MRI/Torcula. This coronal (+90 degrees) partial saturation TR 800, TE 25 msec MRI demonstrates the tentorium cerebelli dividing the occipital lobes from the cerebellar hemispheres. The cranial contents appear to be divided into four quadrants related to the dural sinuses, falx cerebri, and tentorium cerebelli. The low signal arcs about the cerebellum are related to normal spinal fluid within the folia.

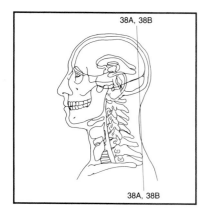

38A, 38B

38A, 38B

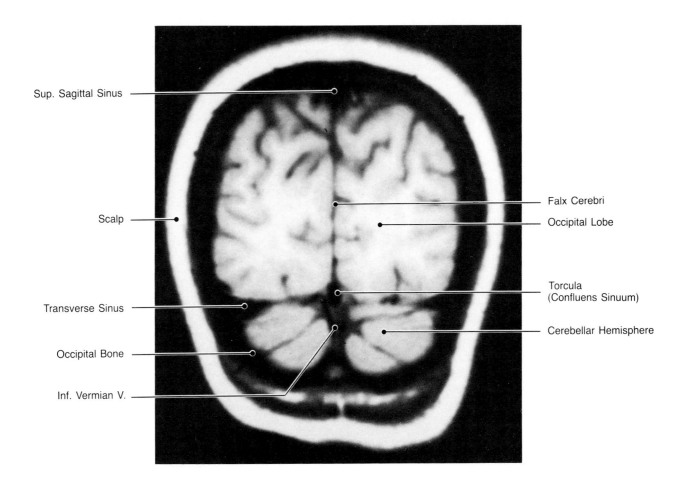

Sup. Sagittal Sinus

Scalp

Transverse Sinus

Occipital Bone

Inf. Vermian V.

Falx Cerebri

Occipital Lobe

Torcula
(Confluens Sinuum)

Cerebellar Hemisphere

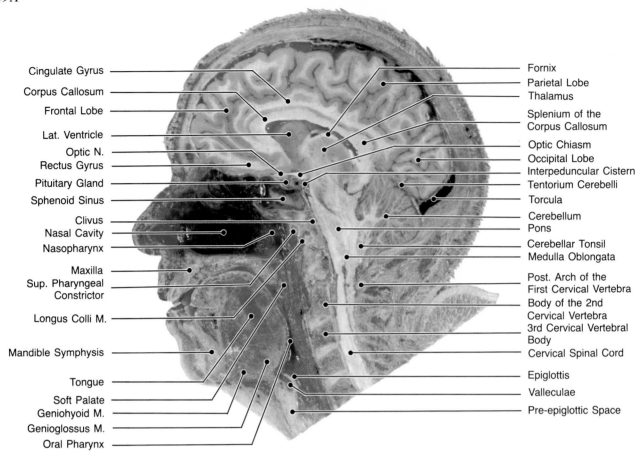

Cingulate Gyrus
Corpus Callosum
Frontal Lobe
Lat. Ventricle
Optic N.
Rectus Gyrus
Pituitary Gland
Sphenoid Sinus
Clivus
Nasal Cavity
Nasopharynx
Maxilla
Sup. Pharyngeal Constrictor
Longus Colli M.
Mandible Symphysis
Tongue
Soft Palate
Geniohyoid M.
Genioglossus M.
Oral Pharynx

Fornix
Parietal Lobe
Thalamus
Splenium of the Corpus Callosum
Optic Chiasm
Occipital Lobe
Interpeduncular Cistern
Tentorium Cerebelli
Torcula
Cerebellum
Pons
Cerebellar Tonsil
Medulla Oblongata
Post. Arch of the First Cervical Vertebra
Body of the 2nd Cervical Vertebra
3rd Cervical Vertebral Body
Cervical Spinal Cord
Epiglottis
Valleculae
Pre-epiglottic Space

39A Sagittal/Anatomic Plate/Midline. This midline sagittal section of the brain demonstrates some of the distortion of the brainstem and cerebellum caused by postmortem brain herniation. The posterior fossa contents are molded to the bony margins with the cerebellar tonsils herniating into the foramen magnum.

39B Sagittal/MRI/Midline. This midline partial saturation TR 800, TE 25 msec MRI demonstrates the fat and scalp structures as high signal regions, whereas the spinal fluid spaces are low signal. A number of the vascular structures in the midline are also identifiable, including the internal cerebral veins lying directly below the body of the fornix, the basilar artery anterior to the pons, and the vein of Galen directly posterior to the splenium of the corpus callosum. This midline section is the ideal plane for evaluation of the corpus callosum, fourth ventricle, brainstem, and cerebellar vermis, since these are all symmetric midline structures.

39C Sagittal/MRI/Midline Magnification. This midline partial saturation TR 600, TE 25 msec MRI demonstrates a number of small structures. The anterior third ventricle with its suprachiasmatic and infundibular recesses is visible. Note that the aqueduct of Sylvius has a slightly lower signal intensity than the third and fourth ventricles, which is related to spinal fluid flow. The flow of the spinal fluid produces a decreased recorded signal, which is similar to that seen in the blood vessels. Within the sella turcica we frequently identify two components; the pituitary gland is seen anteriorly, while a normal small fat pad is usually identifiable in the posterior portion of the sella.

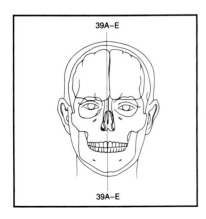

39A–E

39A–E

39B

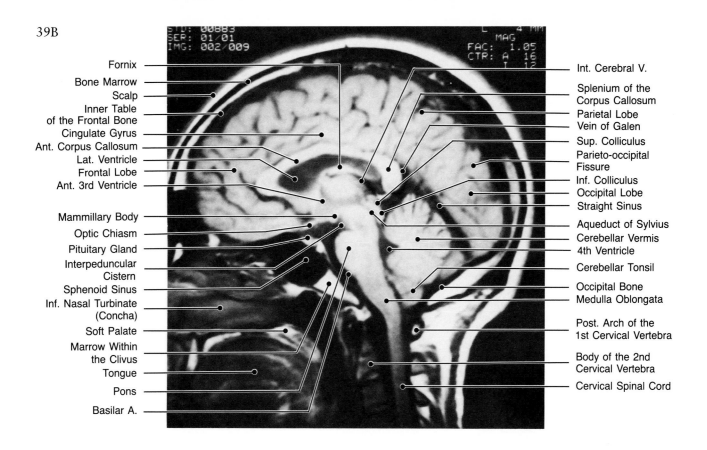

Fornix — Int. Cerebral V.

Bone Marrow — Splenium of the Corpus Callosum

Scalp — Parietal Lobe

Inner Table of the Frontal Bone — Vein of Galen

Cingulate Gyrus — Sup. Colliculus

Ant. Corpus Callosum — Parieto-occipital Fissure

Lat. Ventricle — Inf. Colliculus

Frontal Lobe — Occipital Lobe

Ant. 3rd Ventricle — Straight Sinus

Mammillary Body — Aqueduct of Sylvius

Optic Chiasm — Cerebellar Vermis

Pituitary Gland — 4th Ventricle

Interpeduncular Cistern — Cerebellar Tonsil

Sphenoid Sinus — Occipital Bone

Inf. Nasal Turbinate (Concha) — Medulla Oblongata

Soft Palate — Post. Arch of the 1st Cervical Vertebra

Marrow Within the Clivus — Body of the 2nd Cervical Vertebra

Tongue — Cervical Spinal Cord

Pons

Basilar A.

39C

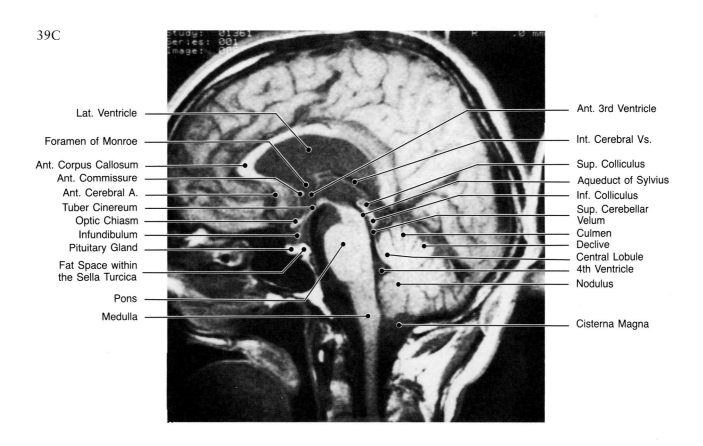

Lat. Ventricle — Ant. 3rd Ventricle

Foramen of Monroe — Int. Cerebral Vs.

Ant. Corpus Callosum — Sup. Colliculus

Ant. Commissure — Aqueduct of Sylvius

Ant. Cerebral A. — Inf. Colliculus

Tuber Cinereum — Sup. Cerebellar Velum

Optic Chiasm — Culmen

Infundibulum — Declive

Pituitary Gland — Central Lobule

Fat Space within the Sella Turcica — 4th Ventricle

Pons — Nodulus

Medulla — Cisterna Magna

105

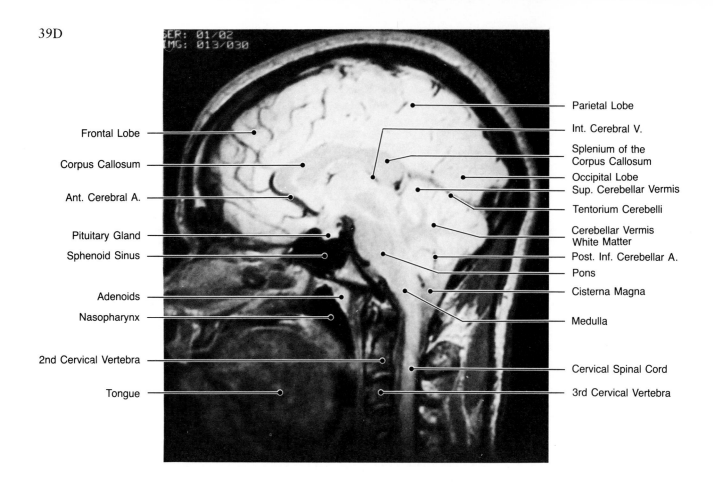

SER: 01/02
IMG: 013/030

Frontal Lobe

Corpus Callosum

Ant. Cerebral A.

Pituitary Gland

Sphenoid Sinus

Adenoids

Nasopharynx

2nd Cervical Vertebra

Tongue

Parietal Lobe

Int. Cerebral V.

Splenium of the
Corpus Callosum

Occipital Lobe
Sup. Cerebellar Vermis

Tentorium Cerebelli

Cerebellar Vermis
White Matter

Post. Inf. Cerebellar A.

Pons

Cisterna Magna

Medulla

Cervical Spinal Cord

3rd Cervical Vertebra

39D Sagittal/MRI/Midline. This TR 2000, TE 25 msec spin echo midline MRI demonstrates distinctly different contrast relationships of the tissues than the shorter repetition time images. The white matter tracts seen in the cerebellar vermis and the corpus callosum demonstrate low signal. The blood vessels, including the anterior cerebral arteries, posterior inferior cerebellar arteries, and internal cerebral veins are clearly visible because they are outlined by adjacent subarachnoid spaces and brain structures. There is a small soft tissue structure in the superior nasopharynx related to adenoid tissue.

39E Sagittal/MRI/Midline. This is a direct sagittal high resolution thin section CT of a cadaver head following the removal of the brain. The sella turcica is well seen, with the dorsum sella forming its posterior margin and the sphenoid sinus forming the sella floor. The planum sphenoidale is seen extending anteriorly from the stella turcica and is in continuity with the roof of the ethmoid air cells forming the floor of the anterior fossa.

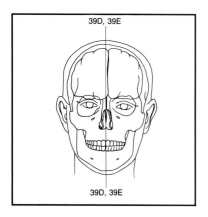

39D, 39E

39D, 39E

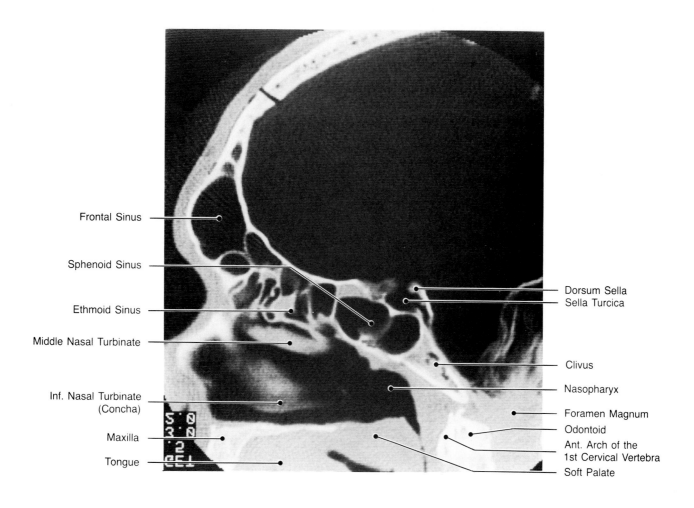

Frontal Sinus

Sphenoid Sinus

Ethmoid Sinus

Middle Nasal Turbinate

Inf. Nasal Turbinate
(Concha)

Maxilla

Tongue

Dorsum Sella
Sella Turcica

Clivus

Nasopharyx

Foramen Magnum

Odontoid

Ant. Arch of the
1st Cervical Vertebra

Soft Palate

40A

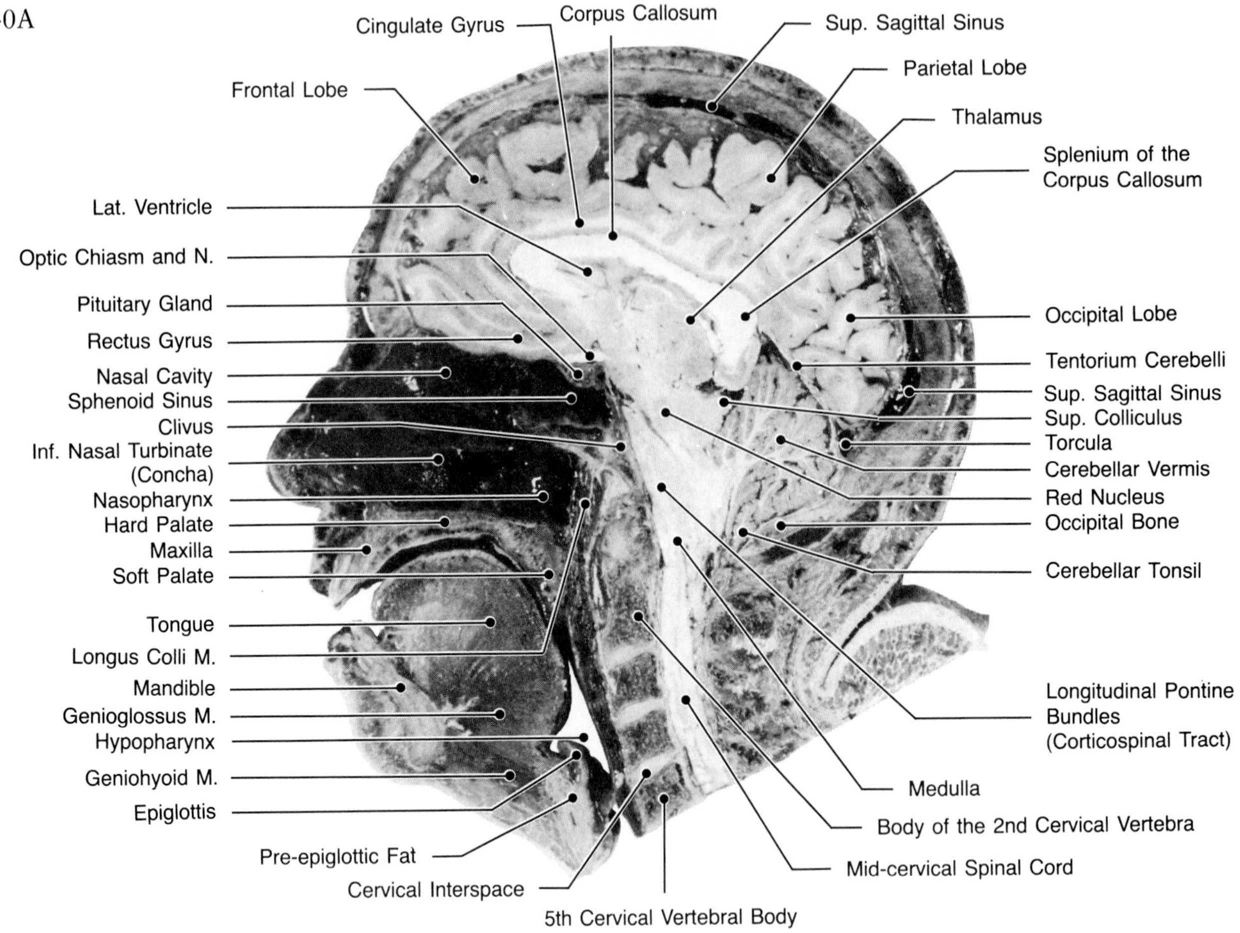

Cingulate Gyrus
Corpus Callosum
Sup. Sagittal Sinus
Frontal Lobe
Parietal Lobe
Thalamus
Splenium of the Corpus Callosum
Lat. Ventricle
Optic Chiasm and N.
Pituitary Gland
Rectus Gyrus
Nasal Cavity
Sphenoid Sinus
Clivus
Inf. Nasal Turbinate (Concha)
Nasopharynx
Hard Palate
Maxilla
Soft Palate
Tongue
Longus Colli M.
Mandible
Genioglossus M.
Hypopharynx
Geniohyoid M.
Epiglottis
Pre-epiglottic Fat
Cervical Interspace
5th Cervical Vertebral Body
Occipital Lobe
Tentorium Cerebelli
Sup. Sagittal Sinus
Sup. Colliculus
Torcula
Cerebellar Vermis
Red Nucleus
Occipital Bone
Cerebellar Tonsil
Longitudinal Pontine Bundles (Corticospinal Tract)
Medulla
Body of the 2nd Cervical Vertebra
Mid-cervical Spinal Cord

40A Sagittal/Anatomic Plate/Paramidline. The cerebellar tonsils and brainstem on this paramidline section are deformed secondary to herniation through the foramen magnum. The radiating appearance of the tongue muscles are well demonstrated on this section, with the most inferior group related to the geniohyoid muscle. The superior sagittal sinus is seen as a vascular cavity enveloped by two layers of dura paralleling the inner table of the skull.

40B Sagittal/MRI/Paramidline. This partial saturation TR 800, TE 25 msec sagittal paramidline MRI demonstrates a number of the vascular structures as low signal serpentine regions, including the posterior cerebral artery in the interpeduncular cistern, the posterior inferior cerebellar artery looping over the superior portion of the cerebellar tonsil, and the vein of Galen draped about the splenium of the corpus callosum. A number of the bony structures are seen as bright regions related to their high fat marrow content.

40C Sagittal/MRI/Paramidline. This spin echo TR 2000, TE 50 msec MRI demonstrates that the fat-containing spaces, such as the clivus and cervical vertebrae, have a low signal related to their T2 properties. Tissues with a high water content, such as the nasal turbinates and spinal fluid spaces, demonstrate rather high signal character. Because of the high signal arising from the brain and spinal fluid spaces, the blood vessels are particularly well outlined.

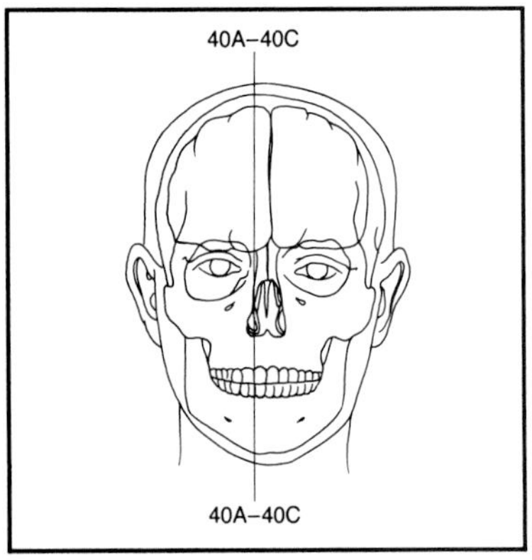

40A–40C
40A–40C

40B

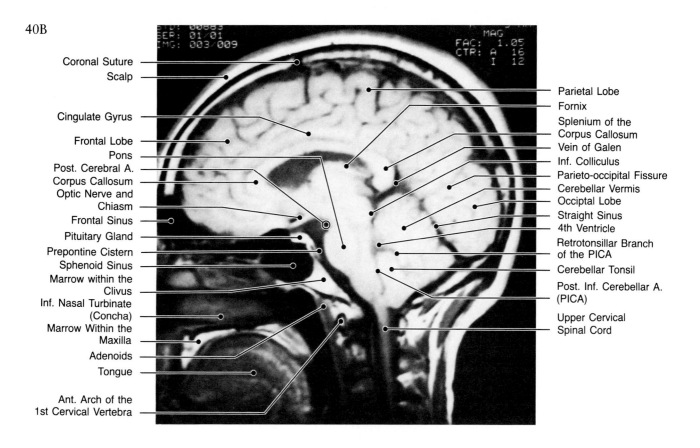

Coronal Suture
Scalp

Cingulate Gyrus
Frontal Lobe
Pons
Post. Cerebral A.
Corpus Callosum
Optic Nerve and Chiasm
Frontal Sinus
Pituitary Gland
Prepontine Cistern
Sphenoid Sinus
Marrow within the Clivus
Inf. Nasal Turbinate (Concha)
Marrow Within the Maxilla
Adenoids
Tongue

Ant. Arch of the 1st Cervical Vertebra

Parietal Lobe
Fornix
Splenium of the Corpus Callosum
Vein of Galen
Inf. Colliculus
Parieto-occipital Fissure
Cerebellar Vermis
Occiptal Lobe
Straight Sinus
4th Ventricle
Retrotonsillar Branch of the PICA
Cerebellar Tonsil
Post. Inf. Cerebellar A. (PICA)
Upper Cervical Spinal Cord

40C

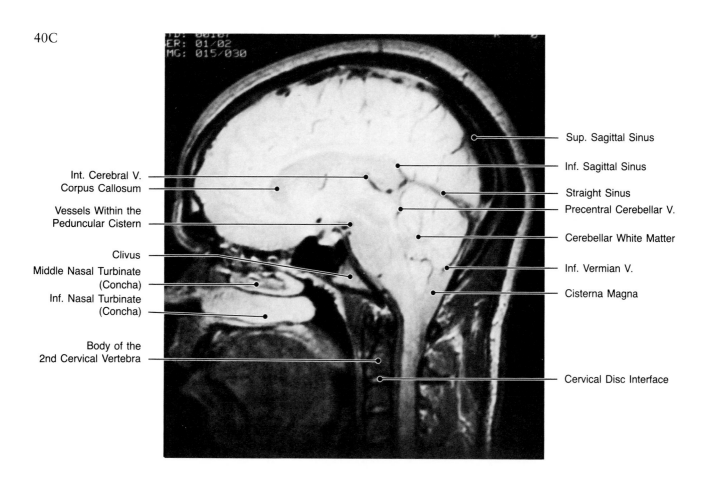

Int. Cerebral V.
Corpus Callosum

Vessels Within the Peduncular Cistern

Clivus
Middle Nasal Turbinate (Concha)
Inf. Nasal Turbinate (Concha)

Body of the 2nd Cervical Vertebra

Sup. Sagittal Sinus

Inf. Sagittal Sinus

Straight Sinus

Precentral Cerebellar V.

Cerebellar White Matter

Inf. Vermian V.

Cisterna Magna

Cervical Disc Interface

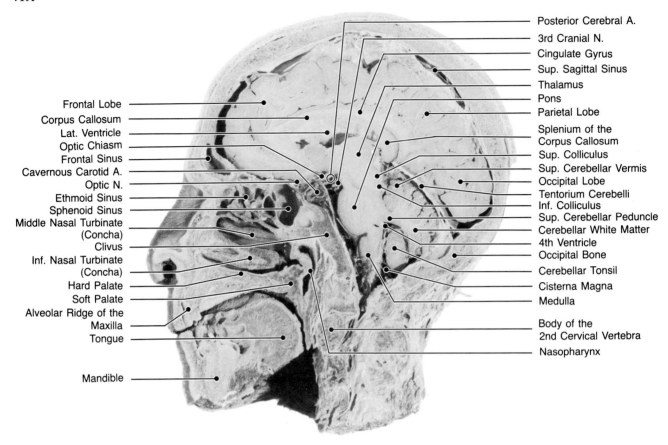

Posterior Cerebral A.
3rd Cranial N.
Cingulate Gyrus
Sup. Sagittal Sinus
Thalamus
Pons
Parietal Lobe
Splenium of the Corpus Callosum
Sup. Colliculus
Sup. Cerebellar Vermis
Occipital Lobe
Tentorium Cerebelli
Inf. Colliculus
Sup. Cerebellar Peduncle
Cerebellar White Matter
4th Ventricle
Occipital Bone
Cerebellar Tonsil
Cisterna Magna
Medulla
Body of the 2nd Cervical Vertebra
Nasopharynx

Frontal Lobe
Corpus Callosum
Lat. Ventricle
Optic Chiasm
Frontal Sinus
Cavernous Carotid A.
Optic N.
Ethmoid Sinus
Sphenoid Sinus
Middle Nasal Turbinate (Concha)
Clivus
Inf. Nasal Turbinate (Concha)
Hard Palate
Soft Palate
Alveolar Ridge of the Maxilla
Tongue
Mandible

41A Sagittal/Anatomic Plate/Paramidline. This sagittal anatomic plate is made slightly off the midline in the region of the interpeduncular cistern and demonstrates a number of structures, including the optic nerve, the posterior cerebral artery, and the third cranial nerve. The fourth ventricle is seen as a triangular spinal fluid space between the cerebellar tonsil posteriorly and the brainstem anteriorly. There are a number of small subarachnoid spaces surrounding the cerebellar tonsil within which lie branches of the posterior inferior cerebellar artery.

41B Sagittal/MRI/Optic Nerve. This sagittal paramidline partial saturation TR 800, TE 25 msec MRI demonstrates the anterior cerebral artery as it crosses over the superior margin of the optic nerve in the suprasellar cistern region. The muscular structures of the posterior nasopharynx, including the longus colli and pharyngeal constrictor muscles, which insert on the skull base, are seen.

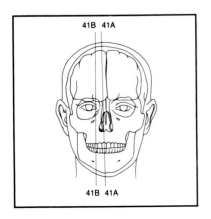

41B 41A

41B 41A

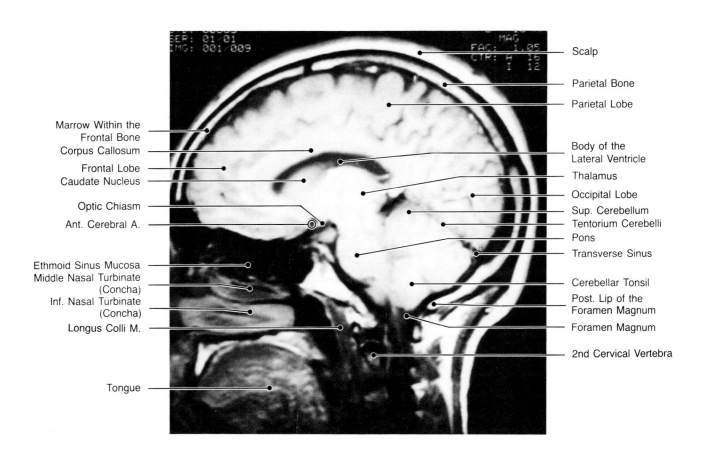

Scalp

Parietal Bone

Parietal Lobe

Marrow Within the
Frontal Bone

Corpus Callosum

Frontal Lobe

Caudate Nucleus

Optic Chiasm

Ant. Cerebral A.

Ethmoid Sinus Mucosa

Middle Nasal Turbinate
(Concha)

Inf. Nasal Turbinate
(Concha)

Longus Colli M.

Tongue

Body of the
Lateral Ventricle

Thalamus

Occipital Lobe

Sup. Cerebellum

Tentorium Cerebelli

Pons

Transverse Sinus

Cerebellar Tonsil

Post. Lip of the
Foramen Magnum

Foramen Magnum

2nd Cervical Vertebra

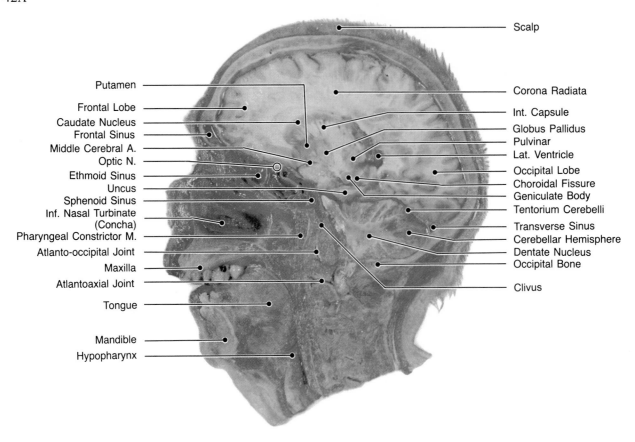

Scalp

Putamen

Frontal Lobe
Caudate Nucleus
Frontal Sinus
Middle Cerebral A.
Optic N.
Ethmoid Sinus
Uncus
Sphenoid Sinus
Inf. Nasal Turbinate
(Concha)
Pharyngeal Constrictor M.
Atlanto-occipital Joint
Maxilla
Atlantoaxial Joint

Tongue

Mandible
Hypopharynx

Corona Radiata

Int. Capsule
Globus Pallidus
Pulvinar
Lat. Ventricle
Occipital Lobe
Choroidal Fissure
Geniculate Body
Tentorium Cerebelli
Transverse Sinus
Cerebellar Hemisphere
Dentate Nucleus
Occipital Bone

Clivus

42A Sagittal/Anatomic Plate/Occipital Condyles. The articulations of the upper cervical spine and occipital condyle are well seen on this sagittal section. Located centrally within the paramidline white matter tracts of the cerebellum is the dentate nucleus. The white matter tracts of the internal capsule are seen coursing between the more peripheral arc-shaped caudate nucleus and the deeper putamen and globus pallidus. The pulvinar lies further posterior and borders the choroidal fissure, which separates the ambient cistern from the lateral ventricle.

42B Sagittal/MRI/Cerebral Peduncle. This sagittal partial saturation TR 800, TE 25 msec MRI is made through the cerebral peduncle of the lateral brainstem. The sphenoid and cavernous portion of the carotid artery are visible looping through the cavernous sinus. The caudate nucleus is seen as a small arc-shaped protuberance in the lateral ventricle.

42C Sagittal/MRI/Cervical Peduncle. This spin echo TR 2000, TE 25 msec MRI demonstrates the white matter tracts as low signal areas outlined by higher signal gray matter structures. The white matter tracts of the cerebellum form a frond-like pattern within the cerebellar hemisphere. A number of the vessels are well seen, including small vessels interposed between the pulvinar of the thalamus and the adjacent temporal lobe. This is the region of the choroidal fissure, where the posterior lateral choroidal artery is located.

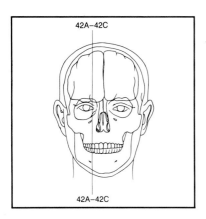

42A–42C

42A–42C

42B

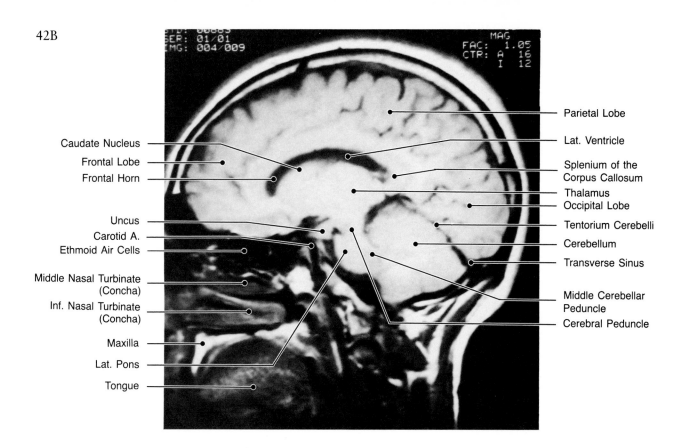

Parietal Lobe

Caudate Nucleus

Frontal Lobe

Frontal Horn

Lat. Ventricle

Splenium of the
Corpus Callosum

Thalamus

Occipital Lobe

Uncus

Carotid A.

Ethmoid Air Cells

Tentorium Cerebelli

Cerebellum

Transverse Sinus

Middle Nasal Turbinate
(Concha)

Inf. Nasal Turbinate
(Concha)

Middle Cerebellar
Peduncle

Cerebral Peduncle

Maxilla

Lat. Pons

Tongue

42C

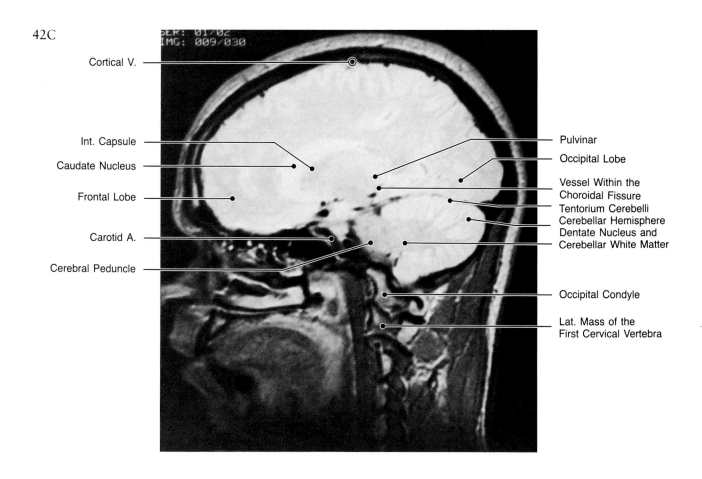

Cortical V.

Int. Capsule

Caudate Nucleus

Frontal Lobe

Carotid A.

Cerebral Peduncle

Pulvinar

Occipital Lobe

Vessel Within the
Choroidal Fissure

Tentorium Cerebelli

Cerebellar Hemisphere

Dentate Nucleus and
Cerebellar White Matter

Occipital Condyle

Lat. Mass of the
First Cervical Vertebra

113

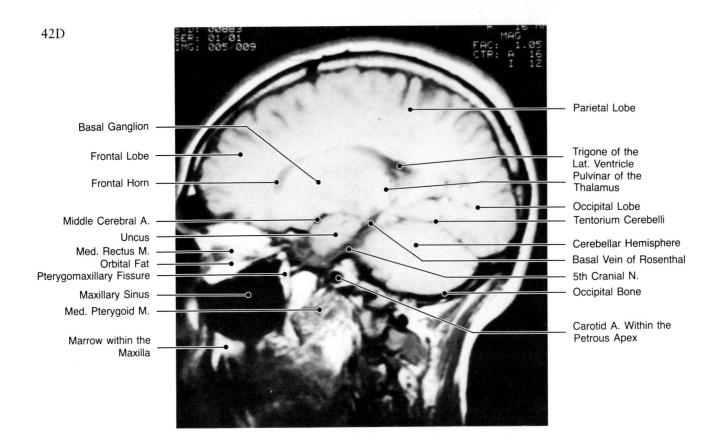

Basal Ganglion

Frontal Lobe

Frontal Horn

Middle Cerebral A.

Uncus

Med. Rectus M.

Orbital Fat

Pterygomaxillary Fissure

Maxillary Sinus

Med. Pterygoid M.

Marrow within the Maxilla

Parietal Lobe

Trigone of the Lat. Ventricle

Pulvinar of the Thalamus

Occipital Lobe

Tentorium Cerebelli

Cerebellar Hemisphere

Basal Vein of Rosenthal

5th Cranial N.

Occipital Bone

Carotid A. Within the Petrous Apex

42D Sagittal/MRI/Uncus. The pterygomaxillary fissure is seen as a bright fat-containing space between the posterior margin of the maxillary sinus and the low signal linear void of the adjacent pterygoid plates on this TR 800, TE 25 msec partial saturation MRI. Small defects in this region are related to branches of the internal maxillary artery. Within the orbit the medial rectus muscle is seen as a linear structure outlined by the adjacent orbital fat. The most medial portion of the temporal lobe (uncus) is seen.

42E Sagittal/MRI/Petrous Apex. This spin echo TR 2000, TE 25 msec MRI demonstrates a number of arterial structures, including the vertebral artery as it ascends through the multiple transverse foramina of the upper cervical spine. The petrous portion of the carotid artery is seen within the triangular low signal void of the temporal bone. The proximal middle cerebral artery in the sylvian fissure is seen just superior to the temporal lobe.

42F Sagittal/CT/Parasagittal Cadaver. This is a direct sagittal CT of a cadaver head following removal of the brain. The articulations of the cervical cranial junction are well seen. The lateral mass of C1 has two hemispheric articular surfaces, concave in configuration, that conform to the occipital condyles superiorly and the articular surface of C2 inferiorly.

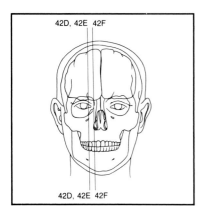

42D, 42E 42F

42D, 42E 42F

42E

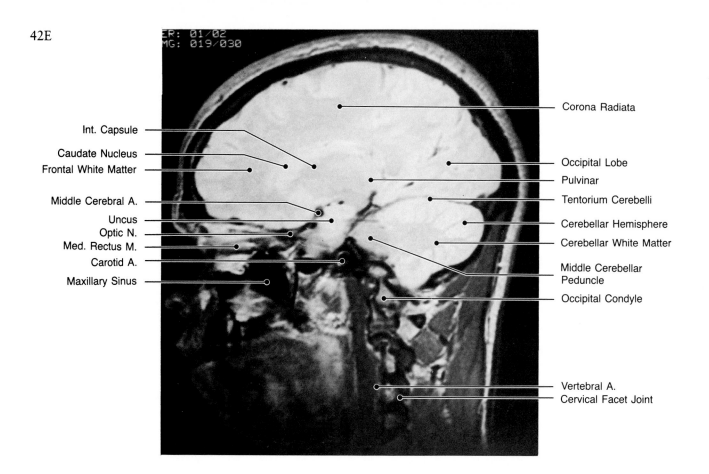

Int. Capsule

Caudate Nucleus

Frontal White Matter

Middle Cerebral A.

Uncus

Optic N.

Med. Rectus M.

Carotid A.

Maxillary Sinus

Corona Radiata

Occipital Lobe

Pulvinar

Tentorium Cerebelli

Cerebellar Hemisphere

Cerebellar White Matter

Middle Cerebellar Peduncle

Occipital Condyle

Vertebral A.

Cervical Facet Joint

42F

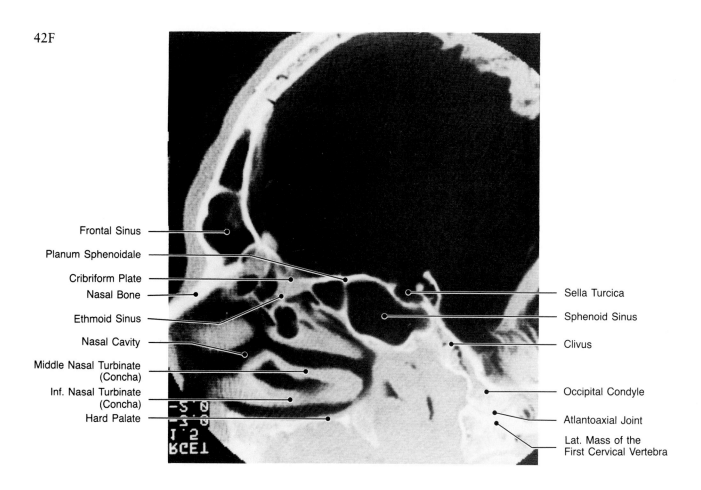

Frontal Sinus

Planum Sphenoidale

Cribriform Plate

Nasal Bone

Ethmoid Sinus

Nasal Cavity

Middle Nasal Turbinate (Concha)

Inf. Nasal Turbinate (Concha)

Hard Palate

Sella Turcica

Sphenoid Sinus

Clivus

Occipital Condyle

Atlantoaxial Joint

Lat. Mass of the First Cervical Vertebra

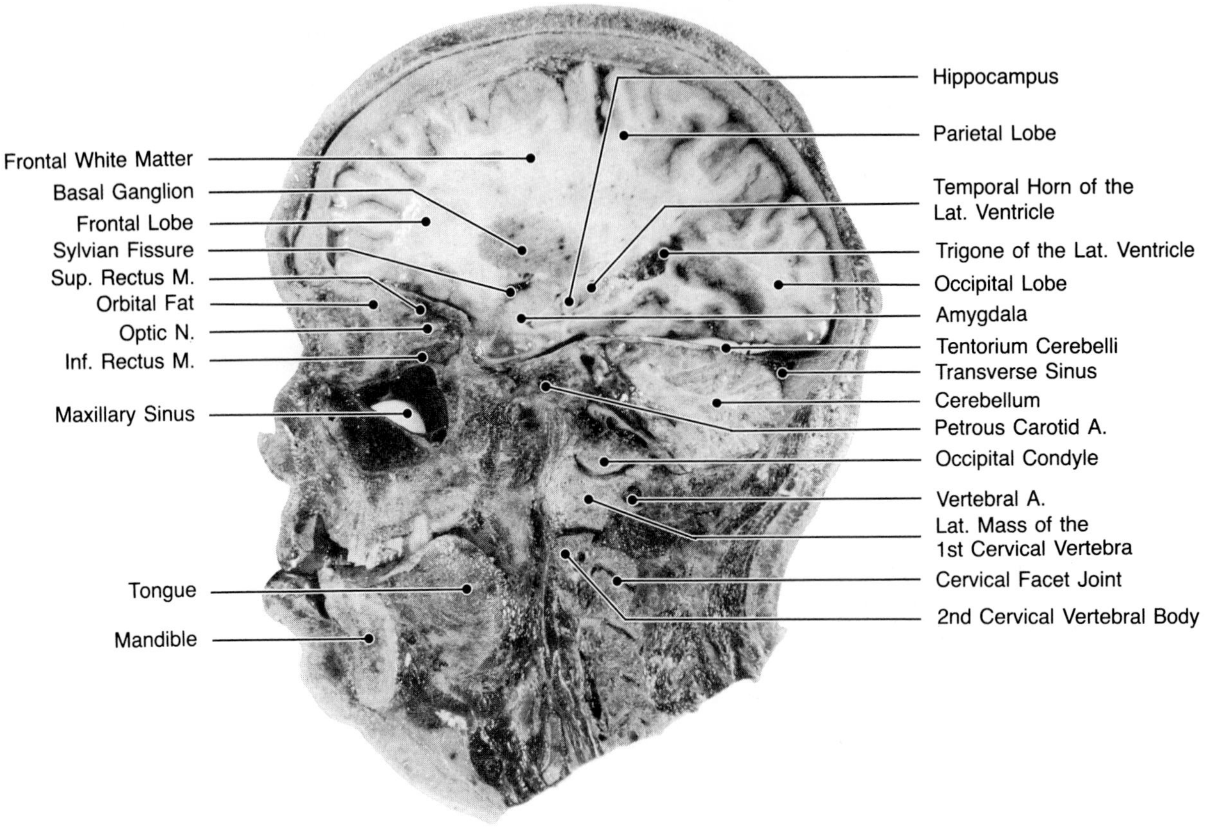
Frontal White Matter
Basal Ganglion
Frontal Lobe
Sylvian Fissure
Sup. Rectus M.
Orbital Fat
Optic N.
Inf. Rectus M.
Maxillary Sinus
Tongue
Mandible
Hippocampus
Parietal Lobe
Temporal Horn of the Lat. Ventricle
Trigone of the Lat. Ventricle
Occipital Lobe
Amygdala
Tentorium Cerebelli
Transverse Sinus
Cerebellum
Petrous Carotid A.
Occipital Condyle
Vertebral A.
Lat. Mass of the 1st Cervical Vertebra
Cervical Facet Joint
2nd Cervical Vertebral Body

43A Sagittal/Anatomic Plate/Temporal Horn. This anatomic sagittal section is made through the occipital condyle. The carotid artery is within the triangle-shaped petrous apex. The superior margin of the temporal bone faces the middle fossa, while the posterior portion faces the cerebellum. At the apex of the orbit the superior and inferior rectus muscles converge, with the optic nerve seen interposed. This section also demonstrates the Y-shaped anterior portion of the temporal horn. The hippocampus folds into the temporal horn, producing this shape.

43B Sagittal/MRI/Optic Nerve. This sagittal partial saturation TR 800, TE 25 msec MRI is made through the mid portion of the orbit. The superior rectus muscle, optic nerve, and inferior rectus muscle are outlined by high intensity fat. The roots of the teeth in the aveolar ridge of the maxilla are also outlined by marrow fat.

43C Sagittal/MRI/Temporal Horn. This sagittal TR 600, TE 25 msec partial saturation MRI is made through the temporal horn. Note that it has a Y-shaped configuration. The central soft tissue density between the two limbs of the subarachnoid space are related to infolding of the hippocampus.

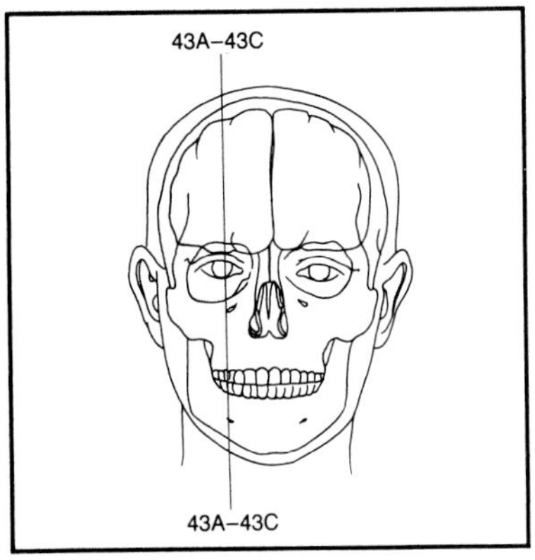
43A–43C
43A–43C

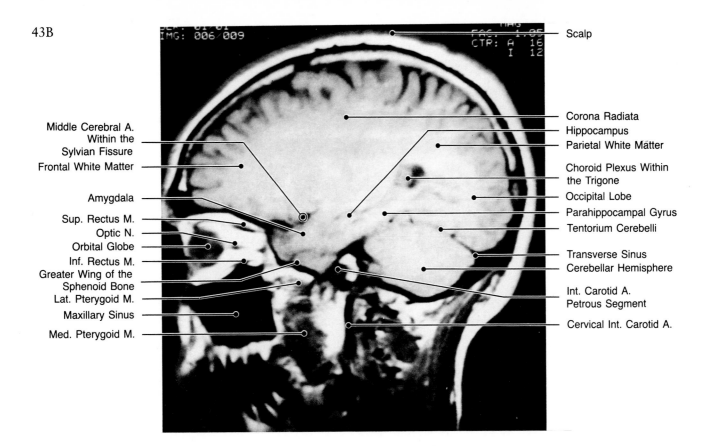

Scalp

Middle Cerebral A.
Within the
Sylvian Fissure
Frontal White Matter

Amygdala

Sup. Rectus M.
Optic N.
Orbital Globe
Inf. Rectus M.
Greater Wing of the
Sphenoid Bone
Lat. Pterygoid M.

Maxillary Sinus

Med. Pterygoid M.

Corona Radiata
Hippocampus
Parietal White Matter

Choroid Plexus Within
the Trigone
Occipital Lobe
Parahippocampal Gyrus
Tentorium Cerebelli

Transverse Sinus
Cerebellar Hemisphere

Int. Carotid A.
Petrous Segment

Cervical Int. Carotid A.

43C

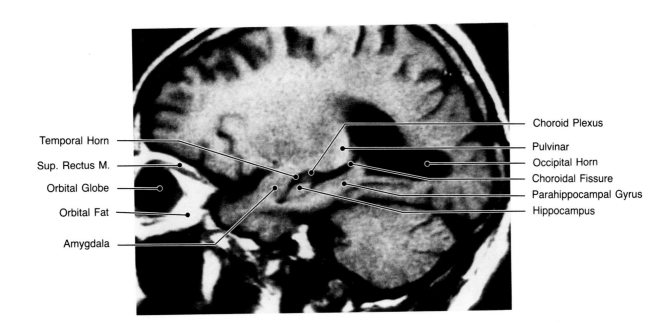

Choroid Plexus

Temporal Horn

Sup. Rectus M.

Orbital Globe

Orbital Fat

Amygdala

Pulvinar
Occipital Horn
Choroidal Fissure
Parahippocampal Gyrus
Hippocampus

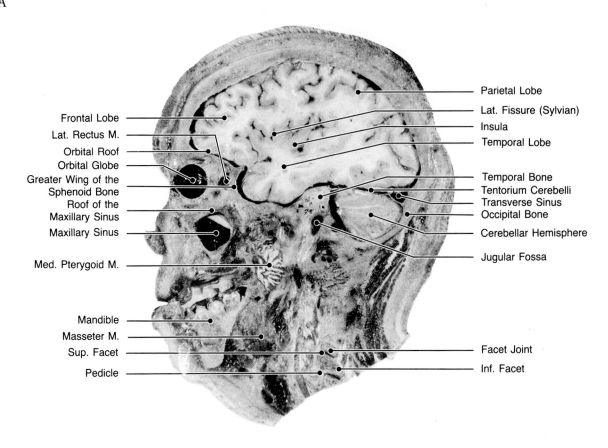

Frontal Lobe
Lat. Rectus M.
Orbital Roof
Orbital Globe
Greater Wing of the Sphenoid Bone
Roof of the Maxillary Sinus
Maxillary Sinus
Med. Pterygoid M.
Mandible
Masseter M.
Sup. Facet
Pedicle

Parietal Lobe
Lat. Fissure (Sylvian)
Insula
Temporal Lobe
Temporal Bone
Tentorium Cerebelli
Transverse Sinus
Occipital Bone
Cerebellar Hemisphere
Jugular Fossa
Facet Joint
Inf. Facet

44A Sagittal/Anatomic Plate/Temporal Lobe. This sagittal anatomic plate is made through the lateral margin of the Sylvian fissure. The Sylvian fissure and insula have a triangular configuration. Middle cerebral artery branches course through this region.

44B Sagittal/MRI/Sylvian Fissure. Multiple branches of the middle cerebral artery in the Sylvian triangle region are seen on this sagittal partial saturation TR 800, TE 25 msec MRI.

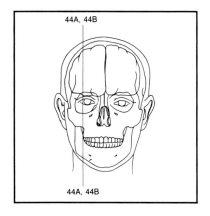

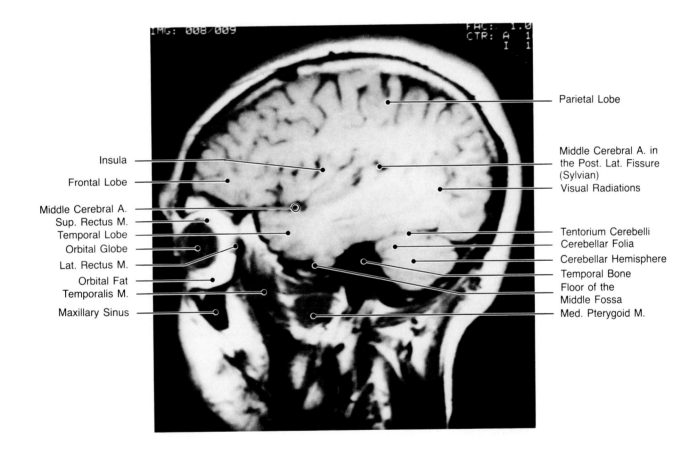

Parietal Lobe

Insula

Frontal Lobe

Middle Cerebral A. in the Post. Lat. Fissure (Sylvian)

Visual Radiations

Middle Cerebral A.
Sup. Rectus M.
Temporal Lobe
Orbital Globe
Lat. Rectus M.
Orbital Fat
Temporalis M.
Maxillary Sinus

Tentorium Cerebelli
Cerebellar Folia
Cerebellar Hemisphere
Temporal Bone
Floor of the
Middle Fossa
Med. Pterygoid M.

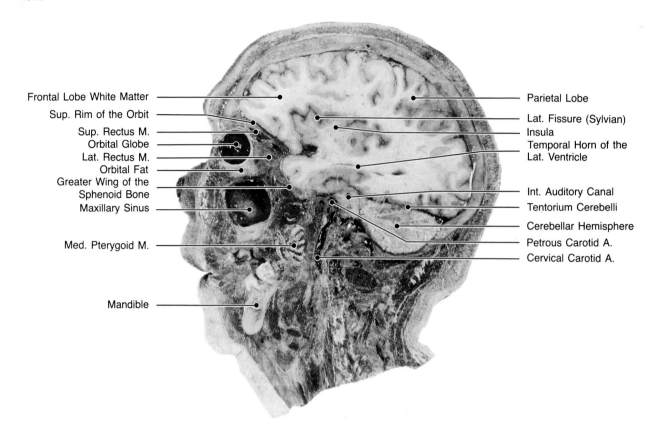

Frontal Lobe White Matter
Sup. Rim of the Orbit
Sup. Rectus M.
Orbital Globe
Lat. Rectus M.
Orbital Fat
Greater Wing of the
Sphenoid Bone
Maxillary Sinus
Med. Pterygoid M.
Mandible

Parietal Lobe
Lat. Fissure (Sylvian)
Insula
Temporal Horn of the
Lat. Ventricle
Int. Auditory Canal
Tentorium Cerebelli
Cerebellar Hemisphere
Petrous Carotid A.
Cervical Carotid A.

45A Sagittal/Anatomic Plate/Lateral Orbit. The temporal bone is seen on this sagittal anatomic plate to have a triangular configuration. In the mid portion of the temporal bone the internal auditory canal is present with the branches of the seventh and eighth cranial nerves. The cervical portion of the internal carotid artery entering the petrous apex is also seen.

45B Sagittal/CT/Carotid Canal. This direct sagittal CT scan of the temporal bone in the sagittal plane was obtained on a cadaver after removal of the brain structures. It is possible to obtain direct sagittal scans on patients, although it is somewhat cumbersome owing to special patient positioning. The close approximation of the converging carotid canal and jugular fossa in the inferior temporal bone is seen. Both the structures lie directly beneath the internal auditory canal. Diverticula of the jugular fossa can extend superiorly toward the internal auditory canal and are occasionally seen as a normal variant.

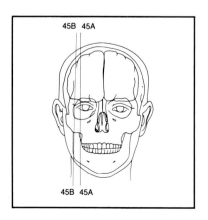

45B 45A

45B 45A

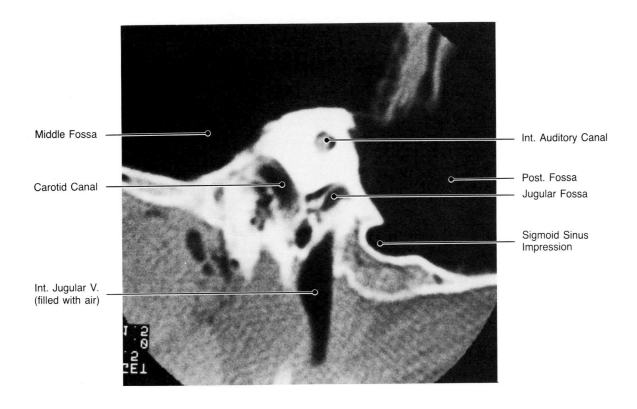

Middle Fossa

Carotid Canal

Int. Jugular V.
(filled with air)

Int. Auditory Canal

Post. Fossa

Jugular Fossa

Sigmoid Sinus
Impression

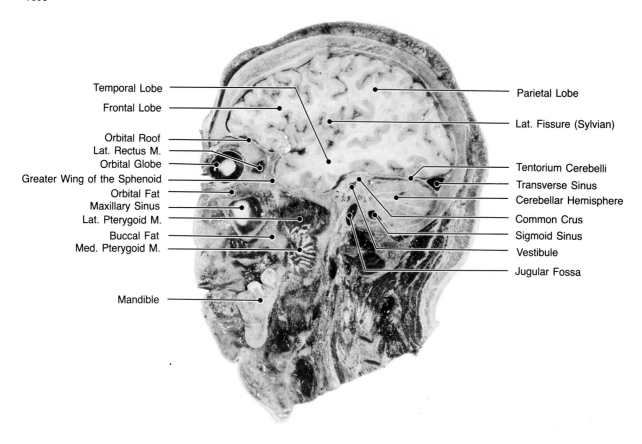

Temporal Lobe —
Frontal Lobe —

Orbital Roof —
Lat. Rectus M. —
Orbital Globe —
Greater Wing of the Sphenoid —
Orbital Fat —
Maxillary Sinus —
Lat. Pterygoid M. —
Buccal Fat —
Med. Pterygoid M. —

Mandible —

— Parietal Lobe

— Lat. Fissure (Sylvian)

— Tentorium Cerebelli
— Transverse Sinus
— Cerebellar Hemisphere
— Common Crus
— Sigmoid Sinus
— Vestibule
— Jugular Fossa

46A Sagittal/Anatomic Plate/Jugular Fossa. A number of the major venous structures of the posterior fossa are well demonstrated on this sagittal anatomic plate. The jugular fossa lies inferior to the vestibule in the inferior temporal bone. The sigmoid sinus is interposed between the anterior inferior portion of the cerebellar hemisphere and the temporal bone. The transverse sinus is a triangular structure within the dural sinuses of the tentorium.

46B Sagittal/MRI/Temporomandibular Joint. This sagittal partial saturation TR 800, TE 25 msec MRI demonstrates a number of the muscles of the face and neck to advantage. The temporalis muscle extends from the lateral margin of the skull between the middle fossa and the zygomatic bone. The masseter muscle arises from the zygomatic arch and descends toward the mandible. The lateral pterygoid muscle inserts on the mandibular ramus.

46C Sagittal/MRI/Carotid Canal Cadaver. This spin echo TR 2000, TE 25 msec MRI is made of an unfixed cadaver head in a sagittal plane. The temporal bone is a triangular low signal region with the internal auditory canal centrally. Note that signal is visible from the region of the carotid canal and the sigmoid sinus related to thrombosed blood. In a normal imaging setting the presence of moving blood produces a signal void in this region, making it difficult to recognize the interface between the inferior temporal bone and these vascular structures.

46D Sagittal/CT/Ossicles. A direct sagittal CT of the molar tooth–shaped ossicles is displayed at bone windows. The anterior ossicle is the long process of the malleus; the long process of the incus is posterior. (From Chakeres DW, and Weider DJ: Computed tomography of the ossicles. Neuroradiology 27:99–107, 1985.)

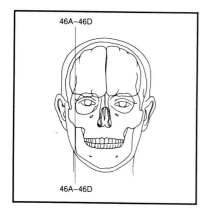

46A–46D

46A–46D

46B

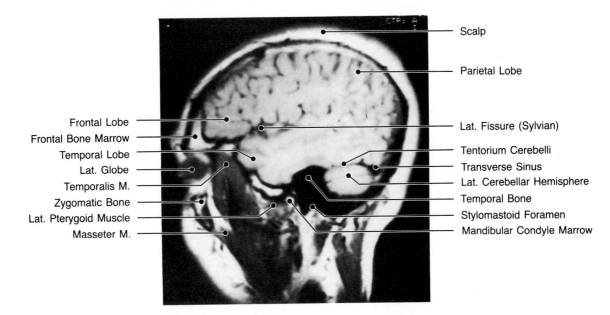

Scalp

Parietal Lobe

Frontal Lobe

Frontal Bone Marrow

Temporal Lobe

Lat. Globe

Temporalis M.

Zygomatic Bone

Lat. Pterygoid Muscle

Masseter M.

Lat. Fissure (Sylvian)

Tentorium Cerebelli

Transverse Sinus

Lat. Cerebellar Hemisphere

Temporal Bone

Stylomastoid Foramen

Mandibular Condyle Marrow

46C

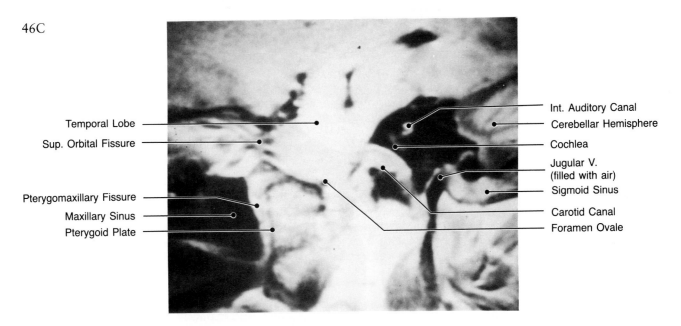

Temporal Lobe

Sup. Orbital Fissure

Pterygomaxillary Fissure

Maxillary Sinus

Pterygoid Plate

Int. Auditory Canal

Cerebellar Hemisphere

Cochlea

Jugular V.
(filled with air)

Sigmoid Sinus

Carotid Canal

Foramen Ovale

46D

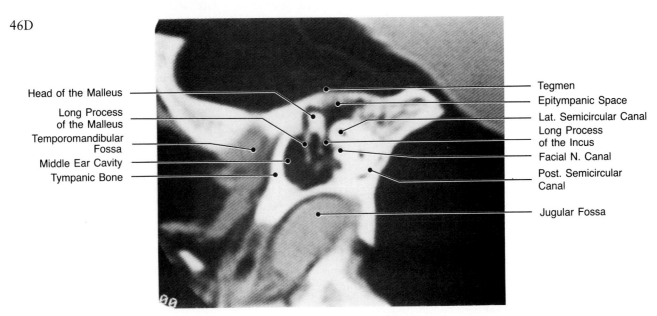

Head of the Malleus

Long Process
of the Malleus

Temporomandibular
Fossa

Middle Ear Cavity

Tympanic Bone

Tegmen

Epitympanic Space

Lat. Semicircular Canal

Long Process
of the Incus

Facial N. Canal

Post. Semicircular
Canal

Jugular Fossa

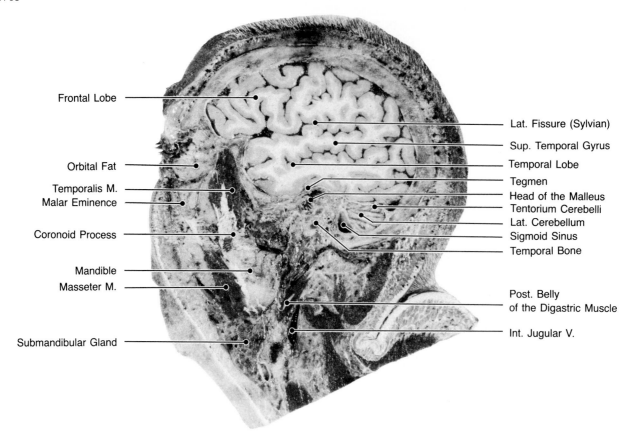

Frontal Lobe

Lat. Fissure (Sylvian)

Sup. Temporal Gyrus

Orbital Fat

Temporal Lobe

Temporalis M.

Tegmen

Malar Eminence

Head of the Malleus

Tentorium Cerebelli

Coronoid Process

Lat. Cerebellum

Sigmoid Sinus

Temporal Bone

Mandible

Masseter M.

Post. Belly
of the Digastric Muscle

Submandibular Gland

Int. Jugular V.

47A Sagittal/Anatomic Plate/Lateral Orbit. The lateral portion of the Sylvian fissure is seen as a spinal fluid space separating the frontal from the temporal lobes on this sagittal section. The superior temporal gyrus parallels the overlying Sylvian fissure.

47B Sagittal/CT/Temporal Bone Cadaver. The descending facial nerve canal is seen arcing posterior to the external auditory canal on this direct sagittal CT of a cadaver temporal bone. The ossicular heads are also seen in the epitympanic space directly above the external auditory canal. Coronal-sagittal reconstructions by computer manipulation of multiple axial sections can generate images of similar configuration though lower resolution.

47C Sagittal/MRI/Descending Facial Nerve Canal. The descending facial nerve canal is seen within the temporal bone on this spin echo TR 2500, TE 25 msec sagittal MRI. Note that there is a small linear defect in the fat at the stylomastoid foramen related to the facial nerve itself. In the facial canal the nerve appears more intense because the adjacent bone produces no signal, while in the fat spaces of the skull base the inverse is true.

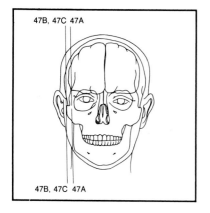

47B, 47C 47A

47B, 47C 47A

47B

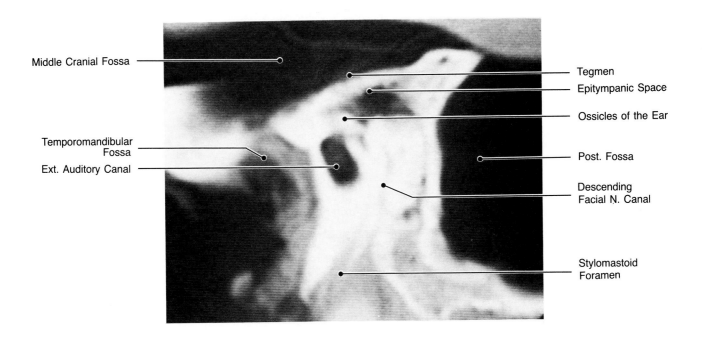

Middle Cranial Fossa

Tegmen

Epitympanic Space

Ossicles of the Ear

Temporomandibular Fossa

Ext. Auditory Canal

Post. Fossa

Descending Facial N. Canal

Stylomastoid Foramen

47C

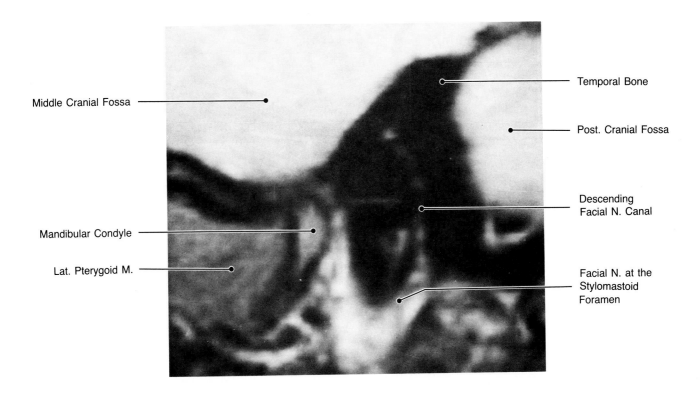

Middle Cranial Fossa

Temporal Bone

Post. Cranial Fossa

Descending Facial N. Canal

Mandibular Condyle

Lat. Pterygoid M.

Facial N. at the Stylomastoid Foramen

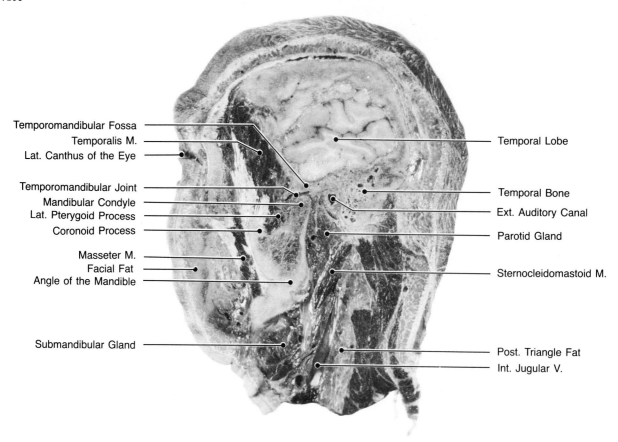

Temporomandibular Fossa
Temporalis M.
Lat. Canthus of the Eye

Temporomandibular Joint
Mandibular Condyle
Lat. Pterygoid Process
Coronoid Process

Masseter M.
Facial Fat
Angle of the Mandible

Submandibular Gland

Temporal Lobe

Temporal Bone
Ext. Auditory Canal

Parotid Gland

Sternocleidomastoid M.

Post. Triangle Fat
Int. Jugular V.

48A Sagittal/Anatomic Plate/Lateral To Orbit. The temporomandibular joint is seen on this sagittal anatomic plate. The rings of the external auditory canal, including the skin and cartilaginous segments, are seen posterior to the temporomandibular joint. A small cartilaginous disc is interposed between the mandibular fossa and the mandibular condyle.

48B Sagittal/MRI/Lateral Temporal Lobe. The superior temporal gyrus is seen at the superior lateral aspect of the temporal lobe on this sagittal spin echo TR 2500, TE 25 msec MRI. A few punctate signal voids are seen just superior to this and are related to branches of the middle cerebral artery exiting the Sylvian fissure to overlie the cerebral hemisphere.

48C Sagittal/MRI/Temporomandibular Joint, Closed Mouth. A surface coil was used to acquire this high resolution sagittal MRI of the temporal bone with a partial saturation technique of TR 600, TE 25 msec. Portions of the mandibular fossa are outlined by marrow cavity and a thin black cortical margin. Between the mandibular head and the fossa is a small low signal area, somewhat rectangular in shape, related to the cartilaginous disc.

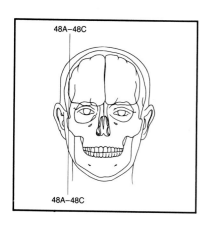

48A–48C

48A–48C

48B

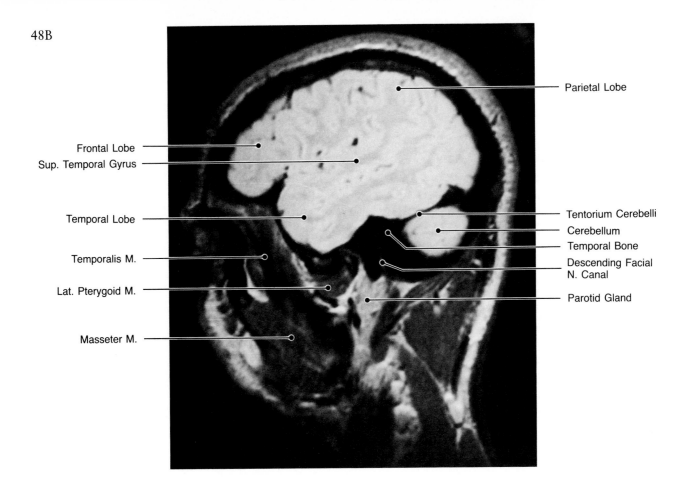

Parietal Lobe

Frontal Lobe

Sup. Temporal Gyrus

Temporal Lobe

Tentorium Cerebelli

Cerebellum

Temporal Bone

Temporalis M.

Descending Facial
N. Canal

Lat. Pterygoid M.

Parotid Gland

Masseter M.

48C

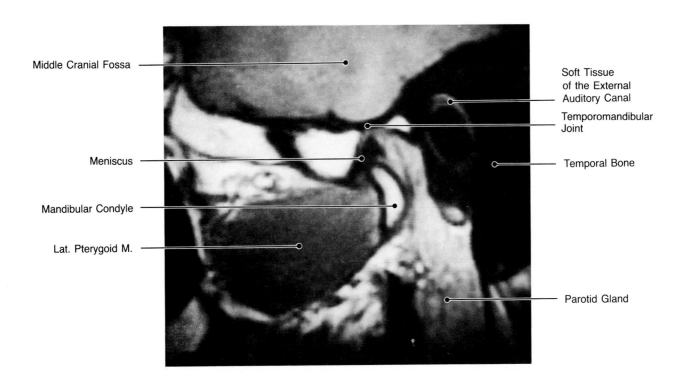

Middle Cranial Fossa

Soft Tissue
of the External
Auditory Canal

Temporomandibular
Joint

Meniscus

Temporal Bone

Mandibular Condyle

Lat. Pterygoid M.

Parotid Gland

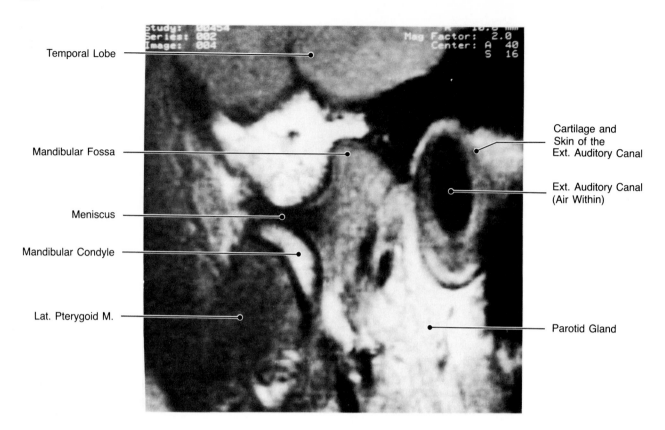

Temporal Lobe

Mandibular Fossa

Meniscus

Mandibular Condyle

Lat. Pterygoid M.

Cartilage and
Skin of the
Ext. Auditory Canal

Ext. Auditory Canal
(Air Within)

Parotid Gland

48D Sagittal/MRI/Temporomandibular Joint, Open Mouth.
A partial saturation TR 600, TE 25 msec sagittal MRI of the
mandible with the mouth open was acquired using a surface
coil. Note that the mandibular condyle has slid anteriorly and
inferiorly in relation to the mandibular fossa. A small low signal region still lies between the condyle and the mandibular
fossa related to the normal movement of the disc.

48E Sagittal/CT/Temporomandibular Joint Cadaver. A direct sagittal CT through the temporomandibular joint is seen
of a cadaver specimen. The mandibular condyle conforms to
the mandibular fossa. The posterior aspect of the fossa is
formed by the tympanic bone.

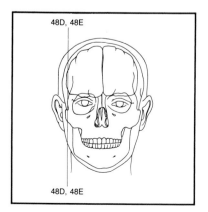

48D, 48E

48D, 48E

Tegmen

Head of the Malleus

Mandibular Fossa

Ext. Auditory Canal

Mandibular Condyle

Mastoid Antrum

Epitympanic Space

Incus

Tympanic Scutum

Sigmoid Sinus Impression

Mastoid Tip

SECTION 2

Spine

Donald W. Chakeres, M.D.
Delmas J. Allen, Ph.D.
A. John Christoforidis, M.D., Ph.D.

The anatomy of the spine on MRI and CT is demonstrated in this chapter. Sagittal anatomic sections are correlated to MR images. The axial anatomic plates of the spine are included in the Head and Neck, Chest, Abdomen, and Pelvis chapters. No coronal images are shown, since they are of limited clinical utility. The series of images begin cephalad and then continue inferiorly. First the sagittal anatomic sections and the comparable MRIs are shown. The respective axial MR images of the cervical, thoracic, and lumbar regions are shown. Finally CT axial images with and without intrathecal contrast similar in level to the MRIs are shown to provide a direct comparison of the anatomic information available with these complementary techniques.

The MR images were made using a number of different pulse sequences and section planes. MRI has the distinct advantage over CT of direct sagittal imaging of the spine; however, the resolution of the bony structures is limited. The relative pixel intensities of one tissue to another change dramatically with different pulse sequences. The shorter TR (<1500 msec), TE (25 msec) images have the greatest resolution. The cerebral spinal fluid (CSF) spaces are low intensity, while the fat spaces are most intense. The late TE (70 msec) images have less resolution because of greater noise, but the CSF and disc interspaces are highlighted as a high signal regions. This technique allows for noninvasive acquisition of information similar to a routine myelogram. MR can provide unique information related to the vertebral interspaces. Normally the discs have a relatively strong signal due to high water content; however, with aging and degeneration the discs lose their water and generate lower signal. MRI is able to delineate not only the external contours of the cord but also the internal anatomy by providing definition of the gray and white matter structures. This is very important in the identification of intramedullary pathology. These advantages account for the profound impact of MRI on the evaluation of spinal pathology.

Axial CT images of the spine are shown at bone and soft tissue windows as well as with and without intrathecal contrast injected for myelography. CT can evaluate the bony changes of the spine in great detail, but there is limited soft tissue detail. Multiple direct axial CT images of the spine can be reformatted by computer manipulation to display sagittal or any other section planes. Intrathecal water soluble contrast helps accentuate the CSF spaces, which may otherwise by marginally seen. Occasionally the contrast may enter the spinal cord itself in pathologic conditions such as syrinx. The cortical bone structures are most dense, and the marrow spaces of the bones are less dense. The CSF spaces are intermediate in density, while the fat spaces (epidural, retroperitoneal) are least dense. The epidural fat acts as an important inherent contrast agent to outline the nerve roots and dural sac.

Bibliography

Gehweiler JA, Osborne RL, Becker RF: *The Radiology of Vertebral Trauma.* Philadelphia, W.B. Saunders, 1980.

Modic MT, Weinstein MA, Pavlicek W, Boumphrey F, Starnes D, Duchesneau PM: Magnetic resonance imaging of the cervical spine: Technical and clinical observations. *AJR* 141:1129–1136, 1983.

Modic MR, Weinstein MA, Pavlicek W, Starnes DL, Duchesneau PM, Boumphrey F, Hardy RJ: Nuclear magnetic resonance imaging of the spine. *Radiology* 148:757–762, 1983.

Modic MT, Pavlicek W, Weinstein MA, Boumphrey F, Ngo F, Hardy R, Duchesneau PM: Magnetic resonance imaging of intervertebral disk disease. *Radiology* 152:103–111, 1984.

Modic MT, Feiglin DH, Piraino DW, Boumphrey F, Weinstein MA, Duchesneau PM, Rehm S: Vertebral osteomyelitis: Assessment using MR. *Radiology* 157:157–166, 1985.

Pech P, Haughton VM: Lumbar intervertebral disk: Correlation MR and anatomic study. *Radiology* 156:699–701, 1985.

Pech P, Haughton VM: CT appearance of unfused ossicles in the lumbar spine. *AJNR* 6:629–631, 1985.

Pech P, Daniels DL, Williams AL, Haughton VM: The cervical neural foramina: Correlation of microtomy and CT anatomy. *Radiology* 155:143–146, 1985.

Pernkopf E: *Atlas of Topographical and Applied Human Anatomy.* 2nd ed. Philadelphia, W.B. Saunders Co., 1964.

Raskin SP, Keating JW: Recognition of lumbar disk disease: Comparison of myelography and computed tomography. *AJR* 139:349–355, 1982.

Schnitzlein HN, Murtagh FR: *Imaging Anatomy of the Head and Spine.* Baltimore, Urban & Schwarzenberg, 1985.

Williams AL, Haughton VM, Meyer GA, Ho KC: Computed tomographic appearance of the bulging annulus. *Radiology* 142:403–408, 1982.

Williams AL, Haughton VM, Daniels DL, Grogan JP: Differential CT diagnosis of extruded nucleus pulposus. *Radiology* 148:141–148, 1983.

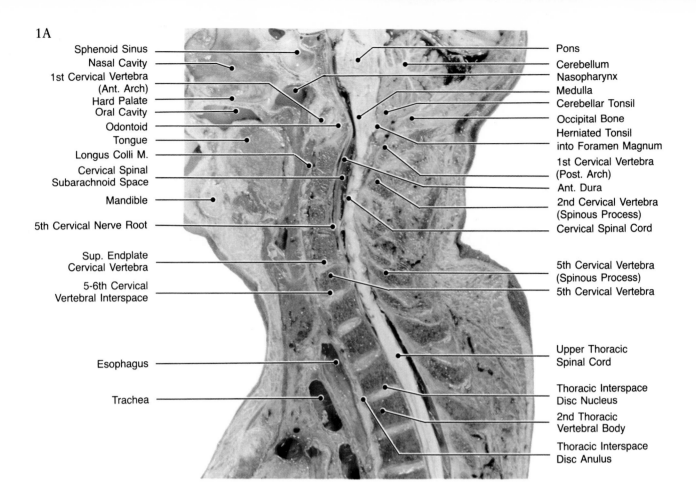

1A Sphenoid Sinus
Nasal Cavity
1st Cervical Vertebra (Ant. Arch)
Hard Palate
Oral Cavity
Odontoid
Tongue
Longus Colli M.
Cervical Spinal Subarachnoid Space
Mandible
5th Cervical Nerve Root
Sup. Endplate Cervical Vertebra
5-6th Cervical Vertebral Interspace
Esophagus
Trachea

Pons
Cerebellum
Nasopharynx
Medulla
Cerebellar Tonsil
Occipital Bone
Herniated Tonsil into Foramen Magnum
1st Cervical Vertebra (Post. Arch)
Ant. Dura
2nd Cervical Vertebra (Spinous Process)
Cervical Spinal Cord
5th Cervical Vertebra (Spinous Process)
5th Cervical Vertebra
Upper Thoracic Spinal Cord
Thoracic Interspace Disc Nucleus
2nd Thoracic Vertebral Body
Thoracic Interspace Disc Anulus

1A Sagittal/Anatomic Plate/Midline Cervical Spine. On this midline sagittal anatomic plate the cerebellar tonsils are herniated into the upper cervical region related to post-mortem brain edema. Because of minor scoliosis the brainstem and thoracic cord are sectioned in the midline, while the surface of the cervical spinal cord is displayed.

1B Sagittal/MRI/Midline Cervical. This spin echo TR 2000, TE 25 msec sagittal MRI is made in the midline cervical region with a body coil. Note that the signal intensity throughout the image is more uniform than that seen when a surface coil is used.

1C Sagittal/MRI/Midline Cervical. This sagittal spin echo TR 2000, TE 50 msec MRI demonstrates similar signal intensities for the CSF and the spinal cord. The CSF spaces are seen as high signal regions rather than low on short TE images.

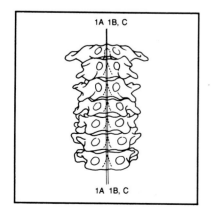

1A 1B, C

1A 1B, C

1B

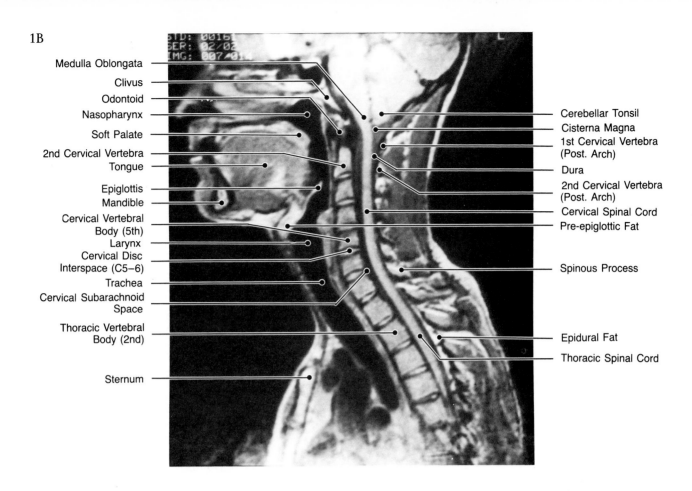

Medulla Oblongata
Clivus
Odontoid
Nasopharynx
Soft Palate
2nd Cervical Vertebra
Tongue
Epiglottis
Mandible
Cervical Vertebral Body (5th)
Larynx
Cervical Disc Interspace (C5–6)
Trachea
Cervical Subarachnoid Space
Thoracic Vertebral Body (2nd)
Sternum

Cerebellar Tonsil
Cisterna Magna
1st Cervical Vertebra (Post. Arch)
Dura
2nd Cervical Vertebra (Post. Arch)
Cervical Spinal Cord
Pre-epiglottic Fat
Spinous Process
Epidural Fat
Thoracic Spinal Cord

1C

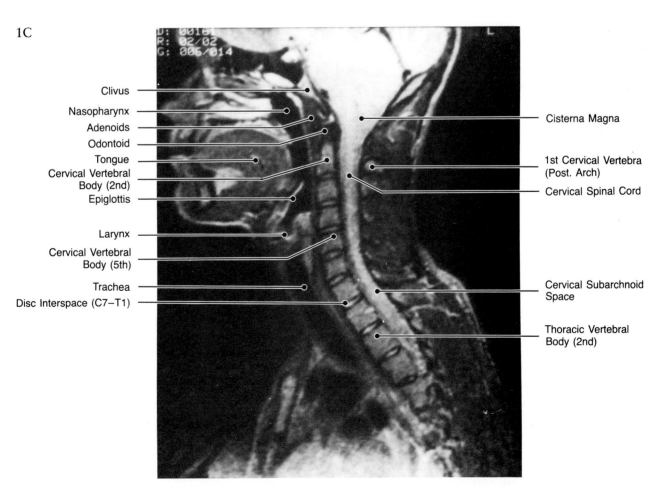

Clivus
Nasopharynx
Adenoids
Odontoid
Tongue
Cervical Vertebral Body (2nd)
Epiglottis
Larynx
Cervical Vertebral Body (5th)
Trachea
Disc Interspace (C7–T1)

Cisterna Magna
1st Cervical Vertebra (Post. Arch)
Cervical Spinal Cord
Cervical Subarachnoid Space
Thoracic Vertebral Body (2nd)

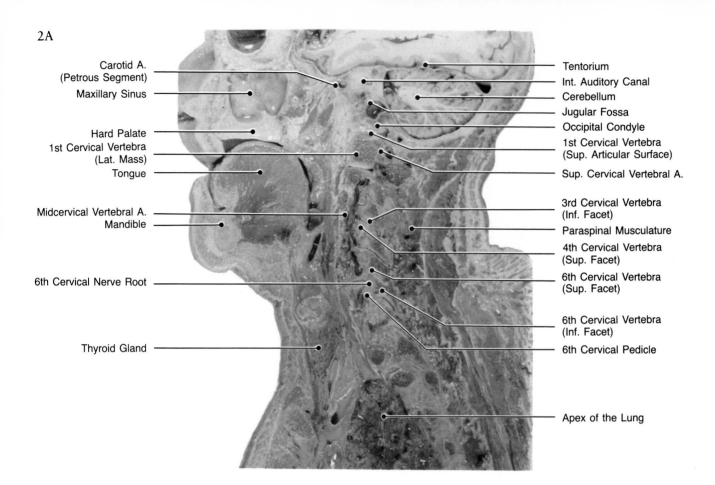

2A

Carotid A. (Petrous Segment)
Maxillary Sinus

Hard Palate
1st Cervical Vertebra (Lat. Mass)
Tongue

Midcervical Vertebral A.
Mandible

6th Cervical Nerve Root

Thyroid Gland

Tentorium
Int. Auditory Canal
Cerebellum
Jugular Fossa
Occipital Condyle
1st Cervical Vertebra (Sup. Articular Surface)
Sup. Cervical Vertebral A.

3rd Cervical Vertebra (Inf. Facet)
Paraspinal Musculature
4th Cervical Vertebra (Sup. Facet)
6th Cervical Vertebra (Sup. Facet)

6th Cervical Vertebra (Inf. Facet)
6th Cervical Pedicle

Apex of the Lung

2A Sagittal/Anatomic Plate/Pedicle of Cervical Spine. This sagittal anatomic section is made slightly off the midline at the level of the facets and pedicles of the cervical spine. The articulations of the inferior facets of the vertebrae above and the superior facets of the vertebrae below are well depicted. The nerve roots exit the neural foramen formed by segments of the bodies, pedicles, and facets of the vertebrae above and below.

2B Sagittal/MRI/Foramen. This sagittal spin echo TR 2000, TE 25 msec MRI demonstrates the neural foramen as high signal circular regions due to their epidural fat content.

2C Sagittal/MRI/Vertebral Artery. This sagittal spin echo TR 2000, TE 25 msec MRI demonstrates the bony structures surrounding the neural foramen.

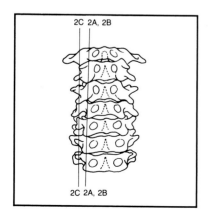

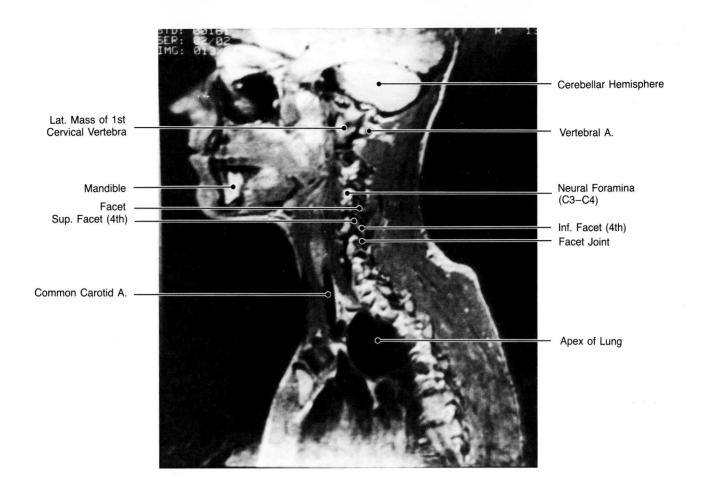

Cerebellar Hemisphere

Lat. Mass of 1st Cervical Vertebra

Vertebral A.

Mandible

Neural Foramina (C3–C4)

Facet

Sup. Facet (4th)

Inf. Facet (4th)

Facet Joint

Common Carotid A.

Apex of Lung

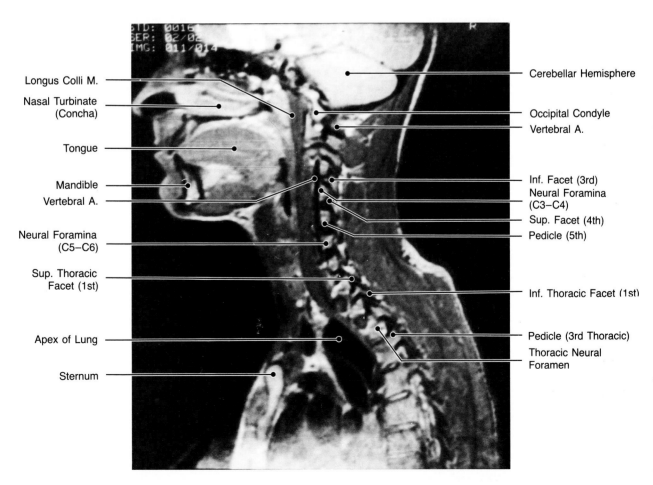

Longus Colli M.

Cerebellar Hemisphere

Nasal Turbinate (Concha)

Occipital Condyle

Vertebral A.

Tongue

Mandible

Inf. Facet (3rd)

Vertebral A.

Neural Foramina (C3–C4)

Sup. Facet (4th)

Neural Foramina (C5–C6)

Pedicle (5th)

Sup. Thoracic Facet (1st)

Inf. Thoracic Facet (1st)

Apex of Lung

Pedicle (3rd Thoracic)

Thoracic Neural Foramen

Sternum

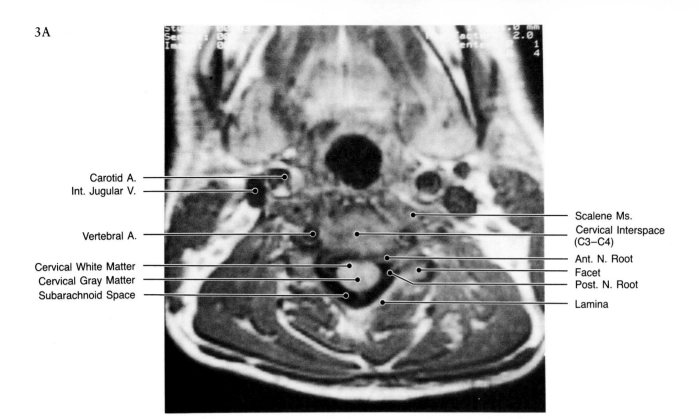

Carotid A.
Int. Jugular V.
Vertebral A.
Cervical White Matter
Cervical Gray Matter
Subarachnoid Space

Scalene Ms.
Cervical Interspace (C3–C4)
Ant. N. Root
Facet
Post. N. Root
Lamina

3A Axial/MRI/Cervical Nerve Roots. This axial partial saturation TR 600, TE 25 msec MRI demonstrates the anterior and posterior nerve roots coursing toward the neural foramen.

3B Axial/CT Bone Windows/Cervical Pedicle. This axial non–contrast-enhanced CT scan demonstrates the complete ring surrounding the spinal canal.

3C Axial/CT/Cervical Pedicle. This axial CT of the cervical spine was obtained after intrathecal injection of contrast. The spinal nerve roots are seen at the lateral margins of the spinal cord.

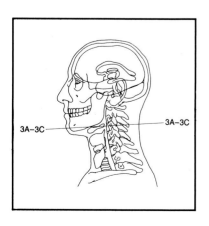

3A–3C   3A–3C

3B

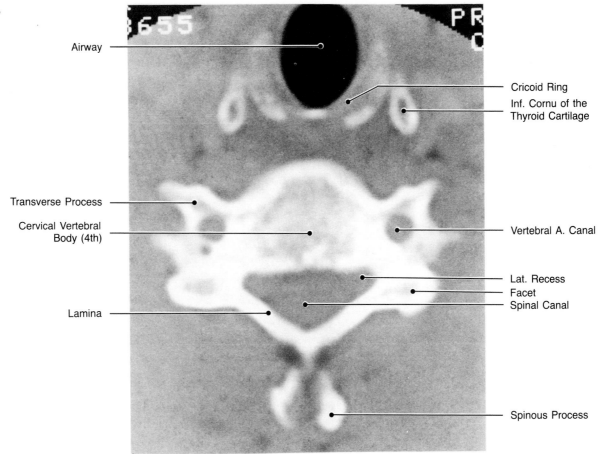

Airway

Cricoid Ring

Inf. Cornu of the Thyroid Cartilage

Transverse Process

Cervical Vertebral Body (4th)

Vertebral A. Canal

Lat. Recess
Facet
Spinal Canal

Lamina

Spinous Process

3C

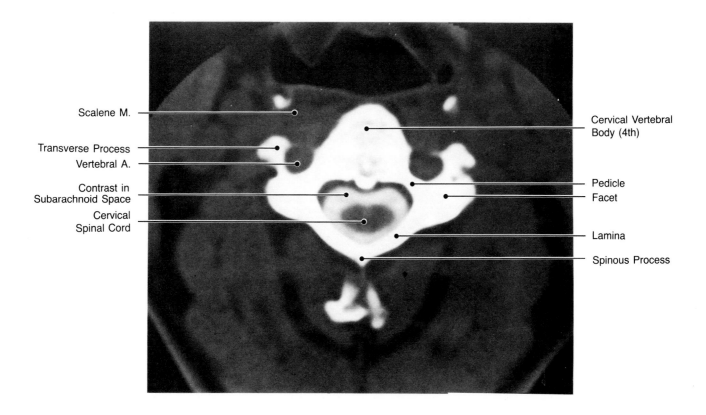

Scalene M.

Cervical Vertebral Body (4th)

Transverse Process
Vertebral A.

Contrast in Subarachnoid Space

Pedicle
Facet

Cervical Spinal Cord

Lamina

Spinous Process

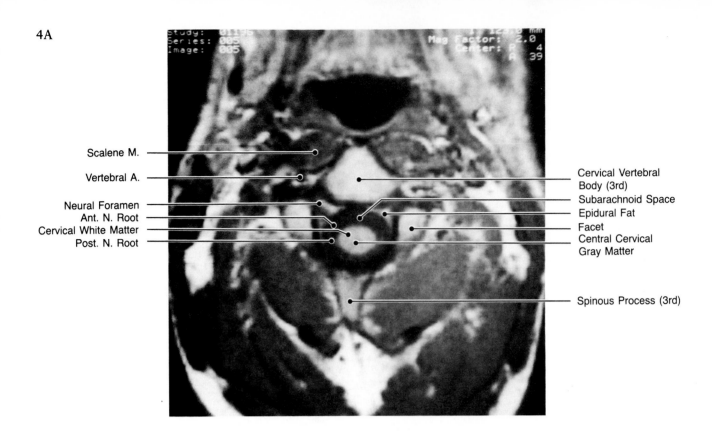

Scalene M. ——

Vertebral A. ——

Neural Foramen ——
Ant. N. Root ——
Cervical White Matter ——
Post. N. Root ——

—— Cervical Vertebral Body (3rd)
—— Subarachnoid Space
—— Epidural Fat
—— Facet
—— Central Cervical Gray Matter

—— Spinous Process (3rd)

4A Axial/MRI/Cervical Neural Foramen. This axial partial saturation TR 600, TE 25 msec MRI is made at the level of the neural foramen. High signal is seen in the vertebral body due to the marrow fat on this pulse sequence.

4B Axial/CT/Neural Foramen. This non–contrast-enhanced axial CT is made at the level of the neural foramen.

4C Axial/CT/Lateral Recess. This axial intrathecal contrast-enhanced CT shows the pointed margins of the anterior lateral subarachnoid space, which are due to the axillary root sheaths that surround the nerve sheaths in the neural foramen.

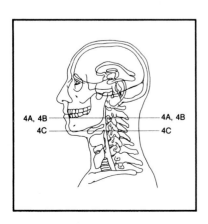

4B

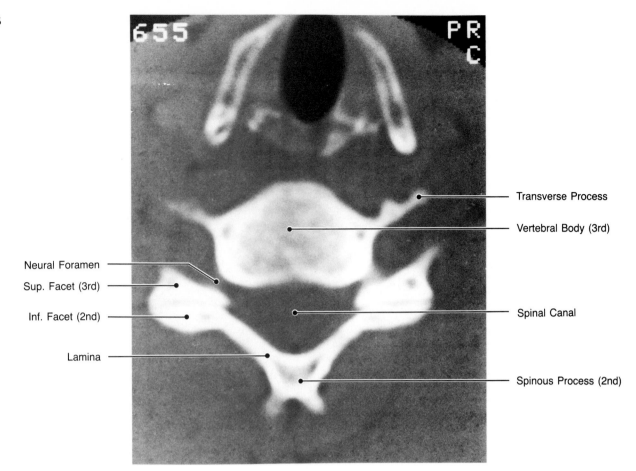

Transverse Process

Vertebral Body (3rd)

Neural Foramen

Sup. Facet (3rd)

Inf. Facet (2nd)

Spinal Canal

Lamina

Spinous Process (2nd)

4C

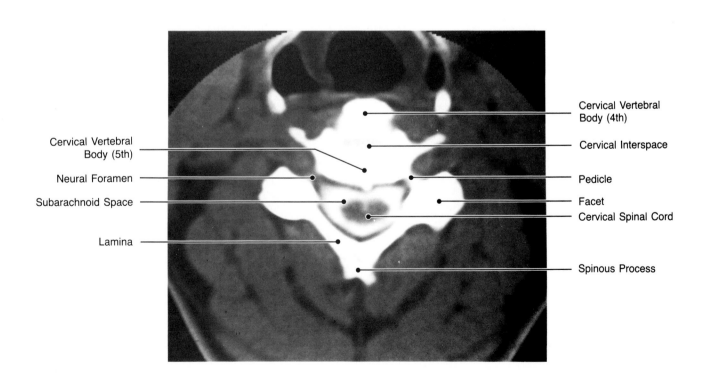

Cervical Vertebral
Body (4th)

Cervical Interspace

Cervical Vertebral
Body (5th)

Neural Foramen

Pedicle

Subarachnoid Space

Facet

Cervical Spinal Cord

Lamina

Spinous Process

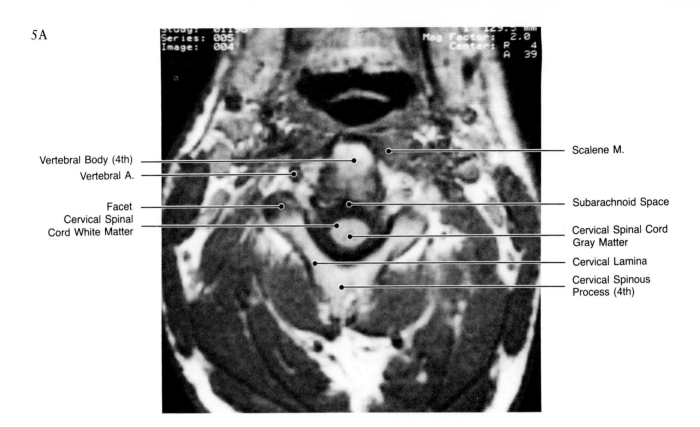

Vertebral Body (4th) —

Vertebral A. —

Facet —
Cervical Spinal
Cord White Matter —

Scalene M.

Subarachnoid Space

Cervical Spinal Cord
Gray Matter

Cervical Lamina

Cervical Spinous
Process (4th)

5A Axial/MRI/Cervical Interspace. This axial partial saturation TR 600, TE 25 msec MRI is made at the level of the interspace. Note that the signal from the disc is less than the marrow fat of the vertebral body.

5B Axial/CT/Cervical Nerve Roots. This axial CT image was made following injection of intrathecal contrast. The small anterior and posterior nerve roots are seen as filling defects within the contrast.

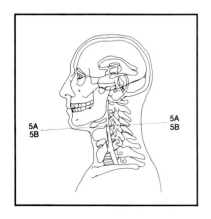

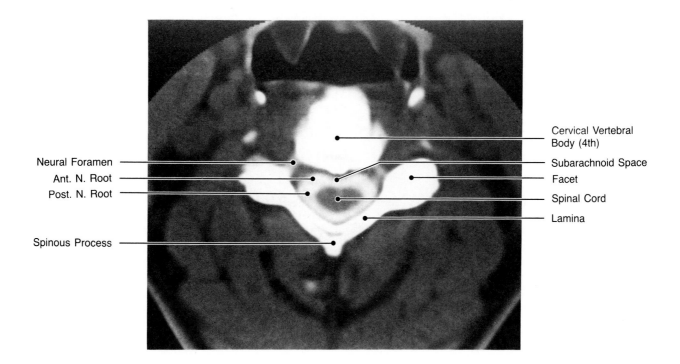

Cervical Vertebral
Body (4th)

Neural Foramen

Subarachnoid Space

Ant. N. Root

Facet

Post. N. Root

Spinal Cord

Lamina

Spinous Process

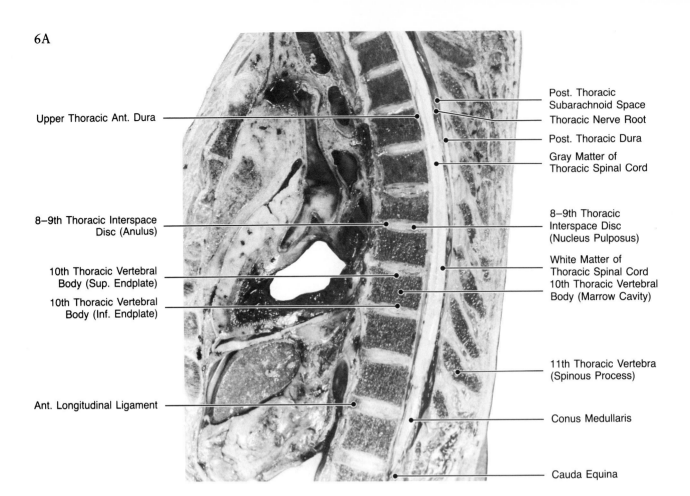

Upper Thoracic Ant. Dura

8–9th Thoracic Interspace Disc (Anulus)

10th Thoracic Vertebral Body (Sup. Endplate)

10th Thoracic Vertebral Body (Inf. Endplate)

Ant. Longitudinal Ligament

Post. Thoracic Subarachnoid Space

Thoracic Nerve Root

Post. Thoracic Dura

Gray Matter of Thoracic Spinal Cord

8–9th Thoracic Interspace Disc (Nucleus Pulposus)

White Matter of Thoracic Spinal Cord

10th Thoracic Vertebral Body (Marrow Cavity)

11th Thoracic Vertebra (Spinous Process)

Conus Medullaris

Cauda Equina

6A Sagittal/Anatomic Plate/Midline Thoracic. This midline sagittal section through the thoracic spine demonstrates the central gray matter structures surrounded by the larger white matter tracts of the spinal cord. There is a minor compression fracture of the eighth thoracic vertebral body evidenced by loss of overall vertebral height as well as a concave superior margin. Small osteophytes are seen from the superior anterior vertebral endplates.

6B Sagittal/MRI/Midline Thoracic. This sagittal spin echo TR 1500, TE 25 msec MR image shows the cervical and thoracic spinal cord. The subarachnoid space is seen as a low signal region around the linear cord.

6C Sagittal/MRI/Midline Thoracic. This spin echo TR 1500, TE 75 msec sagittal MR image demonstrates a nearly isointense signal produced by the spinal cord and subarachnoid space. Artifactual defects can be seen within the spinal fluid spaces related to CSF pulsation.

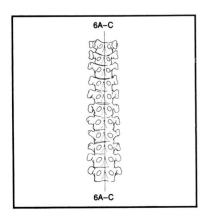

6B

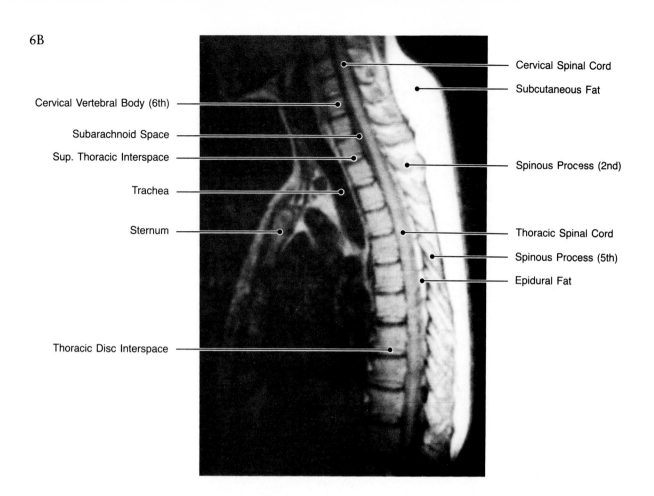

Cervical Vertebral Body (6th) —

Subarachnoid Space —

Sup. Thoracic Interspace —

Trachea —

Sternum —

Thoracic Disc Interspace —

— Cervical Spinal Cord

— Subcutaneous Fat

— Spinous Process (2nd)

— Thoracic Spinal Cord

— Spinous Process (5th)

— Epidural Fat

6C

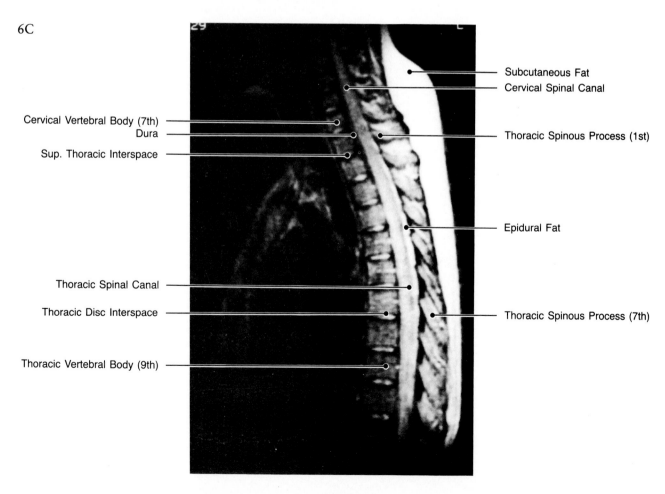

Cervical Vertebral Body (7th) —
Dura —

Sup. Thoracic Interspace —

Thoracic Spinal Canal —

Thoracic Disc Interspace —

Thoracic Vertebral Body (9th) —

— Subcutaneous Fat
— Cervical Spinal Canal

— Thoracic Spinous Process (1st)

— Epidural Fat

— Thoracic Spinous Process (7th)

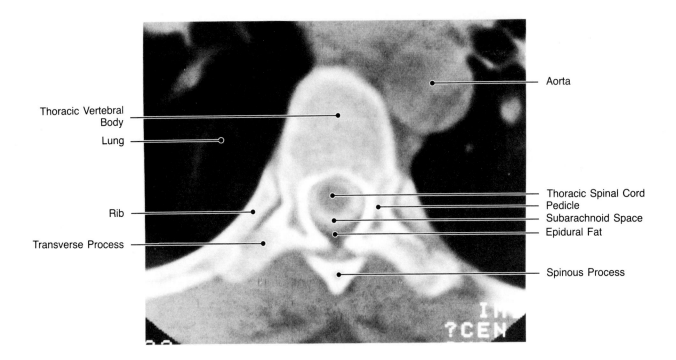

Aorta

Thoracic Vertebral Body

Lung

Thoracic Spinal Cord
Pedicle
Subarachnoid Space
Epidural Fat

Rib

Transverse Process

Spinous Process

7A Axial/CT/Thoracic Spinal Cord. This axial CT demonstrates the thoracic spinal cord outlined by intrathecal contrast.

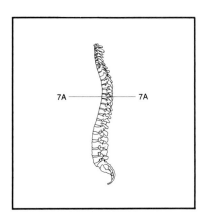

7A — 7A

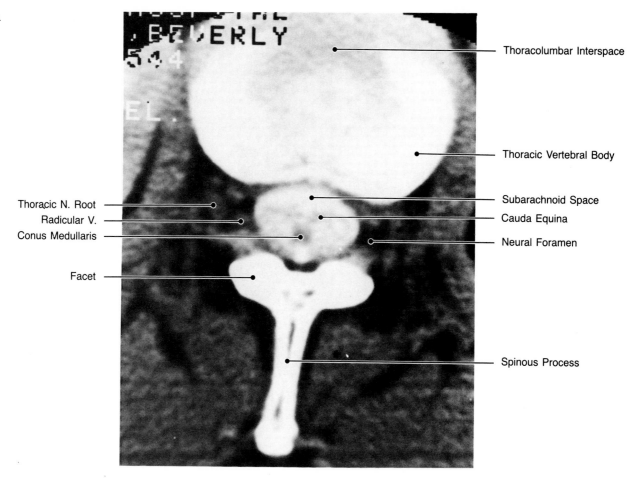

Thoracolumbar Interspace

Thoracic Vertebral Body

Thoracic N. Root

Radicular V.

Conus Medullaris

Facet

Subarachnoid Space

Cauda Equina

Neural Foramen

Spinous Process

8A Axial/CT/Conus. This axial CT scan of the lower thoracic spine shows intrathecal contrast about the conus medullaris and cauda equina.

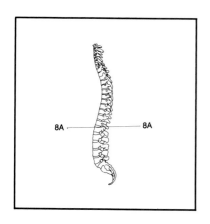

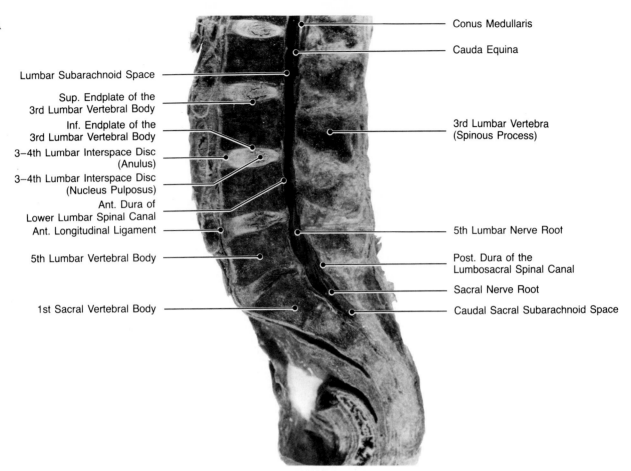

9A

Conus Medullaris

Cauda Equina

Lumbar Subarachnoid Space

Sup. Endplate of the
3rd Lumbar Vertebral Body

3rd Lumbar Vertebra
(Spinous Process)

Inf. Endplate of the
3rd Lumbar Vertebral Body

3–4th Lumbar Interspace Disc
(Anulus)

3–4th Lumbar Interspace Disc
(Nucleus Pulposus)

Ant. Dura of
Lower Lumbar Spinal Canal

Ant. Longitudinal Ligament

5th Lumbar Nerve Root

5th Lumbar Vertebral Body

Post. Dura of the
Lumbosacral Spinal Canal

Sacral Nerve Root

1st Sacral Vertebral Body

Caudal Sacral Subarachnoid Space

9A Sagittal/Anatomic Plate/Midline Lumbar. This midline section of the lumbar spine demonstrates a relatively narrow anterior posterior dimension of the spinal canal. Degenerative changes of the lumbar discs are seen as crevices within the nuclear components of the disc material. The lowest segment of the conus medullaris is seen in conjunction with the cauda equina at the superior lumbar level.

9B Sagittal/MRI/Midline. This midline lumbar sagittal spin echo TR 1500, TE 20 msec image is made with a surface coil under the spine. There is a general decrease in the signal from posterior to anterior. This is related to the decreasing signal reception as the distance is increased from the surface coil.

9C Sagittal/MRI/Midline. This midline sagittal MR spin echo TR 1500, TE 70 msec image demonstrates high signal, particularly from the CSF and discs.

9D Sagittal/MRI/Paramidline. This paramidline sagittal spin echo TR 1500, TE 20 msec MR image demonstrates the fat spaces within the neural foramen and posterior spine to be of high signal intensity. The marrow spaces within the vertebral bodies are less intense due primarily to their distance from the surface coil.

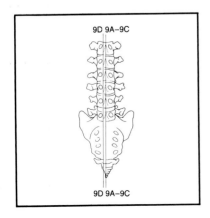

9D 9A–9C

9D 9A–9C

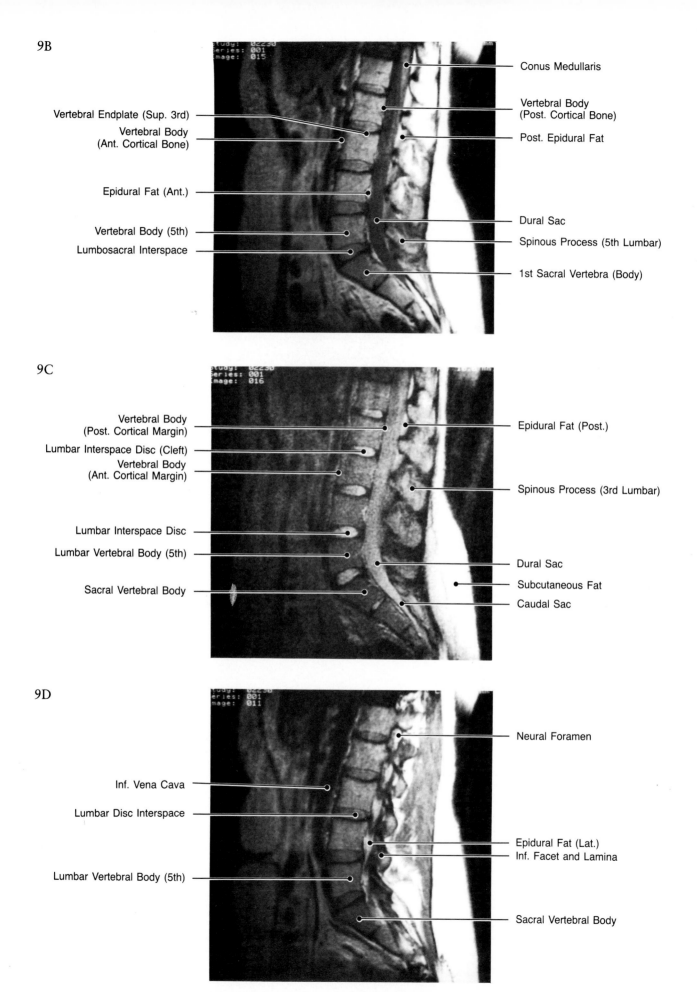

9B

Vertebral Endplate (Sup. 3rd)
Vertebral Body (Ant. Cortical Bone)
Epidural Fat (Ant.)
Vertebral Body (5th)
Lumbosacral Interspace

Conus Medullaris
Vertebral Body (Post. Cortical Bone)
Post. Epidural Fat
Dural Sac
Spinous Process (5th Lumbar)
1st Sacral Vertebra (Body)

9C

Vertebral Body (Post. Cortical Margin)
Lumbar Interspace Disc (Cleft)
Vertebral Body (Ant. Cortical Margin)
Lumbar Interspace Disc
Lumbar Vertebral Body (5th)
Sacral Vertebral Body

Epidural Fat (Post.)
Spinous Process (3rd Lumbar)
Dural Sac
Subcutaneous Fat
Caudal Sac

9D

Inf. Vena Cava
Lumbar Disc Interspace
Lumbar Vertebral Body (5th)

Neural Foramen
Epidural Fat (Lat.)
Inf. Facet and Lamina
Sacral Vertebral Body

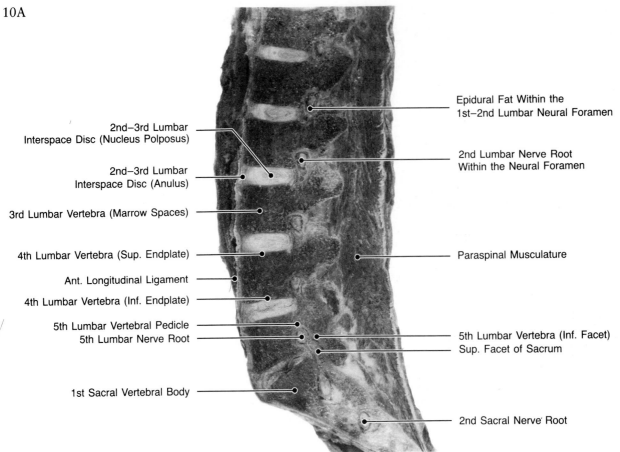

2nd–3rd Lumbar Interspace Disc (Nucleus Polposus)

2nd–3rd Lumbar Interspace Disc (Anulus)

3rd Lumbar Vertebra (Marrow Spaces)

4th Lumbar Vertebra (Sup. Endplate)

Ant. Longitudinal Ligament

4th Lumbar Vertebra (Inf. Endplate)

5th Lumbar Vertebral Pedicle
5th Lumbar Nerve Root

1st Sacral Vertebral Body

Epidural Fat Within the 1st–2nd Lumbar Neural Foramen

2nd Lumbar Nerve Root Within the Neural Foramen

Paraspinal Musculature

5th Lumbar Vertebra (Inf. Facet)
Sup. Facet of Sacrum

2nd Sacral Nerve Root

10A Sagittal/Anatomic Plate/Neural Foramen. The lumbar nerve roots are seen coursing below the pedicles on this paramidline sagittal anatomic plate. Note that the nerve roots exit the foramen at a level above the disc interspace.

10B Sagittal/MRI/Pedicle. This sagittal spin echo TR 1500, TE 20 msec MR image is made through the pedicle. Small radicular veins are seen directly superior to the lumbar nerve root in the foramen.

10C Sagittal/MRI/Neural Foramen. This sagittal spin echo TR 1500, TE 25 msec MR image demonstrates the structures surrounding the neural foramen.

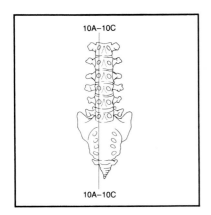

10A–10C

10A–10C

10B

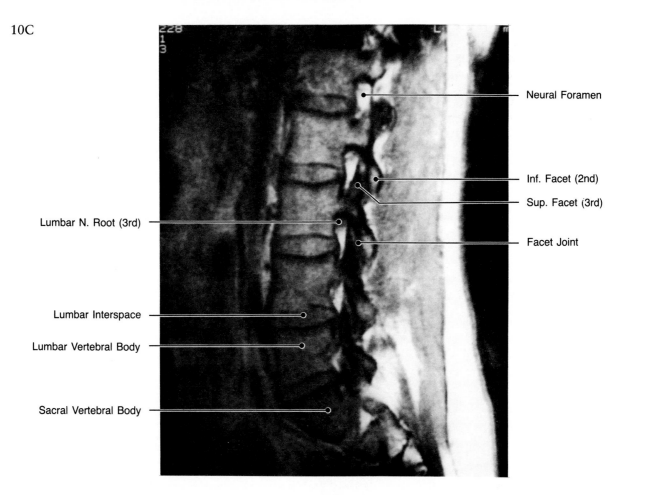

Lumbar Vertebral Body (3rd) —————

Pedicle (3rd)

Radicular V.

Lumbar Disc Interspace —————

Lumbar N. Root
(Within Neural Foramen)

Sup. Facet (5th) —————

Inf. Facet (4th)

Sacral Vertebral Body

10C

Neural Foramen

Inf. Facet (2nd)

Sup. Facet (3rd)

Lumbar N. Root (3rd) —————

Facet Joint

Lumbar Interspace —————

Lumbar Vertebral Body —————

Sacral Vertebral Body —————

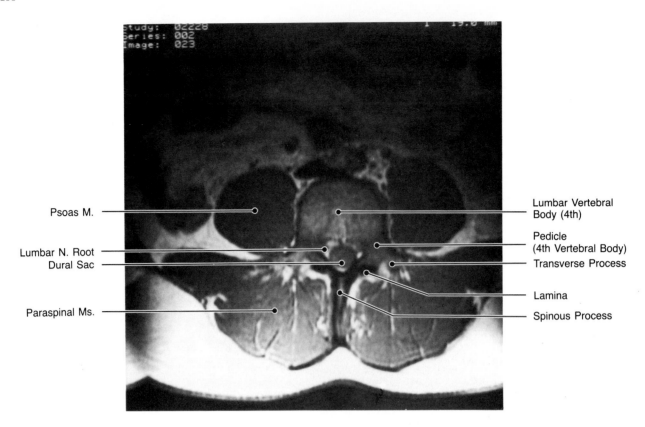

Psoas M.

Lumbar N. Root
Dural Sac

Paraspinal Ms.

Lumbar Vertebral
Body (4th)

Pedicle
(4th Vertebral Body)

Transverse Process

Lamina

Spinous Process

11A Axial/MRI/Pedicle. This axial spin echo TR 1500, TE 20 msec MR image is made through the pedicle. The cortical bone structures are low signal intensity, while the bone containing marrow is of higher signal.

11B Axial/CT/Lumbar Pedicle. This axial non–contrast-enhanced CT scan is made at the level of the pedicle.

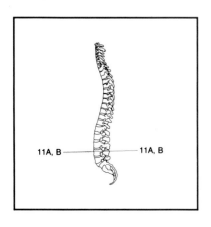

11A, B   11A, B

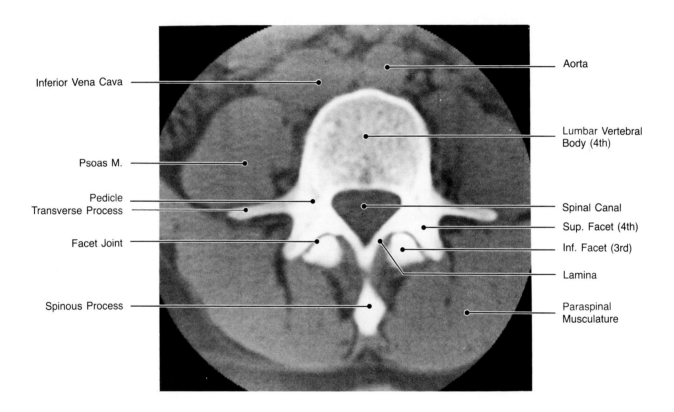

Inferior Vena Cava

Aorta

Psoas M.

Lumbar Vertebral Body (4th)

Pedicle

Transverse Process

Spinal Canal

Sup. Facet (4th)

Facet Joint

Inf. Facet (3rd)

Lamina

Spinous Process

Paraspinal Musculature

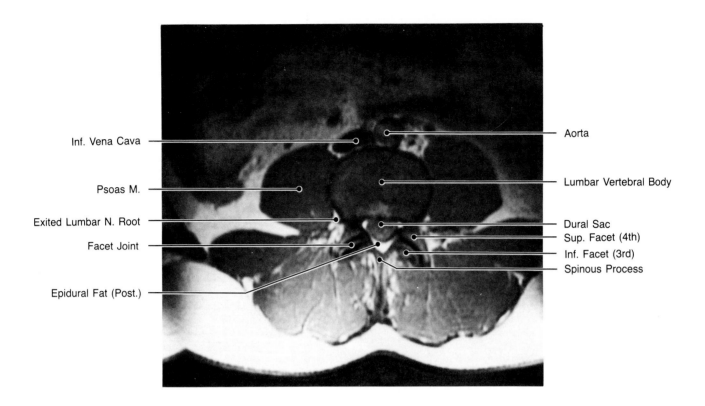

Inf. Vena Cava

Psoas M.

Exited Lumbar N. Root

Facet Joint

Epidural Fat (Post.)

Aorta

Lumbar Vertebral Body

Dural Sac
Sup. Facet (4th)
Inf. Facet (3rd)
Spinous Process

12A Axial/MRI/Facet. This axial spin echo TR 1500, TE 20 msec MR image demonstrates small high signal regions within the spinal canal related to the epidural fat.

12B Axial/CT/Lateral Recess. This axial non–contrast-enhanced CT scan demonstrates the spinal canal to have a triangular configuration. The anterior lateral extensions of the spinal canal are related to the lateral recesses.

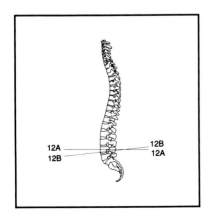

12A
12B

12B
12A

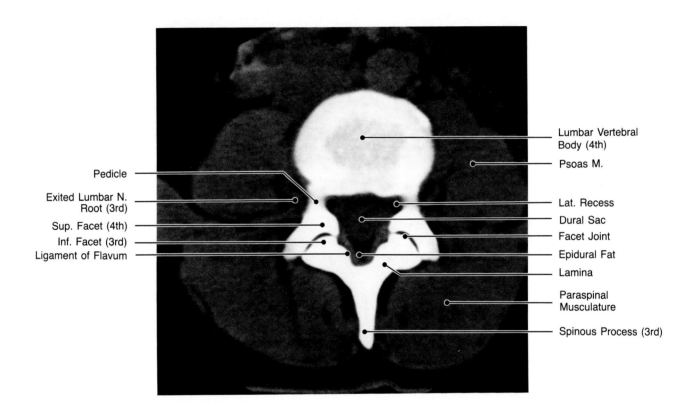

Pedicle

Exited Lumbar N. Root (3rd)

Sup. Facet (4th)

Inf. Facet (3rd)

Ligament of Flavum

Lumbar Vertebral Body (4th)

Psoas M.

Lat. Recess

Dural Sac

Facet Joint

Epidural Fat

Lamina

Paraspinal Musculature

Spinous Process (3rd)

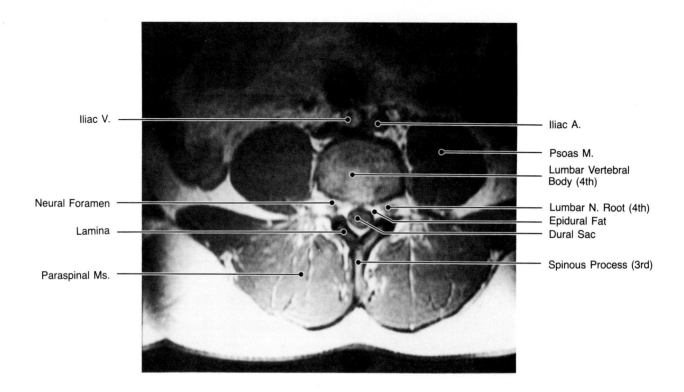

Iliac V.

Iliac A.

Psoas M.

Lumbar Vertebral Body (4th)

Neural Foramen

Lumbar N. Root (4th)

Epidural Fat

Lamina

Dural Sac

Paraspinal Ms.

Spinous Process (3rd)

13A Axial/MRI/Neural Foramen. This axial spin echo TR 1500, TE 20 msec MR image demonstrates the lumbar nerve roots outlined by fat within the neural foramen.

13B Axial/CT/Lumbar Neural Foramen. This axial CT scan without contrast enhancement shows the radicular veins and lumbar nerve roots outlined by fat in the neural foramen.

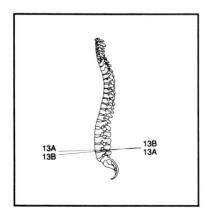

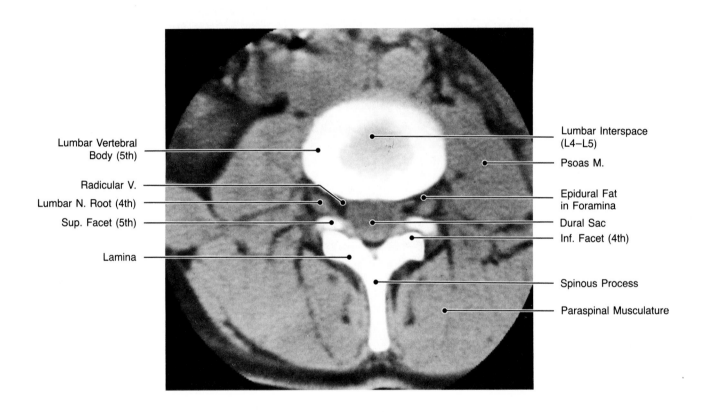

Lumbar Vertebral
Body (5th)

Radicular V.

Lumbar N. Root (4th)

Sup. Facet (5th)

Lamina

Lumbar Interspace
(L4–L5)

Psoas M.

Epidural Fat
in Foramina

Dural Sac

Inf. Facet (4th)

Spinous Process

Paraspinal Musculature

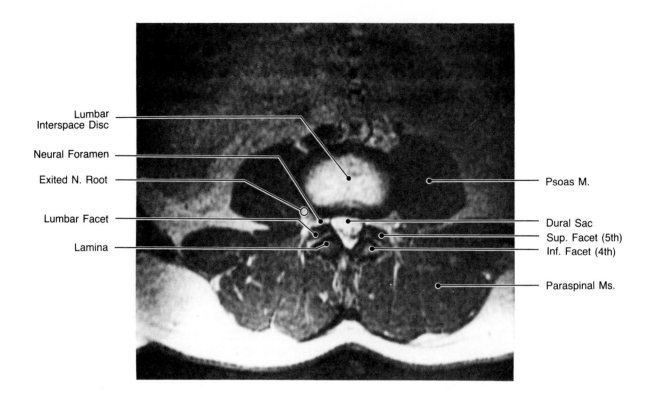

Lumbar Interspace Disc

Neural Foramen

Exited N. Root

Lumbar Facet

Lamina

Psoas M.

Dural Sac
Sup. Facet (5th)
Inf. Facet (4th)

Paraspinal Ms.

14A Axial/MRI/Disc. This axial spin echo TR 1500, TE 70 msec MR image shows the high signal intensity dural sac and lumbar disc. Definition of the dural sac and contents is similar to that seen on an intrathecal contrast–enhanced CT.

14B Axial/CT/Lumbar Interspace. This axial non–contrast-enhanced CT scan is made at the level of the lumbar disc interspace. Note that the lumbar nerve roots have already exited the neural foramen.

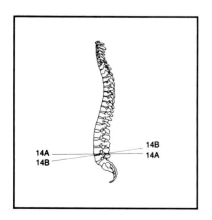

14A

14B

14B

14A

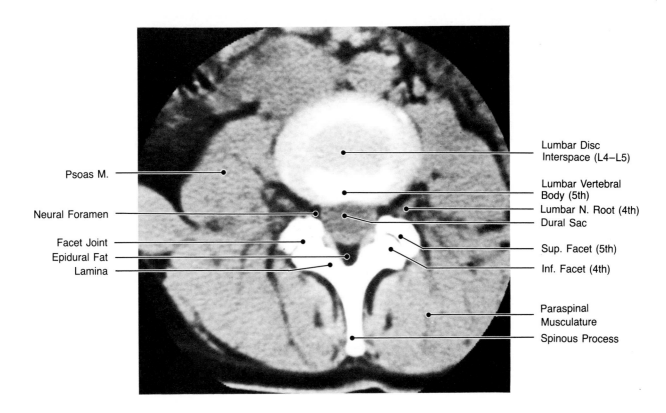

Psoas M.

Neural Foramen

Facet Joint
Epidural Fat
Lamina

Lumbar Disc
Interspace (L4–L5)

Lumbar Vertebral
Body (5th)

Lumbar N. Root (4th)

Dural Sac

Sup. Facet (5th)

Inf. Facet (4th)

Paraspinal
Musculature

Spinous Process

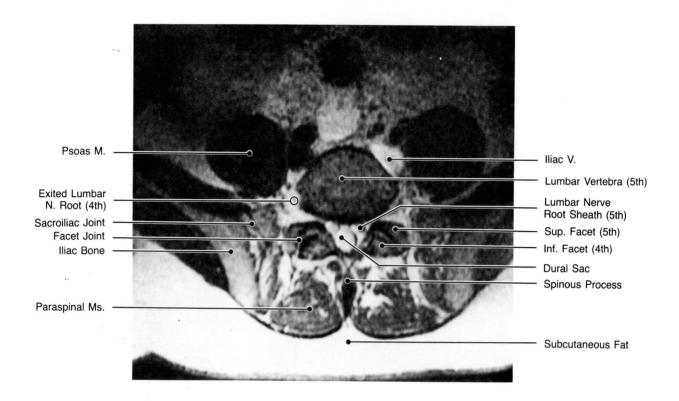

Psoas M.

Exited Lumbar
N. Root (4th)

Sacroiliac Joint
Facet Joint
Iliac Bone

Paraspinal Ms.

Iliac V.

Lumbar Vertebra (5th)

Lumbar Nerve
Root Sheath (5th)

Sup. Facet (5th)

Inf. Facet (4th)

Dural Sac
Spinous Process

Subcutaneous Fat

15A Axial/MRI/Fifth Lumbar Vertebra. This axial spin echo TR 1500, TE 70 msec MR image demonstrates low signal crescents at the left lateral margins of the dural sac and nerve roots. This is due to a chemical shift artifact in the frequency encoded axis, since there are slight differences in the precessional frequency of water and fat.

15B Axial/MRI/Lumbosacral Interspace. This axial spin echo TR 1500, TE 20 msec MR image, which is not parallel to the interspace, demonstrates a low signal band related to the lumbar sacral interspace disc, which is of lower signal than the marrow of the adjacent vertebral bodies.

15C Axial/CT/Sacrum. This axial CT scan through the superior sacrum demonstrates the branching spinal canal cavity. The sacroiliac joints are seen laterally.

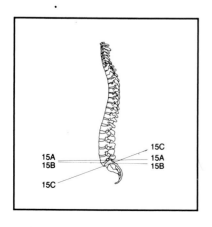

15B

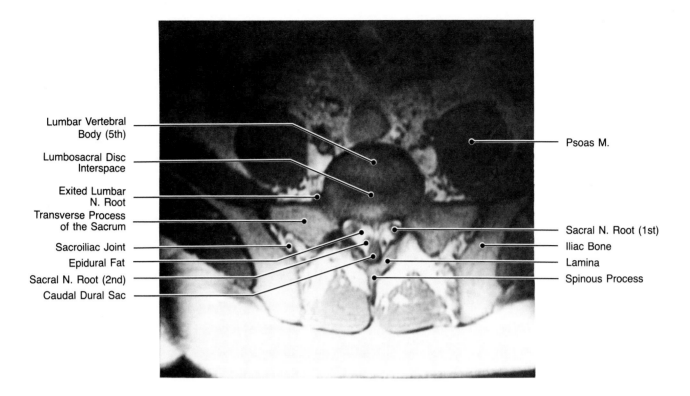

Lumbar Vertebral Body (5th)

Lumbosacral Disc Interspace

Exited Lumbar N. Root

Transverse Process of the Sacrum

Sacroiliac Joint

Epidural Fat

Sacral N. Root (2nd)

Caudal Dural Sac

Psoas M.

Sacral N. Root (1st)

Iliac Bone

Lamina

Spinous Process

15C

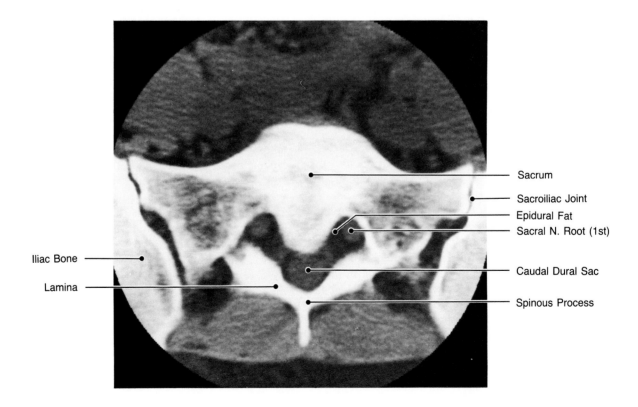

Sacrum

Sacroiliac Joint

Epidural Fat

Sacral N. Root (1st)

Iliac Bone

Lamina

Caudal Dural Sac

Spinous Process

SECTION 3

Thorax

A. J. Christoforidis, M.D., Ph.D.
Charles F. Mueller, M.D.
John A. Negulesco, Ph.D.

The anatomic plates of this section were obtained from four different cadavers. The technique of preparation consisted of freezing the cadavers without embalming. The rationale for not embalming the cadavers was to preserve the natural color of the tissues as much as possible. Axial computed tomographic sections of a fresh cadaver were obtained before the freezing, after placing markers on the skin to accurately indicate the level of the sections. This secured precise correlation of the axial CT sections of the cadaver with the anatomic ones. One, of course, should keep in mind that CT images are two-dimensional representations of three-dimensional anatomic sections of varying thicknesses.

The sectioning in the axial, coronal, and sagittal planes at approximately 2-cm intervals was made 8 to 10 days later using a band saw. The frozen sections were then left to thaw and were subsequently photographed using the technique described in the general introduction.

Axial CT sections were obtained also from living individuals, all of whom were patients referred for different clinical indications. While the cadaver CT sections serve the objective of precise correlation with the anatomic specimen, the CT images of the living individuals provide important information related to the normal physiology and anatomy enhanced by the utilization of intravenous contrast as well as by the introduction of dilute contrast by mouth or rectum. This reflects more realistically the daily experience of the radiologist and the clinician in general. Inherent in this is the anatomic variance, so prevalent in daily clinical experience. The computed tomographic sections of the cadaver, as expected, correspond to an "end expiratory phase," while the examination of the living individual is obtained, as a rule, during the end inspiratory phase. This explains the difference in the relationship of the level of the diaphragm to the degree of imaging of the liver in the sections of the lower thorax in the cadaver

and the living patients. Furthermore, CT windowing is generally performed for and tailored to the specific function and purpose of the examination. Since many varied CT windows (almost an infinite variety) are available, not all CT windows and centers are equivalent. In our living patients, for example, we choose routinely both mediastinal and lung windows to demonstrate possible pathology. In the interest of space we demonstrate on the CT cadaver only the mediastinal window. Because of dependent lung fluid, window level and center variations may occasionally simulate disease where none is present.

Some major anatomicopathologic variations are also present in cadaver sections. Most important in deviation from normal is the cadaver used for coronal sectioning of the chest. Here a pre-existing colon cancer resulted in extensive hepatic metastases and bile-stained ascites and pleural fluid. The diaphragm was elevated by the ascites, and some of the tissues were stained. The pathologic changes do not significantly alter the anatomic relationships. The inclusion of a not uncommon anatomicopathologic clinical case was considered familiar territory, particularly for the clinician.

It should be mentioned that pleural effusion was present on the CT images of the fresh cadaver done before freezing and subsequent sectioning. That pleural effusion was not always in the same location in the anatomic sections. The absence of pleural fluid in some of the anatomic specimens provided additional separation between the planes that actually might prove helpful to the viewer. We doubt that these slight variations of the pleural space will mislead the reader in an evaluation of the correlative anatomy.

Reformatted CT images in coronal or sagittal projections were not included, as these projections are not routinely used in the thorax, as might be the case for other parts of the body. The cadaver images were obtained with a third-generation GE8800 CT scanner. The living patients' CT images were obtained from a fourth-generation CT system, either the 2060 or the 1440 H.P.S. by Technicare.

Magnetic resonance images were obtained in axial, coronal, and sagittal planes. We used a 1.5 Tesla GE Signa unit and partial saturation techniques, TR of 400 to 500 milliseconds and TE of 25 to 30 milliseconds. Other spin echo images, TR 2500 milliseconds and TE 40 to 80 milliseconds, were also obtained. In some cases more excitations and averages than in others were obtained, particularly in the gated studies.

Sagittal sections of the chest, both anatomic and magnetic resonance imaging, are not as thorough, by intention, because it is our belief that sagittal images will principally be used for assessment of mediastinal and paramediastinal information. Middle and lateral lung fields are adequately evaluated with other imaging modalities, including CT and conventional radiography.

We believe that the cardiac gated magnetic resonance images will become routine in the future and more widely available than currently. Respiratory and cardiac gating are available to a limited degree at this time at a few centers. We included pertinent sections of the thoracic cardiovascular system for which MRI is expected to play a major role. We recognize that the rate at which technical advances can and will occur in the near future will contribute to the improvement of the study of the anatomy and physiology of the cardiovascular system.

Bibliography

Bo WJ, Meschan I, Kruegar, WA: *Basic Atlas of Cross-Sectional Anatomy.* Philadelphia, W. B. Saunders Co., 1980.

Frederick HM, Bernardino ME, Baron M; et al: Accuracy of chest computerized tomography in detecting malignant hilar and mediastinal involvement by squamous cell carcinoma of the lung. *Radiology* 156:280, 1985.

Gamsu G, Webb WR, Sheldon P, Kaufman L, Crooks LE, Birnberg FA, Goodman P, Hinchcliffe WA, Hedgecock M. Nuclear Magnetic resonance imaging of the thorax. *Radiology* 147:473, 1983.

Glazer GM: Evaluation of the pulmonary hila by CT. *Radiology* 146:261, 1983.

Glazer GM: Evaluation of the pulmonary hila by CT: Further comment. *Radiology* 146:262, 1983.

Haaga JR, Alfidi RT: *Computed Tomography of the Whole Body.* St. Louis, C.V. Mosby, 1983.

Koritké JG, and Sick H: *Atlas of Sectional Human Anatomy*: by *Frontal, Sagittal and Horizontal Planes.* Baltimore, Urban and Schwarzenberg, 1983.

Levitt RG, Glazer HS, Roper CL, et al: Magnetic resonance imaging of the mediastinal and hilar masses: Comparison with CT. *AJR* 145:9, 1985.

Moss AA, Gamsu, G, Genant H: *Computed Tomography of the Body.* Philadelphia, W.B. Saunders Co., 1983.

Naidich DP, Zerhouni EA, Siegelman SS: *Computed Tomography of the Thorax.* New York, Raven Press, 1984.

Pernkopf E: Atlas of Topographical and Applied Human Anatomy. 2nd ed. Philadelphia, W.B. Saunders Co., 1964.

Vogler, III J, Helms C, Callen P: *Normal variants and pitfalls in imaging,* Philadelphia, W. B. Saunders Company, 1986.

Wagner M, Lawson T: *Segmental Anatomy: Applications to Clinical Medicine.* New York, Macmillan, 1982.

Webb WR, Gamsu G, Stark DD, Moore EH: Magnetic resonance imaging of the normal and abnormal pulmonary hila. 152:89, 1984.

Webb WR, Jensen BG, Gamsu G, et al: Coronal magnetic resonance imaging of the chest: Normal and abnormal. 153:279, 1984.

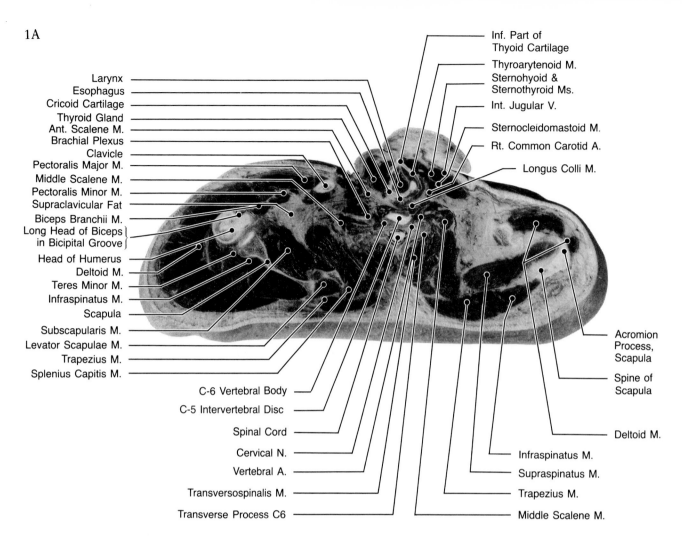

1A

Larynx
Esophagus
Cricoid Cartilage
Thyroid Gland
Ant. Scalene M.
Brachial Plexus
Clavicle
Pectoralis Major M.
Middle Scalene M.
Pectoralis Minor M.
Supraclavicular Fat
Biceps Branchii M.
Long Head of Biceps in Bicipital Groove
Head of Humerus
Deltoid M.
Teres Minor M.
Infraspinatus M.
Scapula
Subscapularis M.
Levator Scapulae M.
Trapezius M.
Splenius Capitis M.

Inf. Part of Thyoid Cartilage
Thyroarytenoid M.
Sternohyoid & Sternothyroid Ms.
Int. Jugular V.
Sternocleidomastoid M.
Rt. Common Carotid A.
Longus Colli M.

Acromion Process, Scapula
Spine of Scapula
Deltoid M.
Infraspinatus M.
Supraspinatus M.
Trapezius M.
Middle Scalene M.

C-6 Vertebral Body
C-5 Intervertebral Disc
Spinal Cord
Cervical N.
Vertebral A.
Transversospinalis M.
Transverse Process C6

1A Axial/Anatomic Plate/Larynx. This and subsequent anatomic sections were obtained with the patients' arms at their sides while associated CT sections were generally performed with the arms over the head to reduce artifact. This section was cut obliquely with the left side higher than the right, which shows the C5 intervertebral disk and C6 vertebral body at the transverse process level. It demonstrates the anterior location of the thyroid laminae, which join and form a typical male angle of 90 degrees (a more obtuse angle of about 120 degrees would be typically female). The posterior left lamina of the thyroid cartilage is separated from the trachea at the larynx by the thyroarytenoid muscle, and because it is slightly higher on the left, a small portion of the left vocal cord is identified. The anteriorly opened C-shaped cricoid cartilage surrounds the posterior tracheal margin and hugs the anterior aspect of the esophagus. Superficial to the thyroid cartilage on the left are the sternohyoid and sternothyroid muscles, and on the right is the thyroid gland. The broad sternocleidomastoid muscle lies just superficial to the large internal jugular vein, which in turns lies just anterior to the left common carotid artery. Only the longus colli muscle separates the cervical spine from the esophagus. This section shows a small fragment of anterior C5 body separated by the whitish intervertebral C5 disk and more posteriorly the C6 vertebral body. On the left side the large C6 nerve root may be seen lying just anterior to the C6 transverse process. Anterior and middle scalene muscles generally lie anterior to the level of the transverse process, while the posterior scalene muscle lies slightly posterior to these and usually just lateral to the transverse process of the lower cervical vertebra. The pectoral girdles are prominent in this section particularly; on the right, showing the glenohumeral joint and the body of the scapula separating subscapularis from infraspinatus and teres minor muscles. Along the medial border of the scapula are the large levator scapulae and splenius capitis muscles.

1B Axial/CT Cadaver/Larynx. This CT cadaver plate is used to correlate with the anatomic section. Because the arms are overhead for this cadaver section, it demonstrates the first rib, the thyroid gland and the trachea. The brachial plexus of nerves is best seen between the scalenus anterior and the scalenus medius muscles.

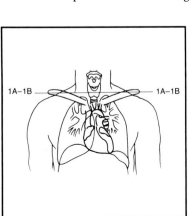

1A-1B   1A-1B

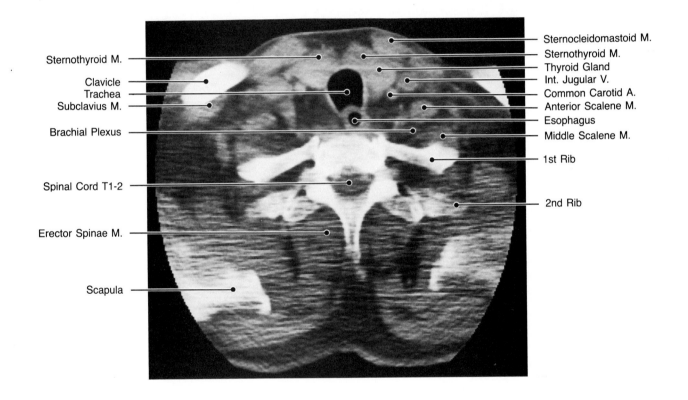

Sternocleidomastoid M.

Sternothyroid M.

Sternothyroid M.

Thyroid Gland

Clavicle

Int. Jugular V.

Trachea

Common Carotid A.

Subclavius M.

Anterior Scalene M.

Esophagus

Brachial Plexus

Middle Scalene M.

1st Rib

Spinal Cord T1-2

2nd Rib

Erector Spinae M.

Scapula

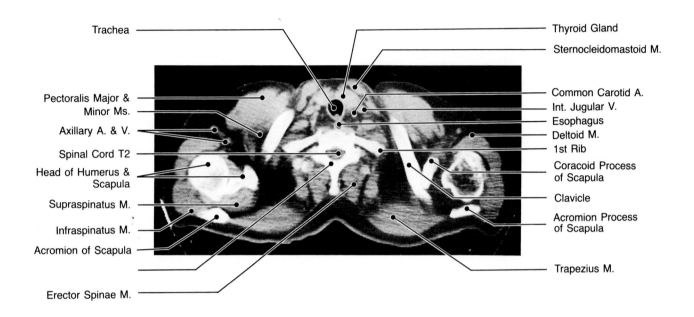

Trachea

Thyroid Gland

Sternocleidomastoid M.

Pectoralis Major & Minor Ms.

Common Carotid A.

Int. Jugular V.

Axillary A. & V.

Esophagus

Deltoid M.

Spinal Cord T2

1st Rib

Head of Humerus & Scapula

Coracoid Process of Scapula

Supraspinatus M.

Clavicle

Infraspinatus M.

Acromion Process of Scapula

Acromion of Scapula

Trapezius M.

Erector Spinae M.

IC, ID Axial/CT Cadaver/Larynx. These two cadaver sections can be correlated with the anatomic section (1A). The brachial plexus is seen between the scalenus anterior and the combined scalenus medius and scalenus posterior muscles. With the arms overhead the relationship of the supraspinatus to the infraspinatus is significantly altered. Common carotid, internal jugular vein, and esophagus are demonstrated at these levels.

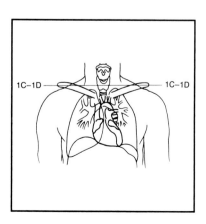

1C–1D

1C–1D

Lt. Thyroid Lobe

Longus Colli M.

Ant. Scalene M.

Middle Scalene M.

Glenoid Labrum (Ant.)

Glenoid Fossa

Shoulder
(Glenohumeral) Joint

Trapezius M.

Lt. Carotid A.

Jugular V.

Pectoralis Major M.

Ant. Scalene M.

Brachial Plexus

Head of Humerus

Acromion Process
of Scapula

Levator Scapulae M.

Deep Cervical M.

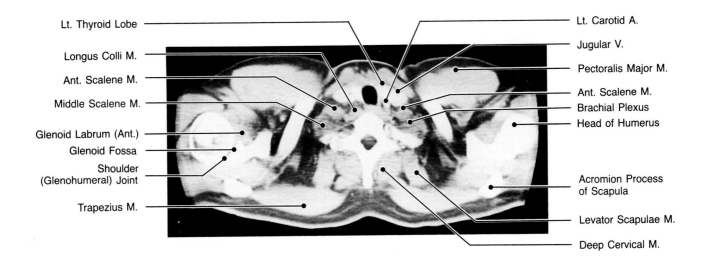

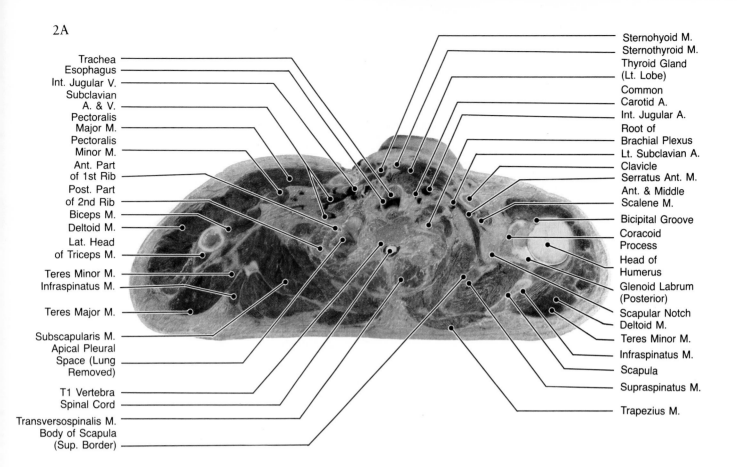

Trachea
Esophagus
Int. Jugular V.
Subclavian
A. & V.
Pectoralis
Major M.
Pectoralis
Minor M.
Ant. Part
of 1st Rib
Post. Part
of 2nd Rib
Biceps M.
Deltoid M.
Lat. Head
of Triceps M.
Teres Minor M.
Infraspinatus M.
Teres Major M.
Subscapularis M.
Apical Pleural
Space (Lung
Removed)
T1 Vertebra
Spinal Cord
Transversospinalis M.
Body of Scapula
(Sup. Border)

Sternohyoid M.
Sternothyroid M.
Thyroid Gland
(Lt. Lobe)
Common
Carotid A.
Int. Jugular A.
Root of
Brachial Plexus
Lt. Subclavian A.
Clavicle
Serratus Ant. M.
Ant. & Middle
Scalene M.
Bicipital Groove
Coracoid
Process
Head of
Humerus
Glenoid Labrum
(Posterior)
Scapular Notch
Deltoid M.
Teres Minor M.
Infraspinatus M.
Scapula
Supraspinatus M.
Trapezius M.

2A Axial/Anatomic Plate/Lung Apex. This Anatomic section is cut so that the specimen's left side is higher than the right and demonstrates the lower pole of the left thyroid gland. The incomplete cartilaginous rings of the trachea are demonstrated at this level as well as the characteristically flattened or concave posterior wall of the trachea, which is in continuity with the anterior wall of the esophagus. Nerve roots of the brachial plexus lie between the anterior and middle scalene muscles, and on the lower right side the tip of the right pleural cavity is exposed just deep to the laterally placed first and second ribs. The subclavian vessels lie between the anterior first rib and the medial aspect of the right clavicle, exposing the favorite approach for the subclavian line placement. The coracoid process of the scapula is identified anteriorly just medial to the bicipital groove of the humerus, and the bony and cartilaginous labra of the posterior glenoid are demon-

strated on the right. The superior margin of the body of the scapula forms the most anterior extent of the large supraspinatus muscle, which extends to the spine of the scapula posteriorly. Posterolateral to the spine are the infraspinatus, teres minor, and deltoid muscles.

2B Axial/CT/Lung Apex. This cadaver CT plate is magnified to show principally the apex of the lung and the first and second ribs. The relationship of the internal jugular vein and carotid artery to the scapula and anterior first rib is well demonstrated.

2C Axial/CT/Lung Apex. CT images of the apices of the lung on a living patient after contrast injection densely opacify the subclavian vessels and carotid arteries. In the axillary fat, the axillary vessels lie among the pectoralis, deltoid, and subscapularis muscles.

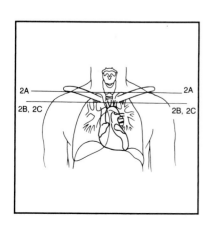

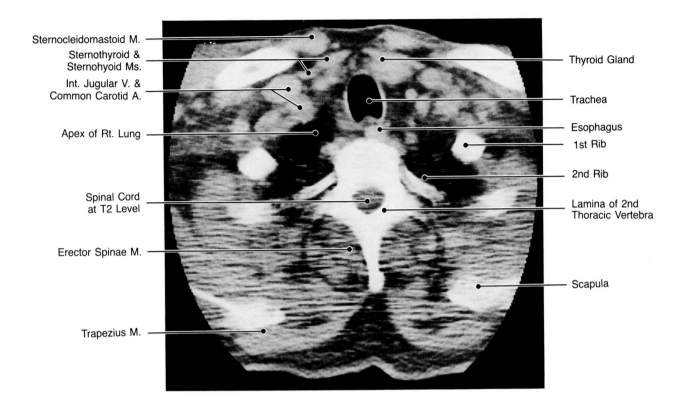

Sternocleidomastoid M.

Sternothyroid & Sternohyoid Ms.

Int. Jugular V. & Common Carotid A.

Apex of Rt. Lung

Spinal Cord at T2 Level

Erector Spinae M.

Trapezius M.

Thyroid Gland

Trachea

Esophagus

1st Rib

2nd Rib

Lamina of 2nd Thoracic Vertebra

Scapula

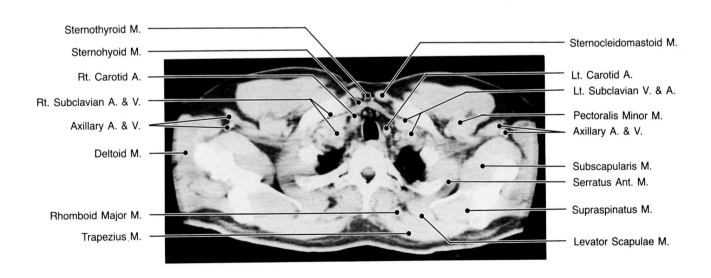

Sternothyroid M.

Sternohyoid M.

Rt. Carotid A.

Rt. Subclavian A. & V.

Axillary A. & V.

Deltoid M.

Rhomboid Major M.

Trapezius M.

Sternocleidomastoid M.

Lt. Carotid A.

Lt. Subclavian V. & A.

Pectoralis Minor M.

Axillary A. & V.

Subscapularis M.

Serratus Ant. M.

Supraspinatus M.

Levator Scapulae M.

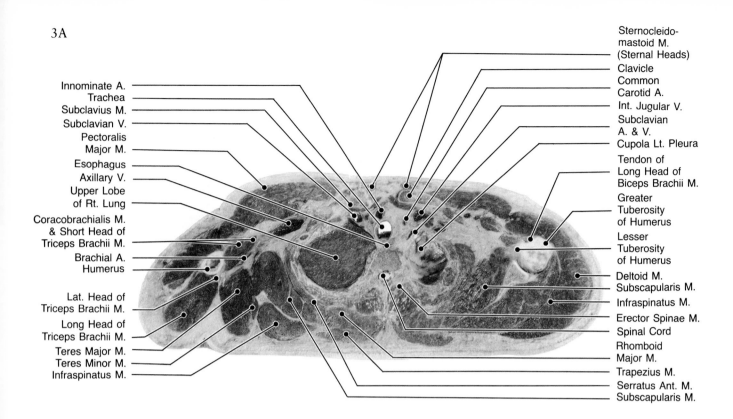

3A

Innominate A.
Trachea
Subclavius M.
Subclavian V.
Pectoralis Major M.
Esophagus
Axillary V.
Upper Lobe of Rt. Lung
Coracobrachialis M. & Short Head of Triceps Brachii M.
Brachial A.
Humerus
Lat. Head of Triceps Brachii M.
Long Head of Triceps Brachii M.
Teres Major M.
Teres Minor M.
Infraspinatus M.

Sternocleido-mastoid M. (Sternal Heads)
Clavicle
Common Carotid A.
Int. Jugular V.
Subclavian A. & V.
Cupola Lt. Pleura
Tendon of Long Head of Biceps Brachii M.
Greater Tuberosity of Humerus
Lesser Tuberosity of Humerus
Deltoid M.
Subscapularis M.
Infraspinatus M.
Erector Spinae M.
Spinal Cord
Rhomboid Major M.
Trapezius M.
Serratus Ant. M.
Subscapularis M.

3A Axial Anatomic Plate/Innominate Artery. The innominate artery is as yet unbranched on this plate and lies just anterior to the trachea and is separated from the right clavicular head by the subclavius muscle. The sternal heads of the sternocleidomastoid lie on the surface of the medial clavicular heads. The left side is higher than the right side, and the left apical pleural surface may be seen, since the lung has been removed. On the right, the right upper lobe is seen as a homogeneous fluid-filled structure post-mortem. Portions of the axillary vein and the subclavian vein are demonstrated on the right, with the distal axillary and proximal brachial artery identified medial to the coracobrachialis and short head of biceps brachii muscles. Teres major and triceps brachii muscles are nearly of equal size at this level on the right. Surrounding the posterior thoracic cage are serratus anterior, subscapularis, rhomboid major, and erector spinae muscles. On the left the

greater and lesser tuberosities of the humerus separate the tendon for the long head of the biceps lying in the bicipital groove of the humerus.

3B, 3C Axial/CT/Common Carotid–Subclavian Vessels. Cadaver CT sections demonstrate the relationship of the first anterior rib and the deep surface of the clavicle and its retroclavicular subclavian vein. Superior mediastinal fat outlines the common carotid and left subclavian arteries. Some lung detail is appreciated due to thickened interstitial structures in the lung.

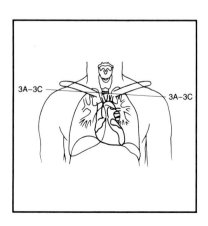

3A–3C   3A–3C

3B

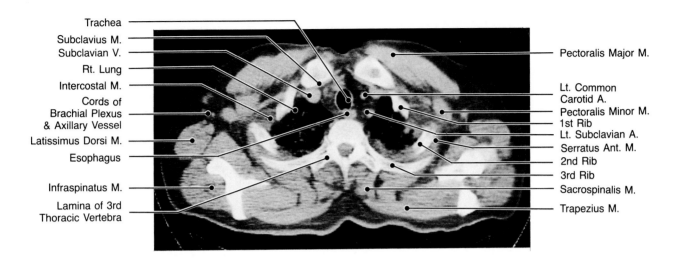

Trachea

Subclavius M.

Subclavian V.

Rt. Lung

Intercostal M.

Cords of
Brachial Plexus
& Axillary Vessel

Latissimus Dorsi M.

Esophagus

Infraspinatus M.

Lamina of 3rd
Thoracic Vertebra

Pectoralis Major M.

Lt. Common
Carotid A.

Pectoralis Minor M.

1st Rib

Lt. Subclavian A.

Serratus Ant. M.

2nd Rib

3rd Rib

Sacrospinalis M.

Trapezius M.

3C

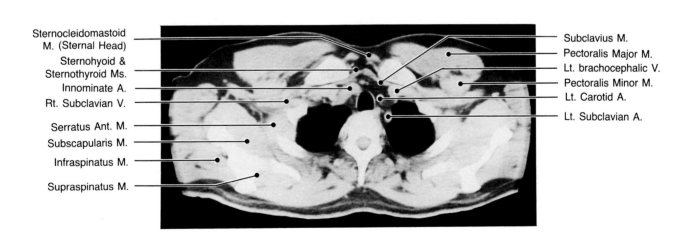

Sternocleidomastoid
M. (Sternal Head)

Sternohyoid &
Sternothyroid Ms.

Innominate A.

Rt. Subclavian V.

Serratus Ant. M.

Subscapularis M.

Infraspinatus M.

Supraspinatus M.

Subclavius M.

Pectoralis Major M.

Lt. brachiocephalic V.

Pectoralis Minor M.

Lt. Carotid A.

Lt. Subclavian A.

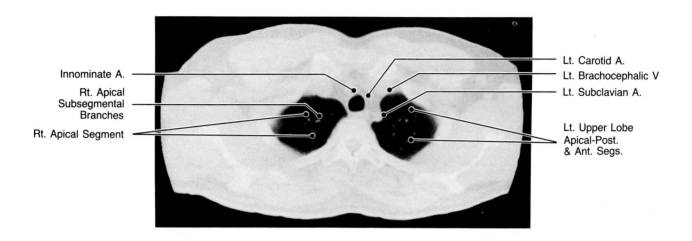

Innominate A. —

Rt. Apical
Subsegmental
Branches

Rt. Apical Segment

Lt. Carotid A.
Lt. Brachocephalic V
Lt. Subclavian A.

Lt. Upper Lobe
Apical-Post.
& Ant. Segs.

3D Axial/CT/Innominate Artery. CT section at mediastinal window at the level of the innominate artery in the living patient with contrast enhancement clearly show the relationships of subclavian, carotid, innominate, and brachiocephalic vessels.

3E Axial/CT/Innominate Artery. CT section at lung window shows the delicate stroma of the air filled bronchi of the apical (right) and the apical-posterior (left) segments, and the feeding arteries and draining veins. Note the left subclavian artery grooving the medial aspect of the left lung apex; at the apex of the lung this groove is called the superior sulcus.

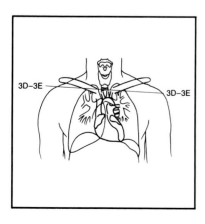

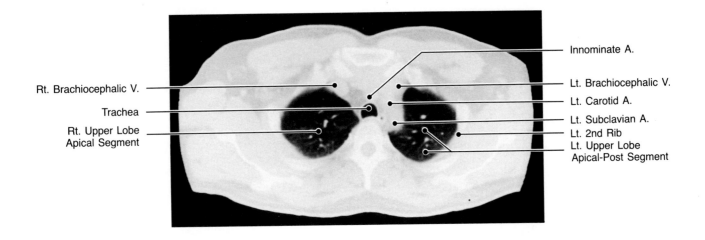

Innominate A.

Lt. Brachiocephalic V.

Lt. Carotid A.

Lt. Subclavian A.

Lt. 2nd Rib

Lt. Upper Lobe
Apical-Post Segment

Rt. Brachiocephalic V.

Trachea

Rt. Upper Lobe
Apical Segment

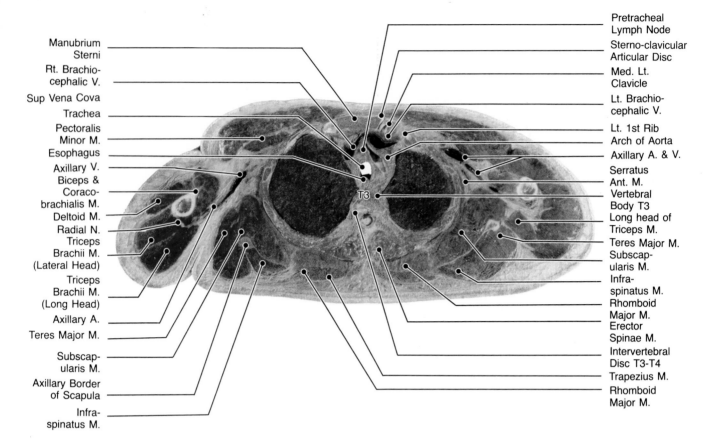

Manubrium Sterni

Rt. Brachio-cephalic V.

Sup Vena Cova

Trachea

Pectoralis Minor M.

Esophagus

Axillary V.

Biceps & Coraco-brachialis M.

Deltoid M.

Radial N.

Triceps Brachii M. (Lateral Head)

Triceps Brachii M. (Long Head)

Axillary A.

Teres Major M.

Subscap-ularis M.

Axillary Border of Scapula

Infra-spinatus M.

Pretracheal Lymph Node

Sterno-clavicular Articular Disc

Med. Lt. Clavicle

Lt. Brachio-cephalic V.

Lt. 1st Rib

Arch of Aorta

Axillary A. & V.

Serratus Ant. M.

Vertebral Body T3

Long head of Triceps M.

Teres Major M.

Subscap-ularis M.

Infra-spinatus M.

Rhomboid Major M.

Erector Spinae M.

Intervertebral Disc T3-T4

Trapezius M.

Rhomboid Major M.

T3

4A Axial/Anatomic Plate/Sternoclavicular Junction. The manubriosternal junction on the left and its intervening articular disk and ligaments are seen. Behind these midline bony structures large brachiocephalic veins from each side drape over the collapsed arch of the aorta. The left side is cut higher and demonstrates axillary artery and vein, while on the right brachial artery and vein are demonstrated. This is also an upper thoracic level with demonstration of a vertebral body, intervertebral disk, and the neural arch elements of the next lower vertebral level. Ribs and intercostal muscles together with the finger-like slips of the serratus anterior, the rhomboid major, and the erector spinae muscles surround the lungs. On the right the teres major, subscapularis, and infraspinatus surround the prominent axillary border of the scapula. On the left side at a slightly higher level the long head of the triceps is seen to arise just lateral to the axillary border of the scapula in an infraglenoidal location.

4B Axial/Cadaver CT/Sternoclavicular Level. This cadaver CT demonstrates the manubrium and clavicular articulations with the stronger posterior and less well-developed anterior sternoclavicular ligaments. The interclavicular ligament may faintly be seen connecting the two posterior sternoclavicular ligaments. This joint with its capsule and articular disk is the only true articulation between the pectoral girdle and the axial skeleton. The preterminal lung fluid is seen in the dependent and, in several areas, nondependent segments of the left and right lungs.

4C, 4D Axial/CT/Sternoclavicular Level. The contrast-enhanced CT images at the sternoclavicular joint in the living patient at mediastinal and lung windows show the vascular structures and upper lobe segmental bronchii and their vascular supply, respectively.

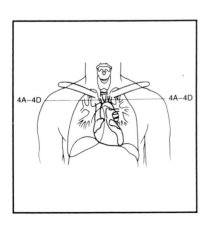

4A–4D   4A–4D

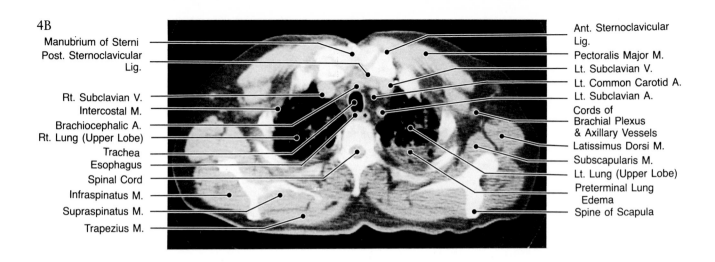

4B

Manubrium of Sterni
Post. Sternoclavicular Lig.

Rt. Subclavian V.
Intercostal M.
Brachiocephalic A.
Rt. Lung (Upper Lobe)
Trachea
Esophagus
Spinal Cord
Infraspinatus M.
Supraspinatus M.
Trapezius M.

Ant. Sternoclavicular Lig.
Pectoralis Major M.
Lt. Subclavian V.
Lt. Common Carotid A.
Lt. Subclavian A.
Cords of Brachial Plexus & Axillary Vessels
Latissimus Dorsi M.
Subscapularis M.
Lt. Lung (Upper Lobe)
Preterminal Lung Edema
Spine of Scapula

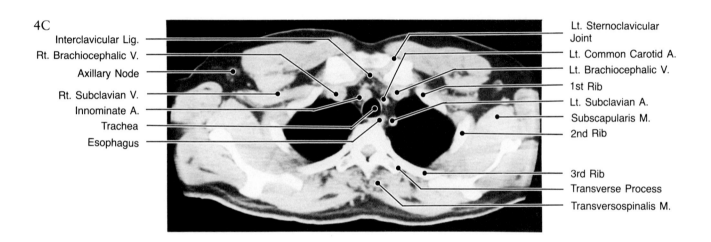

4C

Interclavicular Lig.
Rt. Brachiocephalic V.
Axillary Node
Rt. Subclavian V.
Innominate A.
Trachea
Esophagus

Lt. Sternoclavicular Joint
Lt. Common Carotid A.
Lt. Brachiocephalic V.
1st Rib
Lt. Subclavian A.
Subscapularis M.
2nd Rib

3rd Rib
Transverse Process
Transversospinalis M.

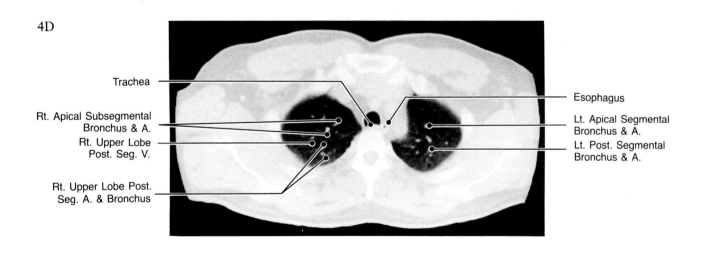

4D

Trachea

Rt. Apical Subsegmental Bronchus & A.
Rt. Upper Lobe Post. Seg. V.

Rt. Upper Lobe Post. Seg. A. & Bronchus

Esophagus

Lt. Apical Segmental Bronchus & A.
Lt. Post. Segmental Bronchus & A.

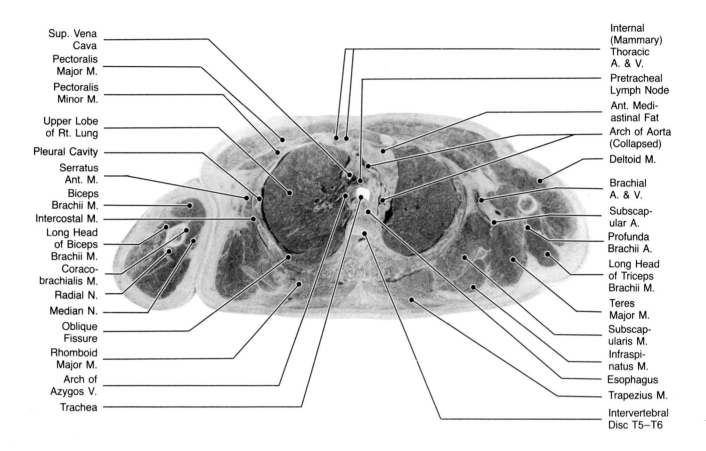

Sup. Vena Cava
Pectoralis Major M.
Pectoralis Minor M.
Upper Lobe of Rt. Lung
Pleural Cavity
Serratus Ant. M.
Biceps Brachii M.
Intercostal M.
Long Head of Biceps Brachii M.
Coracobrachialis M.
Radial N.
Median N.
Oblique Fissure
Rhomboid Major M.
Arch of Azygos V.
Trachea

Internal (Mammary) Thoracic A. & V.
Pretracheal Lymph Node
Ant. Mediastinal Fat
Arch of Aorta (Collapsed)
Deltoid M.
Brachial A. & V.
Subscapular A.
Profunda Brachii A.
Long Head of Triceps Brachii M.
Teres Major M.
Subscapularis M.
Infraspinatus M.
Esophagus
Trapezius M.
Intervertebral Disc T5–T6

5A Axial/Anatomic Plate/Aortic Arch. The collapsed aortic arch from its most anterior to its most posterior extent is demonstrated on this anatomic plate. A small thymic remnant is seen within the anterior mediastinal fat just to the left of the midline and anterior to the arch. The major fissure of each lung is seen, but only a small crescent of the lower lobe lies behind each major fissure.

5B Axial/CT/Aortic Arch. This cadaver CT study demonstrates a great deal of parenchymal lung consolidation from the premorbid event. Ribs 1 through 4 are seen together with pectoralis major and minor and intercostal muscles. Trapezius, supraspinatus, infraspinatus, and teres minor muscles are well demonstrated.

5C, 5D Axial/CT/Aortic Arch. The contrast-enhanced CT images on the living patient demonstrate the origin of the great vessels at mediastinal window and chiefly upper lobe vessels at mediastinal window.

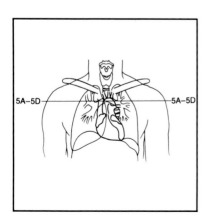

5A–5D 5A–5D

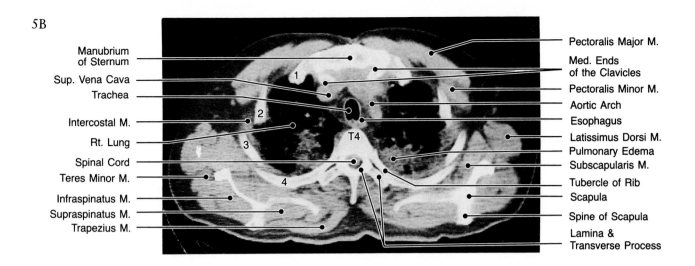

5B

Manubrium of Sternum

Sup. Vena Cava

Trachea

Intercostal M.

Rt. Lung

Spinal Cord

Teres Minor M.

Infraspinatus M.

Supraspinatus M.

Trapezius M.

Pectoralis Major M.

Med. Ends of the Clavicles

Pectoralis Minor M.

Aortic Arch

Esophagus

Latissimus Dorsi M.

Pulmonary Edema

Subscapularis M.

Tubercle of Rib

Scapula

Spine of Scapula

Lamina & Transverse Process

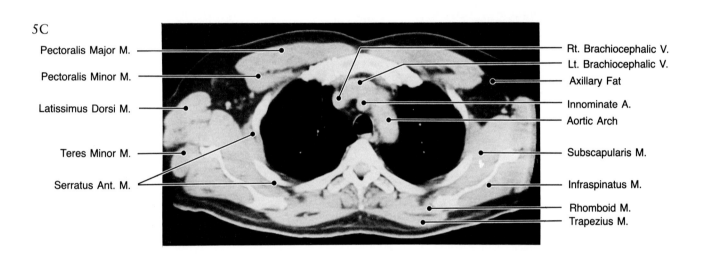

5C

Pectoralis Major M.

Pectoralis Minor M.

Latissimus Dorsi M.

Teres Minor M.

Serratus Ant. M.

Rt. Brachiocephalic V.

Lt. Brachiocephalic V.

Axillary Fat

Innominate A.

Aortic Arch

Subscapularis M.

Infraspinatus M.

Rhomboid M.

Trapezius M.

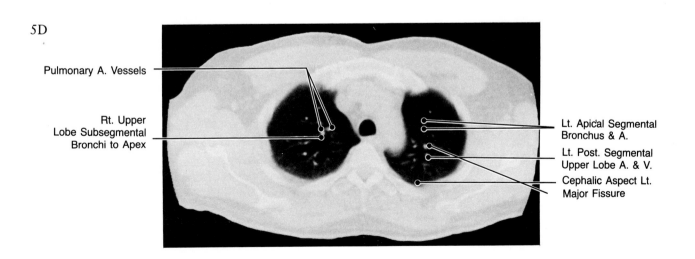

5D

Pulmonary A. Vessels

Rt. Upper Lobe Subsegmental Bronchi to Apex

Lt. Apical Segmental Bronchus & A.

Lt. Post. Segmental Upper Lobe A. & V.

Cephalic Aspect Lt. Major Fissure

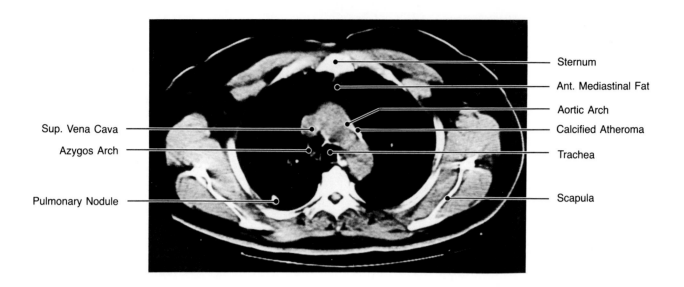

Sternum

Ant. Mediastinal Fat

Aortic Arch

Sup. Vena Cava

Calcified Atheroma

Azygos Arch

Trachea

Pulmonary Nodule

Scapula

5E, 5F, 5G Axial/CT/Pulmonary Nodule Assessment.   Three additional CTs of a patient who has a pulmonary nodule are shown. The patient's anterior mediastinal fat is quite prominent, as are some atherosclerotic aortic arch calcifications. The azyous arch is present as on the anatomic plate at this level. In the region of the posterior aspect of the right lung there is a pulmonary nodule, the density of which approaches that of the adjacent rib. When viewed at lung window the nodule is seen to have a calcified center and a noncalcified rim. That calcification is confirmed on the third plate at lung window where a Hounsfield assay indicates the calcified nidus has a mean Hounsfield value of 317. This indicates a benign granulomatous nodule calcified in its center despite a noncalcified rim.

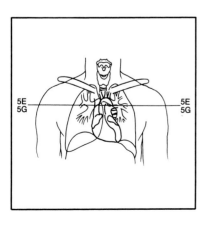

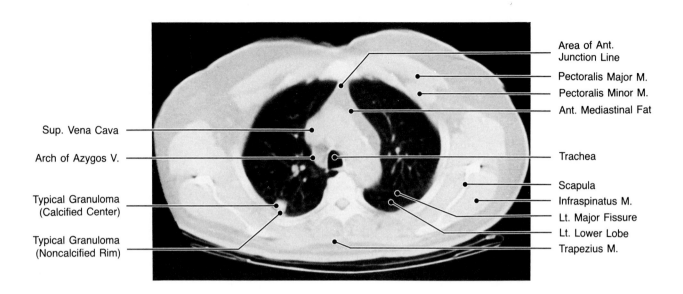

Area of Ant.
Junction Line

Pectoralis Major M.

Pectoralis Minor M.

Ant. Mediastinal Fat

Sup. Vena Cava

Arch of Azygos V.

Trachea

Typical Granuloma
(Calcified Center)

Scapula

Infraspinatus M.

Lt. Major Fissure

Lt. Lower Lobe

Typical Granuloma
(Noncalcified Rim)

Trapezius M.

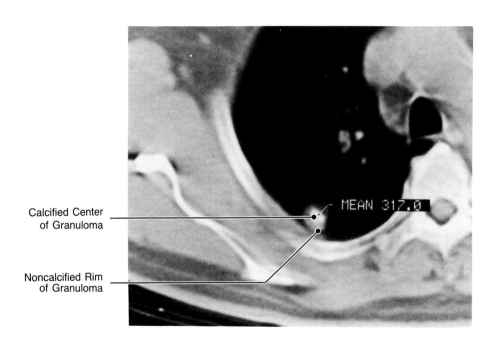

Calcified Center
of Granuloma

MEAN 317.0

Noncalcified Rim
of Granuloma

6A

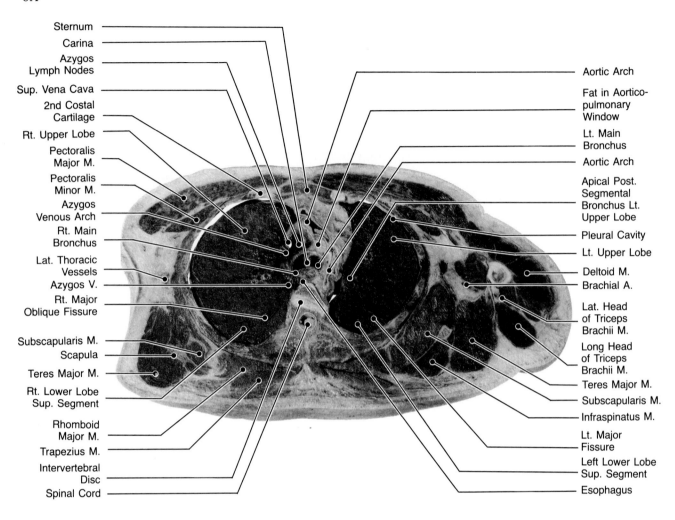

Sternum
Carina
Azygos Lymph Nodes
Sup. Vena Cava
2nd Costal Cartilage
Rt. Upper Lobe
Pectoralis Major M.
Pectoralis Minor M.
Azygos Venous Arch
Rt. Main Bronchus
Lat. Thoracic Vessels
Azygos V.
Rt. Major Oblique Fissure
Subscapularis M.
Scapula
Teres Major M.
Rt. Lower Lobe Sup. Segment
Rhomboid Major M.
Trapezius M.
Intervertebral Disc
Spinal Cord

Aortic Arch
Fat in Aortico-pulmonary Window
Lt. Main Bronchus
Aortic Arch
Apical Post. Segmental Bronchus Lt. Upper Lobe
Pleural Cavity
Lt. Upper Lobe
Deltoid M.
Brachial A.
Lat. Head of Triceps Brachii M.
Long Head of Triceps Brachii M.
Teres Major M.
Subscapularis M.
Infraspinatus M.
Lt. Major Fissure
Left Lower Lobe Sup. Segment
Esophagus

6A Axial/Anatomic Plate/Carina. This anatomic plate demonstrates the central division in the trachea where the two bronchi are separated by the carina into right and left main bronchi. The azygous vein may be seen behind the right main bronchus with its arch extending from posterior to anterior, where it empties into the superior vena cava. The major fissures are seen as crescent-like separations in the posterior lungs. Visceral and parietal pleura can be identified, and a small pleural cavity is appreciated due to minimal atelectasis post-mortem. A few central bronchi and vessels may be seen medially in both lungs. The apical posterior bronchus on the left is particularly prominent. A large fatty collection seen just below the mid portion between the proximal and distal aortic arch represents fat in the aorticopulmonary window region. An azygous lymph node is identified just at the medial aspect of the junction of the azygous vein and the superior vena cava.

6B Axial/CT/Carina. This CT plate from a cadaver corresponds most closely to 6A. It demonstrates the superior vena cava at the level of the aortic arch. Considerable consolidation is present in both lungs. Cords of brachial plexus nerves and axillary vessels are seen in the axillary fat bilaterally between latissimus dorsi and pectoralis muscles. Artifacts can be seen off a central venous catheter in the region of the left subclavian vein.

6C Axial/CT/Carina. Contrast-enhanced CT demonstrates the venous azygous arch at the level of the carina.

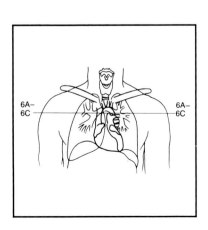

6A–6C

178

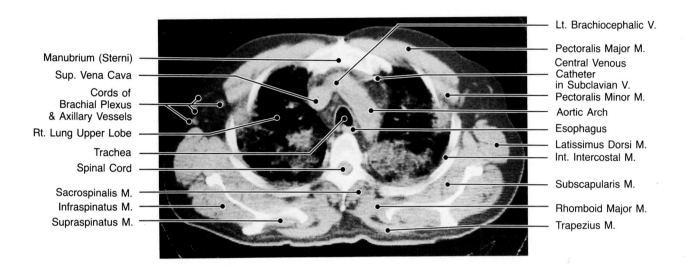

Manubrium (Sterni) —
Sup. Vena Cava —
Cords of Brachial Plexus & Axillary Vessels —
Rt. Lung Upper Lobe —
Trachea —
Spinal Cord —
Sacrospinalis M. —
Infraspinatus M. —
Supraspinatus M. —

— Lt. Brachiocephalic V.
— Pectoralis Major M.
— Central Venous Catheter in Subclavian V.
— Pectoralis Minor M.
— Aortic Arch
— Esophagus
— Latissimus Dorsi M.
— Int. Intercostal M.
— Subscapularis M.
— Rhomboid Major M.
— Trapezius M.

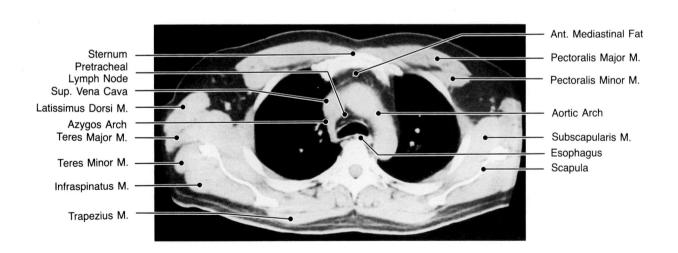

Sternum —
Pretracheal Lymph Node —
Sup. Vena Cava —
Latissimus Dorsi M. —
Azygos Arch —
Teres Major M. —
Teres Minor M. —
Infraspinatus M. —
Trapezius M. —

— Ant. Mediastinal Fat
— Pectoralis Major M.
— Pectoralis Minor M.
— Aortic Arch
— Subscapularis M.
— Esophagus
— Scapula

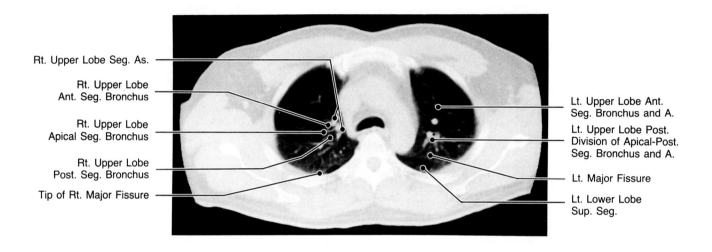

Rt. Upper Lobe Seg. As.

Rt. Upper Lobe
Ant. Seg. Bronchus

Rt. Upper Lobe
Apical Seg. Bronchus

Rt. Upper Lobe
Post. Seg. Bronchus

Tip of Rt. Major Fissure

Lt. Upper Lobe Ant.
Seg. Bronchus and A.

Lt. Upper Lobe Post.
Division of Apical-Post.
Seg. Bronchus and A.

Lt. Major Fissure

Lt. Lower Lobe
Sup. Seg.

6D Axial/CT/Carina. Contrast enhanced CT at lung window. The segmental bronchi to both upper lobes are seen together with their corresponding vessels. Major fissures on the right and left are also faintly seen close to the posterior pleural margins.

6E, 6F Axial/CT/Major Upper Lobe Bronchi and Vessels. The next set of contrast-enhanced CTs on a living patient are at a level 1 cm inferior to the previous level and demonstrate the carina and both right and left main bronchi. Some of the soft tissues averaging the top of the left pulmonary artery with aorticopulmonary fat are seen between the ascending and descending aorta. The azygous vein lies along the posterior aspect of the right main bronchus, while the esophagus lies behind the left main bronchus. The CT printed at lung window demonstrates clearly the anterior and posterior segmental bronchi to the right upper lobe and the intervening right superior pulmonary vein. The trunk of the right pulmonary artery lies anterior to the right upper lobe bronchus and indents it slightly. In the left upper lobe the anterior segmental bronchus and the main left upper bronchus as well as the apical posterior segmental bronchus of the left upper lobe are clearly demonstrated. The accessory posterior upper lobe vein lies just lateral to the left upper lobe bronchus. Both major fissures are demonstrated.

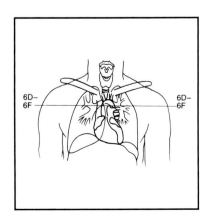

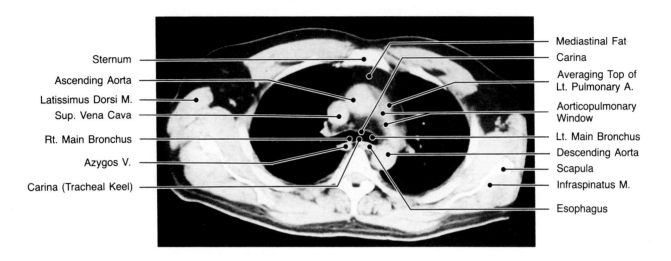

Sternum

Ascending Aorta

Latissimus Dorsi M.

Sup. Vena Cava

Rt. Main Bronchus

Azygos V.

Carina (Tracheal Keel)

Mediastinal Fat

Carina

Averaging Top of Lt. Pulmonary A.

Aorticopulmonary Window

Lt. Main Bronchus

Descending Aorta

Scapula

Infraspinatus M.

Esophagus

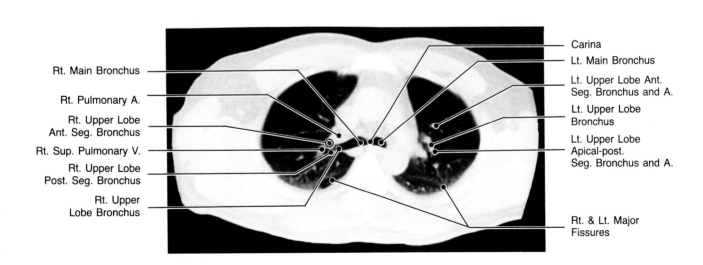

Rt. Main Bronchus

Rt. Pulmonary A.

Rt. Upper Lobe
Ant. Seg. Bronchus

Rt. Sup. Pulmonary V.

Rt. Upper Lobe
Post. Seg. Bronchus

Rt. Upper
Lobe Bronchus

Carina

Lt. Main Bronchus

Lt. Upper Lobe Ant.
Seg. Bronchus and A.

Lt. Upper Lobe
Bronchus

Lt. Upper Lobe
Apical-post.
Seg. Bronchus and A.

Rt. & Lt. Major
Fissures

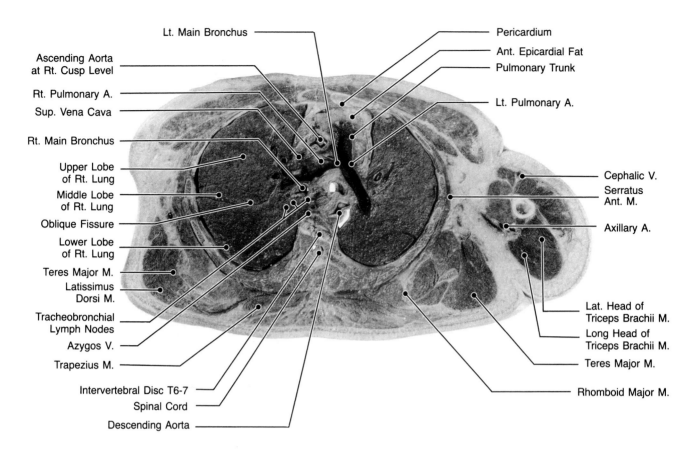

Lt. Main Bronchus

Ascending Aorta at Rt. Cusp Level

Rt. Pulmonary A.

Sup. Vena Cava

Rt. Main Bronchus

Upper Lobe of Rt. Lung

Middle Lobe of Rt. Lung

Oblique Fissure

Lower Lobe of Rt. Lung

Teres Major M.

Latissimus Dorsi M.

Tracheobronchial Lymph Nodes

Azygos V.

Trapezius M.

Intervertebral Disc T6-7

Spinal Cord

Descending Aorta

Pericardium

Ant. Epicardial Fat

Pulmonary Trunk

Lt. Pulmonary A.

Cephalic V.

Serratus Ant. M.

Axillary A.

Lat. Head of Triceps Brachii M.

Long Head of Triceps Brachii M.

Teres Major M.

Rhomboid Major M.

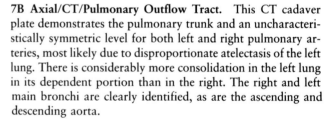

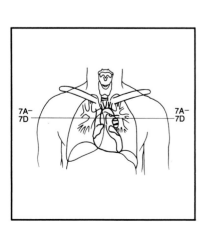

7A Axial/Anatomic Plate/Pulmonary Outflow Tract. This important plate demonstrates the pulmonary outflow tract and the left main and right main pulmonary arteries at a level about 2 cm below the carina, where tracheobronchial lymph nodes predominate. Both the superior vena cava and ascending aorta at the level of the right aortic cusp are collapsed. The major fissure has moved forward to nearly mid lung location from to back on the right lung and elements or the right lower lobe, right middle lobe, and right upper lobe are seen. Some dependent congestion in the right lung is noted posteriorly.

7B Axial/CT/Pulmonary Outflow Tract. This CT cadaver plate demonstrates the pulmonary trunk and an uncharacteristically symmetric level for both left and right pulmonary arteries, most likely due to disproportionate atelectasis of the left lung. There is considerably more consolidation in the left lung in its dependent portion than in the right. The right and left main bronchi are clearly identified, as are the ascending and descending aorta.

7C, 7D Axial/CT/Major Bronchi and Vessels. The CT images with contrast enhancement in the living patient at mediastinal and lung windows demonstrate the immediate subcarinal area and the left main pulmonary artery, which is normally slightly higher than the right. Some of the pulmonary outflow tract and anterior mediastinal fat are also included. The azygous vein lies behind the right main bronchus, and the right pulmonary artery indents the right main bronchus where it becomes right upper lobe bronchus. The lung window image more clearly demonstrates the left upper lobe pulmonary artery and its medially lying left superior pulmonary vein and the apical posterior bronchus of the left upper lobe. The absence of vessels identifies the level of the left major fissure posteriorly. On the right the right main bronchus lies just anterior to the azygous vein, and the right pulmonary truncus anterior indents the right upper lobe bronchus where it gives off the anterior segmental bronchus to the right upper lobe. The continuation of the right main bronchus distal to the take off of the right upper lobe bronchus is more medial and has a thin posterior wall called the bronchus intermedius. The transected vessels in the most dependent portion of the right posterior lung are those from the superior segment of the lower lobe.

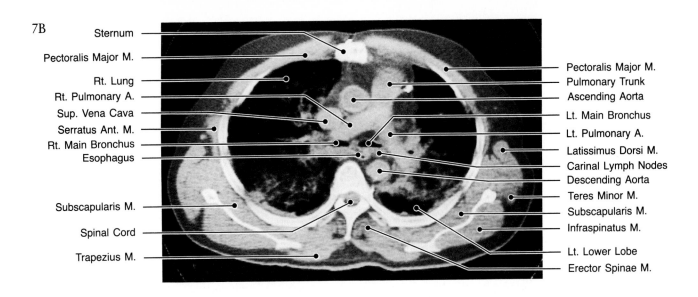

7B

Sternum
Pectoralis Major M.
Rt. Lung
Rt. Pulmonary A.
Sup. Vena Cava
Serratus Ant. M.
Rt. Main Bronchus
Esophagus

Subscapularis M.
Spinal Cord
Trapezius M.

Pectoralis Major M.
Pulmonary Trunk
Ascending Aorta
Lt. Main Bronchus
Lt. Pulmonary A.
Latissimus Dorsi M.
Carinal Lymph Nodes
Descending Aorta
Teres Minor M.
Subscapularis M.
Infraspinatus M.

Lt. Lower Lobe
Erector Spinae M.

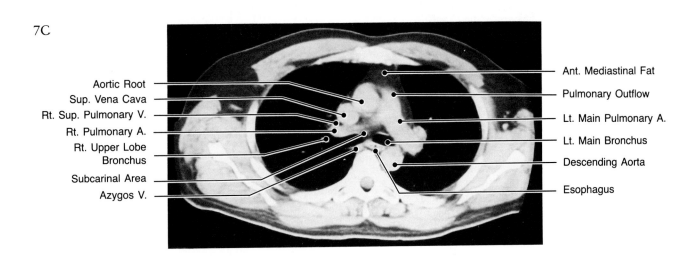

7C

Aortic Root
Sup. Vena Cava
Rt. Sup. Pulmonary V.
Rt. Pulmonary A.
Rt. Upper Lobe
Bronchus
Subcarinal Area
Azygos V.

Ant. Mediastinal Fat
Pulmonary Outflow
Lt. Main Pulmonary A.
Lt. Main Bronchus
Descending Aorta
Esophagus

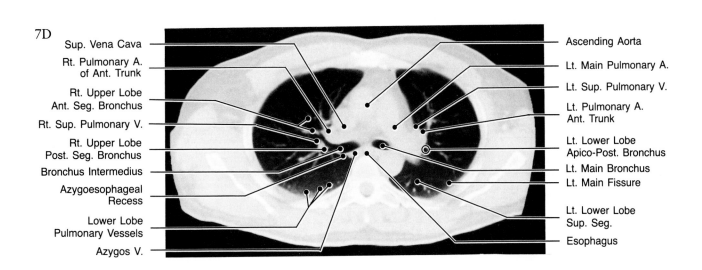

7D

Sup. Vena Cava
Rt. Pulmonary A.
of Ant. Trunk
Rt. Upper Lobe
Ant. Seg. Bronchus
Rt. Sup. Pulmonary V.
Rt. Upper Lobe
Post. Seg. Bronchus
Bronchus Intermedius
Azygoesophageal
Recess
Lower Lobe
Pulmonary Vessels
Azygos V.

Ascending Aorta
Lt. Main Pulmonary A.
Lt. Sup. Pulmonary V.
Lt. Pulmonary A.
Ant. Trunk
Lt. Lower Lobe
Apico-Post. Bronchus
Lt. Main Bronchus
Lt. Main Fissure
Lt. Lower Lobe
Sup. Seg.
Esophagus

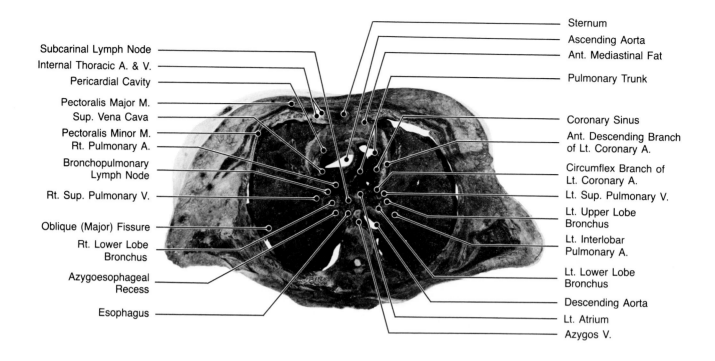

Subcarinal Lymph Node
Internal Thoracic A. & V.
Pericardial Cavity
Pectoralis Major M.
Sup. Vena Cava
Pectoralis Minor M.
Rt. Pulmonary A.
Bronchopulmonary Lymph Node
Rt. Sup. Pulmonary V.
Oblique (Major) Fissure
Rt. Lower Lobe Bronchus
Azygoesophageal Recess
Esophagus

Sternum
Ascending Aorta
Ant. Mediastinal Fat
Pulmonary Trunk
Coronary Sinus
Ant. Descending Branch of Lt. Coronary A.
Circumflex Branch of Lt. Coronary A.
Lt. Sup. Pulmonary V.
Lt. Upper Lobe Bronchus
Lt. Interlobar Pulmonary A.
Lt. Lower Lobe Bronchus
Descending Aorta
Lt. Atrium
Azygos V.

7E Axial/Anatomic Plate/Ascending Aorta and Pulmonary Trunk. A supplemental anatomic plate for further cardiac detail demonstrates the pulmonary trunk and outflow tract and the right-sided adjacent ascending aorta. This anatomic level corresponds about to the 6th thoracic vertebra. Left upper and lower lobe bronchi are identified with their adjacent anterior lobar pulmonary artery lying just lateral. Some dark subcarinal lymph node density is seen between the left and right lower lobe bronchi. The right pulmonary artery lies just anterior to the right lower lobe bronchus, with the right superior pulmonary vein lying just lateral to it.

7F Axial/MRI/Base of Heart. Axial MRI section done at partial saturation technique with cardiac but not respiratory gating demonstrates the right atrium, the root of the aorta, the right ventricle, and the abundant fat lying just above the left ventricular musculature just to the left of the root of the aorta. The left atrium and the adjacent artifacts from the mitral valve leaflet movements are also labeled. The signal void is present wherever flowing blood or air is present.

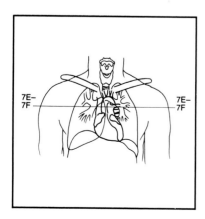

7E–7F

7F

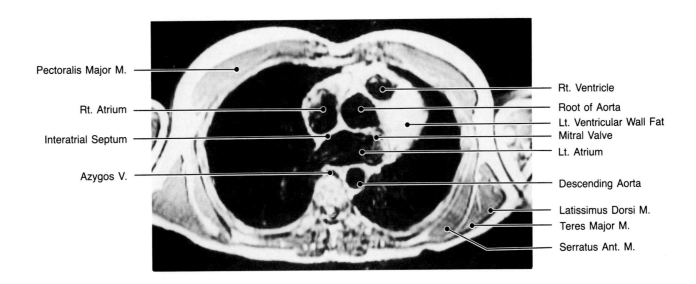

Pectoralis Major M.

Rt. Atrium

Interatrial Septum

Azygos V.

Rt. Ventricle

Root of Aorta

Lt. Ventricular Wall Fat

Mitral Valve

Lt. Atrium

Descending Aorta

Latissimus Dorsi M.

Teres Major M.

Serratus Ant. M.

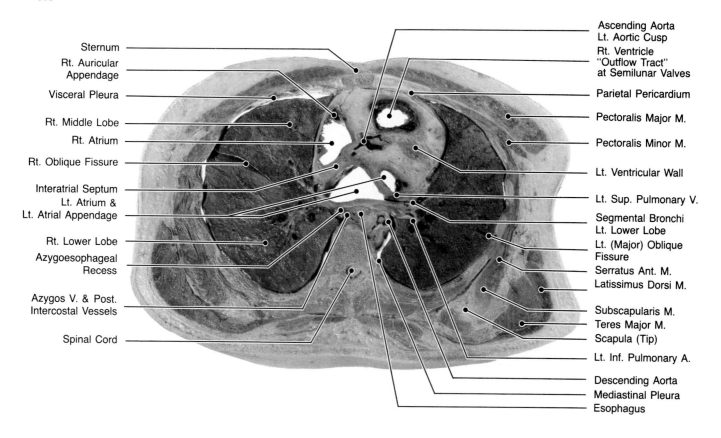

Sternum

Rt. Auricular Appendage

Visceral Pleura

Rt. Middle Lobe

Rt. Atrium

Rt. Oblique Fissure

Interatrial Septum
Lt. Atrium &
Lt. Atrial Appendage

Rt. Lower Lobe

Azygoesophageal Recess

Azygos V. & Post. Intercostal Vessels

Spinal Cord

Ascending Aorta
Lt. Aortic Cusp

Rt. Ventricle "Outflow Tract" at Semilunar Valves

Parietal Pericardium

Pectoralis Major M.

Pectoralis Minor M.

Lt. Ventricular Wall

Lt. Sup. Pulmonary V.

Segmental Bronchi
Lt. Lower Lobe

Lt. (Major) Oblique Fissure

Serratus Ant. M.

Latissimus Dorsi M.

Subscapularis M.

Teres Major M.

Scapula (Tip)

Lt. Inf. Pulmonary A.

Descending Aorta

Mediastinal Pleura

Esophagus

8A Axial/Anatomic Plate/Left Atrium and Left Auricular Appendage. This anatomic plate demonstrates the interatrial septum separating the right and left atria and also the left auricular appendage and still a small portion of the left superior pulmonary vein entering the left atrium. Some of the left ventricular wall is appreciated. The left aortic cusp is seen.

8B Axial/CT/Left Atrium. The cadaver CT plate at about T7 demonstrates the left atrium and right atrium with their interatrial septal component and calcification in the right coronary artery in the right atrioventricular sulcus. Considerable consolidation is seen in the left posterior lung, and some pleural fluid outlines the dependent posterior aspect of each lung.

8C, 8D Axial/CT/Bronchus Intermedius. The CT plates with contrast enhancement in the living patient at mediastinal and lung windows demonstrate the right main pulmonary artery crossing anterior to the bronchus intermedius and left interlobar pulmonary artery lying behind the left main bronchus. When seen at lung window the bronchus intermedius with its thin posterior wall is clearly identified and a void of vessels in the right mid anterior lung indicates the level of the minor fissure. Several arterial and pulmonary venous branches to the left upper lobe and the large left interlobar pulmonary artery are identified separated by the left upper lobe bronchus.

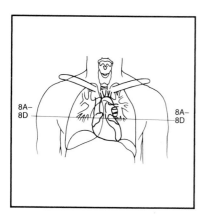

8B

Pectoralis Major M.

Rt. Coronary A.

Rt. Atrium

Serratus Ant. M.
Lt. Atrium
Rt. Pulmonary V.

Esophagus

Infraspinatus M.
Latissimus Dorsi M.

Trapezius M.

Sternum (Body)
Pericardial Fat
Rt. Ventricle
Lingula

Lt. Ventricle

Lt. Pulmonary V.

Segmental
Bronchus & A.
Descending Aorta
Subscapularis M.
Pleural Fluid

Lt. Lower Lobe
(With Bronchopneumonia)

Sacrospinalis M.

8C

Ascending Aorta

Sup. Vena Cava
Sup. Pulmonary V.
Rt. Pulmonary A.

Rt. Bronchus
Intermedius
Latissimus Dorsi M.
Teres Major M.
Subscapularis M.

Azygos V.

Trapezius M.

Pectoralis Major M.

Pulmonary
Outflow Tract

Lt. Pulmonary A.
(Averaging of Fat &
Lt. Pulmonary A.)
Lt. Sup. Pulmonary V.
Lt. Upper Lobe
Bronchus
Lt. Main Bronchus

Lt. Interlobar
Pulmonary A.
Aorta

Esophagus

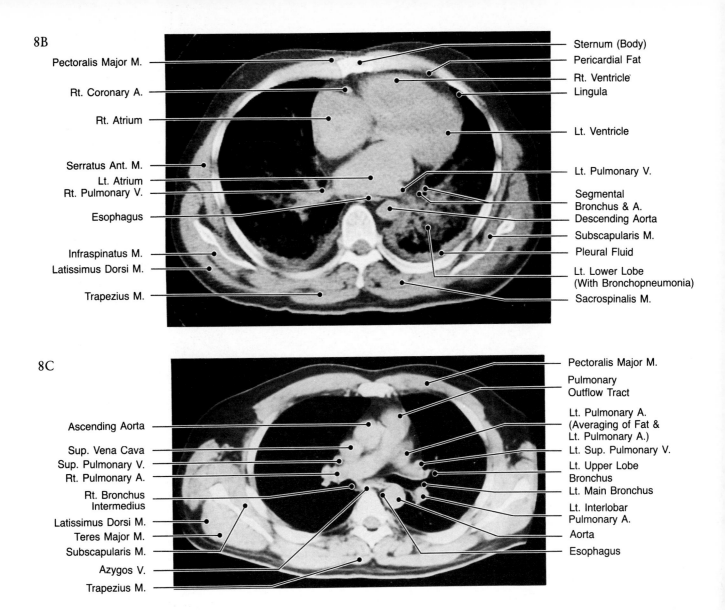

8D

Rt. Sup. Pulmonary V.

Void of Vessels
from Minor Fissure

Interlobar
Rt. Pulmonary A.

Bronchus Intermedius

Post. Wall
Bronchus Intermedius

Rt. Main Pulmonary A.

Lt. Superior
Pulmonary V.

Art. Branches to Lingula

Lt. Upper Lobe
Ant. Seg. A.

Lt. Upper Lobe
Ant. Seg. Bronchus

Lt. Upper Lobe
Bronchus

Lt. Interlobar
Pulmonary A.

Lt. Main Bronchus

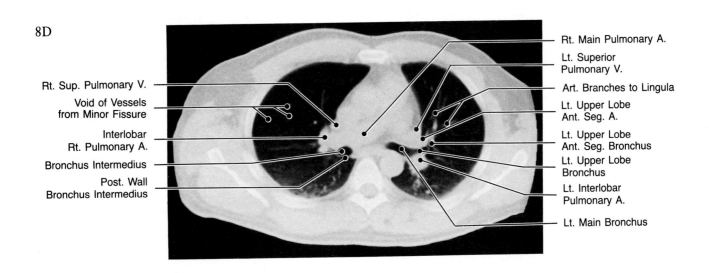

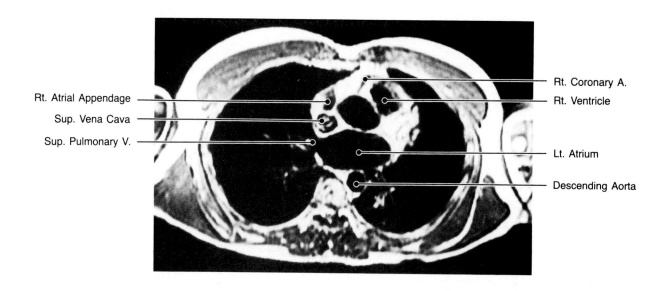

Rt. Coronary A.

Rt. Ventricle

Rt. Atrial Appendage

Sup. Vena Cava

Sup. Pulmonary V.

Lt. Atrium

Descending Aorta

8E Axial/MRI/Interatrial Septum. A Magnetic Resonance Imaging scan of the heart at this level using partial saturation technique with cardiac gating clearly shows the right atrium separated from the left atrium by the interatrial septum at the level of the mitral valve. The right ventricle is demonstrated, but the left ventricle is below the plane of this cut, which is demonstrating abundant fat in the waist of the heart above the top of the left ventricle.

8F Axial/MRI/Root of the Aorta Level. This is a cardiac gated dynamic scan. The root of the aorta, the anterior leaflet of the mitral valve, and the origin of the pulmonary artery are demonstrated. The entrance of the superior vena cava at the level of the right auricular appendage is seen, as well as the origin of the right coronary artery.

8G Axial/MRI. This image demonstrates diastolic atrial filling and systolic ejection into the aorta. Flowing blood is bright electronically enhanced.

8H Axial/MRI. At a level 2 cm above section 8G, this image demonstrates the root of the aorta, the pulmonary outflow tract, and the pulmonary veins.

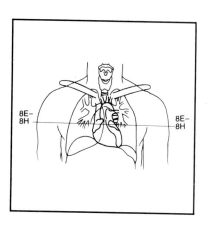

8E– 8H

8E– 8H

8F

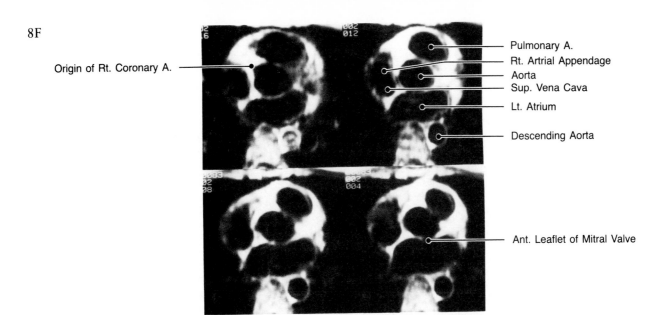

Origin of Rt. Coronary A.

Pulmonary A.
Rt. Artrial Appendage
Aorta
Sup. Vena Cava
Lt. Atrium
Descending Aorta

Ant. Leaflet of Mitral Valve

8G

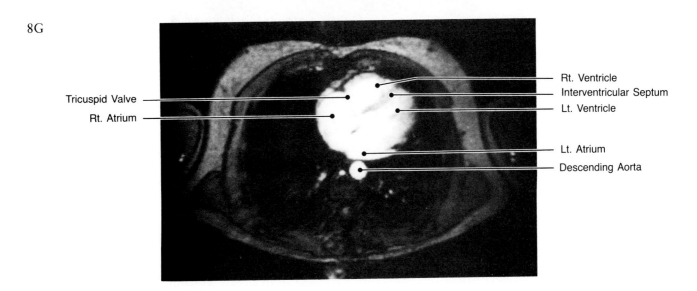

Tricuspid Valve
Rt. Atrium

Rt. Ventricle
Interventricular Septum
Lt. Ventricle

Lt. Atrium
Descending Aorta

8H

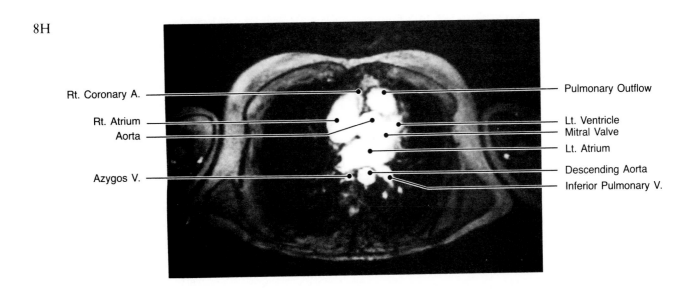

Rt. Coronary A.

Rt. Atrium
Aorta

Azygos V.

Pulmonary Outflow

Lt. Ventricle
Mitral Valve

Lt. Atrium

Descending Aorta
Inferior Pulmonary V.

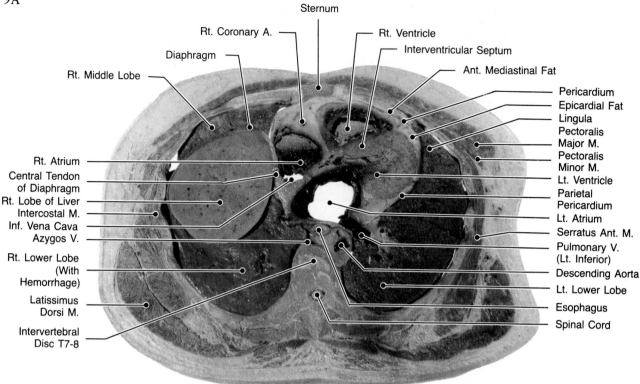

Sternum
Rt. Coronary A.
Rt. Ventricle
Interventricular Septum
Diaphragm
Ant. Mediastinal Fat
Rt. Middle Lobe
Pericardium
Epicardial Fat
Lingula
Pectoralis
Major M.
Rt. Atrium
Pectoralis
Minor M.
Central Tendon
of Diaphragm
Lt. Ventricle
Rt. Lobe of Liver
Parietal
Intercostal M.
Pericardium
Inf. Vena Cava
Lt. Atrium
Azygos V.
Serratus Ant. M.
Pulmonary V.
Rt. Lower Lobe
(Lt. Inferior)
(With
Descending Aorta
Hemorrhage)
Lt. Lower Lobe
Latissimus
Esophagus
Dorsi M.
Spinal Cord
Intervertebral
Disc T7-8

9A Axial/Anatomic Plate/Right Ventricle. The right and left ventricles are both identified on this section; however, the right ventricular lumen and the trabeculated musculature are much more apparent. The right coronary artery is clearly seen in the atrial ventricular fat anteriorly between the right-sided chambers. The thicker myocardium of the left ventricle stands out, as does the interventricular septum. More of the left ventricular lumen will be seen on lower sections. Both atria and their interatrial septum and the draining inferior pulmonary vein on the left are clearly seen. The cardiac structures are invested by the pericardium, which has a thick fibrous outer layer and an inner serous membrane visceral pericardium, leaving a pericardial space between them. Superficial to the parietal pericardium is the anterior mediastinal fat, and deep to the visceral pericardium is the epicardial fat in which the coronary vessels lie. Like the pericardium the pleural space consists of an outer parietal and an inner visceral pleura, re-

sulting in a potential space, the pleural space. It is usual for a small amount of fluid to lubricate the normal motion between visceral and parietal pleural and pericardial surfaces. Where the great vessels enter the heart a continuity between serous pericardium and parietal or fibrous pericardium is present, and at the hilus of the lung continuity between visceral and parietal pleura is present. Where they are adjacent the parietal pericardium and pleura are intimately attached.

9B Axial/CT/Right Ventricle. The CT cadaver section demonstrates the inferior pulmonary vein draining into the left atrium, portions of the interatrial septum, and the homogeneous densities of the right and left ventricles. A small amount of liver and dome of right hemidiaphragm can be seen centrally, opacifying the right lung at this level, T_8.

9C, 9D Axial/CT/Right Ventricle. The contrast-enhanced CTs at mediastinal and lung windows on the living patient clearly demonstrate the left atrium and draining inferior pulmonary veins, the right atrium, and segments of each ventricle. The right coronary artery may be seen rising from the anterior aspect of the root of the aorta. At lung window middle lobe vasculature and both medial and lateral segmental bronchi are identified anterior to the interlobar pulmonary artery, while the right lower lobe bronchus and its superior segmental bronchus and the inferior pulmonary vein are seen behind the interlobar artery. The inferior pulmonary vein is the dominant vascular structure seen on the left, medial and anterior to the left inferior pulmonary artery and lower lobe bronchopulmonary segmental branches. Note that in the living patient the diaphragmatic excursion takes the liver out of the field, whereas in the cadaver case an anatomic section of the liver dome is seen at a much higher level.

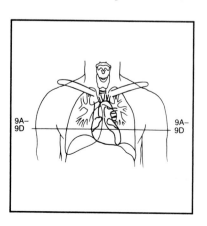

9A–
9D

9A–
9D

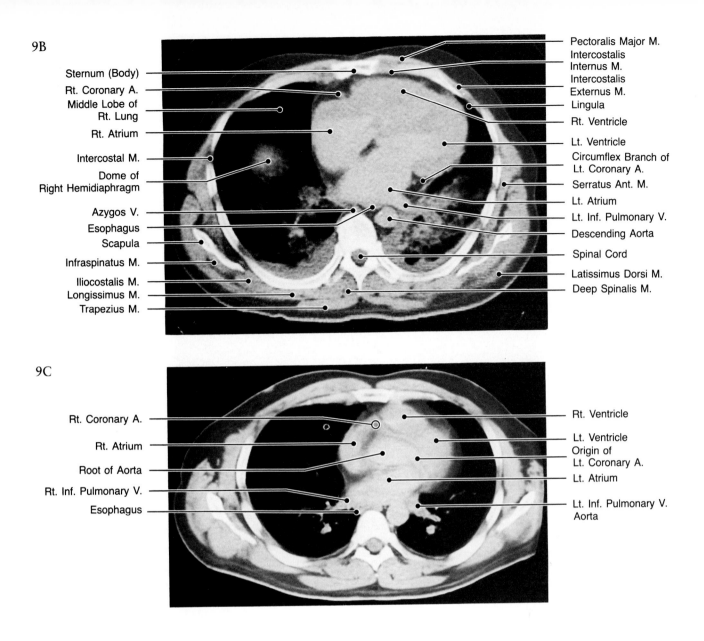

9B

Sternum (Body)
Rt. Coronary A.
Middle Lobe of Rt. Lung
Rt. Atrium
Intercostal M.
Dome of Right Hemidiaphragm
Azygos V.
Esophagus
Scapula
Infraspinatus M.
Iliocostalis M.
Longissimus M.
Trapezius M.

Pectoralis Major M.
Intercostalis Internus M.
Intercostalis Externus M.
Lingula
Rt. Ventricle
Lt. Ventricle
Circumflex Branch of Lt. Coronary A.
Serratus Ant. M.
Lt. Atrium
Lt. Inf. Pulmonary V.
Descending Aorta
Spinal Cord
Latissimus Dorsi M.
Deep Spinalis M.

9C

Rt. Coronary A.
Rt. Atrium
Root of Aorta
Rt. Inf. Pulmonary V.
Esophagus

Rt. Ventricle
Lt. Ventricle
Origin of Lt. Coronary A.
Lt. Atrium
Lt. Inf. Pulmonary V.
Aorta

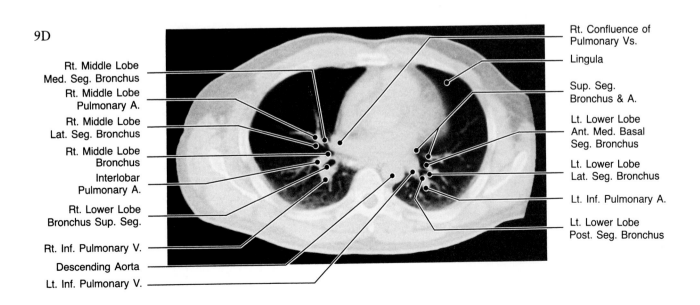

9D

Rt. Middle Lobe Med. Seg. Bronchus
Rt. Middle Lobe Pulmonary A.
Rt. Middle Lobe Lat. Seg. Bronchus
Rt. Middle Lobe Bronchus
Interlobar Pulmonary A.
Rt. Lower Lobe Bronchus Sup. Seg.
Rt. Inf. Pulmonary V.
Descending Aorta
Lt. Inf. Pulmonary V.

Rt. Confluence of Pulmonary Vs.
Lingula
Sup. Seg. Bronchus & A.
Lt. Lower Lobe Ant. Med. Basal Seg. Bronchus
Lt. Lower Lobe Lat. Seg. Bronchus
Lt. Inf. Pulmonary A.
Lt. Lower Lobe Post. Seg. Bronchus

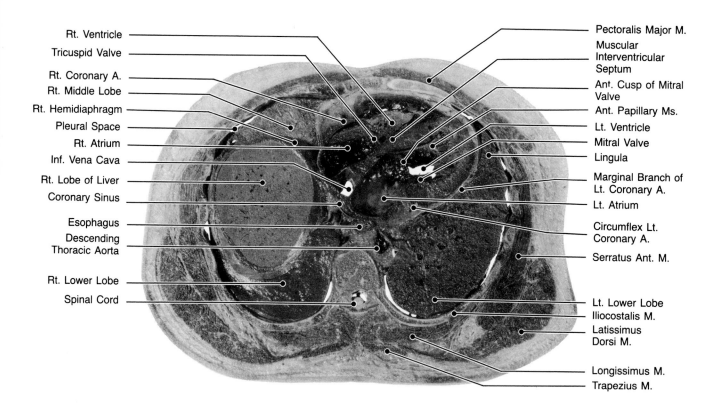

Rt. Ventricle
Tricuspid Valve
Rt. Coronary A.
Rt. Middle Lobe
Rt. Hemidiaphragm
Pleural Space
Rt. Atrium
Inf. Vena Cava
Rt. Lobe of Liver
Coronary Sinus
Esophagus
Descending Thoracic Aorta
Rt. Lower Lobe
Spinal Cord

Pectoralis Major M.
Muscular Interventricular Septum
Ant. Cusp of Mitral Valve
Ant. Papillary Ms.
Lt. Ventricle
Mitral Valve
Lingula
Marginal Branch of Lt. Coronary A.
Lt. Atrium
Circumflex Lt. Coronary A.
Serratus Ant. M.
Lt. Lower Lobe
Iliocostalis M.
Latissimus Dorsi M.
Longissimus M.
Trapezius M.

10A Axial/Anatomic Plate/Atrioventricular Valves. At this level we have more definition of the tricuspid valve and of the mitral valve with the anterior papillary muscle and anterior cusp of the mitral valve demonstrated with the interventricular septum between them. The inferior vena cava appears as a separate structure posterior to the right atrium filled with clotted blood. Both left atrium and left ventricle are shown to communicate through the leaflets of the mitral valve. The coronary sinus may be seen along the posterior aspect of the left atrium coursing toward the right atrium.

10B Axial/CT/Atrioventricular Level. The corresponding cadaver CT section shows some layering of the clotted blood leading to different densities in the cardiac chambers in this early post-mortem state. Hemorrhagic consolidation is seen in the dependent portions of both lungs, a little more obvious on the CT section on the left and a little more obvious on the anatomic plate on the right. The circumflex coronary arterial branch and the right coronary artery in the inferior AV sulcus are seen.

10C, 10D Axial/CT/Atrioventricular Level. CT sections with contrast enhancement corresponding to the anatomic plate in the living patient demonstrates an abundance of pericardial fat surrounding the left ventricular wall and a faint decrease in density at the level of the interventricular muscular septum. At lung window a comparable section demonstrates some dependent congestion in the lung bases posteriorly not seen in the lingula or middle lobe. Such an increase when vessels remain clearly visible is a normal variation and does not represent disease.

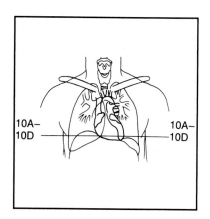

10A–
10D

10A–
10D

10B

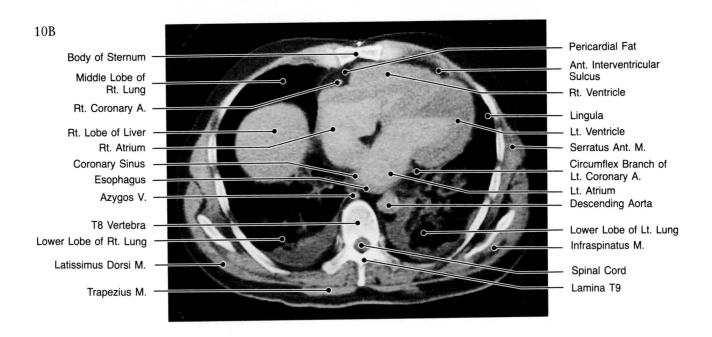

Body of Sternum	Pericardial Fat
Middle Lobe of Rt. Lung	Ant. Interventricular Sulcus
Rt. Coronary A.	Rt. Ventricle
Rt. Lobe of Liver	Lingula
Rt. Atrium	Lt. Ventricle
Coronary Sinus	Serratus Ant. M.
Esophagus	Circumflex Branch of Lt. Coronary A.
Azygos V.	Lt. Atrium
T8 Vertebra	Descending Aorta
Lower Lobe of Rt. Lung	Lower Lobe of Lt. Lung
Latissimus Dorsi M.	Infraspinatus M.
Trapezius M.	Spinal Cord
	Lamina T9

10C

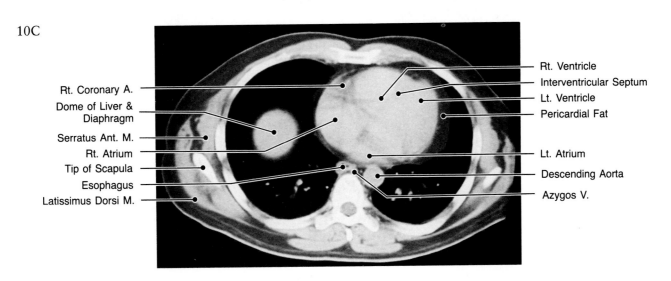

Rt. Coronary A.	Rt. Ventricle
Dome of Liver & Diaphragm	Interventricular Septum
Serratus Ant. M.	Lt. Ventricle
Rt. Atrium	Pericardial Fat
Tip of Scapula	Lt. Atrium
Esophagus	Descending Aorta
Latissimus Dorsi M.	Azygos V.

10D

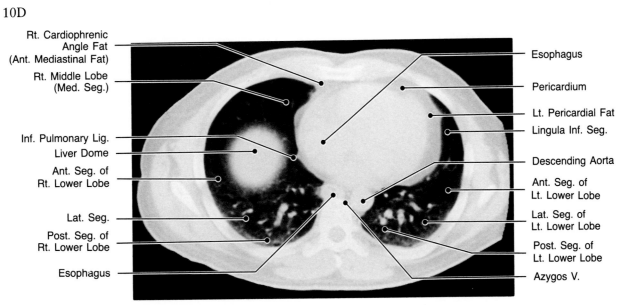

Rt. Cardiophrenic Angle Fat (Ant. Mediastinal Fat)	Esophagus
Rt. Middle Lobe (Med. Seg.)	Pericardium
Inf. Pulmonary Lig.	Lt. Pericardial Fat
Liver Dome	Lingula Inf. Seg.
Ant. Seg. of Rt. Lower Lobe	Descending Aorta
Lat. Seg.	Ant. Seg. of Lt. Lower Lobe
Post. Seg. of Rt. Lower Lobe	Lat. Seg. of Lt. Lower Lobe
Esophagus	Post. Seg. of Lt. Lower Lobe
	Azygos V.

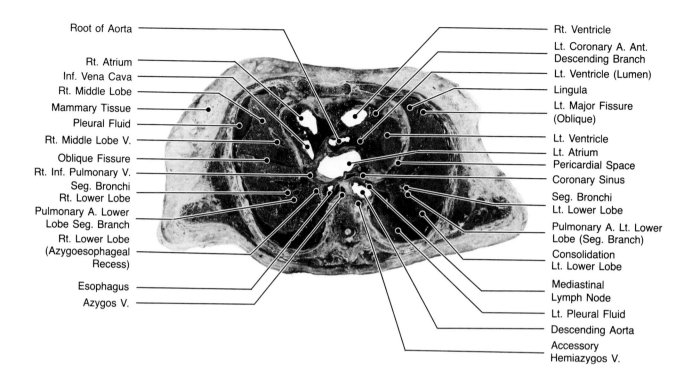

Root of Aorta

Rt. Atrium
Inf. Vena Cava
Rt. Middle Lobe
Mammary Tissue
Pleural Fluid
Rt. Middle Lobe V.

Oblique Fissure
Rt. Inf. Pulmonary V.
Seg. Bronchi
Rt. Lower Lobe
Pulmonary A. Lower
Lobe Seg. Branch
Rt. Lower Lobe
(Azygoesophageal
Recess)

Esophagus
Azygos V.

Rt. Ventricle
Lt. Coronary A. Ant.
Descending Branch
Lt. Ventricle (Lumen)
Lingula
Lt. Major Fissure
(Oblique)
Lt. Ventricle
Lt. Atrium
Pericardial Space
Coronary Sinus
Seg. Bronchi
Lt. Lower Lobe
Pulmonary A. Lt. Lower
Lobe (Seg. Branch)
Consolidation
Lt. Lower Lobe
Mediastinal
Lymph Node
Lt. Pleural Fluid
Descending Aorta
Accessory
Hemiazygos V.

10E Axial/Anatomic Plate/Right Atrium. An additional anatomic plate from a second cadaver whose pleural space contains fluid and whose lungs show focal congestion, particularly in the left lower lobe, and compensatory atelectasis. From within the lumen of the left atrium we can identify the orifice of the inferior pulmonary vein on the right. Bilaterally lower lobe segmental bronchi are demonstrated lying just lateral to their supplying pulmonary arterial branches. Both inferior vena cava and right atrium are seen along the right heart margin.

10E Axial/CT. At lung window this section demonstrates some dependent congestion in the lung bases posteriorly not seen in the lingula on middle lobe. Such an increase when vessels remain clearly visible is a phenomenon related to gravity and does not represent disease.

10F Axial/MRI/Right Atrium. Partial saturation MR image demonstrates the tricuspid and mitral valves, both ventricles, and the interventricular septum. Fat at partial saturation gives an intense signal. The right atrium lies along the right heart margin and receives coronary venous blood via the coronary sinus.

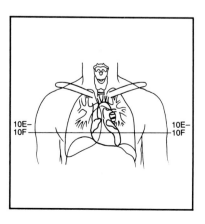

10F

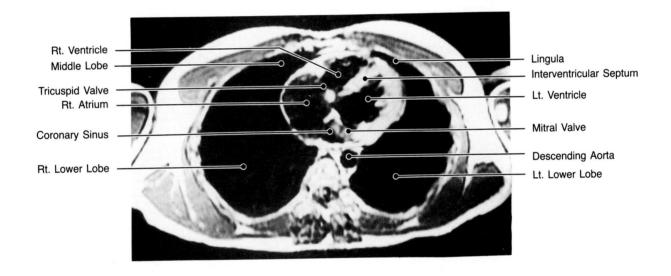

Rt. Ventricle

Middle Lobe

Tricuspid Valve

Rt. Atrium

Coronary Sinus

Rt. Lower Lobe

Lingula

Interventricular Septum

Lt. Ventricle

Mitral Valve

Descending Aorta

Lt. Lower Lobe

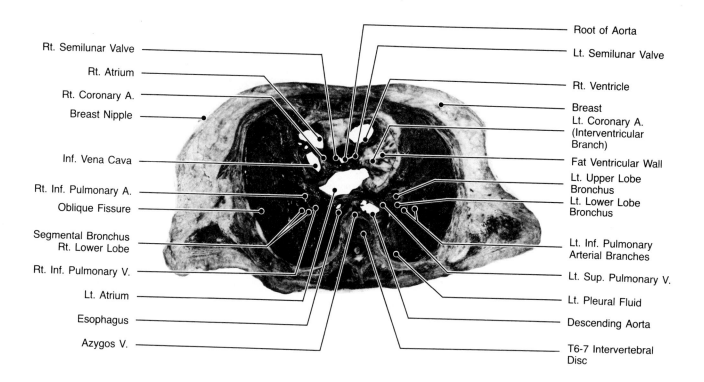

Rt. Semilunar Valve

Rt. Atrium

Rt. Coronary A.

Breast Nipple

Inf. Vena Cava

Rt. Inf. Pulmonary A.

Oblique Fissure

Segmental Bronchus
Rt. Lower Lobe

Rt. Inf. Pulmonary V.

Lt. Atrium

Esophagus

Azygos V.

Root of Aorta

Lt. Semilunar Valve

Rt. Ventricle

Breast

Lt. Coronary A.
(Interventricular
Branch)

Fat Ventricular Wall

Lt. Upper Lobe
Bronchus

Lt. Lower Lobe
Bronchus

Lt. Inf. Pulmonary
Arterial Branches

Lt. Sup. Pulmonary V.

Lt. Pleural Fluid

Descending Aorta

T6-7 Intervertebral
Disc

10G Axial/Anatomic Plate/Left Atrium Supplemental. This plate, 1.5 cm higher than 10E, demonstrates the aortic root and semilunar valves. The confluence of the inferior vena cava to the right atrium is well seen. The left upper and left lower lobe bronchi are demonstrated. Pleural fluid occupies the dependent part of the left hemithorax. The right coronary artery and an interventricular branch of the left coronary artery can be seen.

10H Axial/MRI/Atrioventricular Valves. Partial saturation MR images of the heart, demonstrate the tricuspid and mitral valves, both ventricles and the interventricular septum, and both atria without clear demonstration of the atrial septa. Fat at partial saturation appears as an intensely white signal.

10I Axial/MRI/Dynamic Scan at Atrioventricular Level. Plate 10I is a composite ventricular level dynamic cardiac gated scan of the heart using MRI. Images A and B demonstrate the open and closed tricuspid valve, while images C and D show the progressive contraction of the right ventricle in systole. The mitral valve region appears patent on image A and mitral valve leaflet approximation is seen on images C and D as left ventricular systole progresses. Note the interventricular septal thinning during diastole with filling of the left anterior descending branch of the coronary artery. During progressive contraction the muscular septum thickens and the coronary flow is not appreciated as a flow void in the coronary artery.

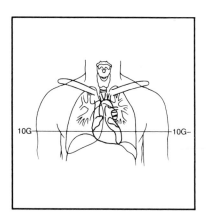

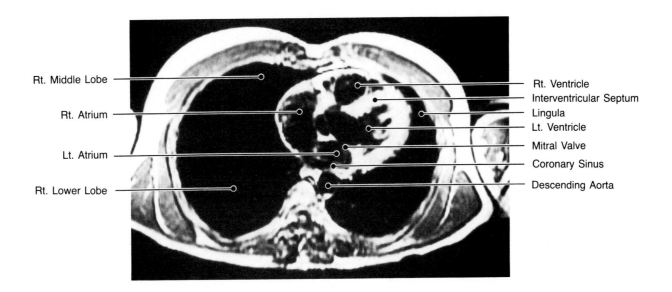

Rt. Middle Lobe

Rt. Atrium

Lt. Atrium

Rt. Lower Lobe

Rt. Ventricle
Interventricular Septum
Lingula
Lt. Ventricle
Mitral Valve
Coronary Sinus
Descending Aorta

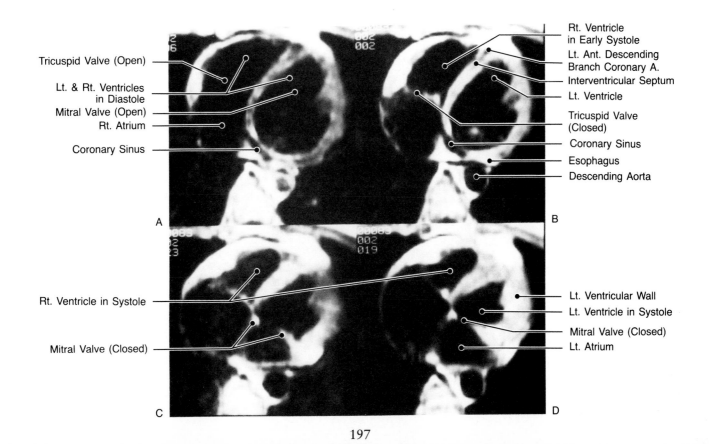

Tricuspid Valve (Open)

Lt. & Rt. Ventricles
in Diastole
Mitral Valve (Open)
Rt. Atrium

Coronary Sinus

Rt. Ventricle in Systole

Mitral Valve (Closed)

Rt. Ventricle
in Early Systole
Lt. Ant. Descending
Branch Coronary A.
Interventricular Septum
Lt. Ventricle
Tricuspid Valve
(Closed)
Coronary Sinus
Esophagus
Descending Aorta

A

B

C

D

Lt. Ventricular Wall
Lt. Ventricle in Systole
Mitral Valve (Closed)
Lt. Atrium

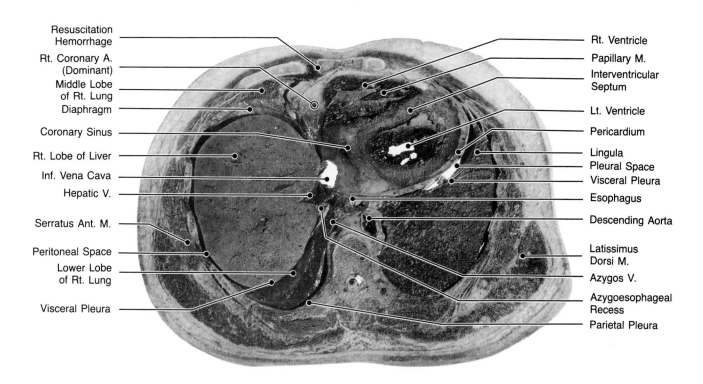

Resuscitation Hemorrhage — Rt. Ventricle

Rt. Coronary A. (Dominant) — Papillary M.

Middle Lobe of Rt. Lung — Interventricular Septum

Diaphragm — Lt. Ventricle

Coronary Sinus — Pericardium

Rt. Lobe of Liver — Lingula

Inf. Vena Cava — Pleural Space

Hepatic V. — Visceral Pleura

— Esophagus

Serratus Ant. M. — Descending Aorta

Peritoneal Space — Latissimus Dorsi M.

Lower Lobe of Rt. Lung — Azygos V.

— Azygoesophageal Recess

Visceral Pleura — Parietal Pleura

11A Axial/Anatomic Plate/Esophagus. The esophagus lies in the same fibrous sheath from the level of the aortic knob, where it attaches to the posterior wall of the aorta, to near the diaphragmatic perforation, where each has its own separate opening. The azygous vein lies along the right posterior wall of the esophagus. The tongue of lung inserting behind the inferior vena cava and heart is frequently referred to as the azygoesophageal recess. Dense consolidation can be seen in the thin crescentic portion of the most dependent right lower lobe.

11B Axial/CT/Coronary Sinus. The cadaver CT section corresponds to the anatomic plate and demonstrates the coronary sinus emptying into the right atrium and the significant bibasilar dependent consolidation in the lungs.

11C, 11D Axial/CT/Coronary Sinus. Contrast-enhanced CT on the living patient at the level of the diaphragmatic dome demonstrates the inferior vena cava posterior to the right atrium and a portion of the coronary sinus along the sulcus between the right atrium and the left ventricle. A small, persistent anatomic structure seen between the azygous vein and the descending aorta is the thoracic duct surrounded by fat. On lung windows dependent edema is identified. Areas for the segmental anatomy of the left and right lower lobes are identified as well.

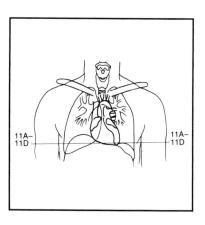

11A–11D   11A–11D

11B

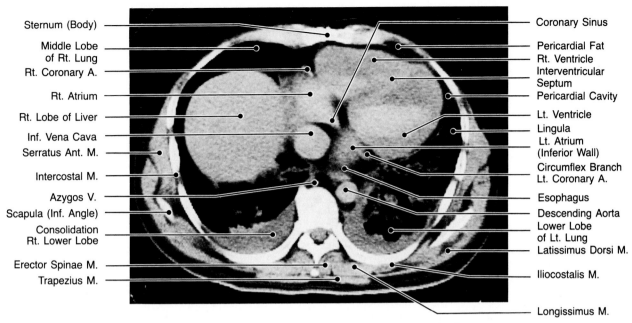

Sternum (Body)
Middle Lobe of Rt. Lung
Rt. Coronary A.
Rt. Atrium
Rt. Lobe of Liver
Inf. Vena Cava
Serratus Ant. M.
Intercostal M.
Azygos V.
Scapula (Inf. Angle)
Consolidation Rt. Lower Lobe
Erector Spinae M.
Trapezius M.

Coronary Sinus
Pericardial Fat
Rt. Ventricle
Interventricular Septum
Pericardial Cavity
Lt. Ventricle
Lingula
Lt. Atrium (Inferior Wall)
Circumflex Branch Lt. Coronary A.
Esophagus
Descending Aorta
Lower Lobe of Lt. Lung
Latissimus Dorsi M.
Iliocostalis M.
Longissimus M.

11C

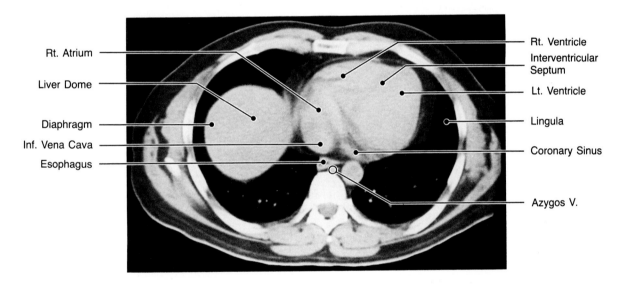

Rt. Atrium
Liver Dome
Diaphragm
Inf. Vena Cava
Esophagus

Rt. Ventricle
Interventricular Septum
Lt. Ventricle
Lingula
Coronary Sinus
Azygos V.

11D

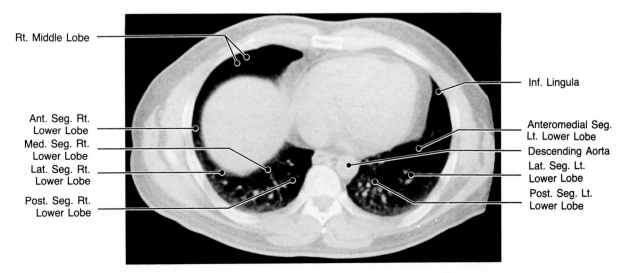

Rt. Middle Lobe
Ant. Seg. Rt. Lower Lobe
Med. Seg. Rt. Lower Lobe
Lat. Seg. Rt. Lower Lobe
Post. Seg. Rt. Lower Lobe

Inf. Lingula
Anteromedial Seg. Lt. Lower Lobe
Descending Aorta
Lat. Seg. Lt. Lower Lobe
Post. Seg. Lt. Lower Lobe

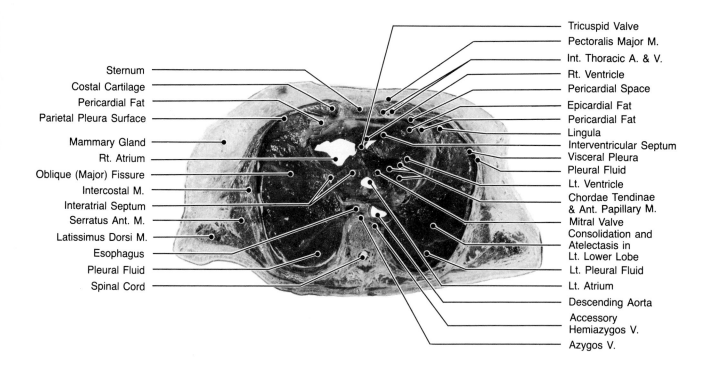

Sternum
Costal Cartilage
Pericardial Fat
Parietal Pleura Surface
Mammary Gland
Rt. Atrium
Oblique (Major) Fissure
Intercostal M.
Interatrial Septum
Serratus Ant. M.
Latissimus Dorsi M.
Esophagus
Pleural Fluid
Spinal Cord

Tricuspid Valve
Pectoralis Major M.
Int. Thoracic A. & V.
Rt. Ventricle
Pericardial Space
Epicardial Fat
Pericardial Fat
Lingula
Interventricular Septum
Visceral Pleura
Pleural Fluid
Lt. Ventricle
Chordae Tendinae
& Ant. Papillary M.
Mitral Valve
Consolidation and
Atelectasis in
Lt. Lower Lobe
Lt. Pleural Fluid
Lt. Atrium
Descending Aorta
Accessory
Hemiazygos V.
Azygos V.

11E Axial/Anatomic Plate/Chordae Tendinae. An additional anatomic plate from a female cadaver, at a higher level, identified by breast tissue demonstrates fluid in the dependent pleural spaces and considerable compensatory atelectasis and blood on the surface of the dependent lower lobes. Well demonstrated in the left ventricular lumen are chordae tendinae and papillary muscles for the anterior leaflet of the mitral valve. A leaflet of the tricuspid valve is also identified between right atrium and right ventricle.

11F Axial/MRI/Coronary Sinus. At a corresponding level a partial saturation cardiac gated MR image demonstrates the coronary sinus, the tricuspid valve, and the interventricular septum.

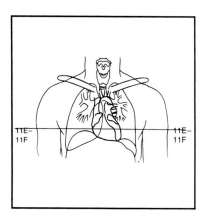

11F

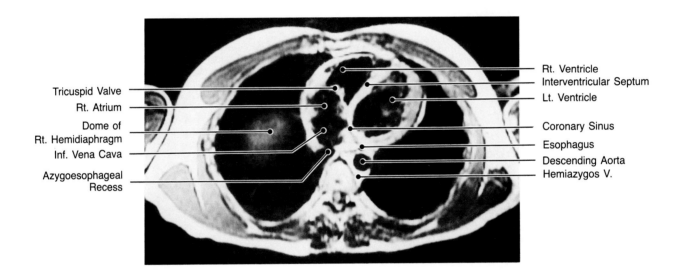

Rt. Ventricle
Interventricular Septum
Lt. Ventricle

Tricuspid Valve
Rt. Atrium

Coronary Sinus

Dome of
Rt. Hemidiaphragm
Inf. Vena Cava

Esophagus
Descending Aorta
Hemiazygos V.

Azygoesophageal
Recess

12A

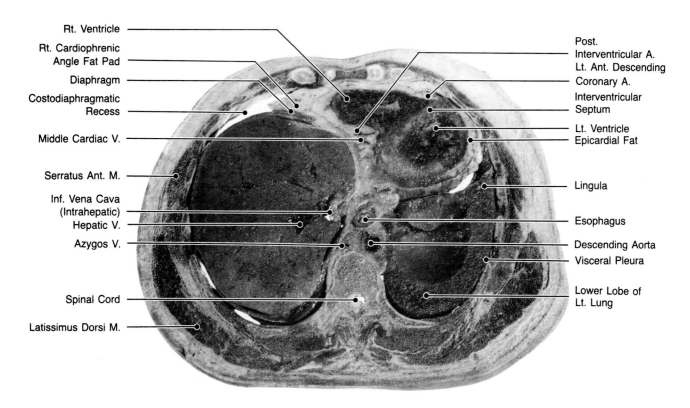

Rt. Ventricle

Rt. Cardiophrenic Angle Fat Pad

Diaphragm

Costodiaphragmatic Recess

Middle Cardiac V.

Serratus Ant. M.

Inf. Vena Cava (Intrahepatic)
Hepatic V.

Azygos V.

Spinal Cord

Latissimus Dorsi M.

Post. Interventricular A.
Lt. Ant. Descending Coronary A.

Interventricular Septum

Lt. Ventricle
Epicardial Fat

Lingula

Esophagus

Descending Aorta
Visceral Pleura

Lower Lobe of Lt. Lung

12A Axial/Anatomic Plate/Diaphragmatic Level. The anatomic plate at the diaphragmatic level consists almost completely of the liver in the right lower thoracic and upper abdominal cavity with muscular diaphragm identified on its anterior surface. Hepatic veins and inferior vena cava appear to be joining at the central portion of the medial aspect of the liver. The aorta and esophagus are now separate, and in fact the right crus of the diaphragm lying just anterior to the azygous vein appears to separate the aorta from the esophagus slightly. At the base of the left lung visceral pleura is seen covering the cephalic aspect of the lower lobe and the caudal aspect of the lingula. Only the two ventricles remain of the four chambers of the heart and abundant cardiophrenic angle and epicardial fat.

12B Axial/CT/Right Coronary Artery. The cadaver CT demonstrates some right coronary artery calcification lying in the posterior descending branch of the right coronary artery just anterior to the middle cardiac vein. Some pleural fluid and adjacent consolidation can hardly be distinguished without contrast in the posterior costophrenic recesses.

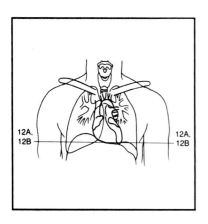

12B

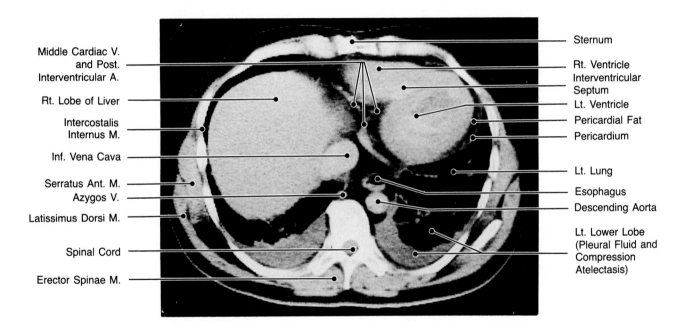

Middle Cardiac V. and Post. Interventricular A.

Rt. Lobe of Liver

Intercostalis Internus M.

Inf. Vena Cava

Serratus Ant. M.

Azygos V.

Latissimus Dorsi M.

Spinal Cord

Erector Spinae M.

Sternum

Rt. Ventricle

Interventricular Septum

Lt. Ventricle

Pericardial Fat

Pericardium

Lt. Lung

Esophagus

Descending Aorta

Lt. Lower Lobe (Pleural Fluid and Compression Atelectasis)

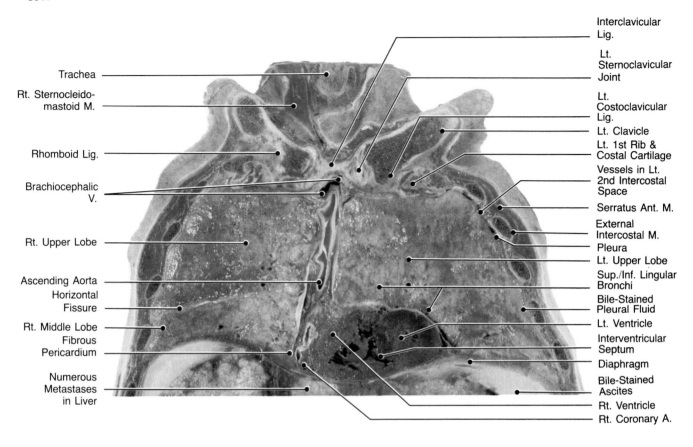

Trachea

Rt. Sternocleido-
mastoid M.

Rhomboid Lig.

Brachiocephalic
V.

Rt. Upper Lobe

Ascending Aorta
Horizontal
Fissure

Rt. Middle Lobe
Fibrous
Pericardium

Numerous
Metastases
in Liver

Interclavicular
Lig.

Lt.
Sternoclavicular
Joint

Lt.
Costoclavicular
Lig.

Lt. Clavicle

Lt. 1st Rib &
Costal Cartilage

Vessels in Lt.
2nd Intercostal
Space

Serratus Ant. M.

External
Intercostal M.

Pleura

Lt. Upper Lobe

Sup./Inf. Lingular
Bronchi

Bile-Stained
Pleural Fluid

Lt. Ventricle

Interventricular
Septum

Diaphragm

Bile-Stained
Ascites

Rt. Ventricle

Rt. Coronary A.

13A Coronal/Anatomic Plate/Right Ventricle. This anatomic plate is through the anterior chest at the level of the right ventricle. A rather long segment of ascending aorta, a portion of the thickly muscular left ventricle, and the interventricular septum are seen. Much of the anterior portion of the right upper lobe is seen separated by the horizontal fissure from the middle lobe. On the left, only the upper lobe is seen with lingular bronchi identified in cross-section just above the left heart margin. Some bile staining of the pleural surface is present on the left in this male with extensive liver metastases from primary colon carcinoma.

13B Coronal/MRI/Right Heart. This coronal magnetic resonance partial saturation image of the anterior chest and upper abdomen demonstrates the right-sided chambers of the heart and main pulmonary artery. Portions of the left ventricle and ascending aorta are also appreciated. Both upper lobes are also demonstrated.

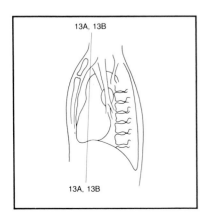

13A, 13B

13A, 13B

13B

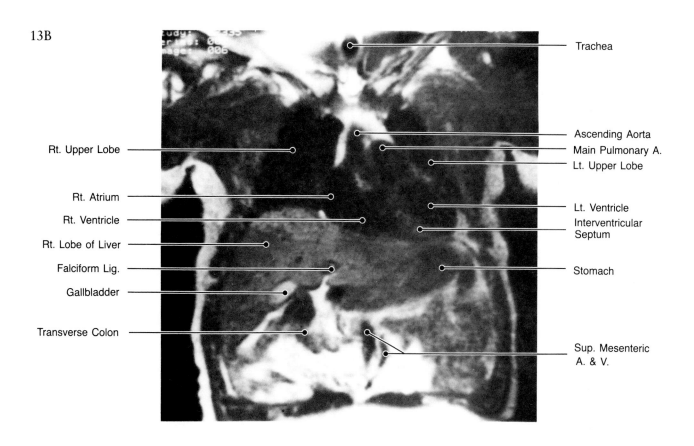

Trachea

Ascending Aorta

Main Pulmonary A.

Lt. Upper Lobe

Rt. Upper Lobe

Rt. Atrium

Rt. Ventricle

Rt. Lobe of Liver

Falciform Lig.

Gallbladder

Transverse Colon

Lt. Ventricle

Interventricular
Septum

Stomach

Sup. Mesenteric
A. & V.

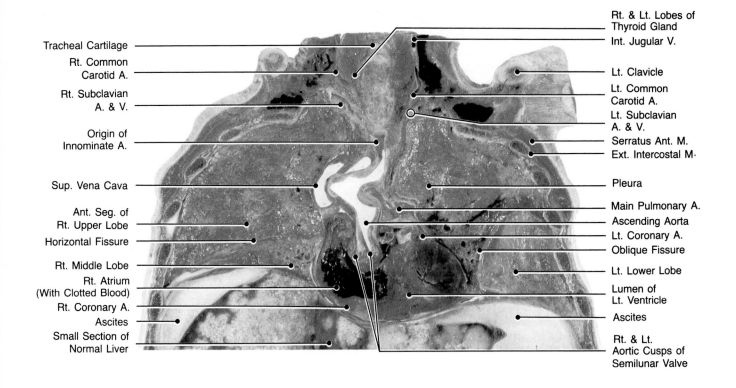

Tracheal Cartilage
Rt. Common Carotid A.
Rt. Subclavian A. & V.
Origin of Innominate A.
Sup. Vena Cava
Ant. Seg. of Rt. Upper Lobe
Horizontal Fissure
Rt. Middle Lobe
Rt. Atrium (With Clotted Blood)
Rt. Coronary A.
Ascites
Small Section of Normal Liver

Rt. & Lt. Lobes of Thyroid Gland
Int. Jugular V.
Lt. Clavicle
Lt. Common Carotid A.
Lt. Subclavian A. & V.
Serratus Ant. M.
Ext. Intercostal M·
Pleura
Main Pulmonary A.
Ascending Aorta
Lt. Coronary A.
Oblique Fissure
Lt. Lower Lobe
Lumen of Lt. Ventricle
Ascites
Rt. & Lt. Aortic Cusps of Semilunar Valve

14A Coronal/Anatomic Plate/Aortic Valve. This anatomic plate demonstrates the ascending aorta and root of the aorta, including right and left aortic cusps. Note the relationships of the aortic root to the right atrium and left ventricle as well as the superior vena cava and main pulmonary artery. The right lung consists chiefly of the upper lobe and a small wedge of middle lobe medially adjacent to the right heart margin. The left oblique or major fissure separates the upper lobe and lingula from the lower lobe. The great vessels in the neck are well demonstrated. Below the diaphragm the liver is replaced by tumor and floats in bile-stained ascites.

14B Coronal/MRI/Aortic Valve. The Coronal MRI section with cardiac gating at the level of the aortic valve demonstrates the same relationships between ascending aorta, pulmonary artery, superior vena cava, and right atrium as seen on the anatomic plate. Because these images were obtained in diastole the left ventricle has a far larger luminal capacity than is seen on the anatomic specimen.

14C Coronal/MRI/Systole. Aorta and ventricles during systolic phase, with flowing blood electronically enhanced to resemble contrast.

14D Coronal/MRI/Diastole. Change in the calibre and size of aorta and left ventricle during diastole. Compare this with 14C.

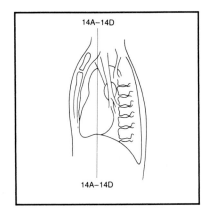

14A–14D

14A–14D

14B

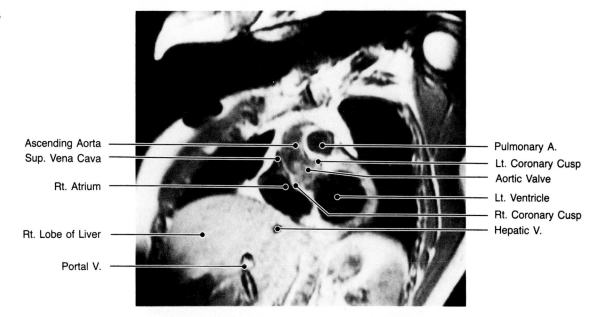

Ascending Aorta — — Pulmonary A.
Sup. Vena Cava — — Lt. Coronary Cusp
— Aortic Valve
Rt. Atrium — — Lt. Ventricle
— Rt. Coronary Cusp
Rt. Lobe of Liver — — Hepatic V.
Portal V. —

14C

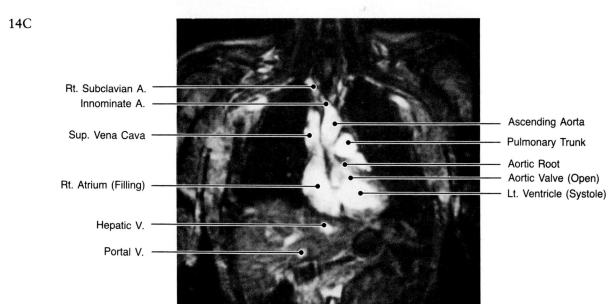

Rt. Subclavian A. —
Innominate A. —
— Ascending Aorta
Sup. Vena Cava — — Pulmonary Trunk
— Aortic Root
— Aortic Valve (Open)
Rt. Atrium (Filling) — — Lt. Ventricle (Systole)
Hepatic V. —
Portal V. —

14D

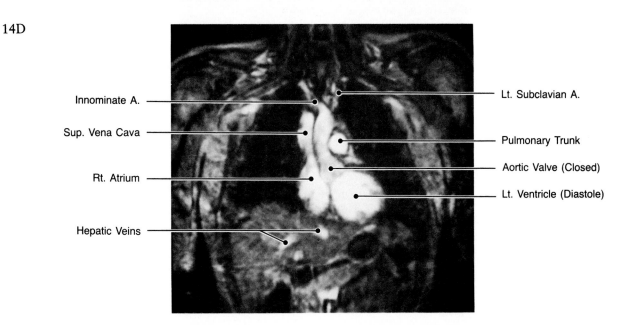

— Lt. Subclavian A.
Innominate A. —
Sup. Vena Cava — — Pulmonary Trunk
— Aortic Valve (Closed)
Rt. Atrium — — Lt. Ventricle (Diastole)
Hepatic Veins —

207

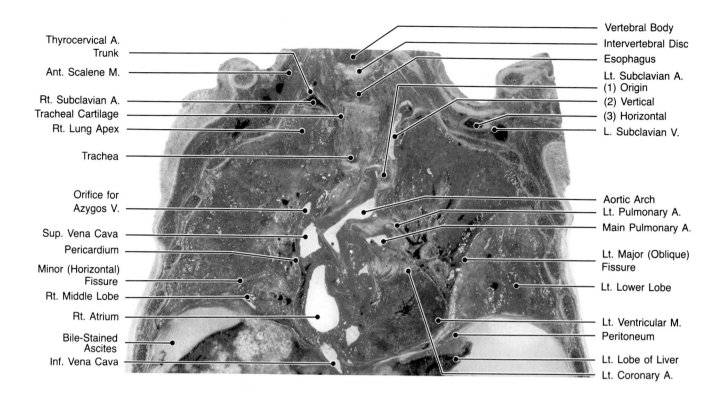

Thyrocervical A. Trunk
Ant. Scalene M.
Rt. Subclavian A.
Tracheal Cartilage
Rt. Lung Apex
Trachea
Orifice for Azygos V.
Sup. Vena Cava
Pericardium
Minor (Horizontal) Fissure
Rt. Middle Lobe
Rt. Atrium
Bile-Stained Ascites
Inf. Vena Cava

Vertebral Body
Intervertebral Disc
Esophagus
Lt. Subclavian A.
(1) Origin
(2) Vertical
(3) Horizontal
L. Subclavian V.
Aortic Arch
Lt. Pulmonary A.
Main Pulmonary A.
Lt. Major (Oblique) Fissure
Lt. Lower Lobe
Lt. Ventricular M.
Peritoneum
Lt. Lobe of Liver
Lt. Coronary A.

15A Coronal/Anatomic Plate/Tracheal Level. This anatomic plate shows the trachea in the midline highlighted by its tracheal cartilages. Above the trachea in sequence lie the collapsed esophagus, the intervertebral disc, and the vertebral body of the upper thoracic spine. To the left of the trachea, the three segments of the left subclavian artery (1, its origin from the aortic arch; 2, its vertical portion, where it indents the medial aspect of the lung extrapleurally; and 3, its horizontal portion lying next to the subclavian vein above the first anterior rib) are seen. The superior vena cava and its orifice for the azygous vein may be seen between the right upper lobe and aortic arch. The right atrium and inferior vena cava complete the right heart margin.

15B, 15C Coronal/MRI/Left Ventricular Level. The MRI studies that are cardiac gated through this anatomic plane demonstrate important differences in left ventricular volume, aortic size, and coronary flow. Also during left ventricular systole better filling of the innominate and left subclavian arteries can be seen with the aortic valve leaflets open. Without electronic manipulation the flowing blood returns no signal.

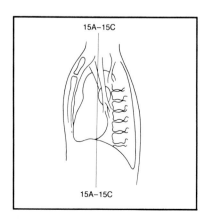

15A–15C

15A–15C

15B

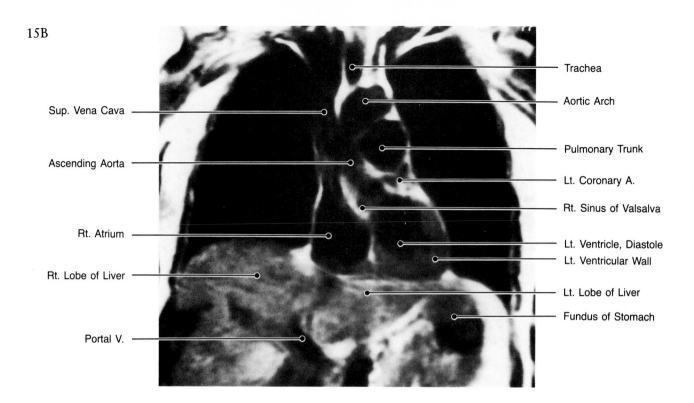

Sup. Vena Cava

Ascending Aorta

Rt. Atrium

Rt. Lobe of Liver

Portal V.

Trachea

Aortic Arch

Pulmonary Trunk

Lt. Coronary A.

Rt. Sinus of Valsalva

Lt. Ventricle, Diastole

Lt. Ventricular Wall

Lt. Lobe of Liver

Fundus of Stomach

15C

Trachea

Innominate A.

Rt. Atrium

Rt. Cardiophrenic
Angle Fat Pad

Rt. Lobe of Liver

Inf. Vena Cava

Lt. Subclavian A.

Ascending Aorta

Pulmonary Trunk

Aortic Valve (Open)

Lt. Ventricle, Systole

Lt. Apex Pericardial
Fat Pad

Lt. Lobe of Liver

Stomach

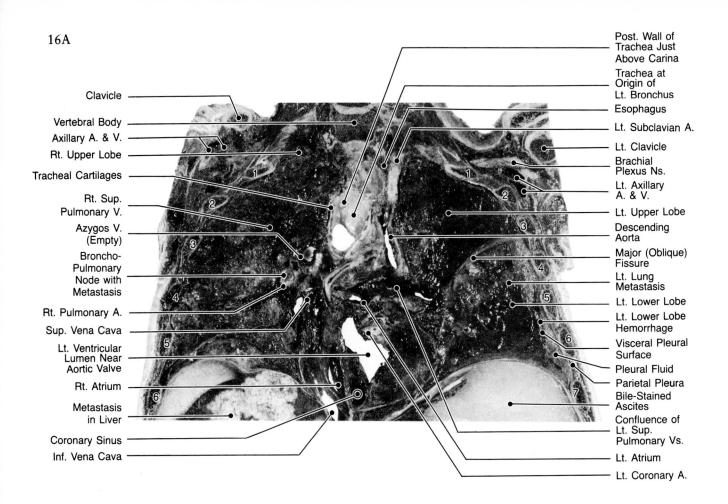

Clavicle

Vertebral Body

Axillary A. & V.

Rt. Upper Lobe

Tracheal Cartilages

Rt. Sup. Pulmonary V.

Azygos V. (Empty)

Broncho-Pulmonary Node with Metastasis

Rt. Pulmonary A.

Sup. Vena Cava

Lt. Ventricular Lumen Near Aortic Valve

Rt. Atrium

Metastasis in Liver

Coronary Sinus

Inf. Vena Cava

Post. Wall of Trachea Just Above Carina

Trachea at Origin of Lt. Bronchus

Esophagus

Lt. Subclavian A.

Lt. Clavicle

Brachial Plexus Ns.

Lt. Axillary A. & V.

Lt. Upper Lobe

Descending Aorta

Major (Oblique) Fissure

Lt. Lung Metastasis

Lt. Lower Lobe

Lt. Lower Lobe Hemorrhage

Visceral Pleural Surface

Pleural Fluid

Parietal Pleura

Bile-Stained Ascites

Confluence of Lt. Sup. Pulmonary Vs.

Lt. Atrium

Lt. Coronary A.

16A Coronal/Anatomic Plate/Tracheal Bifurcation. This anatomic plate demonstrates the central trachea, with a bulge representing the tracheal keel, and the more posterior going left main bronchus and the sectioned lumen of the right main bronchus. To the left of these structures is the esophagus, descending thoracic aorta, and a portion of the left atrium at the superior pulmonary veins. The left coronary artery is calcified and lies surrounded by fat between the left atrium and left ventricular myocardium. The prominent venous structure in the right upper lobe is the right superior pulmonary vein.

16B Coronal/MRI/Tracheal Bifurcation. The MRI coronal section at the level of the tracheal bifurcation shows the right and left main bronchi. To the right of the trachea is the azygous vein, and to the left is the aortic arch; inferior to it is the right main pulmonary artery. This nongated cardiac image done with partial saturation technique demonstrates portions of the inferior vena cava, left atrium, and left ventricle. Below the diaphragm the esophagogastric junction is demonstrated also.

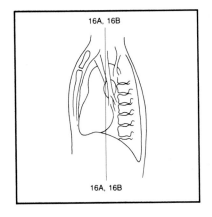

16A, 16B

16A, 16B

16B

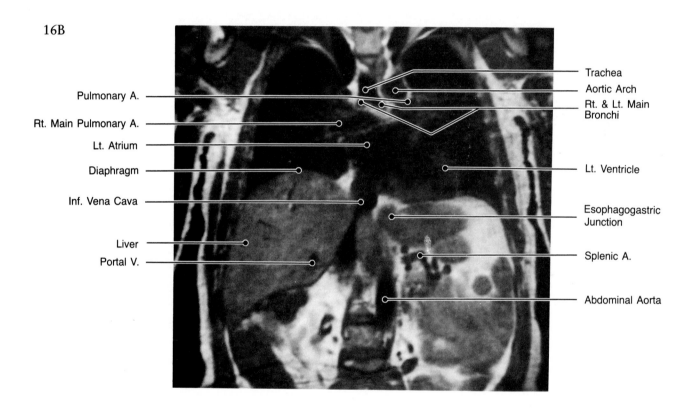

Pulmonary A.

Rt. Main Pulmonary A.

Lt. Atrium

Diaphragm

Inf. Vena Cava

Liver

Portal V.

Trachea

Aortic Arch

Rt. & Lt. Main Bronchi

Lt. Ventricle

Esophagogastric Junction

Splenic A.

Abdominal Aorta

17A

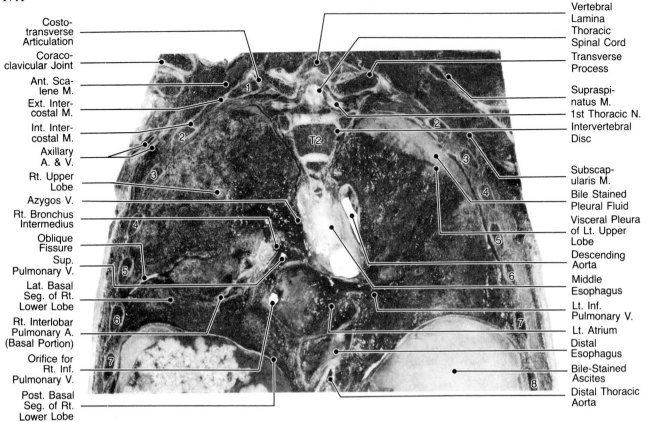

Costo-
transverse
Articulation

Coraco-
clavicular Joint

Ant. Sca-
lene M.

Ext. Inter-
costal M.

Int. Inter-
costal M.

Axillary
A. & V.

Rt. Upper
Lobe

Azygos V.

Rt. Bronchus
Intermedius

Oblique
Fissure

Sup.
Pulmonary V.

Lat. Basal
Seg. of Rt.
Lower Lobe

Rt. Interlobar
Pulmonary A.
(Basal Portion)

Orifice for
Rt. Inf.
Pulmonary V.

Post. Basal
Seg. of Rt.
Lower Lobe

Vertebral
Lamina

Thoracic
Spinal Cord

Transverse
Process

Supraspi-
natus M.

1st Thoracic N.

Intervertebral
Disc

Subscap-
ularis M.

Bile Stained
Pleural Fluid

Visceral Pleura
of Lt. Upper
Lobe

Descending
Aorta

Middle
Esophagus

Lt. Inf.
Pulmonary V.

Lt. Atrium

Distal
Esophagus

Bile-Stained
Ascites

Distal Thoracic
Aorta

17A Coronal/Anatomic Plate/Left Atrium. This plate is through the left atrium and demonstrates right and left pulmonary veins entering it. The descending aorta is seen to the left of the middle portion of the esophagus. A more distal segment of esophagus is seen above the distal thoracic aorta near the bottom of this section. There is bile-stained ascites and left pleural effusion. Intervertebral discs are identified above and below T2 as well as portions of the spinal cord at the T1 level. The major fissure is seen on the right separating upper and lower lobes.

17B Coronal/MRI/Left Atrium. An MRI image in coronal plane shows the left atrium and segments of the aorta, inferior vena cava, and hepatic veins as signal voids. The left pulmonary artery lies above the left main bronchus and the left pulmonary vein.

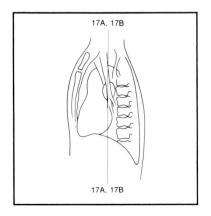

212

17B

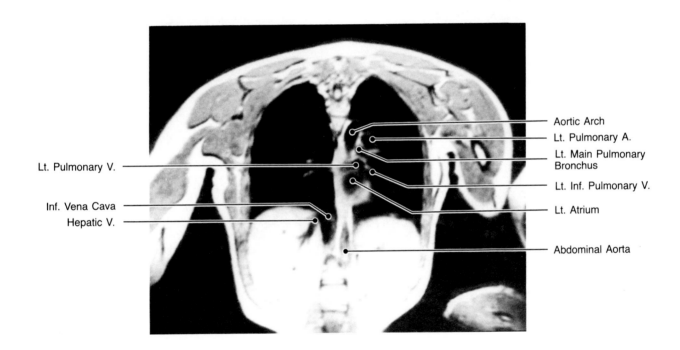

Aortic Arch

Lt. Pulmonary A.

Lt. Main Pulmonary
Bronchus

Lt. Pulmonary V.

Lt. Inf. Pulmonary V.

Inf. Vena Cava

Hepatic V.

Lt. Atrium

Abdominal Aorta

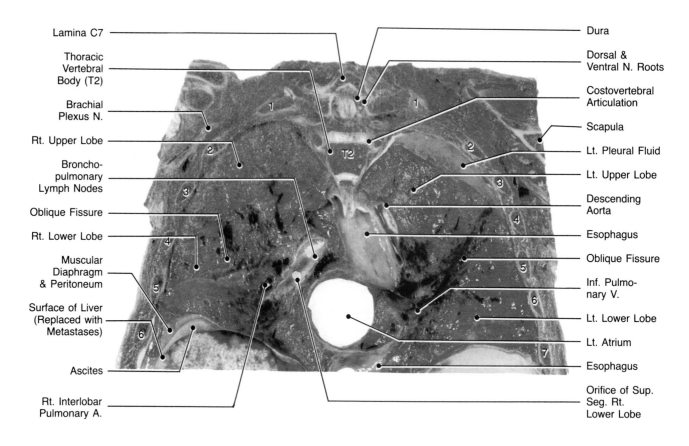

Lamina C7

Thoracic Vertebral Body (T2)

Brachial Plexus N.

Rt. Upper Lobe

Broncho-pulmonary Lymph Nodes

Oblique Fissure

Rt. Lower Lobe

Muscular Diaphragm & Peritoneum

Surface of Liver (Replaced with Metastases)

Ascites

Rt. Interlobar Pulmonary A.

Dura

Dorsal & Ventral N. Roots

Costovertebral Articulation

Scapula

Lt. Pleural Fluid

Lt. Upper Lobe

Descending Aorta

Esophagus

Oblique Fissure

Inf. Pulmonary V.

Lt. Lower Lobe

Lt. Atrium

Esophagus

Orifice of Sup. Seg. Rt. Lower Lobe

18A Coronal/Anatomic Plate/Esophagus. This anatomic plate features the esophagus in the midline with the aorta to the left and the left atrium inferior to it. On the right the bronchus intermedius is seen together with the orifice for the superior segment of the right lower lobe. Just lateral to the airway lie the interlobar pulmonary artery and draining veins. The oblique (major) fissures may be seen on both sides; on the right it runs obliquely superior from the fifth lateral rib towards the right hilum, while on the left it runs from the third intercostal space towards the left infrahilar area.

18B Coronal/MRI/Esophagus and Azygos Vein. The coronal MRI section also demonstrates the esophagus near the esophagogastric junction, where it crosses anterior to the descending thoracic aorta. In the midline is the ascending azygos vein lying just medial to the inferior vena cava. More cephalically, portions of the azygos arch are seen above the right main bronchus.

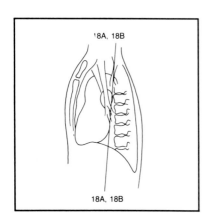

18B

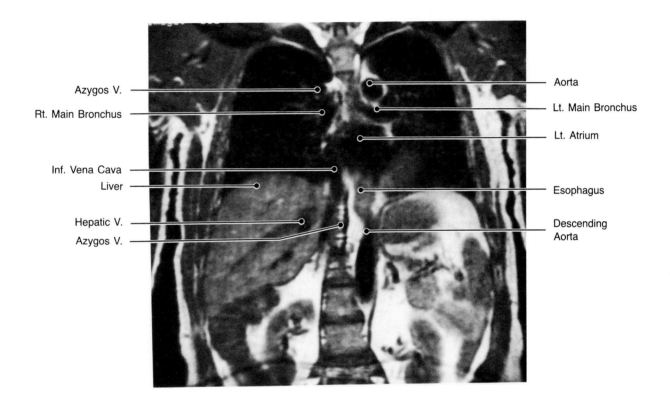

Azygos V.

Rt. Main Bronchus

Inf. Vena Cava

Liver

Hepatic V.

Azygos V.

Aorta

Lt. Main Bronchus

Lt. Atrium

Esophagus

Descending
Aorta

19A

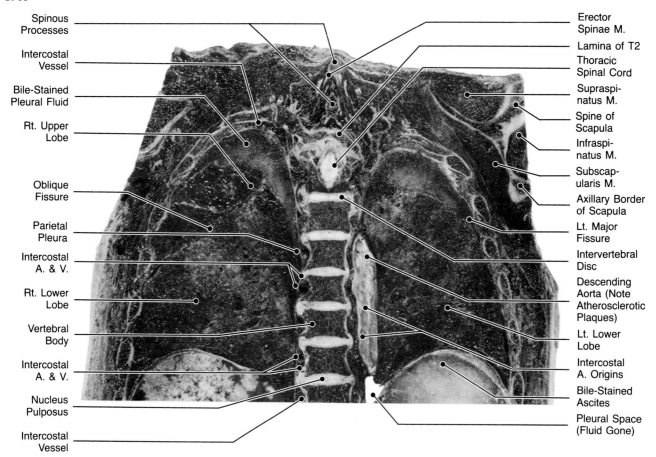

Spinous Processes

Intercostal Vessel

Bile-Stained Pleural Fluid

Rt. Upper Lobe

Oblique Fissure

Parietal Pleura

Intercostal A. & V.

Rt. Lower Lobe

Vertebral Body

Intercostal A. & V.

Nucleus Pulposus

Intercostal Vessel

Erector Spinae M.

Lamina of T2

Thoracic Spinal Cord

Supraspinatus M.

Spine of Scapula

Infraspinatus M.

Subscapularis M.

Axillary Border of Scapula

Lt. Major Fissure

Intervertebral Disc

Descending Aorta (Note Atherosclerotic Plaques)

Lt. Lower Lobe

Intercostal A. Origins

Bile-Stained Ascites

Pleural Space (Fluid Gone)

19A Coronal/Anatomic Plate/Thoracic Spine. This anatomic plate demonstrates the mid thoracic spine from vertebral bodies T4–T9 and their associated interspaces. To the left of the spine is the descending thoracic aorta. Because of dorsal kyphosis at the level of T3 and above we see progressively the spinal cord and more posterior elements, including the lamina and spinous processes of the upper thoracic vertebra.

19B Coronal/MRI/Thoracolumbar Spine. This partial saturation MRI plate demonstrates the descending thoracic aorta to the left of the thoracic spine visualized from about T4 to the lower thoracic and upper lumbar levels. The intervertebral discs and ligamentous structures between the bodies show up as low signal densities like the crus of the diaphragm or the descending aorta. A portion of a middle third of the esophagus is also seen medial to the descending aorta. To the right of the spine at the level of the proximal descending aorta is the arch of the azygos vein. Below the diaphragm the crura of the diaphragm and both kidneys stand out nicely.

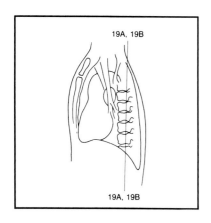

19A, 19B

19A, 19B

19B

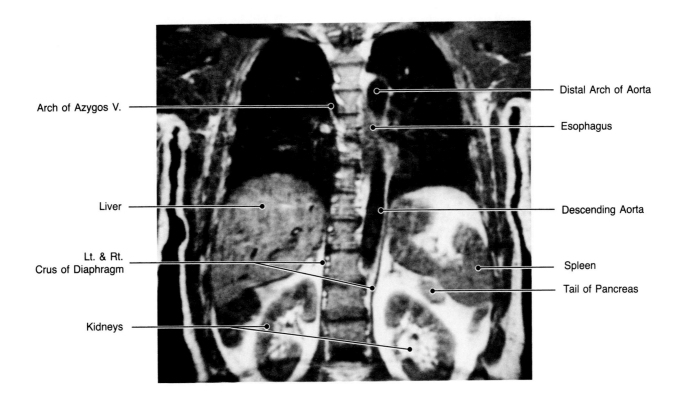

Arch of Azygos V.

Distal Arch of Aorta

Esophagus

Liver

Descending Aorta

Lt. & Rt.
Crus of Diaphragm

Spleen

Tail of Pancreas

Kidneys

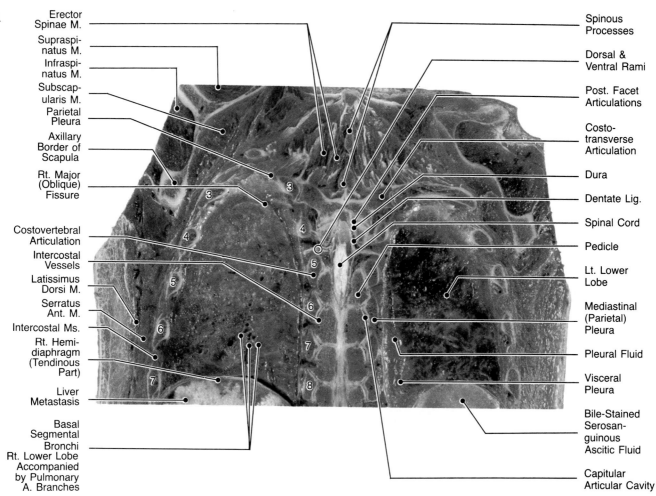

20A

Erector Spinae M.

Supraspi- natus M.

Infraspi- natus M.

Subscap- ularis M.

Parietal Pleura

Axillary Border of Scapula

Rt. Major (Oblique) Fissure

Costovertebral Articulation

Intercostal Vessels

Latissimus Dorsi M.

Serratus Ant. M.

Intercostal Ms.

Rt. Hemi- diaphragm (Tendinous Part)

Liver Metastasis

Basal Segmental Bronchi Rt. Lower Lobe Accompanied by Pulmonary A. Branches

Spinous Processes

Dorsal & Ventral Rami

Post. Facet Articulations

Costo- transverse Articulation

Dura

Dentate Lig.

Spinal Cord

Pedicle

Lt. Lower Lobe

Mediastinal (Parietal) Pleura

Pleural Fluid

Visceral Pleura

Bile-Stained Serosan- guinous Ascitic Fluid

Capitular Articular Cavity

20A Coronal/Anatomic Plate/Spinal Canal. The anatomic section here is principally through the mid thoracic spine at the level of the spinal canal and demonstrates posterior facet articulations, dura, spinal cord, and dorsal as well as ventral thoracic rami, the dentate ligament, the pedicles of the T-spine and costovertebral articulations of ribs 3 through 8. Intercostal vessels may be seen both medially under the origin of the ribs and laterally, where they are protected by the lateral beveled edge of the rib inferiorly.

20B Coronal/MRI/Spinal Canal. This MRI demonstrates the spinal canal and nerves, the pedicles, and at higher levels the more posterior neural arch elements. Below the diaphragm the liver, spleen and retroperitoneal fat around the upper poles of kidneys are seen.

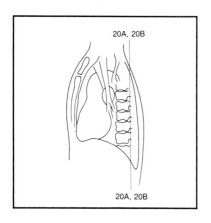

20A, 20B

20A, 20B

20B

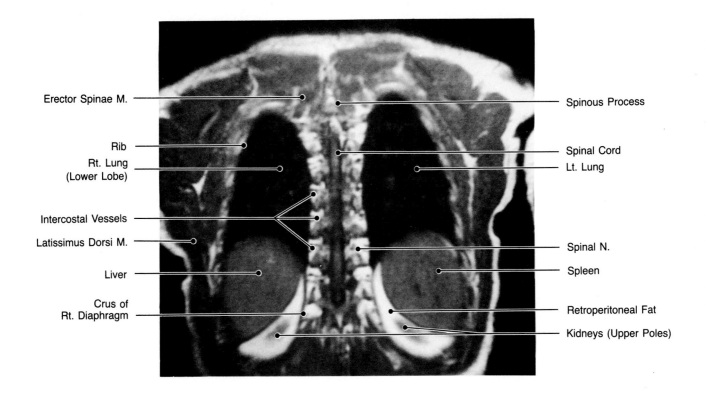

Erector Spinae M.

Rib

Rt. Lung
(Lower Lobe)

Intercostal Vessels

Latissimus Dorsi M.

Liver

Crus of
Rt. Diaphragm

Spinous Process

Spinal Cord

Lt. Lung

Spinal N.

Spleen

Retroperitoneal Fat

Kidneys (Upper Poles)

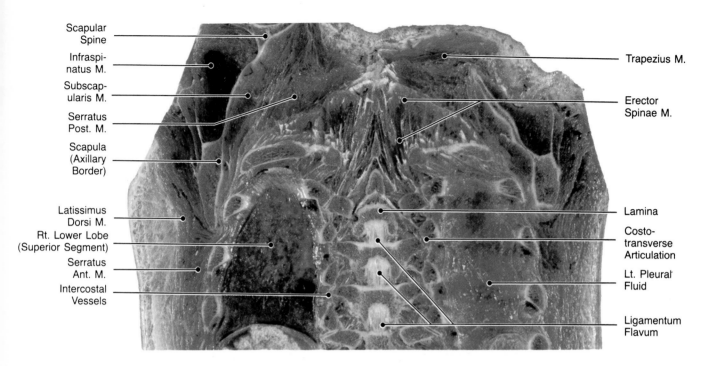

Scapular Spine

Infraspinatus M.

Subscapularis M.

Serratus Post. M.

Scapula (Axillary Border)

Latissimus Dorsi M.

Rt. Lower Lobe (Superior Segment)

Serratus Ant. M.

Intercostal Vessels

Trapezius M.

Erector Spinae M.

Lamina

Costotransverse Articulation

Lt. Pleural Fluid

Ligamentum Flavum

21A Coronal/Anatomic Plate/Posterior Thoracic Cage. This anatomic level shows only a small portion of the most posterior right lung and the pleural fluid in the dependent left pleural space. Between the lamina the prominent yellowish ligamentous structure, the ligamentum flavum, is seen. The posterior medial spinal muscles include the erector spinae group and more superficially the posterior serratus and trapezius muscles. Laterally the deep pectoral musculature around the scapula is demonstrated.

21B Coronal/MRI/Posterior Thoracic Cage. The MRI of the posterior thorax shows some spinal canal and ligaments. The lower lobes, subdiaphragmatic fat and liver, spleen, and right adrenal gland are visualized.

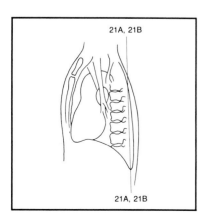

21A, 21B

21A, 21B

21B

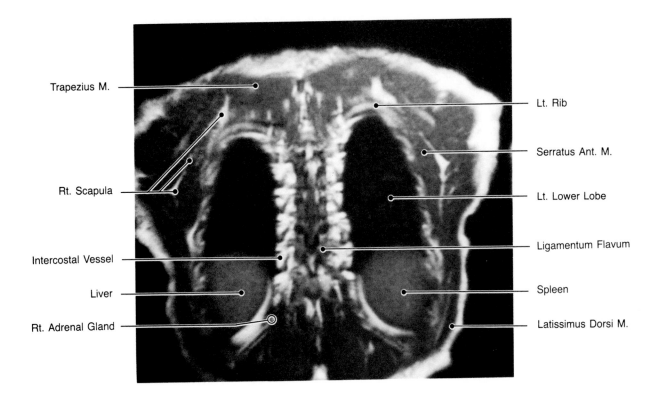

Trapezius M.

Rt. Scapula

Intercostal Vessel

Liver

Rt. Adrenal Gland

Lt. Rib

Serratus Ant. M.

Lt. Lower Lobe

Ligamentum Flavum

Spleen

Latissimus Dorsi M.

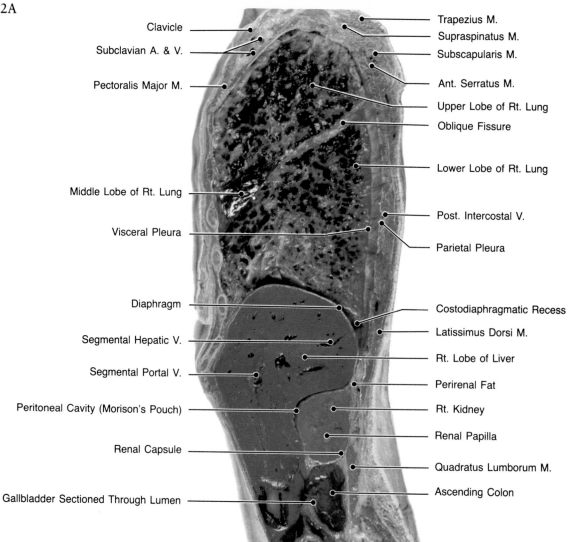

Clavicle

Subclavian A. & V.

Pectoralis Major M.

Middle Lobe of Rt. Lung

Visceral Pleura

Diaphragm

Segmental Hepatic V.

Segmental Portal V.

Peritoneal Cavity (Morison's Pouch)

Renal Capsule

Gallbladder Sectioned Through Lumen

Trapezius M.

Supraspinatus M.

Subscapularis M.

Ant. Serratus M.

Upper Lobe of Rt. Lung

Oblique Fissure

Lower Lobe of Rt. Lung

Post. Intercostal V.

Parietal Pleura

Costodiaphragmatic Recess

Latissimus Dorsi M.

Rt. Lobe of Liver

Perirenal Fat

Rt. Kidney

Renal Papilla

Quadratus Lumborum M.

Ascending Colon

22A Sagittal/Anatomic Plate/Right Lung. This anatomic plate 10 cm to the right of midline demonstrates the right upper lobe separated from the right middle lobe by the minor fissure and the upper and middle lobes separated by the major fissure from the lower lobe. The middle lobe appears small because of the short distance lateral to the right hilum. The right lung shows anthracotic staining. Below the diaphragm this section also includes the right lobe of the liver and Morison's pouch, separating the liver from the kidney. Below the right liver edge are the wall and lumen of the emptied gallbladder.

22B Sagittal/MRI/Right Lung. This sagittal MRI through the thorax and upper abdomen utilizing the "exorcist" for respiratory gating demonstrates a few of the right hilar structures. Flow voids from the moving blood and air within the airways of the right lung are not well discriminated. Flow voids in the abdomen are principally the inferior vena cava and portal venous structures.

22C Sagittal/MRI/Right Lung. This MRI image was obtained with cardiac gating to improve the visualization of the cardiac chambers and great vessels of the right hilar section. Prominent on this image is the inferior vena cava and right atrium but well visualized also are superior pulmonary veins draining into left atrium and the prominent right pulmonary artery lying anterior to the right main bronchial structures.

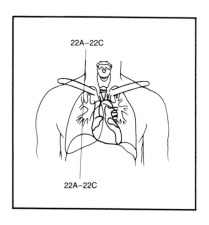

22A–22C

22A–22C

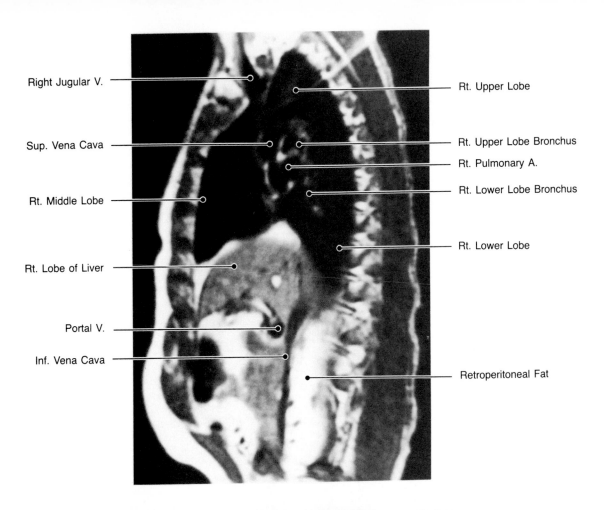

Right Jugular V.

Sup. Vena Cava

Rt. Middle Lobe

Rt. Lobe of Liver

Portal V.

Inf. Vena Cava

Rt. Upper Lobe

Rt. Upper Lobe Bronchus

Rt. Pulmonary A.

Rt. Lower Lobe Bronchus

Rt. Lower Lobe

Retroperitoneal Fat

22C

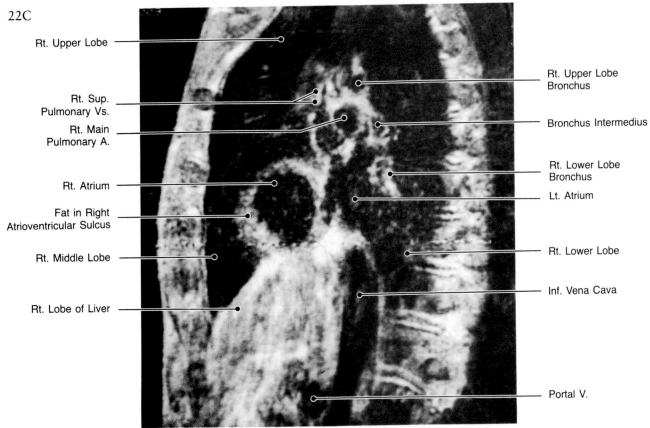

Rt. Upper Lobe

Rt. Sup.
Pulmonary Vs.

Rt. Main
Pulmonary A.

Rt. Atrium

Fat in Right
Atrioventricular Sulcus

Rt. Middle Lobe

Rt. Lobe of Liver

Rt. Upper Lobe
Bronchus

Bronchus Intermedius

Rt. Lower Lobe
Bronchus

Lt. Atrium

Rt. Lower Lobe

Inf. Vena Cava

Portal V.

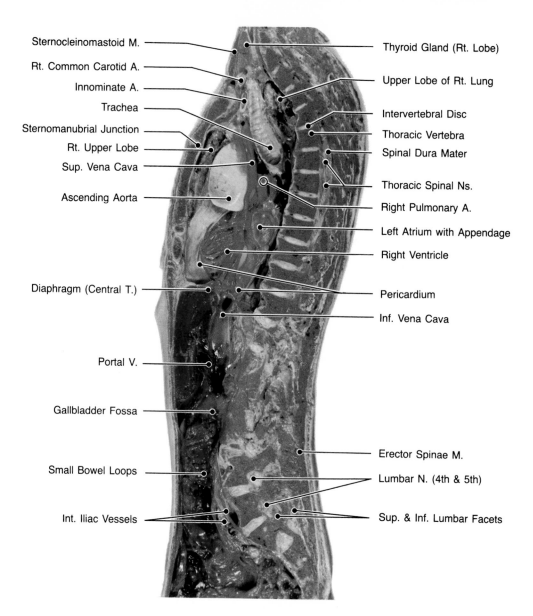

Sternocleinomastoid M. — Thyroid Gland (Rt. Lobe)

Rt. Common Carotid A. — Upper Lobe of Rt. Lung

Innominate A. —

Trachea — Intervertebral Disc

Sternomanubrial Junction — Thoracic Vertebra

Rt. Upper Lobe — Spinal Dura Mater

Sup. Vena Cava —

Ascending Aorta — Thoracic Spinal Ns.

Right Pulmonary A.

Left Atrium with Appendage

Right Ventricle

Diaphragm (Central T.) —

Pericardium

Inf. Vena Cava

Portal V. —

Gallbladder Fossa —

Erector Spinae M.

Small Bowel Loops — Lumbar N. (4th & 5th)

Int. Iliac Vessels — Sup. & Inf. Lumbar Facets

23A Sagittal/Anatomic Plate/Right Paramedial Section. This section to the right of the midline in a slightly scoliotic patient demonstrates the right wall of the trachea and right-sided cardiac and great vessel structures. Mid thoracic levels demonstrate dura mater and spinal nerves because of a slight dextroscoliotic curve. Below the diaphragm the inferior vena cava, portal vein, and fossa of the gallbladder are seen. Slight compensatory scoliosis of the upper lumbar spine is also noted.

23B Sagittal/MRI/Right Paramedial Section. This respiratory gated MRI image at partial saturation technique clearly demonstrates the trachea and portion of the ascending aorta lying anterior to the right pulmonary artery. The spinal cord may be seen within the spinal canal highlighted anteriorly by the posterior longitudinal ligament. Below the diaphragm the liver, portal vein, and gas-filled stomach are identified.

23C Sagittal/MRI/Right Paramedial Section. This MRI cardiac gated image demonstrates the ascending aorta down to the aortic root lying anterior to the left atrium and the right pulmonary artery. Separating the left atrium from the left ventricle is the mitral valve, and the pericardium and epicardial fat may be seen anterior to the right ventricle.

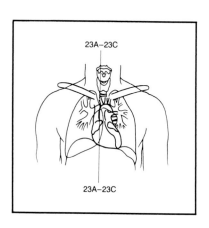

23A–23C

23A–23C

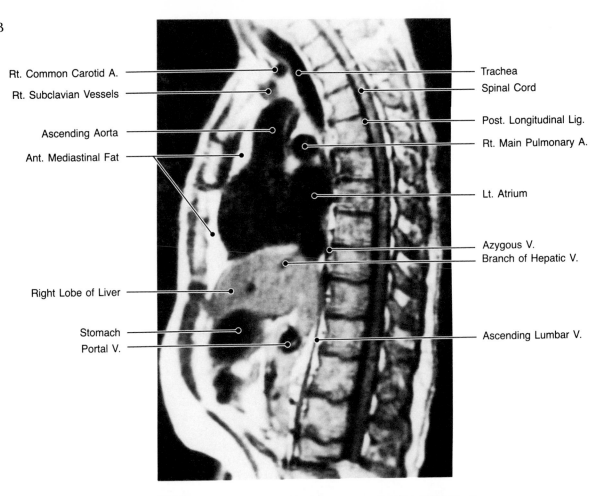

Rt. Common Carotid A. — Trachea

Rt. Subclavian Vessels — Spinal Cord

Ascending Aorta — Post. Longitudinal Lig.

Ant. Mediastinal Fat — Rt. Main Pulmonary A.

— Lt. Atrium

— Azygous V.
— Branch of Hepatic V.

Right Lobe of Liver

Stomach — Ascending Lumbar V.

Portal V.

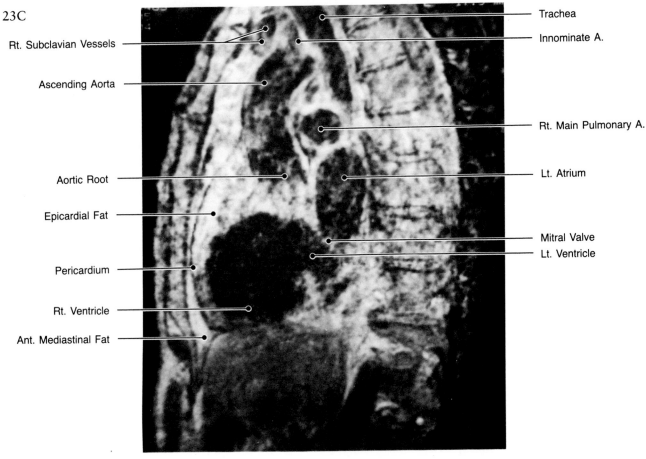

— Trachea

Rt. Subclavian Vessels — Innominate A.

Ascending Aorta

— Rt. Main Pulmonary A.

Aortic Root — Lt. Atrium

Epicardial Fat

Pericardium — Mitral Valve
— Lt. Ventricle

Rt. Ventricle

Ant. Mediastinal Fat

24A

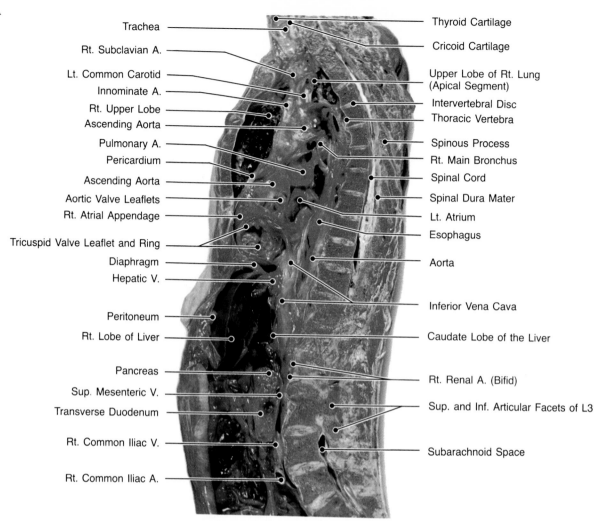

Trachea	Thyroid Cartilage
Rt. Subclavian A.	Cricoid Cartilage
Lt. Common Carotid	Upper Lobe of Rt. Lung (Apical Segment)
Innominate A.	Intervertebral Disc
Rt. Upper Lobe	Thoracic Vertebra
Ascending Aorta	Spinous Process
Pulmonary A.	Rt. Main Bronchus
Pericardium	Spinal Cord
Ascending Aorta	Spinal Dura Mater
Aortic Valve Leaflets	Lt. Atrium
Rt. Atrial Appendage	Esophagus
Tricuspid Valve Leaflet and Ring	Aorta
Diaphragm	
Hepatic V.	Inferior Vena Cava
Peritoneum	Caudate Lobe of the Liver
Rt. Lobe of Liver	
Pancreas	Rt. Renal A. (Bifid)
Sup. Mesenteric V.	Sup. and Inf. Articular Facets of L3
Transverse Duodenum	
Rt. Common Iliac V.	Subarachnoid Space
Rt. Common Iliac A.	

24A Sagittal/Anatomic Plate/Tricuspid and Aortic Valves. This anatomic plate in a patient with slight thoracic dextroscoliosis still remains slightly to the right of midline and demonstrates the atherosclerotic ascending aorta and two of the great vessels above with aortic valve leaflets below and a nearly complete demonstration of the aortic ring lying above the larger tricuspid valve ring and leaflets inferior and anterior to it. Behind the tricuspid valve lie the inferior vena cava and esophagus. Just above left atrium the pulmonary artery on the right is identified just anterior to the esophagus, and above the pulmonary artery is the right main bronchus. Above the top of the aortic arch a small triangle of right apical lung is identified. The diaphragm is incompletely seen because the inferior vena cava is in its intrahepatic portion at the diaphragm. Just ante-

rior to the cava are the quadrate and right lobes of the liver. Because of a levoscoliotic curve in the upper lumbar spine on this section the last vertebral body to be seen in continuity with the thoracic spine is T12 and the first lumbar vertebra to be seen with its body complete is L4.

24B Sagittal/MRI/Tricuspid and Aortic Valves. This respiratory gated MRI study demonstrates superior and inferior vena cava with the azygous vein entering the posterior wall of the superior vena cava at the level of the ascending aorta. The retrosternal signal void is a combination of right upper lobe anterior segment and ascending aorta. Below the diaphragm the liver may be seen separated into its right lobe and its quadrate lobe by the portal vein. Below the liver the duodenum is identified just anterior superior to the pancreas.

24C Sagittal/MRI/Tricuspid and Aortic Valves. This respiratory gated magnetic resonance image more clearly demonstrates the aortic root surrounded anteriorly by the right ventricle, superiorly by the right pulmonary artery, and posteriorly by the left atrium. In this patient none of the superior or inferior vena cava is identified, but the distal aortic arch and descending thoracic aorta rim the posterior heart.

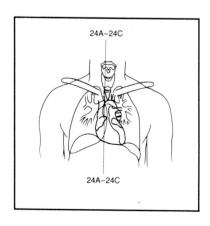

226

Jugular V.

Superior Vena Cava

Ascending Aorta

Rt. Upper Lobe

Rt. Lobe of Liver

Duodenum

Pancreas

Apex of Rt. Lung

Azygos Arch

Rt. Main Bronchus

Rt. Pulmonary A.

Intrahepatic Portion of the Inferior Vena Cava

Quadrate Lobe of Liver

Portal V.

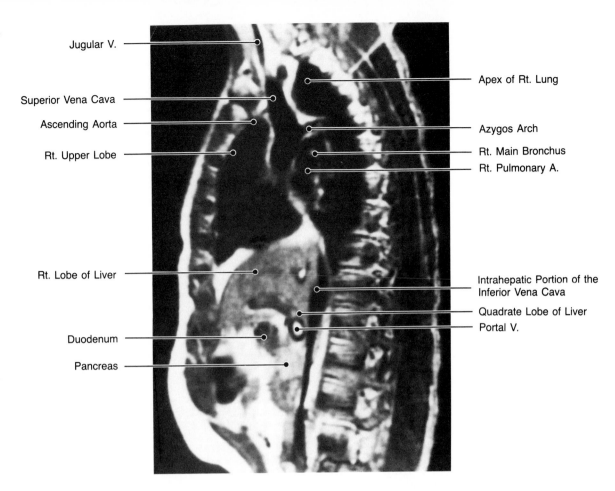

Aortic Arch

Ant. Mediastinal Fat

Main Pulmonary A.

Aortic Root

Rt. Ventricle

Lt. Lobe of Liver

Splenic V.

Pancreas

Stomach

Lt. Upper Lobe

Aorticopulmonary Fat

Lt. Main Bronchus

Lt. Atrium

Descending Aorta

Interventricular Sulcus Fat

Vertebral Vs.

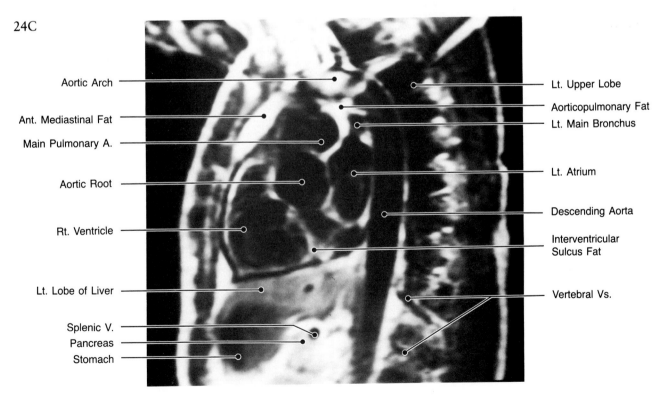

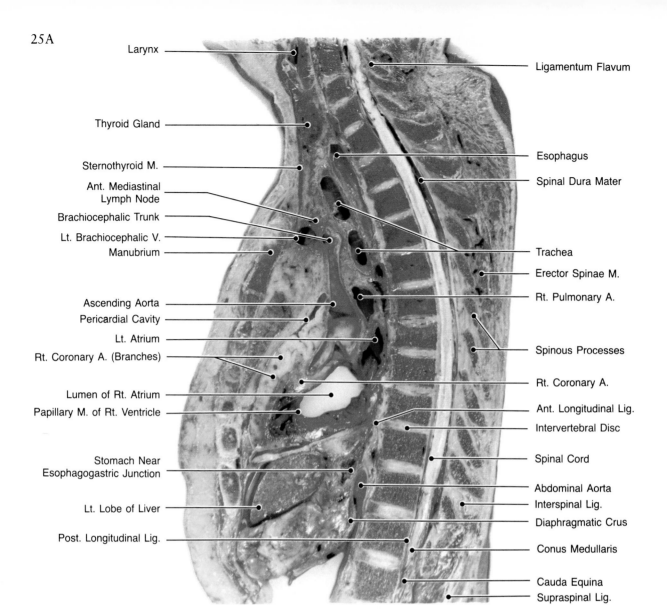

25A

Larynx

Thyroid Gland

Sternothyroid M.

Ant. Mediastinal
Lymph Node

Brachiocephalic Trunk

Lt. Brachiocephalic V.

Manubrium

Ascending Aorta

Pericardial Cavity

Lt. Atrium

Rt. Coronary A. (Branches)

Lumen of Rt. Atrium

Papillary M. of Rt. Ventricle

Stomach Near
Esophagogastric Junction

Lt. Lobe of Liver

Post. Longitudinal Lig.

Ligamentum Flavum

Esophagus

Spinal Dura Mater

Trachea

Erector Spinae M.

Rt. Pulmonary A.

Spinous Processes

Rt. Coronary A.

Ant. Longitudinal Lig.

Intervertebral Disc

Spinal Cord

Abdominal Aorta

Interspinal Lig.

Diaphragmatic Crus

Conus Medullaris

Cauda Equina

Supraspinal Lig.

25A Sagittal/Anatomic Plate/Midline. This mid sagittal section through a nonembalmed cadaver includes the spinal cord from C6 through the conus medullaris and demonstrates both trachea and esophagus at the thoracic inlet with thyroid gland and a portion of larynx and thyroid cartilage. Abundant pericardial fat surrounds the anterior surface of the right ventricle, including the branches of the right coronary artery. The apex of the right ventricle with a portion of the sectioned right papillary muscle lies just anterior to the sectioned lumen of the right atrium. Above this cavity is the ascending aorta anteri-

orly and left atrium posteriorly. Below the diaphragm we see a small portion of the stomach near the esophagogastric junction and a small portion of liver.

25B Sagittal/MRI/Midline. Using respiratory gating this magnetic resonance image also sections the spinal cord from the lower cervical levels to the cauda equina. The trachea is identified at and within the thoracic inlet. Anterior to it lies the great vessels, most prominently the ascending aorta, which lies just anterior to the right main pulmonary artery. The locations for the ventricles and left atrium can also be ascertained but are not well separated from the right lower lobe. Below the diaphragm the left lobe of the liver and the caudate lobe are demonstrated, with the portal vein demonstrating a flow void behind the stomach, which contains gas.

25C Sagittal/MRI/Midline. This cardiac gated MRI better differentiates the cardiac chambers and in particular identifies the aortic and mitral valve levels. Just anterior to the carina above the right pulmonary artery is aorticopulmonary window fat. Epicardial fat may also be seen along the anterior pericardium. Below the diaphragm the celiac trunk and stomach are outlined by signal voids.

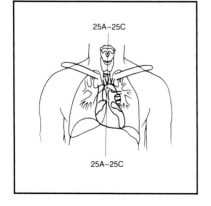

25A–25C

25A–25C

25B

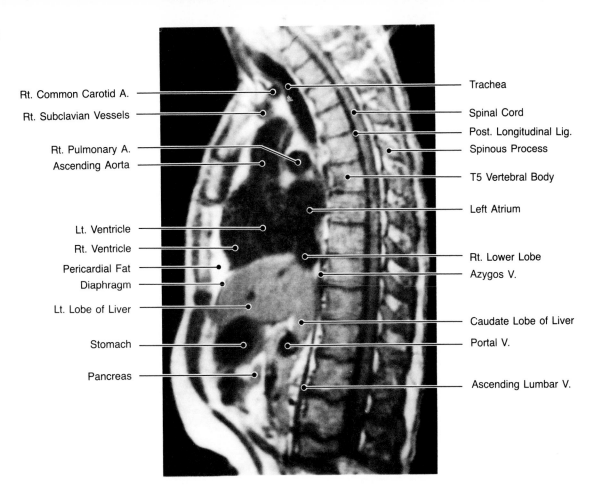

Rt. Common Carotid A.

Rt. Subclavian Vessels

Rt. Pulmonary A.

Ascending Aorta

Lt. Ventricle

Rt. Ventricle

Pericardial Fat

Diaphragm

Lt. Lobe of Liver

Stomach

Pancreas

Trachea

Spinal Cord

Post. Longitudinal Lig.

Spinous Process

T5 Vertebral Body

Left Atrium

Rt. Lower Lobe

Azygos V.

Caudate Lobe of Liver

Portal V.

Ascending Lumbar V.

25C

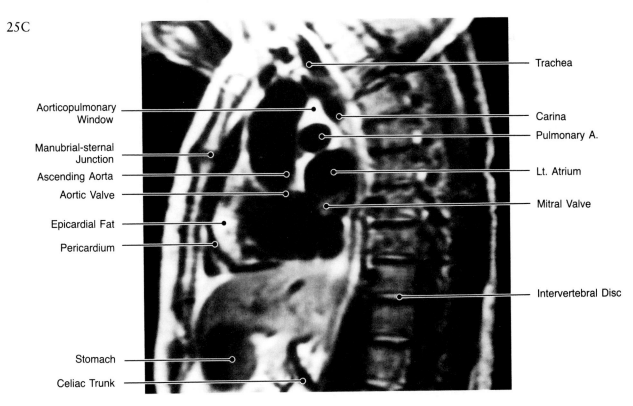

Aorticopulmonary Window

Manubrial-sternal Junction

Ascending Aorta

Aortic Valve

Epicardial Fat

Pericardium

Stomach

Celiac Trunk

Trachea

Carina

Pulmonary A.

Lt. Atrium

Mitral Valve

Intervertebral Disc

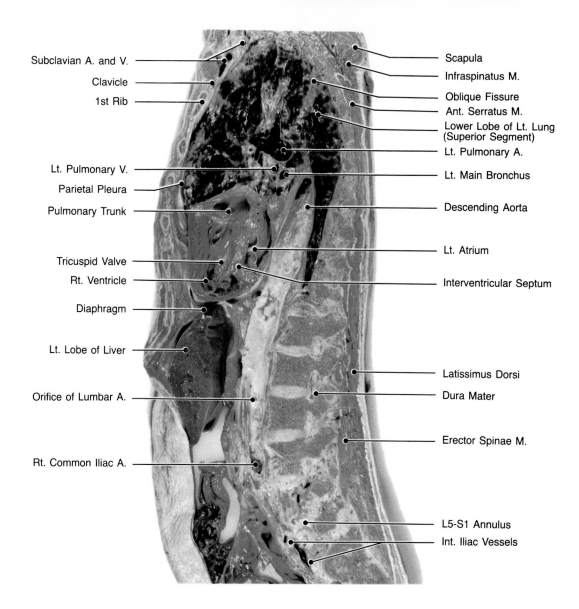

Subclavian A. and V.
Clavicle
1st Rib
Lt. Pulmonary V.
Parietal Pleura
Pulmonary Trunk
Tricuspid Valve
Rt. Ventricle
Diaphragm
Lt. Lobe of Liver
Orifice of Lumbar A.
Rt. Common Iliac A.

Scapula
Infraspinatus M.
Oblique Fissure
Ant. Serratus M.
Lower Lobe of Lt. Lung (Superior Segment)
Lt. Pulmonary A.
Lt. Main Bronchus
Descending Aorta
Lt. Atrium
Interventricular Septum
Latissimus Dorsi
Dura Mater
Erector Spinae M.
L5-S1 Annulus
Int. Iliac Vessels

26A Sagittal/Anatomic Plate/Descending Aorta Level. Because of this patient's levoscoliotic curve in the upper lumbar spine the descending thoracic and upper abdominal aorta lie well to the left of the midline. In the thorax the major fissure is seen to separate left upper and left lower lobes and can be followed inferiorly to the main pulmonary artery on the left and its anterior trunk. Below the left pulmonary artery is the left pulmonary vein; slightly posteriorly and inferior to the vein is the left main bronchus. Within the pericardium the anterior trabeculated right ventricle and the pulmonary trunk can be seen anterior to the interventricular septum. The abdominal aorta is quite atherosclerotic, particularly at the origin of the right common iliac artery displaced off the midline by the scoliosis. The left lobe of the liver is well seen.

26B Sagittal/MRI/Descending Aorta Level. This respiratory gated MRI study at partial saturation technique demonstrates principally the descending thoracic aorta; just within its anterior curvature is the left main bronchus. Anterior to the aorta and left main bronchus is the signal void, consisting of heart, principally the left-sided chambers, and lingula. Behind the aorta the nondescript signal void is of the left lower lobe and corresponds with the anatomic plate. Below the diaphragm the anterior signal void of the stomach is in close approximation with the transverse colon beneath it. Behind the stomach within intermediate density signals is the pancreas with small bowel loops posteroinferior to it.

26C Sagittal/MRI/Descending Aorta Level. This partial saturation MRI done with cardiac gating demonstrates the distal thoracic arch and descending thoracic aorta with pulmonary trunk anterior and inferior to it. Because of the cardiac gating the pulmonary valve and the region of the mitral valve are identified. A small portion of left upper lobe is seen lying anterior to the pericardium of the pulmonary outflow tract.

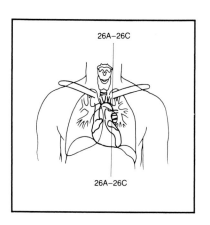

26A–26C

26A–26C

26B

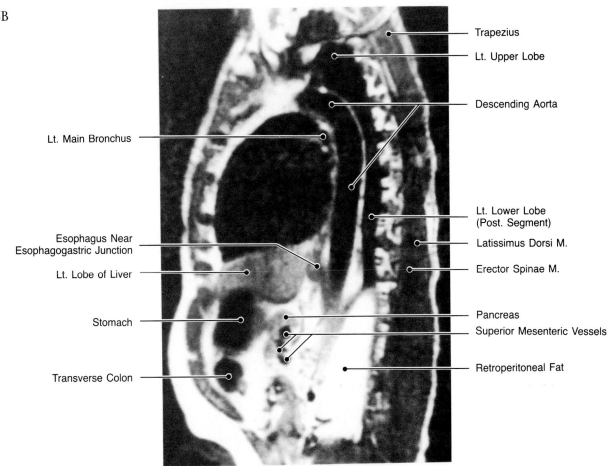

Trapezius

Lt. Upper Lobe

Descending Aorta

Lt. Main Bronchus

Lt. Lower Lobe
(Post. Segment)

Latissimus Dorsi M.

Esophagus Near
Esophagogastric Junction

Erector Spinae M.

Lt. Lobe of Liver

Stomach

Pancreas

Superior Mesenteric Vessels

Transverse Colon

Retroperitoneal Fat

26C

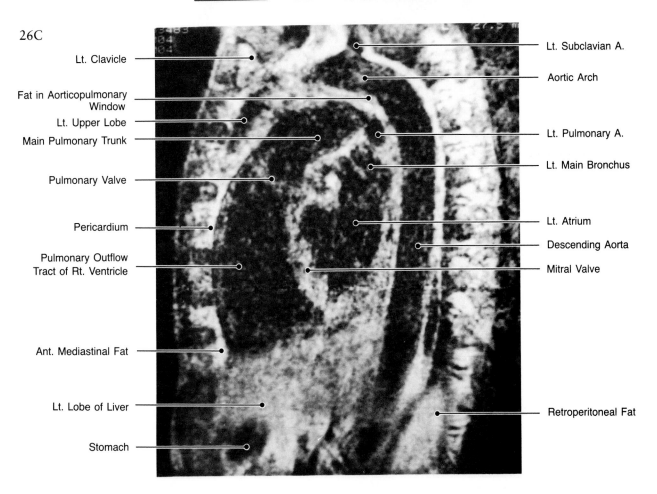

Lt. Clavicle

Lt. Subclavian A.

Fat in Aorticopulmonary
Window

Aortic Arch

Lt. Upper Lobe

Main Pulmonary Trunk

Lt. Pulmonary A.

Pulmonary Valve

Lt. Main Bronchus

Pericardium

Lt. Atrium

Pulmonary Outflow
Tract of Rt. Ventricle

Descending Aorta

Mitral Valve

Ant. Mediastinal Fat

Lt. Lobe of Liver

Retroperitoneal Fat

Stomach

27A

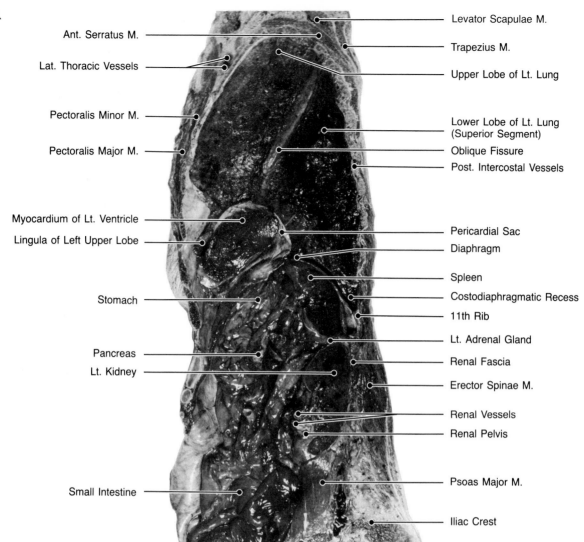

Levator Scapulae M.

Ant. Serratus M.

Trapezius M.

Lat. Thoracic Vessels

Upper Lobe of Lt. Lung

Pectoralis Minor M.

Lower Lobe of Lt. Lung
(Superior Segment)

Pectoralis Major M.

Oblique Fissure

Post. Intercostal Vessels

Myocardium of Lt. Ventricle

Pericardial Sac

Lingula of Left Upper Lobe

Diaphragm

Spleen

Costodiaphragmatic Recess

Stomach

11th Rib

Lt. Adrenal Gland

Pancreas

Renal Fascia

Lt. Kidney

Erector Spinae M.

Renal Vessels

Renal Pelvis

Psoas Major M.

Small Intestine

Iliac Crest

27A Sagittal/Anatomic Plate/Left Lung. This parasagittal section taken 10 cm to the left of the midline demonstrates anthracotic lung of the left upper and left lower lobe separated by the major fissure, which descends towards the diaphragm anteriorly. Before reaching the diaphragm it encounters the pericardium surrounding the left ventricular myocardium, without any left ventricular lumen demonstrated at this level. Below the diaphragm the spleen and left kidney occupy the posterior aspect of the upper abdomen, while the stomach with its characteristic rugae can be seen just anterior to the small portion of retrogastric pancreatic body.

27B Sagittal/MRI/Left Lung. This respiratory gated MRI, slightly closer to the midline than the anatomic plate, demonstrates the left pulmonary artery and left main bronchus with fat above the pulmonary artery in the aorticopulmonary window. The esophagus near the esophagogastric junction lies just posterior to a prominent crus of diaphragm and to the left lobe of the liver. The distal abdominal aorta blends with left common iliac vein posteriorly, while stomach and transverse colon are anterior. Between the vessels and these viscera lies the pancreas and superior mesenteric vein.

27C Sagittal/MRI/Left Lung. This magnetic resonance image with cardiac gating demonstrates to advantage the posterior left ventricular wall, the inferior left ventricular myocardium, the left ventricular lumen, and the more anterior interventricular septum. Below the diaphragm the left lobe of liver lies anterior in the abdomen, while the spleen and its hilar splenic vessels lie posteriorly. Below the splenic structures the left kidney is seen surrounded by retroperitoneal fat.

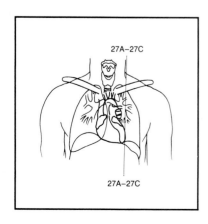

27A–27C

27A–27C

27B

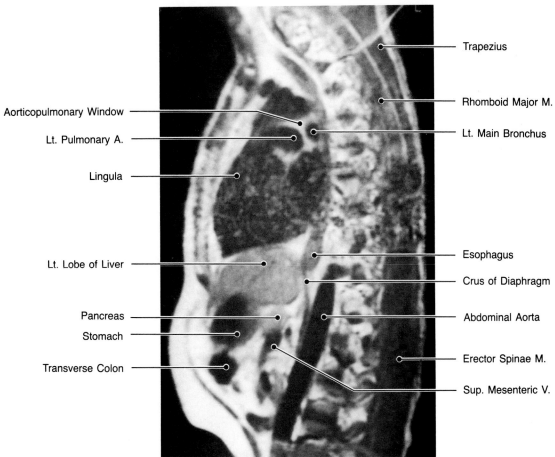

Aorticopulmonary Window

Lt. Pulmonary A.

Lingula

Lt. Lobe of Liver

Pancreas

Stomach

Transverse Colon

Trapezius

Rhomboid Major M.

Lt. Main Bronchus

Esophagus

Crus of Diaphragm

Abdominal Aorta

Erector Spinae M.

Sup. Mesenteric V.

27C

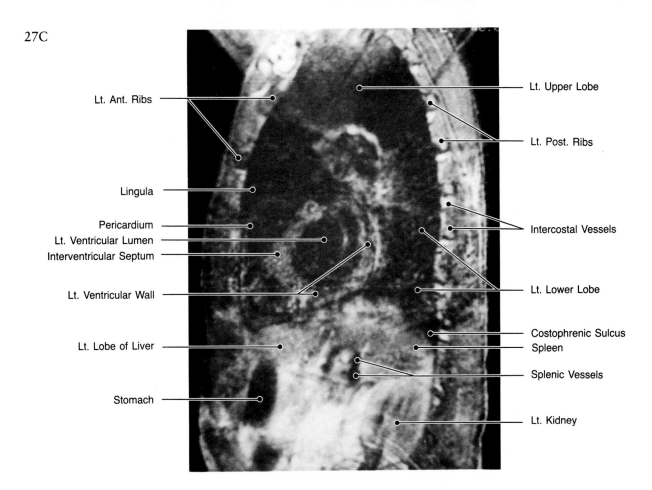

Lt. Ant. Ribs

Lingula

Pericardium

Lt. Ventricular Lumen

Interventricular Septum

Lt. Ventricular Wall

Lt. Lobe of Liver

Stomach

Lt. Upper Lobe

Lt. Post. Ribs

Intercostal Vessels

Lt. Lower Lobe

Costophrenic Sulcus

Spleen

Splenic Vessels

Lt. Kidney

SECTION 4

Abdomen

Charles F. Mueller, M.D.
A. J. Christoforidis, M.D., Ph.D.
John A. Negulesco, Ph.D.

The preparation of the anatomic sections has been described in the general Introduction. All the axial anatomic plates were correlated accurately with computed tomographic sections of the fresh cadaver. Skin markers, placed before the CT examination, at 2-cm intervals, indicated the exact levels to be sectioned.

As expected, post-mortem changes alter the imaging quality without interfering to any significant degree with the important anatomic relationships of the organs.

In this section, as in others, the correlation of the anatomy was made primarily with CT and magnetic resonance images from living individuals. In order to secure accuracy in this correlation, sections from numerous normal clinical examinations were reviewed. The ever present variations due to the difference in body construction of the patients did not compromise the accuracy of the comparative study of the anatomic, computed tomographic, and magnetic resonance imaging. One has to bear in mind also the differences related to functional alterations on the anatomic position of the organs, particularly of the upper abdomen. These factors involve primarily the respiration, resulting in varying degrees of diaphragmatic excursion in different individuals, transmitted mainly to the upper abdominal organs. Intestinal motility is also a significant factor, particularly in the imaging of the small bowel. The volume of the slices of the CT and magnetic resonance images explains minor discrepancies with the photographs of the anatomic plates. The use of contrast medium in the vascular system and in the gastrointestinal tract highlights some of the important anatomic structures in the computed tomographic examinations. Each axial anatomic plate is accompanied by at least one computed tomographic and a corresponding magnetic resonance image. These images were selected, as mentioned above, from different patients. Special effort was made to select the same anatomic level as accurately as possible.

The coronal and sagittal sections are correlated with gated magnetic resonance images. No corresponding CT images are included, as coronal and sagittal imaging is not used for practical purposes in clinical medicine. The reasons are primarily technical. Computer-assisted reconstruction, here as well as in other parts of the body, has limited utility.

As in the imaging presented in other sections, the CT examinations of the patients were performed with fourth-generation 1440 or 2060Q Technicare units. For the cadaver CT sections an 8800 GE unit was used.

All the magnetic resonance examinations were performed with a 1.5 Tesla unit. The factors used in these images were mainly partial saturation with TR varying from 400 to 2000 msec and with a TE ranging from 20 to 80 msec.

A small number of interesting pathologic cases are included in this section. These include a case of chronic pancreatitis with pancreatic pseudocyst, a cirrhotic liver and ampullary carcinoma with extrahepatic and intrahepatic bile duct dilatation, fatty liver, and a case of lymphoma with splenomegaly, retrocrural lymphadenopathy, and ascites, among others.

Duplication of anatomy was avoided by omitting some of the lower abdominal sections demonstrating mainly intestinal loops and omentum.

Bibliography

Bo WJ, Meschan I, Krueger, WA: *Basic Atlas of Cross-Sectional Anatomy.* Philadelphia, W.B. Saunders Co., 1980.

Buonocore E, Borkowski GP, Pavlicek W, et al: NMR imaging of the abdomen technical considerations: *AJR* 140:1217, 1983.

Edelman RR, McFarland E, Stark DD, Ferrucci JR, Jr, Simeone JF, Wismer G, White EM, Rosen BR, Brady TJ: Surface coil MR imaging of abdominal viscera. I. Theory, technique and initial results. *AJR* 157:425, 1985.

Federle MP: Acute abdomen: Computed tomography. *Radiology* 5:307, 1985.

Haaga JR, Alfidi RJ (eds): *Computed Tomography of the Whole Body.* St. Louis, C.V. Mosby, 1983.

Koritké JG, Sick H: *Atlas of Sectional Human Anatomy.* Munich, Urban and Schwarzenberg, 1983.

Love L, Meyers MA, Churchill RJ, et al: Pictorial essay: Computed tomography of extraperitoneal spaces. *AJR* 136:781, 1981.

Moss AA, Gamsu G, Genant HK: *Computed Tomography of the Body.* Philadelphia, W.B. Saunders Co., 1983.

Pernkopf E: *Atlas of Topographical and Applied Human Anatomy.* 2nd ed. Philadelphia, W.B. Saunders Co., 1964.

Wagner M, Lawson T: *Segmental Anatomy: Applications to Clinical Medicine.* New York, Macmillan, 1982.

Zylak CJ, Pallie W: Correlative anatomy and computed tomography: a module on the pancreas and posterior abdominal wall. *Radiology* 1:61, 1981.

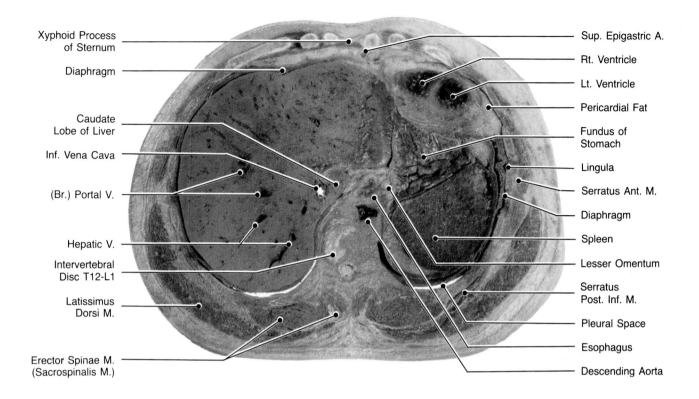

1A Axial/Anatomic Plate/Liver. This anatomic plate is made through a mid portion of the liver and demonstrates the caudate lobe and the intrahepatic inferior vena cava. This level, about 1 cm caudad to anatomic plate 12A of Section 3, where the hepatic veins enter the cava, demonstrates both hepatic and portal veins. This lower section, through the heart, includes interventricular septum, both ventricles, and the expected amount of pericardial fat.

1B Axial/CT/Liver. The cadaver CT plate 1B shows fluid in the pleural spaces, which was not retained at the time the abdomen was sectioned and hence is not seen on the corresponding anatomic plate. Note is made of the faint increased density in the liver from the inferior vena cava because of the clotted blood that is within it.

1C Axial/CT/Liver and Esophagogastric Junction. Additional CT plate 1C from a living patient demonstrates contrast within the stomach and also the esophagogastric junction. In the spleen are rounded calcifications due to granulomatous disease that is now healed. Because contrast has been added intravenously the azygous vein stands out surrounded by retrocrural fat, while the vascular structures are isodense with the liver.

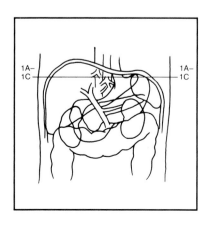

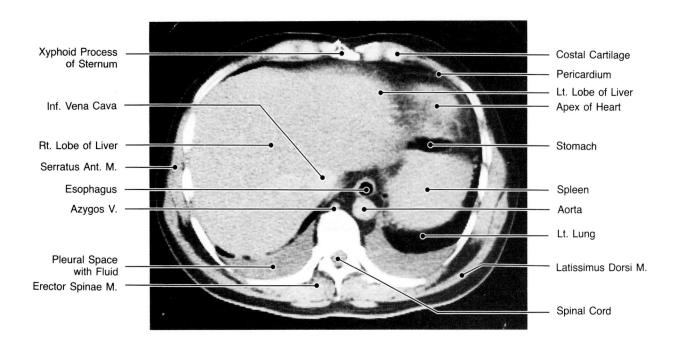

Xyphoid Process of Sternum — Costal Cartilage — Pericardium — Lt. Lobe of Liver — Apex of Heart — Inf. Vena Cava — Rt. Lobe of Liver — Stomach — Serratus Ant. M. — Esophagus — Spleen — Azygos V. — Aorta — Lt. Lung — Pleural Space with Fluid — Erector Spinae M. — Latissimus Dorsi M. — Spinal Cord

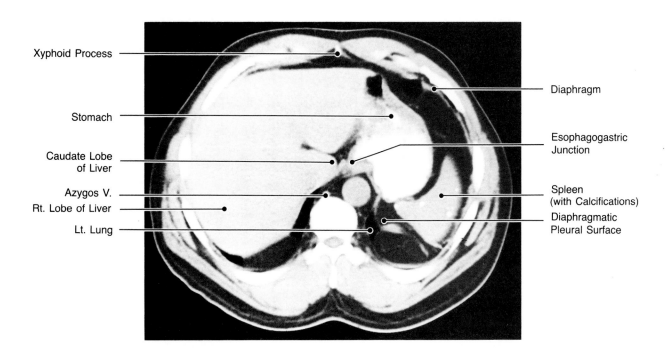

Xyphoid Process — Diaphragm — Stomach — Esophagogastric Junction — Caudate Lobe of Liver — Azygos V. — Spleen (with Calcifications) — Rt. Lobe of Liver — Lt. Lung — Diaphragmatic Pleural Surface

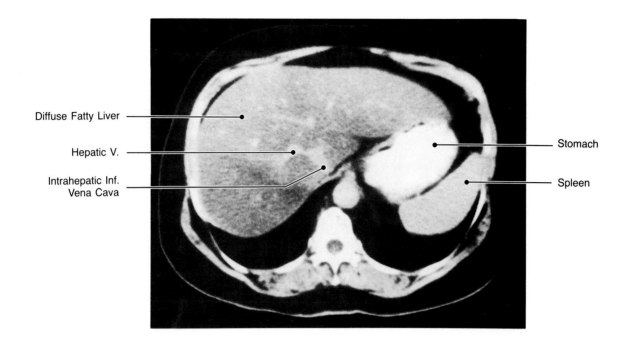

Diffuse Fatty Liver

Hepatic V.

Intrahepatic Inf.
Vena Cava

Stomach

Spleen

1D Axial/CT/Fatty Liver Demonstration. Plate 1D on a different patient demonstrates reversal of liver and spleen densities because this patient has a fatty liver. Contrast enhancement still further outlines the intrahepatic portal and hepatic venous systems, the inferior vena cava, and the aorta.

1E Axial/CT/Intrahepatic Biliary Dilitation. Plate 1E demonstrates abnormally dilated intrahepatic bile ducts as hypodense structures on the background of a normally opacifying liver in a patient with a distal common bile duct carcinoma and obstructed intrahepatic radicles.

1F Axial/MRI/Liver. Plate 1F is a T1 weighted spin echo magnetic resonance image in a respiratory gated study to demonstrate the liver with the intrahepatic flow voids of portal vein and inferior vena cava; fat surrounds the falciform ligament and splenic vessels. Note how fatty structures such as subcutaneous fat and fat in the falciform ligament region provide contrast for the muscles and vessels.

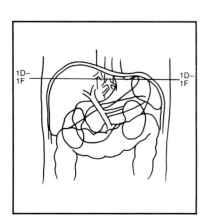

1E

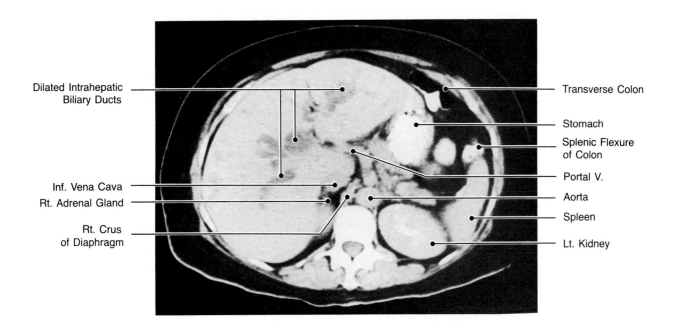

Dilated Intrahepatic Biliary Ducts

Inf. Vena Cava

Rt. Adrenal Gland

Rt. Crus of Diaphragm

Transverse Colon

Stomach

Splenic Flexure of Colon

Portal V.

Aorta

Spleen

Lt. Kidney

1F

Lt. Lobe of Liver

Fat & Falciform Lig.

Inf. Vena Cava

Portal V.

Rt. Lobe of Liver

Azygos V.

Vertebral Body

Intercostal Fat

Rib

Erector Spinae M.

Lamina

Rectus Abdominis M.

Stomach (Fundus)

Omentum

Splenic Flexure of Colon

Aorta

Splenic V.

Splenic Hilus

Splenic A.

Latissimus Dorsi M.

Spleen

Lt. Crus of Diaphragm

Multifidus M.

Spinal Canal

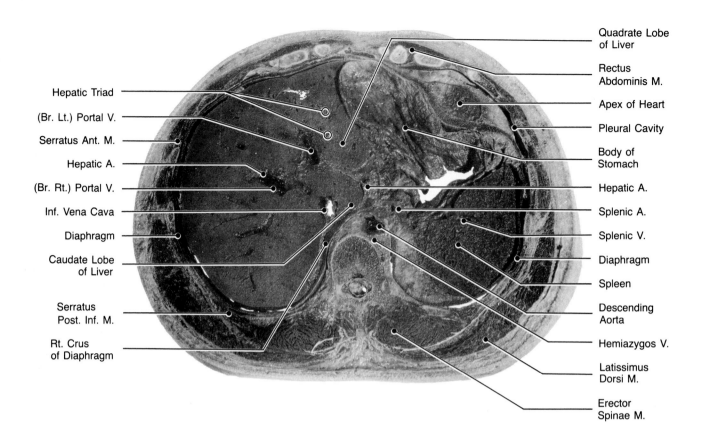

Quadrate Lobe of Liver

Rectus Abdominis M.

Apex of Heart

Pleural Cavity

Body of Stomach

Hepatic A.

Splenic A.

Splenic V.

Diaphragm

Spleen

Descending Aorta

Hemiazygos V.

Latissimus Dorsi M.

Erector Spinae M.

Hepatic Triad

(Br. Lt.) Portal V.

Serratus Ant. M.

Hepatic A.

(Br. Rt.) Portal V.

Inf. Vena Cava

Diaphragm

Caudate Lobe of Liver

Serratus Post. Inf. M.

Rt. Crus of Diaphragm

2A Axial/Anatomic Plate/Splenic and Hepatic Artery. Anatomic plate 2A demonstrates the two major branches of the celiac axis from the aorta, namely the splenic artery and the hepatic artery. In the liver to the left of the axis of the inferior vena cava and portal vein lie the quadrate (anteriorly) and caudate (posteromedially) lobes. In several areas the hepatic triad, consisting of an intrahepatic bile duct, portal vein branch, and hepatic arterial branch can be seen. This is the last anatomic level to show the heart and pericardial fat.

2B Axial /CT/Celiac Axis. The cadaver CT plate 2B shows the origin of the celiac axis and the relationships of the aorta, the crus of the diaphragm, and the hemiazygous vein. Portal venous structures within the liver and the intrahepatic portion of the inferior vena cava appear as slightly higher density structures on the background of the nonenhanced liver. Note that without contrast the liver is of higher density than the spleen.

2C Axial/CT/Fatty Liver. CT plate 2C is from the same patient as plate 1D and demonstrates at this level the portal vein entering the fatty liver. This plate also demonstrates the left renal vein entering the inferior vena cava and the relationship of the superior mesenteric vein and artery to the portal vein. The fatty liver causes the reversal in densities that are normally encountered between liver and spleen.

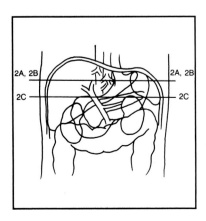

2B

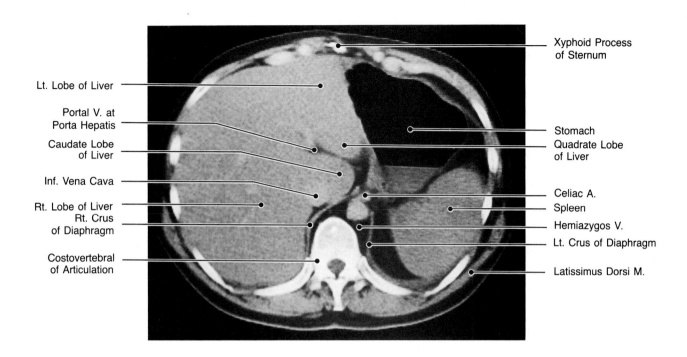

Xyphoid Process of Sternum

Lt. Lobe of Liver

Portal V. at Porta Hepatis

Caudate Lobe of Liver

Inf. Vena Cava

Rt. Lobe of Liver
Rt. Crus of Diaphragm

Costovertebral of Articulation

Stomach
Quadrate Lobe of Liver

Celiac A.
Spleen
Hemiazygos V.
Lt. Crus of Diaphragm

Latissimus Dorsi M.

2C

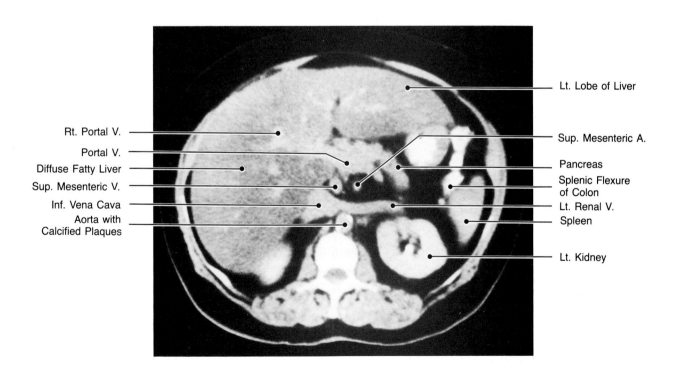

Rt. Portal V.

Portal V.

Diffuse Fatty Liver

Sup. Mesenteric V.

Inf. Vena Cava

Aorta with Calcified Plaques

Lt. Lobe of Liver

Sup. Mesenteric A.

Pancreas

Splenic Flexure of Colon

Lt. Renal V.

Spleen

Lt. Kidney

2D

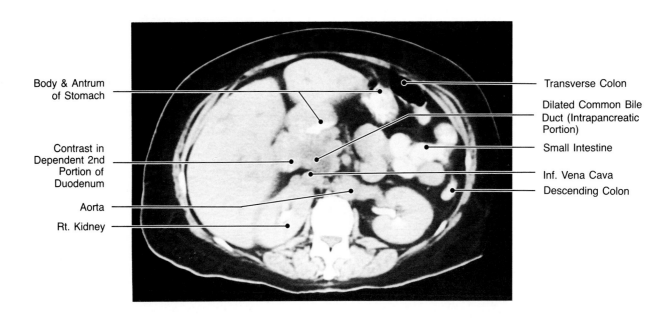

Body & Antrum
of Stomach

Contrast in
Dependent 2nd
Portion of
Duodenum

Aorta

Rt. Kidney

Transverse Colon

Dilated Common Bile
Duct (Intrapancreatic
Portion)

Small Intestine

Inf. Vena Cava

Descending Colon

2D Axial/CT/Biliary Ductal Dilitation. CT plate 2D, from the same patient as plate 1E, demonstrates the dilated common bile duct as a hypodense structures in this patient with periampullary carcinoma.

2E Axial/MRI/Splenic and Portal Vessels. Plate 2E, from the same patient as 1F, is a respiratory gated T1 weighted spin echo magnetic resonance image demonstrating flow voids in the aorta, inferior vena cava, splenic vessels, and portal vein.

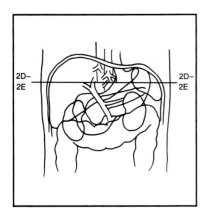

2D–
2E

2D–
2E

2E

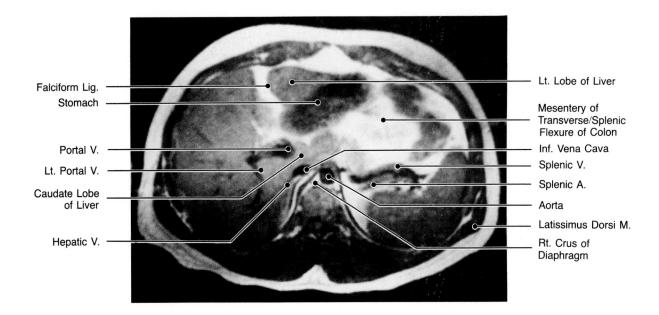

Falciform Lig.

Stomach

Portal V.

Lt. Portal V.

Caudate Lobe
of Liver

Hepatic V.

Lt. Lobe of Liver

Mesentery of
Transverse/Splenic
Flexure of Colon

Inf. Vena Cava

Splenic V.

Splenic A.

Aorta

Latissimus Dorsi M.

Rt. Crus of
Diaphragm

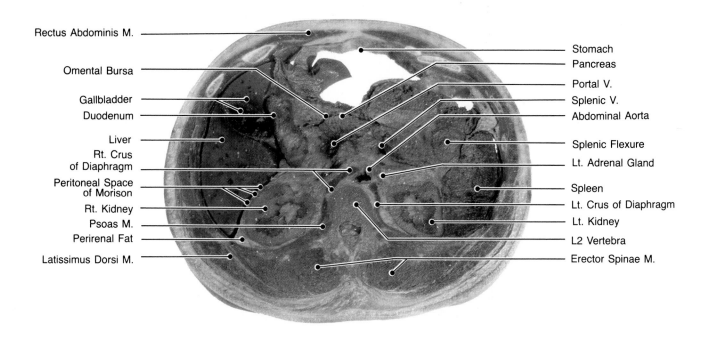

Rectus Abdominis M.

Omental Bursa

Gallbladder

Duodenum

Liver

Rt. Crus
of Diaphragm

Peritoneal Space
of Morison

Rt. Kidney

Psoas M.

Perirenal Fat

Latissimus Dorsi M.

Stomach

Pancreas

Portal V.

Splenic V.

Abdominal Aorta

Splenic Flexure

Lt. Adrenal Gland

Spleen

Lt. Crus of Diaphragm

Lt. Kidney

L2 Vertebra

Erector Spinae M.

3A Axial/Anatomic Plate/Duodenal Level. Anatomic plate 3A shows the duodenum lying medial to the liver and the gallbladder behind the distal stomach; here it appears to extend from its intraperitoneal to its retroperitoneal location. Lesser omentum and the slit-like omental bursa separate the back wall of the stomach from the pancreas in the midline. Just behind the pancreas to the right of the midline is the large portal vein and to the left of the midline is the splenic vein. The spleen and splenic flexure of the colon lie to the left and posterior to the stomach. A recess of peritoneum covering the posterior aspect of the liver and the anterior perirenal fascia is called Morison's pouch. This is the most dependent portion of the peritoneal space in the upper abdomen.

3B Axial/CT/Duodenal Level. The cadaver CT plate 3B at the same level as the anatomic plate demonstrates an air fluid level in the proximal duodenum just medial to the gallbladder.

3C Axial/CT/Lymphoma Retrocrural Level. Plate 3C is from a different patient with lymphoma, retrocrural lymphadenopathy, and ascites. The ascites surrounds the spleen and liver in the greater peritoneal space and enters the potential omental bursa, filling the lesser sac via the foramen of Winslow.

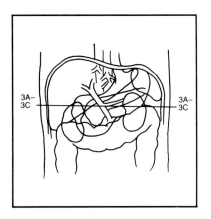

3A–
3C

3A–
3C

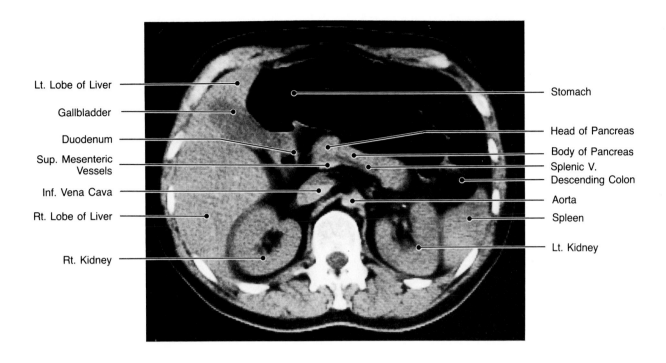

Lt. Lobe of Liver
Gallbladder
Duodenum
Sup. Mesenteric Vessels
Inf. Vena Cava
Rt. Lobe of Liver
Rt. Kidney

Stomach
Head of Pancreas
Body of Pancreas
Splenic V.
Descending Colon
Aorta
Spleen
Lt. Kidney

3C

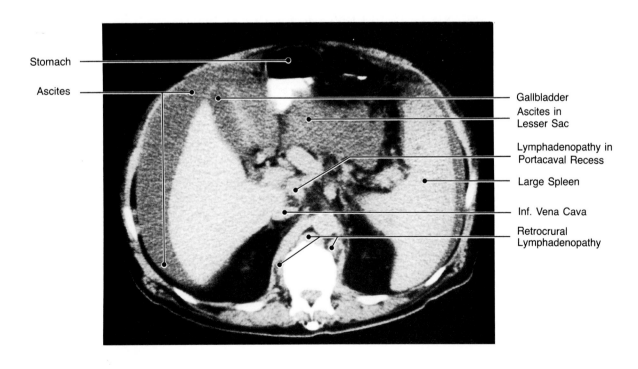

Stomach
Ascites

Gallbladder
Ascites in Lesser Sac
Lymphadenopathy in Portacaval Recess
Large Spleen
Inf. Vena Cava
Retrocrural Lymphadenopathy

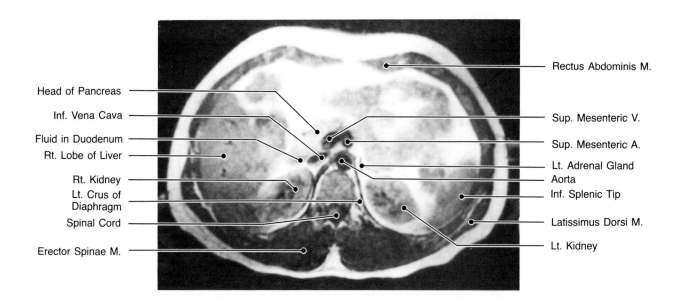

Head of Pancreas

Inf. Vena Cava

Fluid in Duodenum

Rt. Lobe of Liver

Rt. Kidney

Lt. Crus of Diaphragm

Spinal Cord

Erector Spinae M.

Rectus Abdominis M.

Sup. Mesenteric V.

Sup. Mesenteric A.

Lt. Adrenal Gland

Aorta

Inf. Splenic Tip

Latissimus Dorsi M.

Lt. Kidney

3D; 3E Axial/MRI/Duodenal Level. Plates 3D and 3E are T1 weighted spin echo images from a normal volunteer for respiratory gated magnetic resonance of the upper abdomen. Moving blood, which returns no signal, results in the signal voids of the aorta, mesenteric vessels, and inferior vena cava. Fluid identifies the descending duodenum, and just medial to this is the pancreas. At L1-L2 level the cranial and caudal branches of a cephalic loop of superior mesenteric artery can be seen. Intrinsic motion in the small bowel largely blurs the definition of this structure.

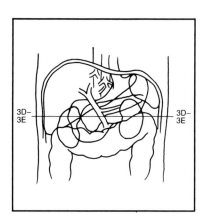

3D–3E

3D–3E

3E

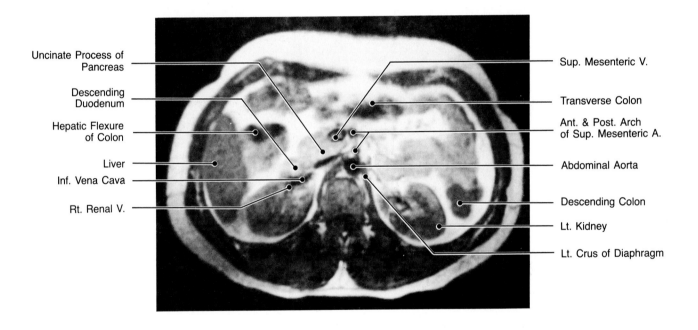

Uncinate Process of Pancreas

Descending Duodenum

Hepatic Flexure of Colon

Liver

Inf. Vena Cava

Rt. Renal V.

Sup. Mesenteric V.

Transverse Colon

Ant. & Post. Arch of Sup. Mesenteric A.

Abdominal Aorta

Descending Colon

Lt. Kidney

Lt. Crus of Diaphragm

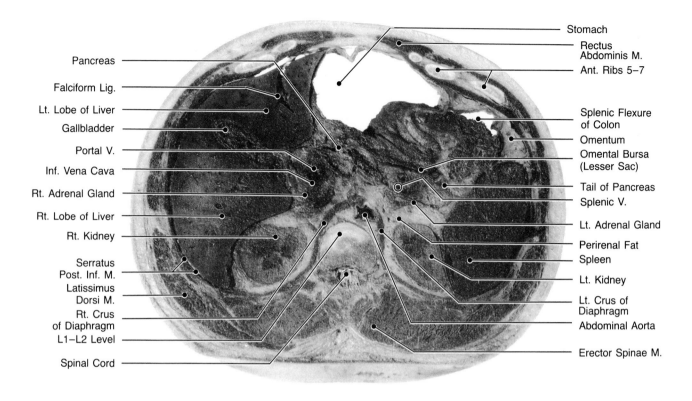

Pancreas

Falciform Lig.

Lt. Lobe of Liver

Gallbladder

Portal V.

Inf. Vena Cava

Rt. Adrenal Gland

Rt. Lobe of Liver

Rt. Kidney

Serratus
Post. Inf. M.
Latissimus
Dorsi M.
Rt. Crus
of Diaphragm
L1–L2 Level

Spinal Cord

Stomach

Rectus
Abdominis M.

Ant. Ribs 5–7

Splenic Flexure
of Colon

Omentum

Omental Bursa
(Lesser Sac)

Tail of Pancreas

Splenic V.

Lt. Adrenal Gland

Perirenal Fat

Spleen

Lt. Kidney

Lt. Crus of
Diaphragm

Abdominal Aorta

Erector Spinae M.

4A Axial/Anatomic Plate/Pancreas, Upper Renal Levels, and Adrenal Glands. Anatomic plate 4A shows kidney parenchyma differentiating intrarenal pyramids and more peripheral cortical medullary parenchyma. At this level the gallbladder fossa and gallbladder are seen as well as some residual falciform ligament in the left lobe of the liver. Portions of the right and left adrenal glands may be seen in the anterior perirenal fat. The right crus of the diaphragm is significantly larger than the left and overrides the anterior portion of the right wall of the aorta, hence separating aorta from cava. A portion of the pancreatic tail lies near the splenic hilus, and on its medial aspect a portion of the splenic vein can be seen. Just anterior to the pancreatic tissue is the slit-like omental bursa. The extrahepatic portal vein is the dominant structure lying just anterior to the inferior vena cava and medial to the gallbladder neck.

4B Axial/CT/Portocaval Recess. The cadaver CT plate 4B demonstrates a slight increase in density in the inferior vena cava and portal vein with a segment of hepatic artery lying in the portocaval recess. More of the pancreas can be seen just anterior to the portal and splenic veins, and a splenic artery calcification is present.

4C Axial/CT/Adrenal Glands. Plate 4C demonstrates the adrenal glands bilaterally. The right adrenal is clearly suprarenal, while the left adrenal lies anterior to the left kidney in its upper pole.

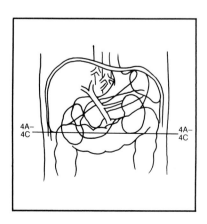

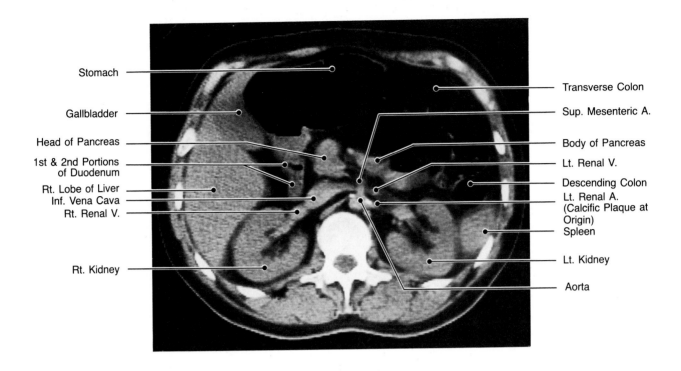

Stomach

Gallbladder

Head of Pancreas

1st & 2nd Portions
of Duodenum

Rt. Lobe of Liver
Inf. Vena Cava
Rt. Renal V.

Rt. Kidney

Transverse Colon

Sup. Mesenteric A.

Body of Pancreas

Lt. Renal V.

Descending Colon

Lt. Renal A.
(Calcific Plaque at
Origin)
Spleen

Lt. Kidney

Aorta

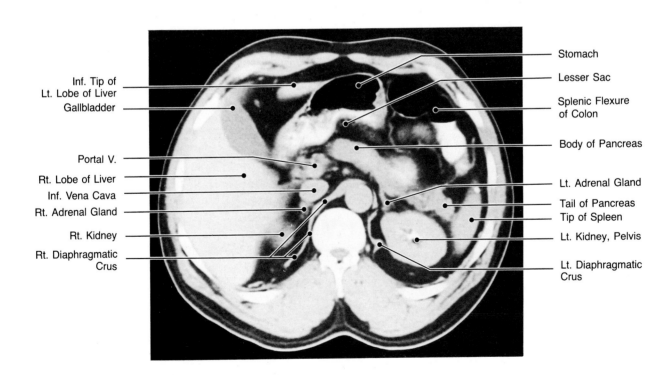

Inf. Tip of
Lt. Lobe of Liver

Gallbladder

Portal V.

Rt. Lobe of Liver

Inf. Vena Cava

Rt. Adrenal Gland

Rt. Kidney

Rt. Diaphragmatic
Crus

Stomach

Lesser Sac

Splenic Flexure
of Colon

Body of Pancreas

Lt. Adrenal Gland

Tail of Pancreas
Tip of Spleen

Lt. Kidney, Pelvis

Lt. Diaphragmatic
Crus

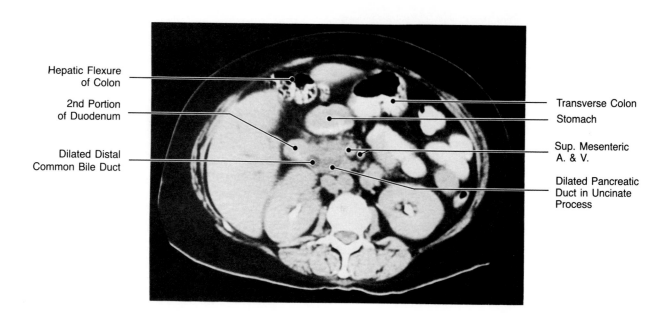

Hepatic Flexure of Colon

2nd Portion of Duodenum

Dilated Distal Common Bile Duct

Transverse Colon

Stomach

Sup. Mesenteric A. & V.

Dilated Pancreatic Duct in Uncinate Process

5D, 5E Axial/CT/Renal Vein. These are additional lower axial plates in the patient (same as Figs. 1E, 2D) who demonstrated intrahepatic biliary radicle dilatation and marked common bile duct dilatation. D demonstrates the dilated distal common bile duct and the double duct sign of the adjacent dilated pancreatic duct in the uncinate process secondary to obstruction from a periampullary carcinoma, which may be seen on E in the duodenum.

5F Axial/MRI/Renal Vein. This magnetic resonance image was made with respiratory gating using a T1 weighted spin echo sequence. Just to the right of the superior mesenteric artery and vein lies the head and uncinate process of the pancreas. Note the slitlike inferior vena cava in this normal patient.

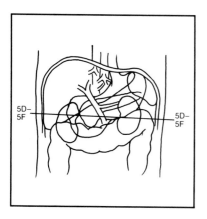

6B

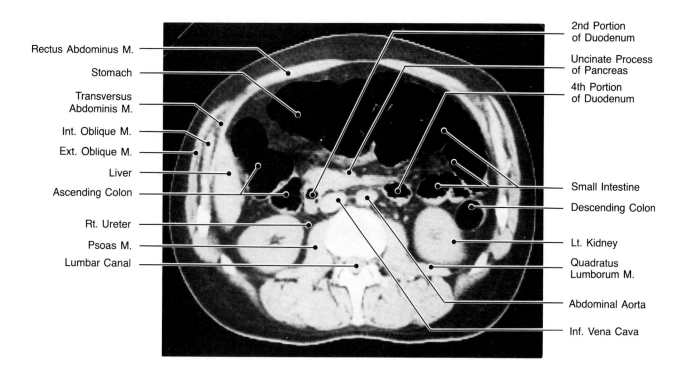

Rectus Abdominus M.

Stomach

Transversus Abdominis M.

Int. Oblique M.

Ext. Oblique M.

Liver

Ascending Colon

Rt. Ureter

Psoas M.

Lumbar Canal

2nd Portion of Duodenum

Uncinate Process of Pancreas

4th Portion of Duodenum

Small Intestine

Descending Colon

Lt. Kidney

Quadratus Lumborum M.

Abdominal Aorta

Inf. Vena Cava

6C

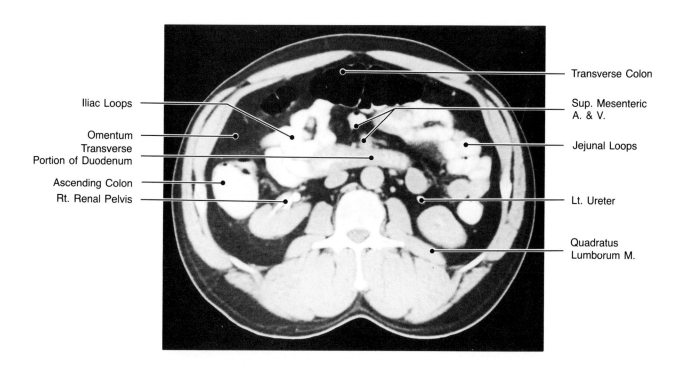

Iliac Loops

Omentum

Transverse Portion of Duodenum

Ascending Colon

Rt. Renal Pelvis

Transverse Colon

Sup. Mesenteric A. & V.

Jejunal Loops

Lt. Ureter

Quadratus Lumborum M.

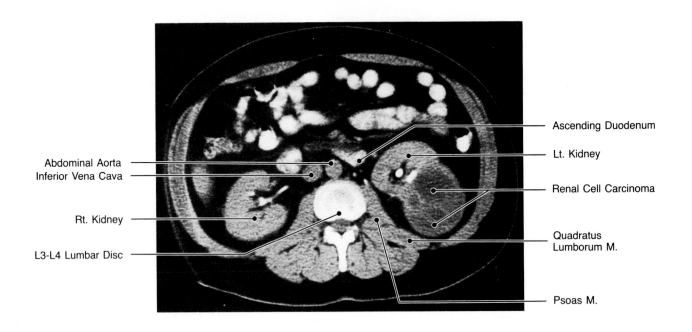

Ascending Duodenum

Lt. Kidney

Renal Cell Carcinoma

Quadratus Lumborum M.

Psoas M.

Abdominal Aorta

Inferior Vena Cava

Rt. Kidney

L3-L4 Lumbar Disc

6D Axial/CT/Renal Tumor. Plate 6D is a contrast-enhanced CT study of the lower kidneys demonstrating a 5 × 6 cm inhomogeneous rim-enhancing tumor from the posterior aspect of the left kidney. The collecting systems do not appear to be significantly distorted. No evidence of metastatic disease from this renal cell carcinoma was demonstrated at this level.

6E Axial/MRI/Kidney Level. Plate 6E is a magnetic resonance image done with partial saturation technique through the lower pole of each kidney. This plate demonstrates the abdominal aorta and inferior vena cava and the uncinate process of the pancreas extending to the left of the abdominal aorta yet still behind the superior mesenteric artery and vein. To the left of these structures is the ascending duodenum lying near the anterior pararenal fat. To the right of the L3 in retroperitoneal fat is the L3 lumbar nerve.

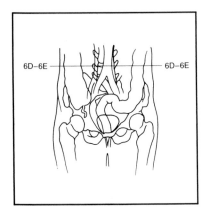

6D-6E   6D-6E

6E

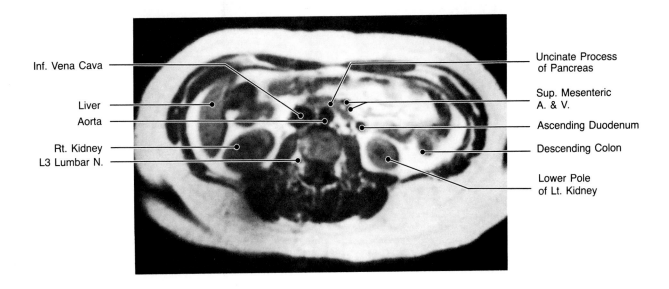

Inf. Vena Cava

Liver

Aorta

Rt. Kidney

L3 Lumbar N.

Uncinate Process
of Pancreas

Sup. Mesenteric
A. & V.

Ascending Duodenum

Descending Colon

Lower Pole
of Lt. Kidney

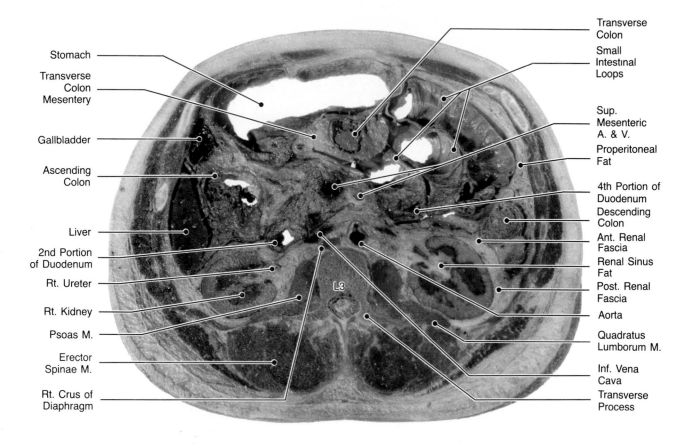

Stomach
Transverse Colon Mesentery
Gallbladder
Ascending Colon
Liver
2nd Portion of Duodenum
Rt. Ureter
Rt. Kidney
Psoas M.
Erector Spinae M.
Rt. Crus of Diaphragm

Transverse Colon
Small Intestinal Loops
Sup. Mesenteric A. & V.
Properitoneal Fat
4th Portion of Duodenum
Descending Colon
Ant. Renal Fascia
Renal Sinus Fat
Post. Renal Fascia
Aorta
Quadratus Lumborum M.
Inf. Vena Cava
Transverse Process

L3

7A Axial/Anatomic Plate/Renal Sinus. Anatomic plate 7A at the level of the renal sinus shows anterior and posterior renal shoulder separated by renal sinus fat; on the right, the right ureter can be seen lying centrally in the more medial perinephric extension of renal sinus fat. Anterior and posterior fascia join laterally and the posterior pararenal fat continues on laterally and anteriorly as properitoneal fat. The fourth portion of the duodenum to the left of the superior mesenteric artery and vein may be seen going from its retroperitoneal to its intraperitoneal continuation as jejunum.

7B Axial/CT/Renal Sinus. Cadaver CT plate 7B demonstrates gas in the fourth portion of the duodenum extending into proximal jejunum intraperitoneally. Small intestinal loops, most of them jejunal lie anterior to the descending colon in the retroperitoneal space. The renal sinus fat lends contrast to the vessels of the renal hilus.

7C Axial/CT/Renal Sinus. Plate 7C living patient CT with intravenous and oral and rectal contrast demonstrate parallel renal veins on the right entering the inferior vena cava and renal pelvis with dense contrast in it on the left. Ascending transverse and descending colon rim the lateral and anterior peritoneal surfaces and contain the small bowel loops. A small portion of uncinate process of pancreas is seen medial to the descending duodenum and behind the superior mesenteric artery and vein. The transverse stomach in this patient lies higher than at this level.

7D Axial/MRI/Renal Sinus. Plate 7D, the T1 weighted MRI image shows renal sinus fat with bright signal centrally in the kidneys. Note the superior mesenteric vessels and uncinate process of pancreas.

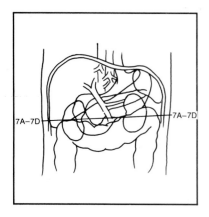

7A–7D   7A–7D

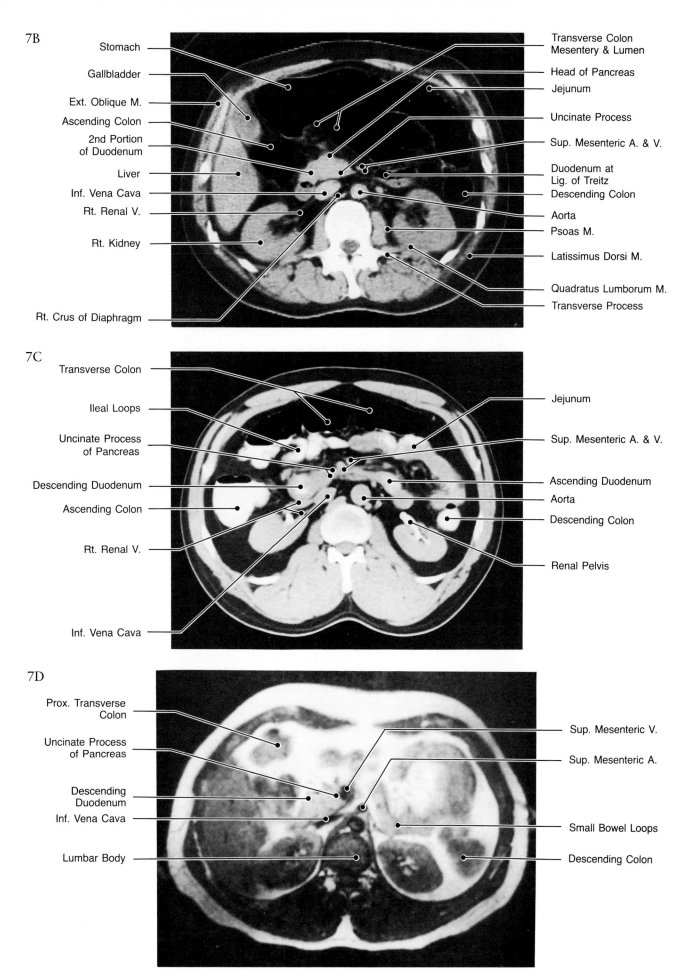

7B

Stomach

Gallbladder

Ext. Oblique M.

Ascending Colon

2nd Portion
of Duodenum

Liver

Inf. Vena Cava

Rt. Renal V.

Rt. Kidney

Rt. Crus of Diaphragm

Transverse Colon
Mesentery & Lumen

Head of Pancreas

Jejunum

Uncinate Process

Sup. Mesenteric A. & V.

Duodenum at
Lig. of Treitz

Descending Colon

Aorta

Psoas M.

Latissimus Dorsi M.

Quadratus Lumborum M.

Transverse Process

7C

Transverse Colon

Ileal Loops

Uncinate Process
of Pancreas

Descending Duodenum

Ascending Colon

Rt. Renal V.

Inf. Vena Cava

Jejunum

Sup. Mesenteric A. & V.

Ascending Duodenum

Aorta

Descending Colon

Renal Pelvis

7D

Prox. Transverse
Colon

Uncinate Process
of Pancreas

Descending
Duodenum

Inf. Vena Cava

Lumbar Body

Sup. Mesenteric V.

Sup. Mesenteric A.

Small Bowel Loops

Descending Colon

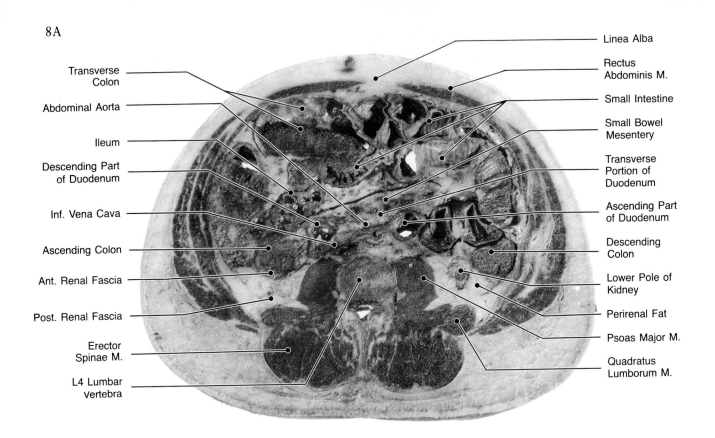

Linea Alba

Rectus Abdominis M.

Small Intestine

Small Bowel Mesentery

Transverse Portion of Duodenum

Ascending Part of Duodenum

Descending Colon

Lower Pole of Kidney

Perirenal Fat

Psoas Major M.

Quadratus Lumborum M.

Transverse Colon

Abdominal Aorta

Ileum

Descending Part of Duodenum

Inf. Vena Cava

Ascending Colon

Ant. Renal Fascia

Post. Renal Fascia

Erector Spinae M.

L4 Lumbar Vertebra

8A Axial/Anatomic Plate/Transverse Duodenum. This anatomic plate demonstrates the lumina for the distal descending and proximal ascending part of the duodenum. Some mucosa of the transverse (third portion) of the duodenum is seen connecting the two just anterior to the aorta and inferior vena cava. Quadratus lumborum muscles lying just lateral to the L4 lumbar transverse processes are largest just above the iliac crest. Anterior to these muscles is the lush retroperitoneal fat divided by posterior and anterior pararenal fascia from perirenal fat and kidney.

8B Axial/CT/Transverse Duodenum. This cadaver CT demonstrates the horizontal portion of the duodenum lying just anterior to the aorta and inferior vena cava. Anterior to the retroperitoneal fat lie the ascending and descending colon. The lateral abdominal wall demonstrates external oblique, internal oblique, and transverse abdominal musculature. Properitoneal fat represents the anterior extension of posterior pararenal fat.

8C Axial/CT/Transverse Duodenum. This contrast-enhanced study in the living patient below the kidneys demonstrates both ureters lying on the anterolateral aspects of the psoas muscles and very well developed quadratus lumborum muscles. The erector spinae muscle groups are also well developed in this patient.

8D Axial/MRI/L4. This MRI demonstrates the anatomy at the level of L4.

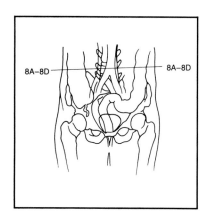

8A–8D   8A–8D

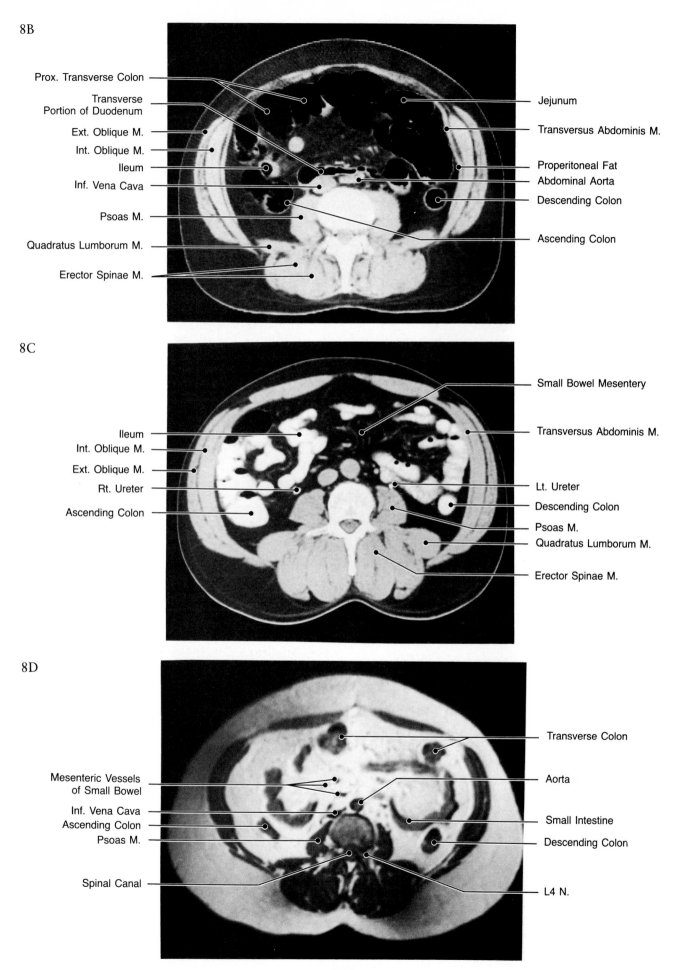

8B

Prox. Transverse Colon

Transverse
Portion of Duodenum

Ext. Oblique M.

Int. Oblique M.

Ileum

Inf. Vena Cava

Psoas M.

Quadratus Lumborum M.

Erector Spinae M.

Jejunum

Transversus Abdominis M.

Properitoneal Fat

Abdominal Aorta

Descending Colon

Ascending Colon

8C

Ileum

Int. Oblique M.

Ext. Oblique M.

Rt. Ureter

Ascending Colon

Small Bowel Mesentery

Transversus Abdominis M.

Lt. Ureter

Descending Colon

Psoas M.

Quadratus Lumborum M.

Erector Spinae M.

8D

Mesenteric Vessels
of Small Bowel

Inf. Vena Cava

Ascending Colon

Psoas M.

Spinal Canal

Transverse Colon

Aorta

Small Intestine

Descending Colon

L4 N.

Small Bowel Mesentery

Transverse Duodenum

Bifurcation of Abdominal Aorta

Ascending Colon

Rt. Ureter

Inf. Vena Cava

Iliacus M.

Gluteus Medius M.

Intervertebral Disc L4-L5

L5 Lumbar Body

Small Intestine

Lt. Ureter

Transversus Abdominis M.

Descending Colon

Ext. Oblique M.

Int. Oblique M.

Psoas Major M.

Ilium

Erector Spinae M.

9A Axial/Anatomic Plate/Aortic Bifurcation. Anatomic Plate 9A demonstrates the iliac bone and the L4-5 lumbar disk. The psoas major muscles have become quite rounded, and the ureters are seen on the anterior aspect of each psoas. Anterior to the retroperitoneum in the midline is the great mass of mesentery of small bowel, which is largely fatty, containing a few major vessels. The iliacus muscle is seen on the anterior surface of the iliac bone on the right.

9B Axial/CT/Aortic Bifurcation. Cadaver CT plate 9B demonstrates the aortic bifurcation and the still unbifurcated inferior vena cava. Both iliac crests are seen.

9C Axial/CT/Aortic Bifurcation. Plate 9C is a CT from a living patient and demonstrates two common iliac arteries with the as yet unbifurcated inferior vena cava behind the right common iliac artery. Because intravenous contrast has been used both ureters are identified.

9D Axial/MRI/Aortic Bifurcation. Partial saturation magnetic resonance imaging at the same level shows the common iliac arteries and an as yet undivided inferior vena cava. The quadratus lumborum muscle is large and seen just above the iliac crest. Abundant subcutaneous and mesenteric fat is seen around and within the muscles of the abdomen.

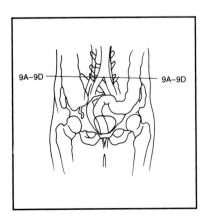

9A–9D

9A–9D

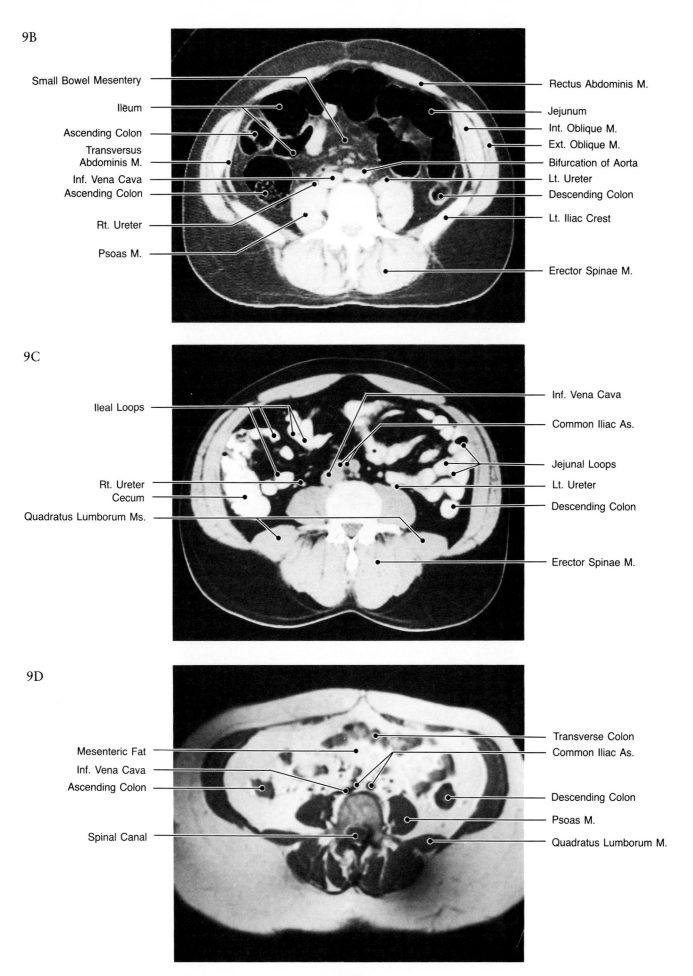

9B

Small Bowel Mesentery

Ileum

Ascending Colon

Transversus Abdominis M.

Inf. Vena Cava

Ascending Colon

Rt. Ureter

Psoas M.

Rectus Abdominis M.

Jejunum

Int. Oblique M.

Ext. Oblique M.

Bifurcation of Aorta

Lt. Ureter

Descending Colon

Lt. Iliac Crest

Erector Spinae M.

9C

Ileal Loops

Rt. Ureter

Cecum

Quadratus Lumborum Ms.

Inf. Vena Cava

Common Iliac As.

Jejunal Loops

Lt. Ureter

Descending Colon

Erector Spinae M.

9D

Mesenteric Fat

Inf. Vena Cava

Ascending Colon

Spinal Canal

Transverse Colon

Common Iliac As.

Descending Colon

Psoas M.

Quadratus Lumborum M.

267

10A

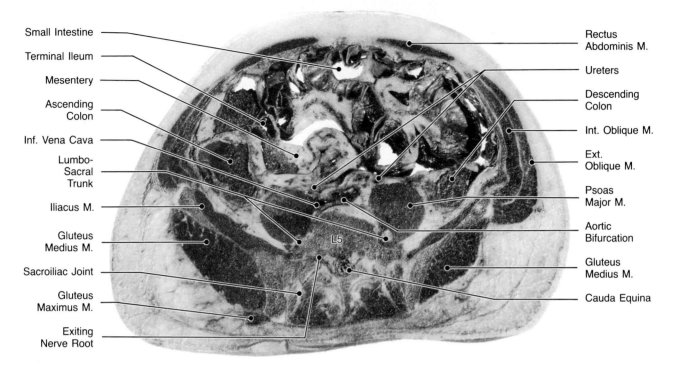

Small Intestine

Terminal Ileum

Mesentery

Ascending Colon

Inf. Vena Cava

Lumbo-Sacral Trunk

Iliacus M.

Gluteus Medius M.

Sacroiliac Joint

Gluteus Maximus M.

Exiting Nerve Root

Rectus Abdominis M.

Ureters

Descending Colon

Int. Oblique M.

Ext. Oblique M.

Psoas Major M.

Aortic Bifurcation

Gluteus Medius M.

Cauda Equina

L5

10A Axial/Anatomic Plate/L5 Level. Anatomic plate 10A demonstrates the L5 vertebral body and ascending and descending colon and in the center the mesentery for small bowel and the small bowel itself. Anterior to the iliac crest are the iliacus muscles and posterior to it are the gluteus medius muscles.

10B Axial/CT/L5 Level. A cadaver CT plate 10B demonstrates the L5 vertebral body and similar levels of ascending colon, descending colon, and small intestine. Abundant mesenteric fat is identified anterior to the bifurcating abdominal aorta.

10C Axial/CT/L4-L5 Level. CT plate 10C, in the living patient, just below the aortic bifurcation demonstrates the bowel, with contrast enhancement, as on the cadaver plates. Note the unusual elongation of the spinal canal in anterior-posterior direction; this is the result of spondylolisthesis at L4-L5.

10D Axial/MRI/L5 Level. Magnetic resonance imaging plate 10D using partial saturation technique in an obese patient at the level of L5 shows an L5 exiting nerve root on each side of the midline ventral to the spinal cord. The common iliac vessels have bifurcated and migrated to their locations along the anteromedial aspects of the psoas major muscles.

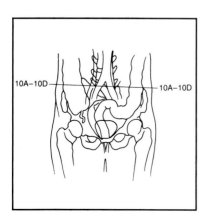

10A–10D   10A–10D

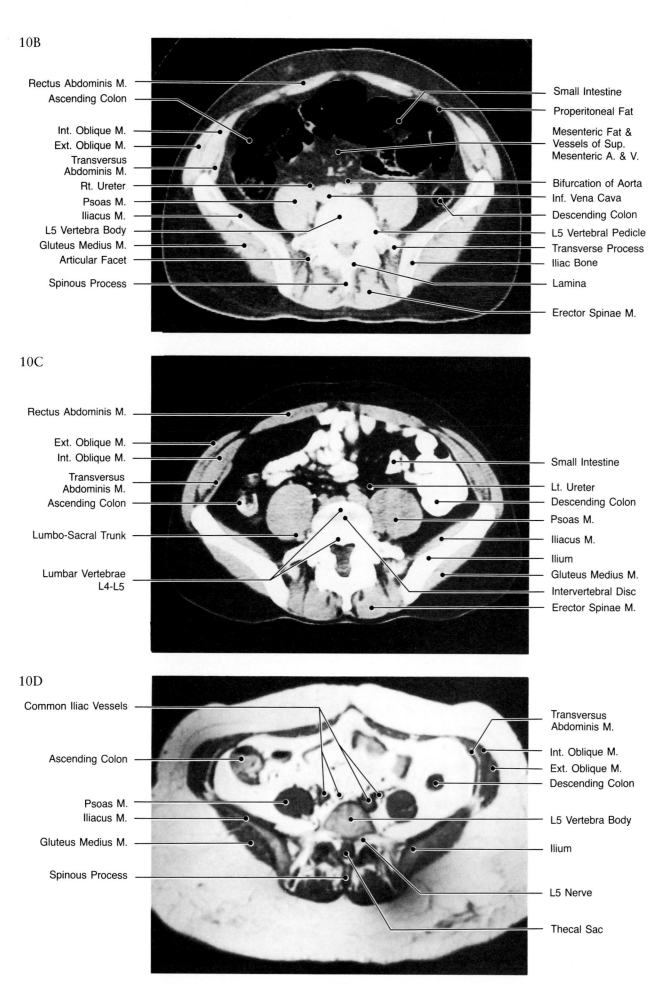

10B

Rectus Abdominis M.
Ascending Colon
Int. Oblique M.
Ext. Oblique M.
Transversus Abdominis M.
Rt. Ureter
Psoas M.
Iliacus M.
L5 Vertebra Body
Gluteus Medius M.
Articular Facet
Spinous Process

Small Intestine
Properitoneal Fat
Mesenteric Fat & Vessels of Sup. Mesenteric A. & V.
Bifurcation of Aorta
Inf. Vena Cava
Descending Colon
L5 Vertebral Pedicle
Transverse Process
Iliac Bone
Lamina
Erector Spinae M.

10C

Rectus Abdominis M.
Ext. Oblique M.
Int. Oblique M.
Transversus Abdominis M.
Ascending Colon
Lumbo-Sacral Trunk
Lumbar Vertebrae L4-L5

Small Intestine
Lt. Ureter
Descending Colon
Psoas M.
Iliacus M.
Ilium
Gluteus Medius M.
Intervertebral Disc
Erector Spinae M.

10D

Common Iliac Vessels
Ascending Colon
Psoas M.
Iliacus M.
Gluteus Medius M.
Spinous Process

Transversus Abdominis M.
Int. Oblique M.
Ext. Oblique M.
Descending Colon
L5 Vertebra Body
Ilium
L5 Nerve
Thecal Sac

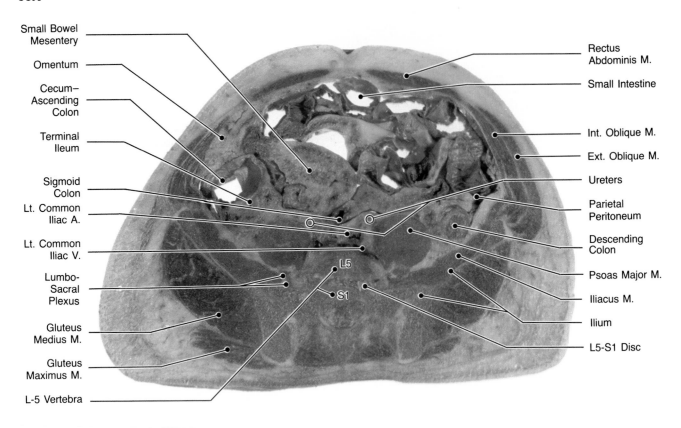

Small Bowel Mesentery

Omentum

Cecum–Ascending Colon

Terminal Ileum

Sigmoid Colon

Lt. Common Iliac A.

Lt. Common Iliac V.

Lumbo-Sacral Plexus

Gluteus Medius M.

Gluteus Maximus M.

L-5 Vertebra

Rectus Abdominis M.

Small Intestine

Int. Oblique M.

Ext. Oblique M.

Ureters

Parietal Peritoneum

Descending Colon

Psoas Major M.

Iliacus M.

Ilium

L5-S1 Disc

11A Axial/Anatomic Plate/L5-S1 Disk. Anatomic plate 11A demonstrates portions of the L5 vertebral body anteriorly and of the neural arch of S1 posteriorly, including lumbar plexus nerves and spinal nerve L5 on the right. Together the psoas and iliacus form a large comma-shaped structure lying in the fossa between L5 and the iliac wing. Between the two psoas muscles are the common iliac arteries and veins. The left iliac vein forms a crescentic density over the anterior L5 vertebral body as it courses from a right to left of midline location. Many loops of small bowel demonstrating valvulae conniventes and the small bowel mesentery can be seen. Cecum and terminal ileum abut at this level in the right lower quadrant.

11B Axial/Cadaver CT/L5-S1 Level. Plate 11B from the cadaver depicts the false pelvis and shows portions of both L5 and S1 vertebrae without a good delineation of the intervening disks because of the window and volume averaging.

11C Axial/CT/Umbilical Level. The CT image from the living patient with contrast in the large and small bowel demonstrates terminal ileal loops in the right lower quadrant and jejunal loops on the left. As with the anatomic section this section clearly shows the umbilicus and well-developed abdominal wall musculature. Disk and lumbar vertebra lie immediately behind the inferior vena cava and the common iliac arteries. Just lateral to these vessels are the ureters.

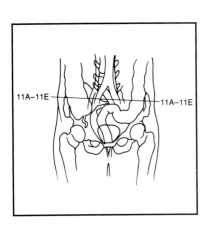

11A–11E

11A–11E

Common Iliac V.
(Rt. & Lt.)

Rt. Ureter

Iliocecal Valve Level

Cecum

Rt. Psoas M.

Iliacus M.

S1 Body

Erector Spinae M.

Common Iliac As.

Small Intestine

Ext. Oblique M.

Int. Oblique M.

Transversus
Abdominis M.

Ant. Sup. Iliac Spine

Descending Colon

Lt. Ilium

Gluteus Medius M.

Gluteus Maximus M.

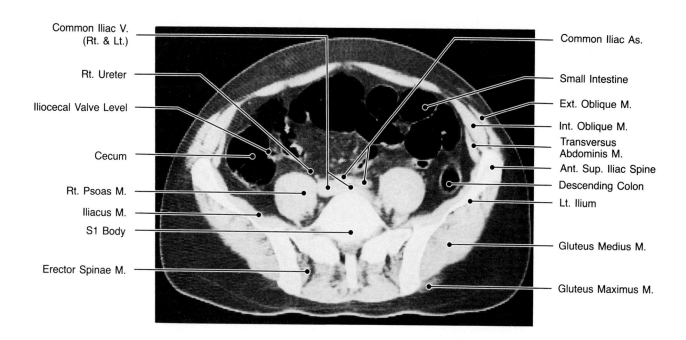

Umbilicus

High Sigmoid
Colon Loop

Rt. Ureter
Inf. Vena Cava
(Formation)

Elongated
Anteroposterior
Dimension of
Spinal Canal

Small Bowel Loops

Descending Colon

Lt. Ureter

Lt. Common Iliac A.

Intervertebral Disc

Lumbar-Sacral Plexus

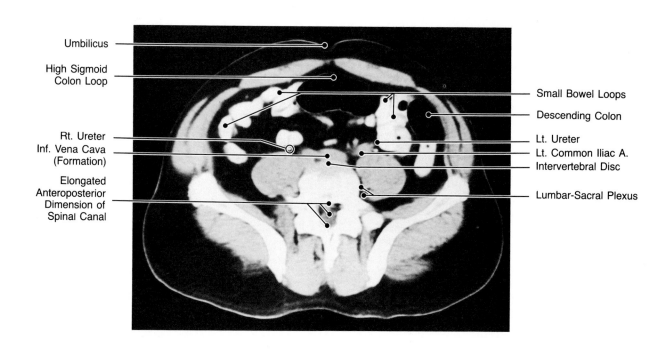

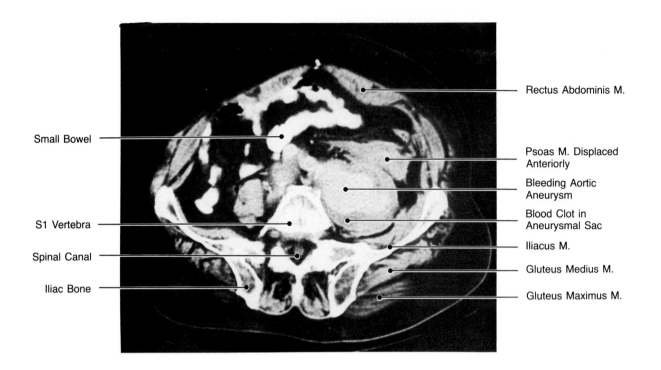

Rectus Abdominis M.

Small Bowel

Psoas M. Displaced Anteriorly

Bleeding Aortic Aneurysm

Blood Clot in Aneurysmal Sac

S1 Vertebra

Iliacus M.

Spinal Canal

Gluteus Medius M.

Iliac Bone

Gluteus Maximus M.

11D Axial/CT/Abdominal Aortic Aneurysm. Plate 11D is an abdominal CT with contrast enhancement on an 80-year-old male with an abdominal aortic aneurysm extending into the left common iliac artery. A 6 × 9 cm mass lateral to the left iliac artery represents the central unclotted (whiter) and the more peripheral clotted aneurysm, which was actively bleeding. There is extension of the bleeding into the left psoas muscle separating the psoas from the iliacus portions.

11E Axial/MRI/L5-S1 Level. MRI partial saturation technique, image 11E, in a patient with abundant mesenteric and subcutaneous fat shows the normal relations of the common iliac vessels at their bifurcations to the psoas muscles. L5 intervertebral disk and S1 vertebra are seen in the midline with fat surrounding the nerves in the sacral exit foramina.

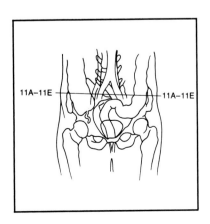

11A–11E   11A–11E

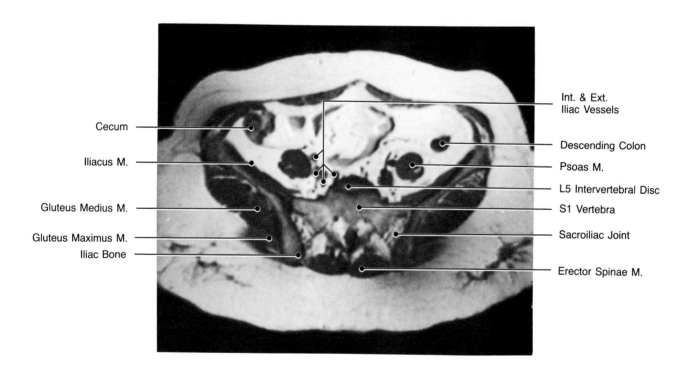

Cecum

Iliacus M.

Gluteus Medius M.

Gluteus Maximus M.

Iliac Bone

Int. & Ext.
Iliac Vessels

Descending Colon

Psoas M.

L5 Intervertebral Disc

S1 Vertebra

Sacroiliac Joint

Erector Spinae M.

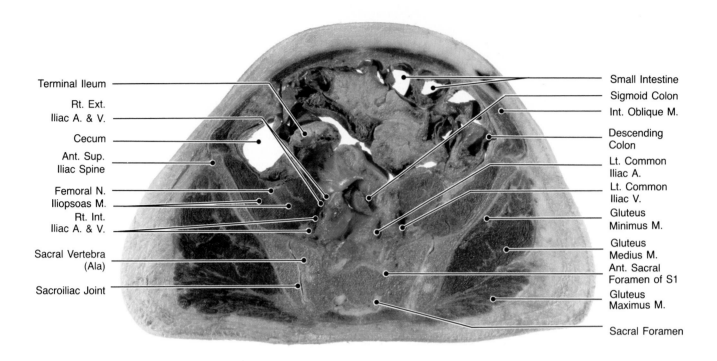

Terminal Ileum

Rt. Ext.
Iliac A. & V.

Cecum

Ant. Sup.
Iliac Spine

Femoral N.
Iliopsoas M.
Rt. Int.
Iliac A. & V.

Sacral Vertebra
(Ala)

Sacroiliac Joint

Small Intestine

Sigmoid Colon

Int. Oblique M.

Descending
Colon

Lt. Common
Iliac A.

Lt. Common
Iliac V.

Gluteus
Minimus M.

Gluteus
Medius M.

Ant. Sacral
Foramen of S1

Gluteus
Maximus M.

Sacral Foramen

12A Axial/Anatomic Plate/Iliac Bifurcation Level.
Anatomic plate 12A sections the right external and internal
iliac arteries and veins. These vessels still abut the most medial
psoas margin. Anterior and posterior sacral foramina are also
identified at this level. The iliac bone has lengthened and
broadened posteriorly to form the sacroiliac articulation and
anteriorly to form the superior iliac spine. The femoral nerve is
identified lying in the notch created by the union of the iliacus
and psoas muscles and is surrounded by fat.

12B Axial/CT/S2-S3. Cadaver CT plate 12B demonstrates
the cecum, descending and sigmoid colon, and the central
small bowel loops. Anterior and posterior sacral foramina of
S2 and S3 are identified. The external abdominal oblique muscle at this level continues on as the external oblique aponeurosis, and the internal oblique muscle belly and transverse abdominis muscle belly remain.

12C Axial/CT/Iliac Bifurcation Level. Plate 12C, the CT
plate from the living patient with contrast enhancement shows
the gastrointestinal structures of the region filled with contrast. Internal and external iliac vessels and anterior and posterior sacral foramina are seen.

12D Axial/MRI/Iliac Bifurcation. Plate 12D is the magnetic
resonance image at the corresponding level. This is the level of
the sacral fossa, and on the left the external iliac vessels are
clearly seen while on the right the internal iliac vessels are also
identified. Descending colon and a small segment of sigmoid
colon are surrounded by mesenteric fat. Fat also surrounds the
nerves of the sacrum in their exit foramina.

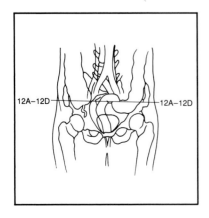

12A–12D   12A–12D

12B

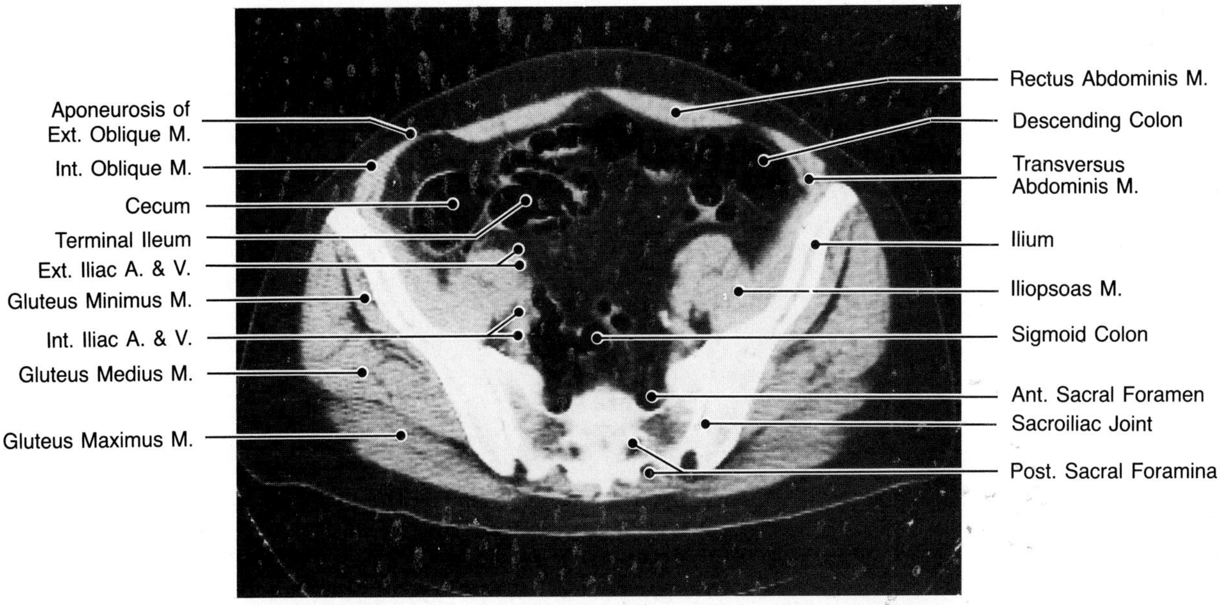

Aponeurosis of Ext. Oblique M.
Int. Oblique M.
Cecum
Terminal Ileum
Ext. Iliac A. & V.
Gluteus Minimus M.
Int. Iliac A. & V.
Gluteus Medius M.
Gluteus Maximus M.

Rectus Abdominis M.
Descending Colon
Transversus Abdominis M.
Ilium
Iliopsoas M.
Sigmoid Colon
Ant. Sacral Foramen
Sacroiliac Joint
Post. Sacral Foramina

12C

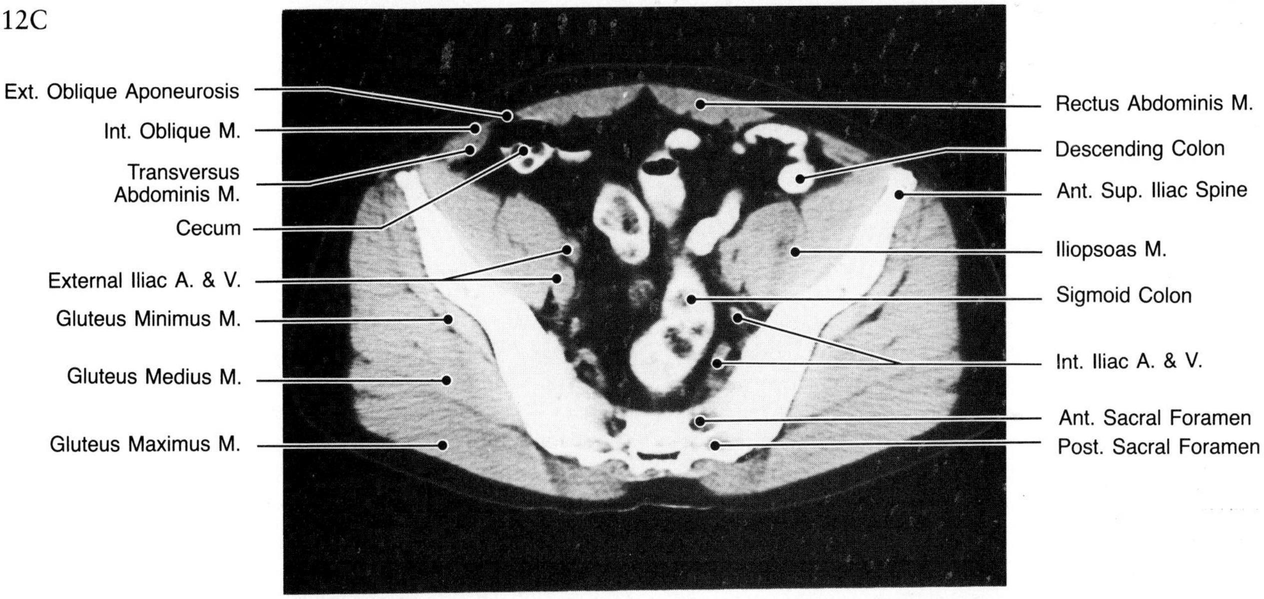

Ext. Oblique Aponeurosis
Int. Oblique M.
Transversus Abdominis M.
Cecum
External Iliac A. & V.
Gluteus Minimus M.
Gluteus Medius M.
Gluteus Maximus M.

Rectus Abdominis M.
Descending Colon
Ant. Sup. Iliac Spine
Iliopsoas M.
Sigmoid Colon
Int. Iliac A. & V.
Ant. Sacral Foramen
Post. Sacral Foramen

12D

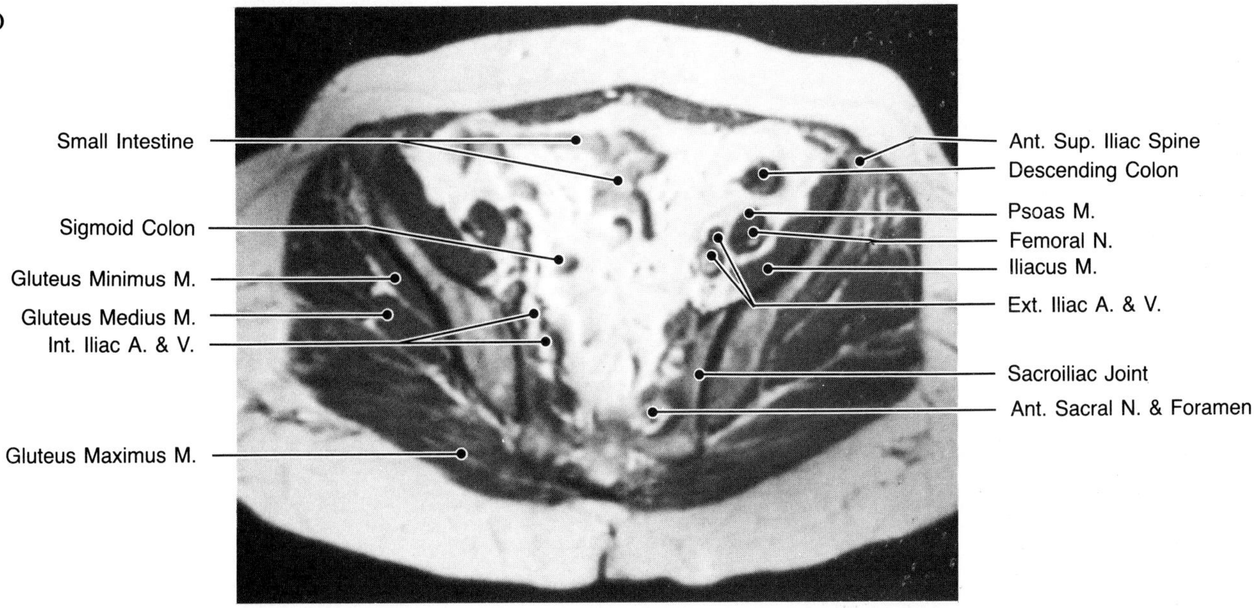

Small Intestine
Sigmoid Colon
Gluteus Minimus M.
Gluteus Medius M.
Int. Iliac A. & V.
Gluteus Maximus M.

Ant. Sup. Iliac Spine
Descending Colon
Psoas M.
Femoral N.
Iliacus M.
Ext. Iliac A. & V.
Sacroiliac Joint
Ant. Sacral N. & Foramen

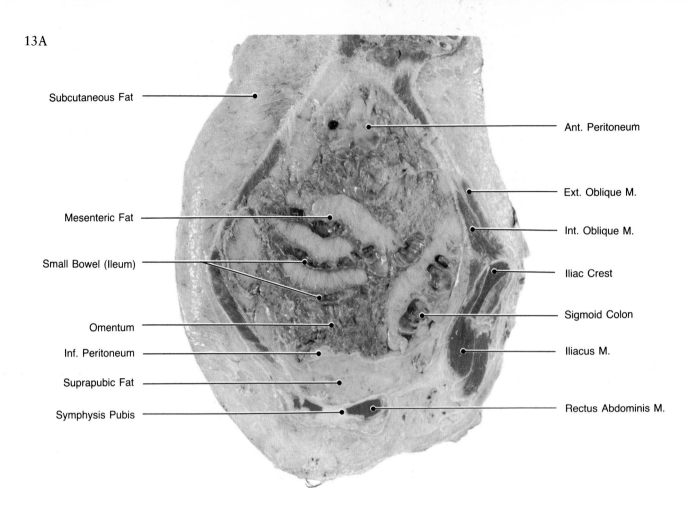

Subcutaneous Fat

Mesenteric Fat

Small Bowel (Ileum)

Omentum

Inf. Peritoneum

Suprapubic Fat

Symphysis Pubis

Ant. Peritoneum

Ext. Oblique M.

Int. Oblique M.

Iliac Crest

Sigmoid Colon

Iliacus M.

Rectus Abdominis M.

13A Coronal/Peritoneal Cavity, Anterior Part. The anatomic section through this obese patient from just below the xiphoid to the anterior symphysis pubis, cuts through the abdominal wall musculature and the left anterior superior iliac spine. Within the peritoneal contour lies the lacy pattern of omentum and the more obviously fatty pattern of the mesentery separating loops of small intestine and colon. Extraperitoneal fat above the symphysis and subcutaneous fat along the lateral aspects of the upper abdomen are abundant.

13B Coronal/MRI/Peritoneal Cavity, Anterior Part. This magnetic resonance image, also from an obese patient, demonstrates abundant omental fat on both sides of the body of the stomach and clearly demonstrates the vascular voids of superior mesenteric artery and vein. Mesenteric fat is also seen following from these vessels to the small bowel. The gallbladder rests in the fossa of the right lobe of the liver, and some intrahepatic portal venous structures are seen.

13C Coronal/MRI/Peritoneal Cavity, Anterior Part. At a level 3 cm posterior on the same patient as seen in the previous plate we can identify the superior mesenteric venous confluence with the splenic vein to form the extrahepatic portal vein. A small portion of the pancreas and that portion of the C loop to the right of the head of the pancreas are demonstrated.

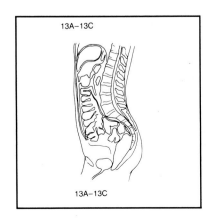

13A–13C

13A–13C

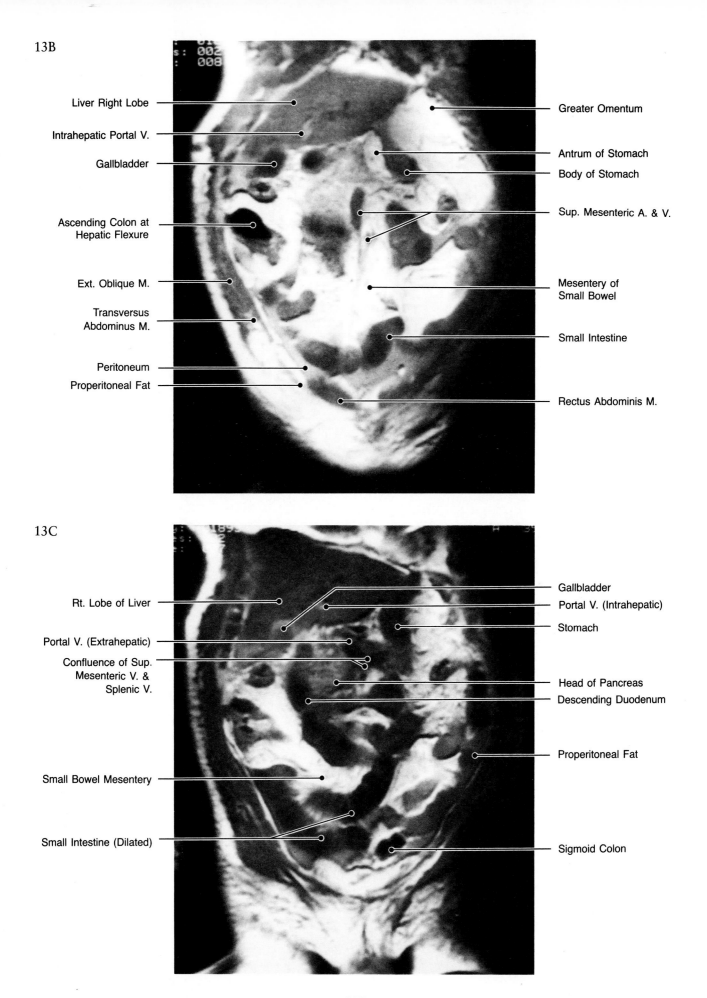

13B

Liver Right Lobe

Intrahepatic Portal V.

Gallbladder

Ascending Colon at
Hepatic Flexure

Ext. Oblique M.

Transversus
Abdominus M.

Peritoneum

Properitoneal Fat

Greater Omentum

Antrum of Stomach

Body of Stomach

Sup. Mesenteric A. & V.

Mesentery of
Small Bowel

Small Intestine

Rectus Abdominis M.

13C

Rt. Lobe of Liver

Portal V. (Extrahepatic)

Confluence of Sup.
Mesenteric V. &
Splenic V.

Small Bowel Mesentery

Small Intestine (Dilated)

Gallbladder

Portal V. (Intrahepatic)

Stomach

Head of Pancreas

Descending Duodenum

Properitoneal Fat

Sigmoid Colon

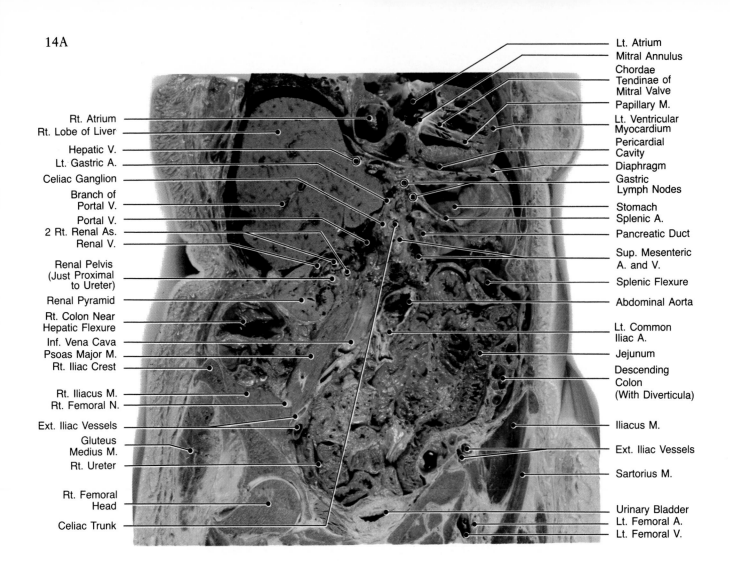

Rt. Atrium
Rt. Lobe of Liver

Hepatic V.
Lt. Gastric A.

Celiac Ganglion

Branch of
Portal V.

Portal V.
2 Rt. Renal As.
Renal V.

Renal Pelvis
(Just Proximal
to Ureter)

Renal Pyramid

Rt. Colon Near
Hepatic Flexure

Inf. Vena Cava

Psoas Major M.
Rt. Iliac Crest

Rt. Iliacus M.
Rt. Femoral N.

Ext. Iliac Vessels

Gluteus
Medius M.

Rt. Ureter

Rt. Femoral
Head

Celiac Trunk

Lt. Atrium
Mitral Annulus
Chordae
Tendinae of
Mitral Valve
Papillary M.
Lt. Ventricular
Myocardium
Pericardial
Cavity
Diaphragm
Gastric
Lymph Nodes
Stomach
Splenic A.
Pancreatic Duct
Sup. Mesenteric
A. and V.
Splenic Flexure
Abdominal Aorta
Lt. Common
Iliac A.
Jejunum
Descending
Colon
(With Diverticula)
Iliacus M.
Ext. Iliac Vessels
Sartorius M.
Urinary Bladder
Lt. Femoral A.
Lt. Femoral V.

14A Coronal/Anatomic Plate/Abdominal Viscera. This impressive coronal section through the mitral valve and chambers of the heart includes papillary muscle of the left ventricle and chorda tendineae and mitral valve. Below the diaphragm the intraperitoneal structures that are seen include the gastric rugae of the stomach and the cardia and upper body region, the ascending and descending colon, including diverticula in the descending colon, and a number of small bowel loops and mesentery. Central to the small bowel loops the abdominal aorta at its bifurcation demonstrates considerable atherosclerotic change, while the smooth-walled inferior vena cava nestles between the distal abdominal aorta and the right psoas major muscle. These viscera are demonstrated sweeping

slightly to the left because of a rotoscoliosis convexity to the left of the upper lumbar spine. As a result, to the left of the spine the pancreas and pancreatic duct are demonstrated, while to the right of the spine the right kidney and its renal pedicle, including vessels, and pelvis are demonstrated. Note the two right renal arteries.

14B Coronal/MRI/Abdominal Viscera. This partial saturation magnetic resonance image demonstrates the retroperitoneal structures to advantage. These include the abdominal aorta and several of its branches. The inferior vena cava is also demonstrated to the right of the aorta. Iliacus and psoas muscles and the femoral nerve lying in the groove created at their junction are seen. The urinary bladder returns little signal, while femoral heads and other marrow fat–containing bones of the pelvis return a bright signal as does fat elsewhere. The ascending colon to hepatic flexure and some ill-defined small intestinal loops are demonstrated.

14C Coronal/MRI/Abdominal Viscera. This partial saturation magnetic resonance image in a different patient demonstrates only the upper lumbar portion of the abdominal aorta and vessels of both renal pedicles. Arising just proximal to the left renal artery is the left superior mesenteric artery. On the right the colon is seen from appendix to hepatic flexure, whereas on the left only splenic flexure and sigmoid colon are identified with certainty; centrally small bowel loops intervene.

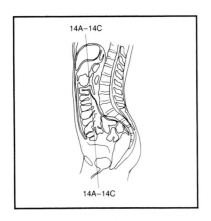

14A–14C

14A–14C

14B

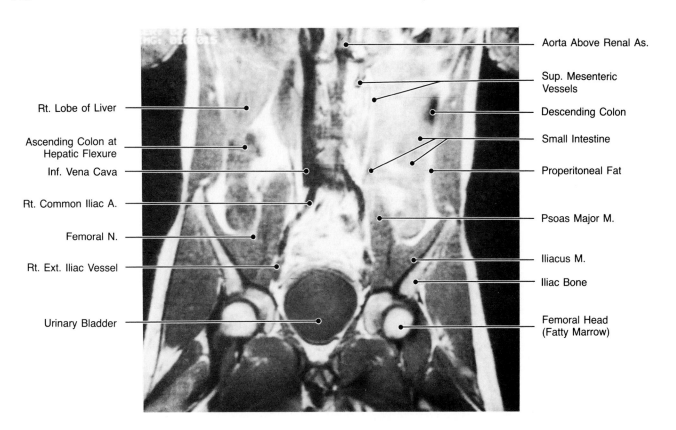

Rt. Lobe of Liver

Ascending Colon at
Hepatic Flexure

Inf. Vena Cava

Rt. Common Iliac A.

Femoral N.

Rt. Ext. Iliac Vessel

Urinary Bladder

Aorta Above Renal As.

Sup. Mesenteric
Vessels

Descending Colon

Small Intestine

Properitoneal Fat

Psoas Major M.

Iliacus M.

Iliac Bone

Femoral Head
(Fatty Marrow)

14C

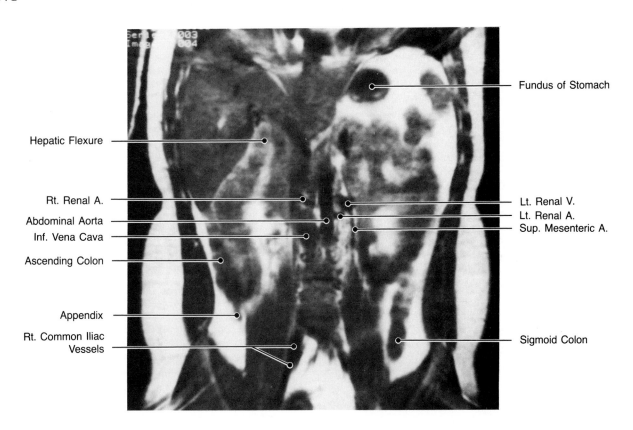

Hepatic Flexure

Rt. Renal A.

Abdominal Aorta

Inf. Vena Cava

Ascending Colon

Appendix

Rt. Common Iliac
Vessels

Fundus of Stomach

Lt. Renal V.

Lt. Renal A.

Sup. Mesenteric A.

Sigmoid Colon

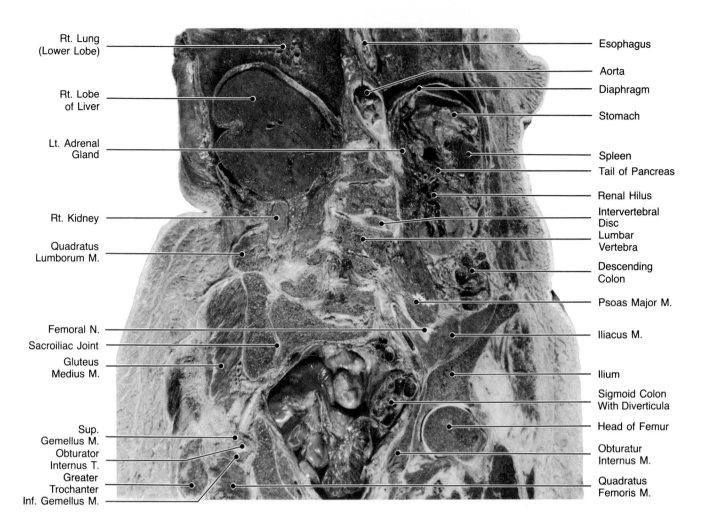

Rt. Lung (Lower Lobe) — Esophagus
Rt. Lobe of Liver — Aorta
Lt. Adrenal Gland — Diaphragm
Rt. Kidney — Stomach
Quadratus Lumborum M. — Spleen
Femoral N. — Tail of Pancreas
Sacroiliac Joint — Renal Hilus
Gluteus Medius M. — Intervertebral Disc
Sup. Gemellus M. — Lumbar Vertebra
Obturator Internus T. — Descending Colon
Greater Trochanter — Psoas Major M.
Inf. Gemellus M. — Iliacus M. — Ilium — Sigmoid Colon With Diverticula — Head of Femur — Obturatur Internus M. — Quadratus Femoris M.

15A Coronal/Anatomic Plate/Retroperitoneum, Anterior Part. Although this anatomic plate contains some intraperitoneal structures, it is chiefly retroperitoneal in its orientation. An oblique section of the abdominal aorta coursing to the left as it enters the abdomen parallels the levorotoscoliotic curve of the lumbar spine. Just lateral to the abdominal aorta on the left lie the adrenal gland, the stomach, and the spleen. Between the adrenal and the spleen and below the stomach is the tail of the pancreas. The left renal hilus and lower pole of the left kidney lying against the lateral aspect of psoas major muscle lie in a bed of retroperitoneal fat. On the concave side of this lumbar scoliosis only the smallest tip of the posterior right kidney is seen lying between the quadratus lumborum and psoas major muscles and within retroperitoneal fat. The relationships of superior and inferior gemellus muscles to the tendon of the obturator internus muscle, the quadratus femoris muscle, and the greater trochanter of the right femur are also demonstrated.

15B Coronal/MRI/Retroperitoneum, Anterior Part. This and the next magnetic resonance image are from the same patient and demonstrate sequentially lower portions and continuity of aorta and inferior vena caval branches, renal pelvis and renal hilar structures, and adrenal gland on the left. The stomach near the esophagogastric junction lies above the splenic vein, which is just above the body of the pancreas. Small intestinal loops are blurred out, but ascending colon may be seen.

15C Coronal/MRI/Retroperitoneum, Anterior Part. This magnetic resonance partial saturation image slightly posterior to the previous plate demonstrates more of the thoracoabdominal aorta, and the relationship of the crus of the left hemidiaphragm to the psoas major muscle is clarified. More of right kidney and descending colon is demonstrated to the splenic flexure. The medial limb of the left adrenal gland accompanies this cut 0.5 cm posterior to the previous image, demonstrating the lateral left adrenal limb. The renal pelvis and kidney on the right match with the renal pelvis and right ureter seen in the more posterior, previous cut.

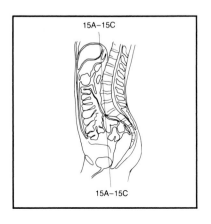

15A–15C

15A–15C

15B

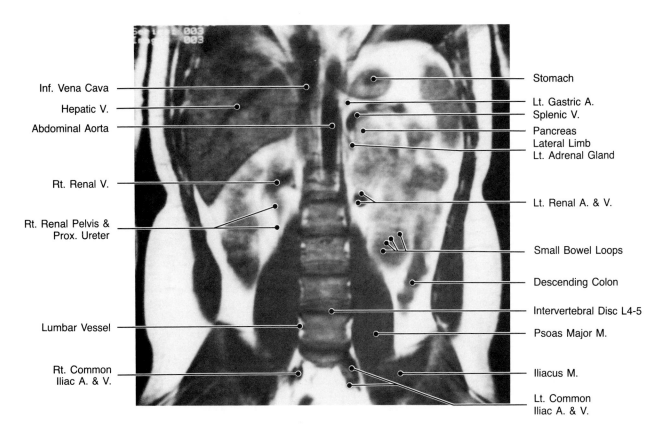

Inf. Vena Cava

Hepatic V.

Abdominal Aorta

Rt. Renal V.

Rt. Renal Pelvis &
Prox. Ureter

Lumbar Vessel

Rt. Common
Iliac A. & V.

Stomach

Lt. Gastric A.
Splenic V.

Pancreas
Lateral Limb
Lt. Adrenal Gland

Lt. Renal A. & V.

Small Bowel Loops

Descending Colon

Intervertebral Disc L4-5

Psoas Major M.

Iliacus M.

Lt. Common
Iliac A. & V.

15C

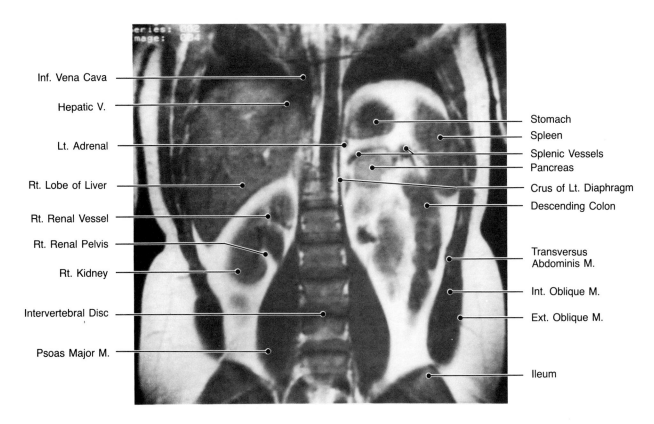

Inf. Vena Cava

Hepatic V.

Lt. Adrenal

Rt. Lobe of Liver

Rt. Renal Vessel

Rt. Renal Pelvis

Rt. Kidney

Intervertebral Disc

Psoas Major M.

Stomach
Spleen
Splenic Vessels
Pancreas
Crus of Lt. Diaphragm
Descending Colon

Transversus
Abdominis M.

Int. Oblique M.

Ext. Oblique M.

Ileum

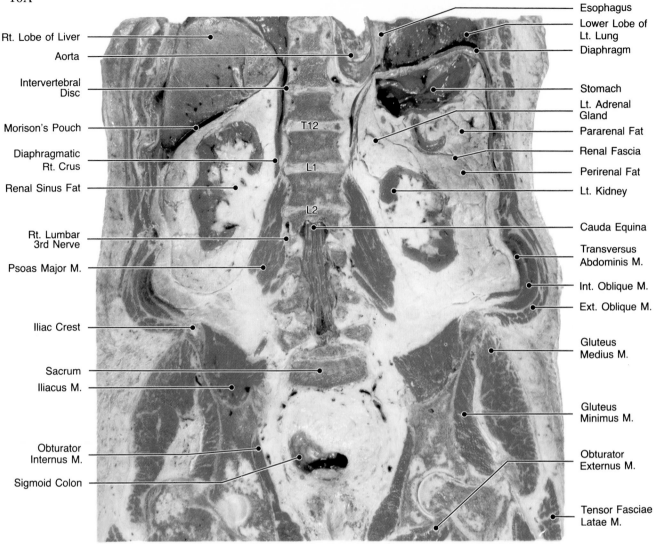

Rt. Lobe of Liver

Aorta

Intervertebral Disc

Morison's Pouch

Diaphragmatic Rt. Crus

Renal Sinus Fat

Rt. Lumbar 3rd Nerve

Psoas Major M.

Iliac Crest

Sacrum

Iliacus M.

Obturator Internus M.

Sigmoid Colon

T12

L1

L2

Esophagus

Lower Lobe of Lt. Lung

Diaphragm

Stomach

Lt. Adrenal Gland

Pararenal Fat

Renal Fascia

Perirenal Fat

Lt. Kidney

Cauda Equina

Transversus Abdominis M.

Int. Oblique M.

Ext. Oblique M.

Gluteus Medius M.

Gluteus Minimus M.

Obturator Externus M.

Tensor Fasciae Latae M.

16A Coronal/Anatomic Plate/Retroperitoneum, posterior part. From a cadaver different from that of previous anatomic plates, this plate demonstrates a straight lumbar spine with cauda equina, dura, and nerve rootlets seen exiting from L3 to the sacrum. The right diaphragmatic crus extends down to the L1 interspace and appears contiguous with the psoas major at the top of L2. Lateral to the psoas the renal capsule bounded by renal fascia containing perinephric fat and kidney is demonstrated, with posterior perirenal fat extending from the psoas major around and laterally to become preperitoneal fat, where it abuts the peritoneum below the liver tip on the right. On the left the adrenal gland lies within the renal capsule lateral to the left psoas muscle posterior to the abdominal aorta and medial to splenic vasculature. Scarring is noted on the lateral aspect of the left kidney.

16B Coronal/MRI/Retroperitoneum, Posterior Part. This magnetic resonance image at partial saturation demonstrates a small portion of stomach and spleen in the left upper quadrant and does not include the pelvis, as is seen on the next magnetic resonance image. The relationship of the thoracolumbar vertebrae and the cauda equina is, however, similar to the anatomic plate. Both kidneys are surrounded by perinephric fat, and both crura of the diaphragm can be seen down to their termination in continuity with the upper most fibers of the psoas major muscles.

16C Coronal/MRI/Retroperitoneum, Posterior Part. This magnetic resonance plate almost perfectly matches the anatomic plate, particularly for the cauda equina and the right L3 lumbar nerve. The renal outlines demonstrate clear differentiation of renal cortex from renal medulla, and on the right the renal pelvis is clearly seen. Only the descending colon is clearly identified as an intraperitoneal structure on this section.

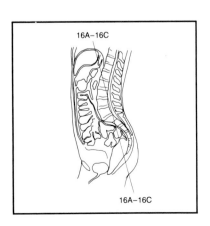

16A–16C

16A–16C

16B

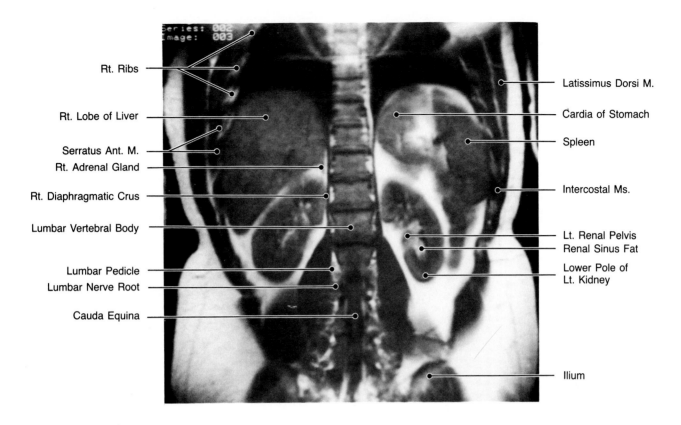

Rt. Ribs

Rt. Lobe of Liver

Serratus Ant. M.
Rt. Adrenal Gland

Rt. Diaphragmatic Crus

Lumbar Vertebral Body

Lumbar Pedicle
Lumbar Nerve Root

Cauda Equina

Latissimus Dorsi M.

Cardia of Stomach

Spleen

Intercostal Ms.

Lt. Renal Pelvis
Renal Sinus Fat

Lower Pole of
Lt. Kidney

Ilium

16C

Rt. Lobe of Liver

Rt. Renal Pelvis

Rt. L3 Pedicle
Rt. L3 N.

Rt. Sacral Ala

Obturator Internus M.

Descending Colon

Lt. Kidney (Medulla)

Lt. Kidney (Cortex)

Lt. Psoas Major M.

Cauda Equina

Lt. Sacroiliac Joint

Sigmoid Colon

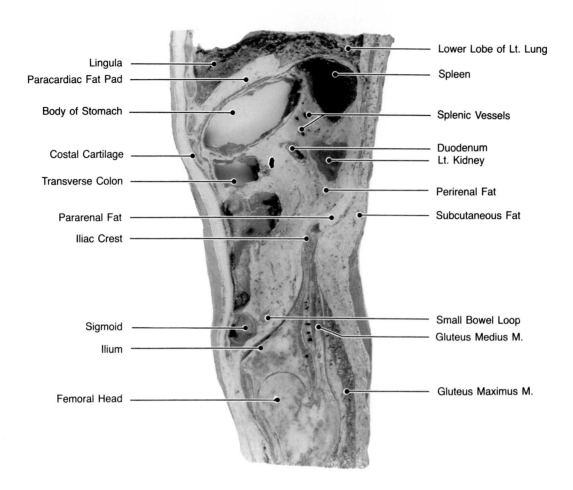

Lingula
Paracardiac Fat Pad
Body of Stomach
Costal Cartilage
Transverse Colon
Pararenal Fat
Iliac Crest
Sigmoid
Ilium
Femoral Head

Lower Lobe of Lt. Lung
Spleen
Splenic Vessels
Duodenum
Lt. Kidney
Perirenal Fat
Subcutaneous Fat
Small Bowel Loop
Gluteus Medius M.
Gluteus Maximus M.

17A Sagittal/Anatomic Plate/Spleen. This section is 10 cm left of midline. The sagittal abdomen in this well-preserved 65-year-old male demonstrates pericardial fat and the lingula of the left upper lobe as well as more posteriorly the lower lobe above the hemidiaphragm. Below the hemidiaphragm the oval lumen of the body of the stomach lies anterior to the spleen and duodenum. The lateral part of the left kidney can be seen posteriorly. Just below the body of the stomach lies the transverse colon, small bowel loops, and a portion of the sigmoid colon. The posterior abdomen consists chiefly of mesenteric fat and retroperitoneal fat and subcutaneous fat.

17B Sagittal/MRI/Spleen. The magnetic resonance image in this young living patient demonstrates the diaphragm and many of the same infradiaphragmatic structures shown in the anatomic sagittal section. Intrinsic motion in the small bowel that exceeds that in the large bowel generally gives it a grayer appearance, while the intraluminal gas yields a darker contrast to the colon and stomach.

17C Sagittal/MRI/Spleen. This magnetic resonance image 3 cm further to the left demonstrates the bulk of the spleen, smaller portion of the stomach, and transverse colon and several small intestinal loops of the left midabdomen. Just posterior to these small bowel loops the descending colon lies just anterior to the retroperitoneal fat.

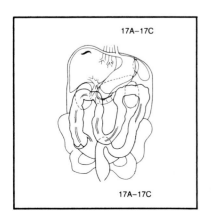

17A–17C

17A–17C

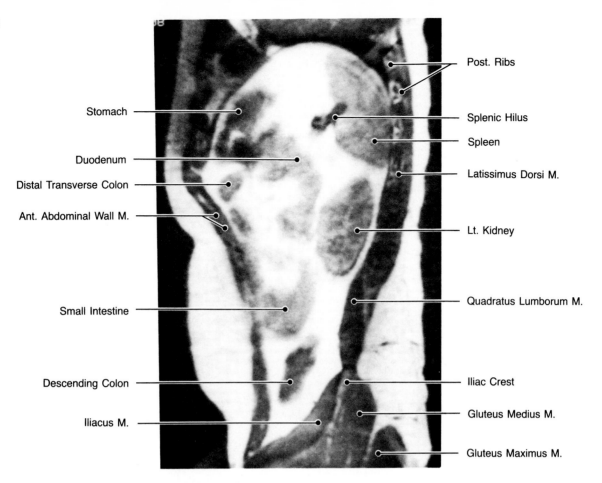

Stomach

Duodenum

Distal Transverse Colon

Ant. Abdominal Wall M.

Small Intestine

Descending Colon

Iliacus M.

Post. Ribs

Splenic Hilus

Spleen

Latissimus Dorsi M.

Lt. Kidney

Quadratus Lumborum M.

Iliac Crest

Gluteus Medius M.

Gluteus Maximus M.

17C

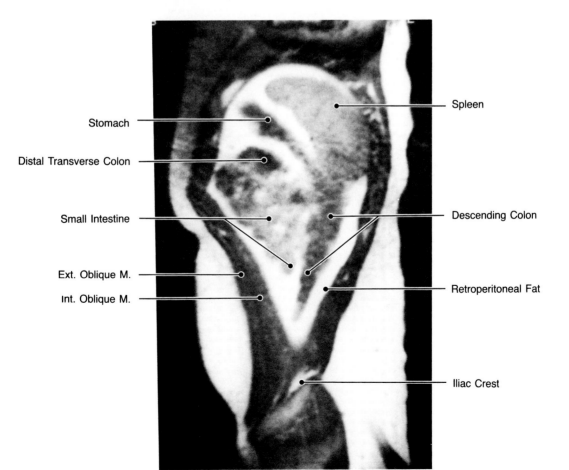

Stomach

Distal Transverse Colon

Small Intestine

Ext. Oblique M.

Int. Oblique M.

Spleen

Descending Colon

Retroperitoneal Fat

Iliac Crest

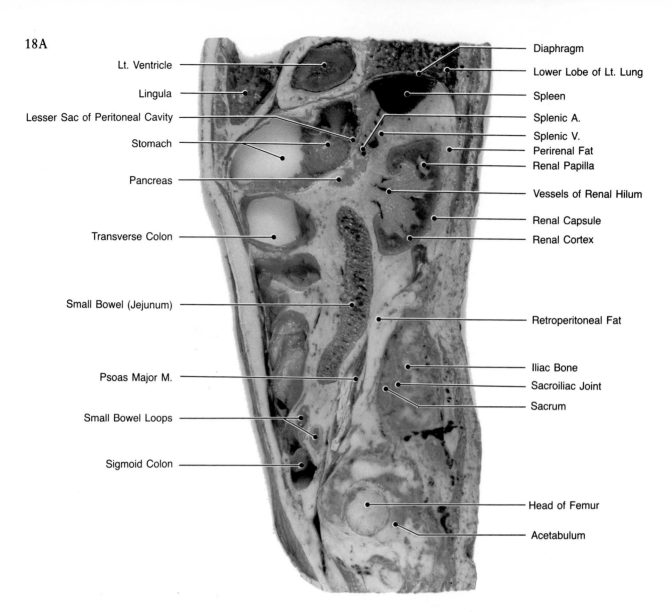

18A — Lt. Ventricle, Lingula, Lesser Sac of Peritoneal Cavity, Stomach, Pancreas, Transverse Colon, Small Bowel (Jejunum), Psoas Major M., Small Bowel Loops, Sigmoid Colon

Diaphragm, Lower Lobe of Lt. Lung, Spleen, Splenic A., Splenic V., Perirenal Fat, Renal Papilla, Vessels of Renal Hilum, Renal Capsule, Renal Cortex, Retroperitoneal Fat, Iliac Bone, Sacroiliac Joint, Sacrum, Head of Femur, Acetabulum

18A Sagittal/Anatomic Plate/Left Kidney. This section is 5 cm left of midline. Above the diaphragm on this anatomic plate we can identify, from anterior to posterior, lingula, pericardium, and portions of left ventricular myocardium and left lower lobe. Below the diaphragm, in addition to stomach and spleen, a central cut through the renal sinus demonstrates renal cortex, papilla, and the hilar vasculature. Valvuli coniventes of the jejunum are nicely demonstrated on the vertical loop of small bowel in the midabdomen. The left sacroiliac joint, the acetabulum, and femoral head are also identified posterior to psoas major and retroperitoneal fat.

18B Sagittal/MRI/Left Kidney. This magnetic resonance image 5 cm from the middle also demonstrates stomach, spleen, and hilar structures of the left kidney. Perinephric fat highlights the visualized portion of proximal left ureter. The posterior abdominal wall musculature includes the erector spinae muscles and quadratus lumborum, psoas, and iliacus muscles. Intrinsic movement of large and small bowel typically blur their outlines.

18C Sagittal/MRI/Left Kidney. This magnetic resonance imaging about 7 cm from the midline demonstrates the renal hilus in this young male patient. Splenic artery and vein can be seen on the ventral surface of the spleen.

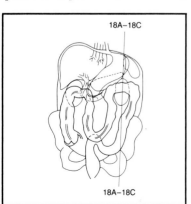

18A–18C

18A–18C

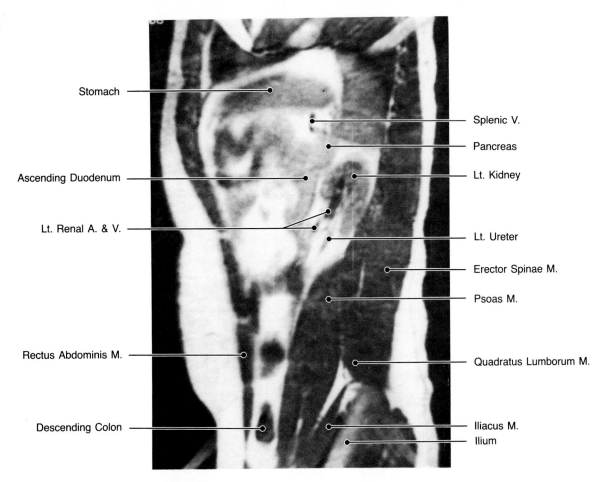

Stomach

Ascending Duodenum

Lt. Renal A. & V.

Rectus Abdominis M.

Descending Colon

Splenic V.

Pancreas

Lt. Kidney

Lt. Ureter

Erector Spinae M.

Psoas M.

Quadratus Lumborum M.

Iliacus M.
Ilium

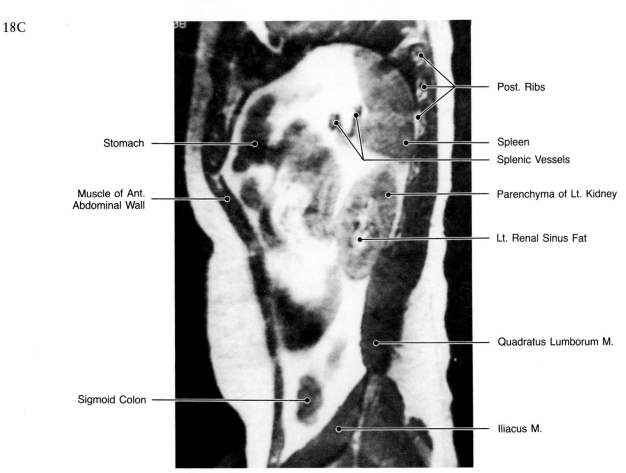

Stomach

Muscle of Ant.
Abdominal Wall

Sigmoid Colon

Post. Ribs

Spleen
Splenic Vessels

Parenchyma of Lt. Kidney

Lt. Renal Sinus Fat

Quadratus Lumborum M.

Iliacus M.

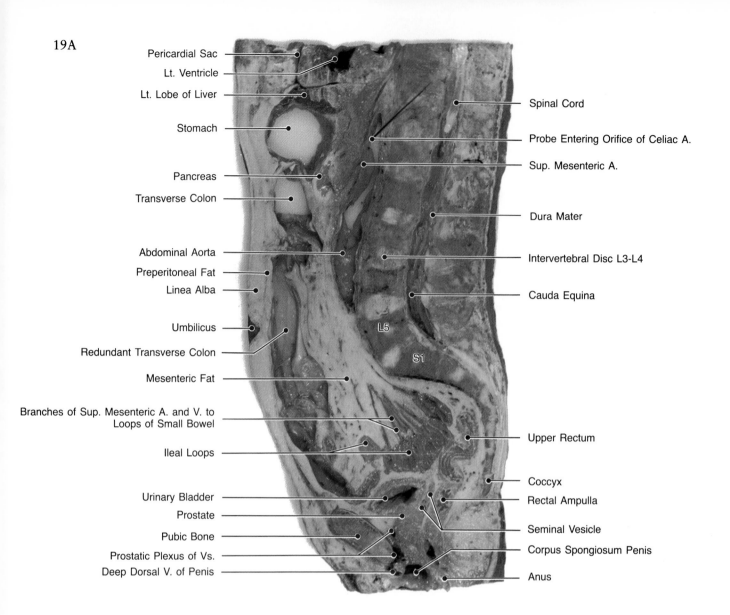

Pericardial Sac
Lt. Ventricle
Lt. Lobe of Liver
Stomach
Pancreas
Transverse Colon
Abdominal Aorta
Preperitoneal Fat
Linea Alba
Umbilicus
Redundant Transverse Colon
Mesenteric Fat
Branches of Sup. Mesenteric A. and V. to Loops of Small Bowel
Ileal Loops
Urinary Bladder
Prostate
Pubic Bone
Prostatic Plexus of Vs.
Deep Dorsal V. of Penis

Spinal Cord
Probe Entering Orifice of Celiac A.
Sup. Mesenteric A.
Dura Mater
Intervertebral Disc L3-L4
Cauda Equina
L5
S1
Upper Rectum
Coccyx
Rectal Ampulla
Seminal Vesicle
Corpus Spongiosum Penis
Anus

19A Sagittal/Anatomic Plate/Abdominal Aorta. This section is 2 centimeters to the left of the midline. Principal structures identified on this anatomic section include the abdominal aorta, with a probe entering the orifice of the celiac artery opposite L1 and the actual orifice of the superior mesenteric artery about 2 cm caudad to it. The intervertebral discs are nicely demonstrated, and portions of a number of spinal nerve roots are identified within the spinal canal. Small bowel mesentery is abundantly seen, with a classic demonstration of the vascular supply to the small bowel loops in the pelvis. Also

identified in the pelvis are the bladder, the seminal vesicles, and a portion of the left lobe of the prostate as well as the venous structures in this male from the prostatic veins to those of the penis. In a presacral and precoccygeal level we can see upper rectum, lower rectum, and even the left lateral wall of the anus. In the anterior abdomen the stomach lies above the transverse colon, which lies above a long redundant loop of transverse colon extending into the pelvis (see 22D)

19B Sagittal/MRI/Aortic Branches. This MRI section and the next are at the midline and demonstrate the vascular structures and gas-filled bowel loops as signal voids; prominent among these are splenic artery and vein, renal artery and vein, left common iliac artery and vein, a number of ascending lumbar veins, and the gas-filled stomach.

19C Sagittal/MRI/Abdominal Aorta. This section is 1 centimeter nearer the midline the abdominal aorta, the celiac trunk and splenic vein and superior mesenteric artery which arches superiorly before turning caudad are demonstrated. The bowel loops, stomach, and sigmoid colon are signal voids (black), while the grayer loops are those small bowel loops moving during the examination. Of intervening density, approaching fat density, is the pancreas.

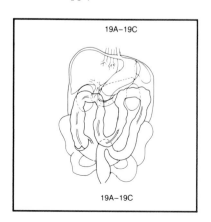

19A–19C

19A–19C

19B

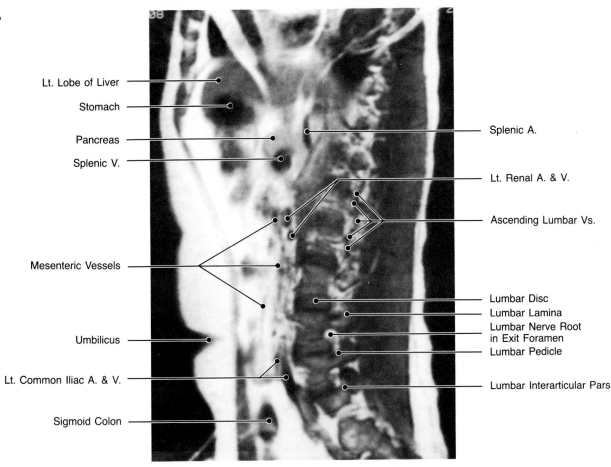

Lt. Lobe of Liver

Stomach

Pancreas

Splenic V.

Splenic A.

Lt. Renal A. & V.

Ascending Lumbar Vs.

Mesenteric Vessels

Lumbar Disc

Lumbar Lamina

Lumbar Nerve Root
in Exit Foramen

Lumbar Pedicle

Umbilicus

Lt. Common Iliac A. & V.

Lumbar Interarticular Pars

Sigmoid Colon

19C

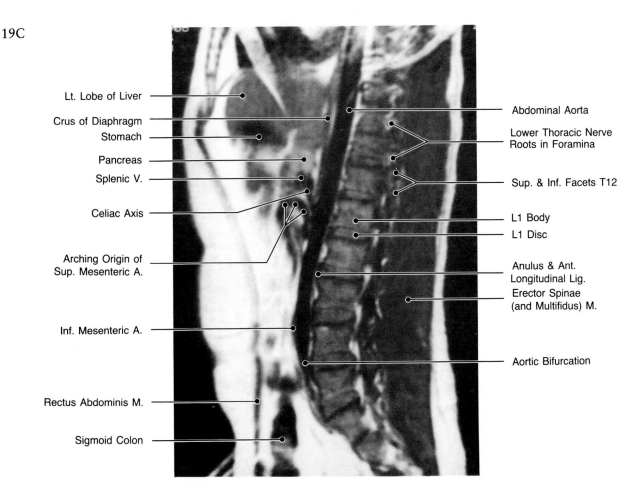

Lt. Lobe of Liver

Crus of Diaphragm

Stomach

Pancreas

Splenic V.

Celiac Axis

Abdominal Aorta

Lower Thoracic Nerve
Roots in Foramina

Sup. & Inf. Facets T12

L1 Body

L1 Disc

Arching Origin of
Sup. Mesenteric A.

Anulus & Ant.
Longitudinal Lig.

Erector Spinae
(and Multifidus) M.

Inf. Mesenteric A.

Aortic Bifurcation

Rectus Abdominis M.

Sigmoid Colon

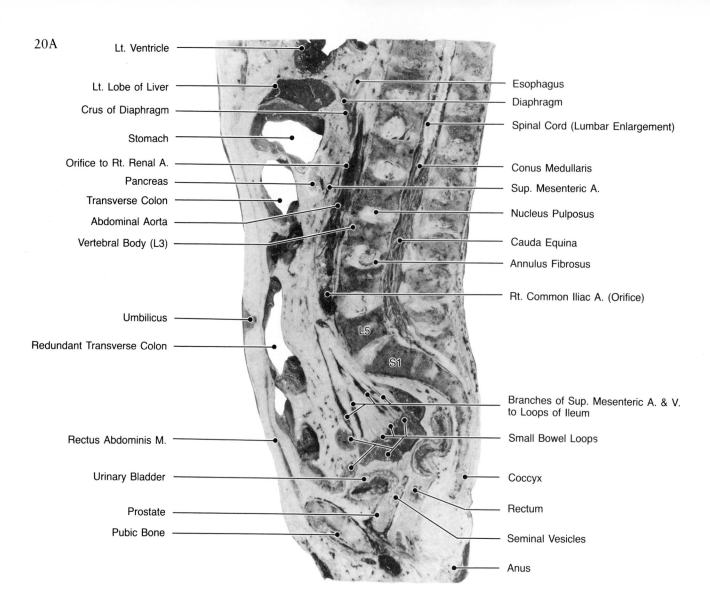

20A

Lt. Ventricle

Lt. Lobe of Liver

Crus of Diaphragm

Stomach

Orifice to Rt. Renal A.

Pancreas

Transverse Colon

Abdominal Aorta

Vertebral Body (L3)

Umbilicus

Redundant Transverse Colon

Rectus Abdominis M.

Urinary Bladder

Prostate

Pubic Bone

Esophagus

Diaphragm

Spinal Cord (Lumbar Enlargement)

Conus Medullaris

Sup. Mesenteric A.

Nucleus Pulposus

Cauda Equina

Annulus Fibrosus

Rt. Common Iliac A. (Orifice)

L5

S1

Branches of Sup. Mesenteric A. & V. to Loops of Ileum

Small Bowel Loops

Coccyx

Rectum

Seminal Vesicles

Anus

20A Sagittal/Anatomic Plate/Midline. The discs are well prepared with good visualization of the nucleus pulposus and anulus fibrosus in this midline section. The spinal cord is identified with the conus medullaris and termination in the filum and the numerous nerve roots of the cauda equina. Since we are sectioning the right wall of the abdominal aorta, the orifice at L1 interspace is to the right renal artery and at the bifurcation is to the common right iliac artery. A small portion of the distal esophagus is seen just above the diaphragm behind the heart and small remaining left lobe of the liver. Posterior to stomach and transverse colon is the pancreas, which lies just

anterior to the superior mesenteric artery. More of the redundant transverse colon is identified in the lower midabdomen anteriorly. Urinary bladder, prostate, and seminal vesicles are again identified anterior to the rectum.

20B Sagittal/MRI/Midline. This magnetic resonance imaging from the living patient seen in Figures 17 through 19 demonstrates the abdominal aorta and right renal artery as on the anatomic specimen. In this young patient the intervertebral disc elements are well demonstrated, and CSF is seen down to the thecal caudal sac at S2. Pancreas lies just anterior to superior mesenteric vessels, while stomach and transverse colon lie just anterior to them.

20C Sagittal/MRI/Midline. This magnetic resonance image demonstrates more clearly the small intestine without the blurring because it was performed on a cadaver. Mesentric vessels and the pancreas have the same anatomic relationship as on the living patient. This section is about 1 cm to the right of the midline and demonstrates the origin of the common iliac artery and vein on the right. A portion of the urinary bladder and right lobe of the prostate are also appreciated.

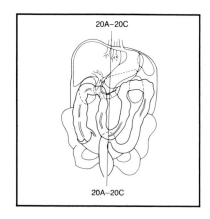

20A–20C

20A–20C

20B

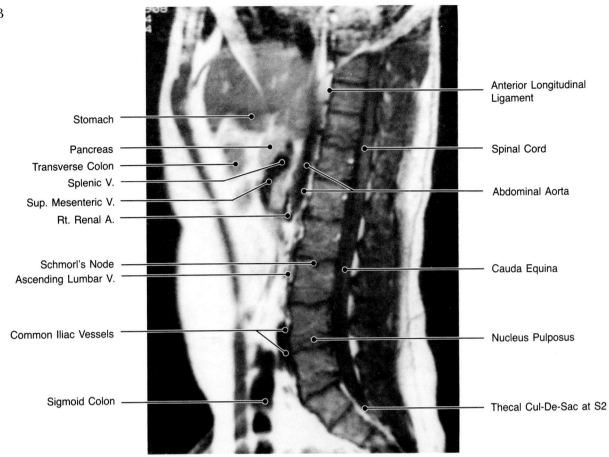

Stomach

Pancreas

Transverse Colon

Splenic V.

Sup. Mesenteric V.

Rt. Renal A.

Schmorl's Node

Ascending Lumbar V.

Common Iliac Vessels

Sigmoid Colon

Anterior Longitudinal Ligament

Spinal Cord

Abdominal Aorta

Cauda Equina

Nucleus Pulposus

Thecal Cul-De-Sac at S2

20C

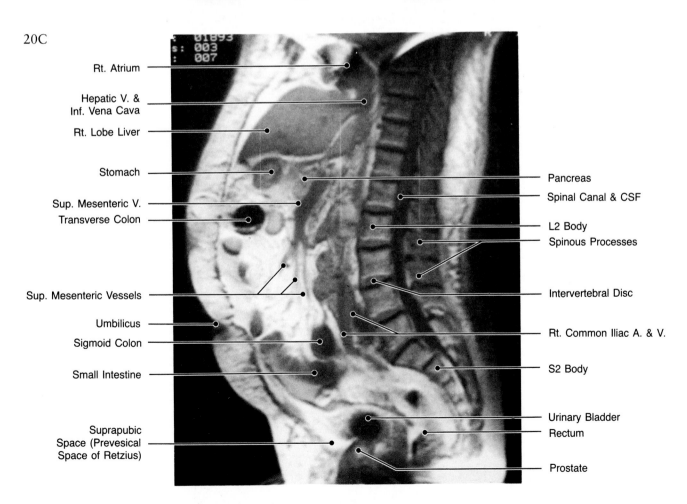

Rt. Atrium

Hepatic V. & Inf. Vena Cava

Rt. Lobe Liver

Stomach

Sup. Mesenteric V.

Transverse Colon

Sup. Mesenteric Vessels

Umbilicus

Sigmoid Colon

Small Intestine

Suprapubic Space (Prevesical Space of Retzius)

Pancreas

Spinal Canal & CSF

L2 Body

Spinous Processes

Intervertebral Disc

Rt. Common Iliac A. & V.

S2 Body

Urinary Bladder

Rectum

Prostate

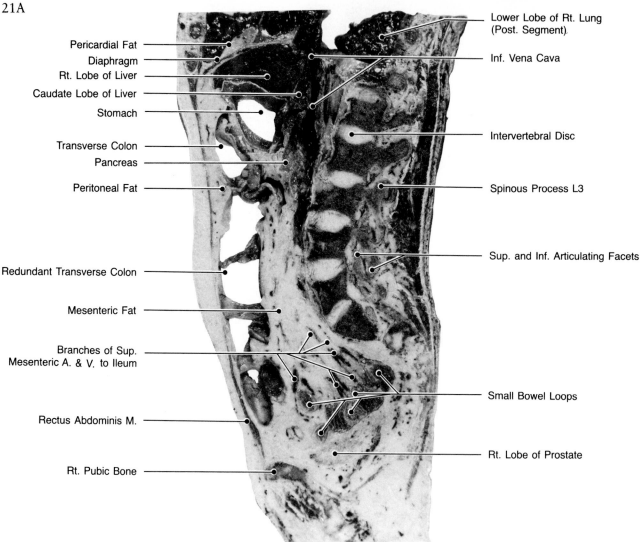

21A

Lower Lobe of Rt. Lung (Post. Segment)

Pericardial Fat

Diaphragm

Rt. Lobe of Liver

Caudate Lobe of Liver

Stomach

Transverse Colon

Pancreas

Peritoneal Fat

Redundant Transverse Colon

Mesenteric Fat

Branches of Sup. Mesenteric A. & V. to Ileum

Rectus Abdominis M.

Rt. Pubic Bone

Inf. Vena Cava

Intervertebral Disc

Spinous Process L3

Sup. and Inf. Articulating Facets

Small Bowel Loops

Rt. Lobe of Prostate

21A Sagittal/Anatomic Plate/Inferior Vena Cava. Two centimeters to the right of the midline the inferior vena cava appears collapsed on this anatomic specimen as it courses through its intrahepatic portion towards the right atrium. Both right lobe and caudate lobe of the liver are identified. Below the stomach the lumen of the redundant transverse colon may be seen. Mesenteric blood supply to ileal loops in the pelvis is also demonstrated.

21B Sagittal/MRI/Inferior Vena Cava. The dominant structure in this magnetic resonance image of this young adult male is the inferior vena cava seen from its right common iliac origin to the right atrium. Prominently seen in the exit foramina of the lumbar spine are the lumbar nerve roots themselves. These are highlighted by the fat in these foramina. Stomach and transverse colon maintain their same relationship as seen in Figures 17 through 20 in this living patient, and the pancreatic head and the portal vein lie just posterior to them.

21C Sagittal/MRI/Inferior Vena Cava. This is the second magnetic resonance image from our cadaver patient seen in Figure 20C. It demonstrates the venous structures from external iliac vein through the inferior vena cava and into its intrahepatic portion. Below the stomach and transverse colon are a number of small bowel loops surrounded by the bright T1-weighted signal from mesenteric and omental fat.

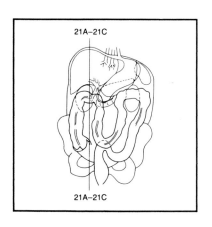

21A–21C

21A–21C

21B

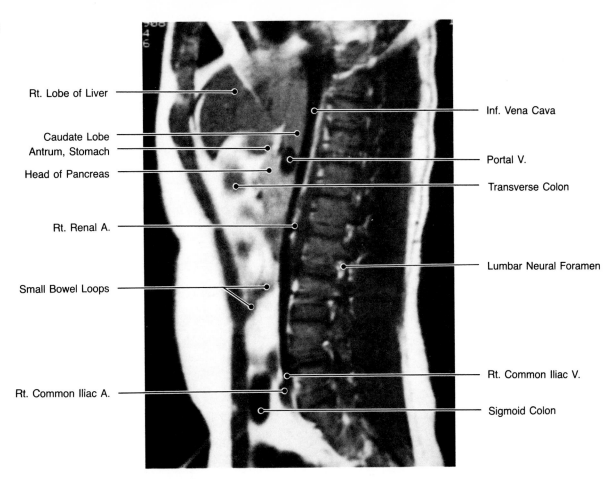

Rt. Lobe of Liver

Caudate Lobe
Antrum, Stomach

Head of Pancreas

Rt. Renal A.

Small Bowel Loops

Rt. Common Iliac A.

Inf. Vena Cava

Portal V.

Transverse Colon

Lumbar Neural Foramen

Rt. Common Iliac V.

Sigmoid Colon

21C

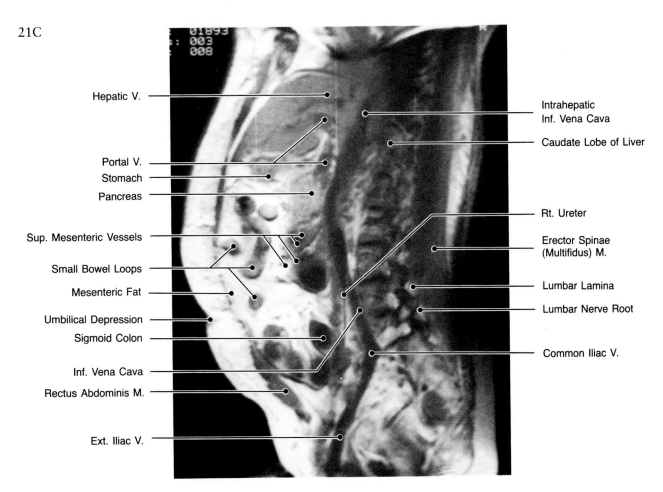

Hepatic V.

Portal V.
Stomach
Pancreas

Sup. Mesenteric Vessels

Small Bowel Loops

Mesenteric Fat

Umbilical Depression

Sigmoid Colon

Inf. Vena Cava

Rectus Abdominis M.

Ext. Iliac V.

Intrahepatic
Inf. Vena Cava

Caudate Lobe of Liver

Rt. Ureter

Erector Spinae
(Multifidus) M.

Lumbar Lamina

Lumbar Nerve Root

Common Iliac V.

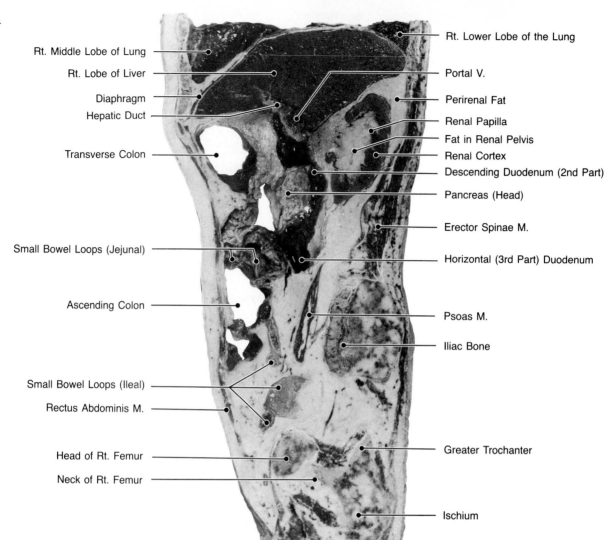

Rt. Middle Lobe of Lung

Rt. Lobe of Liver

Diaphragm

Hepatic Duct

Transverse Colon

Small Bowel Loops (Jejunal)

Ascending Colon

Small Bowel Loops (Ileal)

Rectus Abdominis M.

Head of Rt. Femur

Neck of Rt. Femur

Rt. Lower Lobe of the Lung

Portal V.

Perirenal Fat

Renal Papilla

Fat in Renal Pelvis

Renal Cortex

Descending Duodenum (2nd Part)

Pancreas (Head)

Erector Spinae M.

Horizontal (3rd Part) Duodenum

Psoas M.

Iliac Bone

Greater Trochanter

Ischium

22A Sagittal/Anatomic Plate/Right Kidney. Five centimeters to the right of midline the portal vein and hepatic duct are demonstrated. This section is at the level of the right kidney. The duodenum encircles the back wall of the head of the pancreas. The second portion is seen in its entirety, stained with bile. Ilial loops in the low abdomen and pelvis approach the region of the cecum and ascending colon.

22B Sagittal/MRI/Right Kidney. This and the subsequent MR specimen span the distance from the previous living MR patients images to 5 cm to the right of midline. The portal vein is a large signal void between quadrate and caudal lobes of the

liver and just above the antrum and duodenum and pancreas. Portions of the upper pole of the right kidney and some of the hilar structures of the kidney are demonstrated as well. In the lower abdomen the erector spinae and psoas muscles predominate.

22C Sagittal/MRI/Right Kidney. This section is through the main right lobe of the liver and shows the central portal venous structure at the hilus. The antrum of the stomach and duodenum lying on the anterior surface of the upper pole of the right kidney are demonstrated. Intervening loops of the hepatic flexure of the colon lie above the small bowel which is seen as a grayish density in the mid abdomen. The anterior location of the ascending colon in the right lower quadrant is demonstrated.

22D Sagittal/Supplemental Anatomic Plate/3 cm to the right of midline. This plate clearly identifies the transverse colon with its redundant mid abdominal loop extending toward the pelvis. This plate is included to demonstrate, as indicated in the line drawing, the redundant transverse colonic loop. On the anatomic plate the appendices epiploicae of the transverse colon are also well demonstrated together with the inferior transverse portion of this redundant transverse colon. In the pelvis are a number of small bowel loops along with small bowel mesentery. The bile-stained descending duodenum and right ala of the sacrum are also identified in this right parasagittal plane.

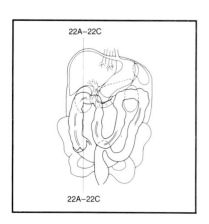

22A–22C

22A–22C

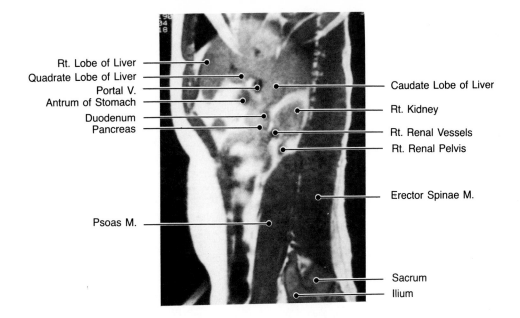

22B

Rt. Lobe of Liver
Quadrate Lobe of Liver
Portal V.
Antrum of Stomach
Duodenum
Pancreas

Caudate Lobe of Liver
Rt. Kidney
Rt. Renal Vessels
Rt. Renal Pelvis

Erector Spinae M.

Psoas M.

Sacrum
Ilium

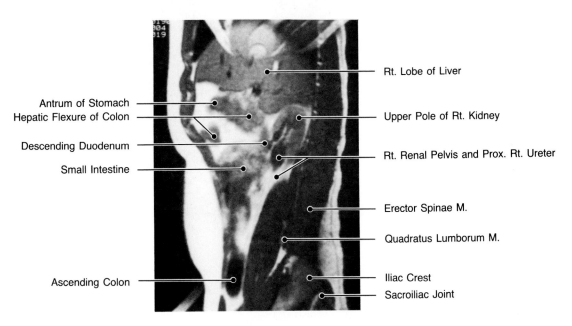

22C

Antrum of Stomach
Hepatic Flexure of Colon

Descending Duodenum

Small Intestine

Rt. Lobe of Liver

Upper Pole of Rt. Kidney

Rt. Renal Pelvis and Prox. Rt. Ureter

Erector Spinae M.

Quadratus Lumborum M.

Ascending Colon

Iliac Crest
Sacroiliac Joint

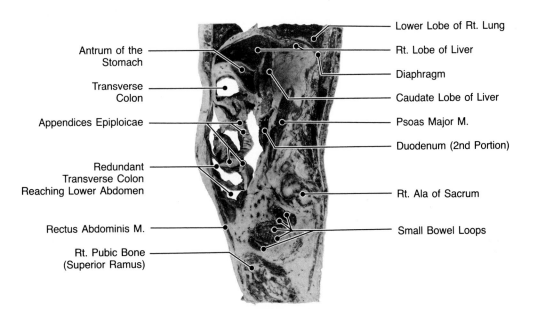

22D

Antrum of the
Stomach

Transverse
Colon

Appendices Epiploicae

Redundant
Transverse Colon
Reaching Lower Abdomen

Rectus Abdominis M.

Rt. Pubic Bone
(Superior Ramus)

Lower Lobe of Rt. Lung
Rt. Lobe of Liver
Diaphragm
Caudate Lobe of Liver
Psoas Major M.
Duodenum (2nd Portion)

Rt. Ala of Sacrum

Small Bowel Loops

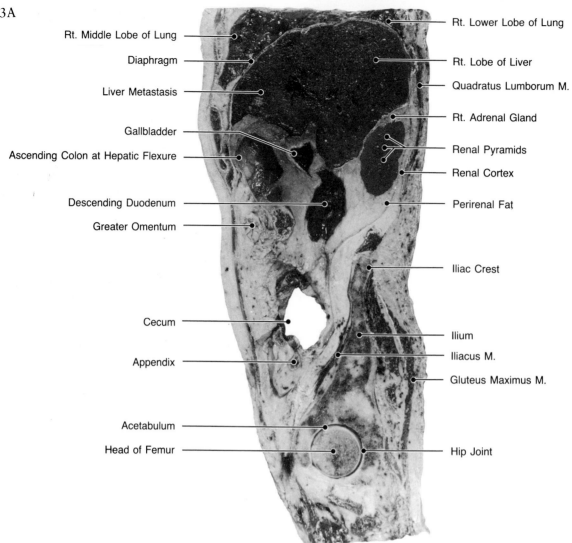

Rt. Middle Lobe of Lung

Diaphragm

Liver Metastasis

Gallbladder

Ascending Colon at Hepatic Flexure

Descending Duodenum

Greater Omentum

Cecum

Appendix

Acetabulum

Head of Femur

Rt. Lower Lobe of Lung

Rt. Lobe of Liver

Quadratus Lumborum M.

Rt. Adrenal Gland

Renal Pyramids

Renal Cortex

Perirenal Fat

Iliac Crest

Ilium

Iliacus M.

Gluteus Maximus M.

Hip Joint

23A Sagittal/Anatomic Plate/Hepatic Hilus. This anatomic plate, 10 cm right of midline, demonstrates the bile-stained second portion of duodenum and the gallbladder. Considerable bile staining from the sectioning of the gallbladder can be seen in the hepatic hilar area. In the right lower quadrant the cecum and appendix are identified surrounded by omentum anteriorly, mesentery superiorly, and retroperitoneal fat posteriorly. A small portion of the lateral aspect of the right kidney demonstrates renal pyramids and cortex as well as the perinephric fat.

23B Sagittal/MRI/Hepatic Hilus. This MRI plate is adjacent to the one seen in Figure 22C and demonstrates the body of the liver and the central portal vein signal void. Hepatic flexure of colon lies beneath the right lobe of the liver, its anterior extent going toward the transverse colon while its posterior extent proceeds to the ascending colon. Portions of the right renal hilus and kidney can be seen together with the relationships of the psoas iliacus and quadratus lumborum muscles to the iliac bone.

23C Sagittal/MRI/Hepatic Hilus. This section of the abdomen demonstrates the full anterior to posterior dimension of the right lobe of the liver with the right portal vein branching in it. Small bowel loops and the cecal tip are seen in the right mid and lower abdomen. The hepatic flexure side of the transverse colon is seen beneath the anterior portion of the liver.

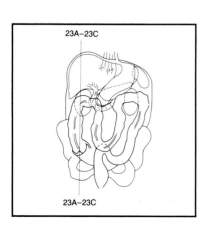

23A–23C

23A–23C

23B

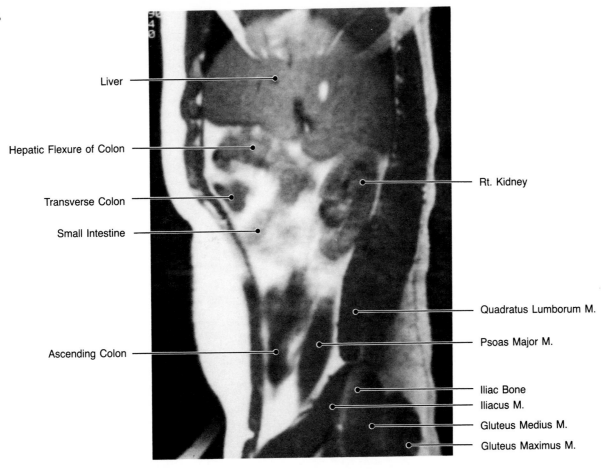

Liver

Hepatic Flexure of Colon

Transverse Colon

Small Intestine

Ascending Colon

Rt. Kidney

Quadratus Lumborum M.

Psoas Major M.

Iliac Bone
Iliacus M.
Gluteus Medius M.
Gluteus Maximus M.

23C

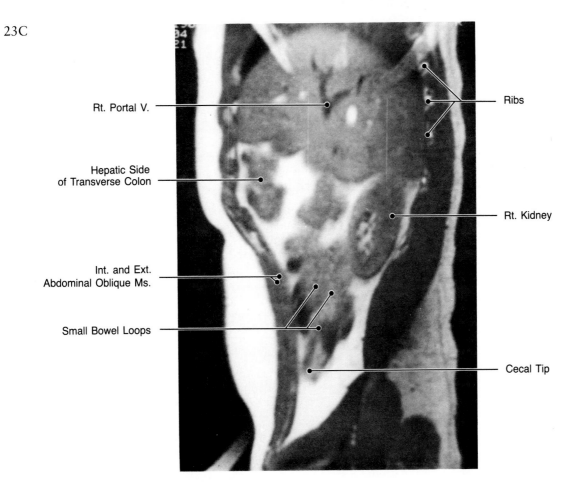

Rt. Portal V.

Hepatic Side
of Transverse Colon

Int. and Ext.
Abdominal Oblique Ms.

Small Bowel Loops

Ribs

Rt. Kidney

Cecal Tip

SECTION 5

Pelvis

Charles Mueller, M. D.
A. J. Christoforidis, M. D., Ph. D.
John Negulesco, Ph. D.

A total of six cadavers were used for the study of the pelvis, three males and three females.

The preparation of the anatomic specimen was made according to the technique described in the Introduction. All the axial anatomic sections were accompanied by at least one computed tomographic and one magnetic resonance image. These images were selected from a number of clinical cases. In selecting the sections emphasis was placed on those that would correspond as accurately as possible to the anatomic preparations. In the axial male pelvis the anatomic sections were made following the CT imaging of the cadaver. Accuracy of this correlation was secured by placing skin markers before the CT examination as a guide to the sectioning that took place following the freezing of the cadaver, as explained in the Introduction. The cadaver CT images were obtained with an 8800 GE unit, while all the CT images of the living subjects were made with either a fourth-generation 1440 or a 2060Q CT Technicare unit. The MRI images were obtained with the 1.5 Tesla GE unit, using mainly partial saturation with TR 400 to 2000 msec while the TE ranged from 20 to 80 msec.

Sagittal and coronal anatomic sections are correlated with magnetic resonance images. As known, these planes are not used as a rule, for direct imaging in computed tomography. The direct three-dimensional imaging capability of the magnetic resonance is of distinct advantage. The lack of any significant reflection of the respiration on the pelvic organs makes respiratory gating not mandatory, although desirable. Actually all our pelvic magnetic resonance images were obtained originally without the benefit of respiratory gating. The sagittal sections of the male pelvis were obtained from a patient who had large amount of ascites as a result of extensive liver metastases with significant jaundice. This resulted in the staining of the ascitic fluid. The presence of ascites displaces the small bowel loops as expected. We considered the inclusion of these anatomic pathologic changes of interest to the clinician, as they did not interfere to any significant degree with the study of the anatomy.

The "monotony" of the normal anatomy was interrupted by a few cases with pathology related to organs and structures of the pelvis. Among them was a case of carcinoma of the urinary bladder (Fig. 6D), a case of carcinoma of the cervix extending to the uterus (Fig. 11C), and a metastatic osteolytic involvement of the iliac bone (Fig. 11D) complementing the CT images of normal subjects, while a left ovarian cyst is shown in one of the magnetic resonance sections (Fig. 10C).

It will be superfluous to remind the reader of the occasional discrepancy between the "surface" photography of the anatomic specimen and the two-dimensional rendition of the volume image of the computed tomographic and magnetic resonance section.

Bibliography

Bo WJ, Meschan I, Krueger, WA: *Basic Atlas of Cross-Sectional Anatomy*. Philadelphia, W. B. Saunders Co., 1980.

Butler H, Bryan PJ, LiPuma JP, et al: Magnetic resonance imaging of the abnormal female pelvis. *AJR: 143*:1259, 1984.

Deutsch AL, Gosink BB: Normal female pelvic anatomy. *Semin Roentgenol 17*:241, 1982.

Gross BH, Moss AA, Mihara K, et al: Review: Computed tomography of gynecologic diseases. *AJR 141*:765, 1983.

Haaga JR. Alfidi RT: Computed Tomography of the Whole Body. St. Louis, C. V. Mosby, 1983.

Hricak H, Alpers C, Crooks LE, et al. Magnetic resonance imaging of the female pelvis, initial experience. *AJR 141*:1119, 1983.

Hricak H, Williams RD, Spring DB, et al: Anatomy and pathology of the male pelvis by magnetic resonance imaging. *AJR 141*:1101, 1983.

Koritké JG, Sick H: *Atlas of Sectional Human Anatomy:*

Frontal, Sagittal and Horizontal Planes. Baltimore, Urban and Schwarzenberg, 1983.

Moss A, Garmsu G, Genant HK: *Computed Tomography of the Body:* Philadelphia, W. B. Saunders, 1983.

Pernkopf E: Atlas of Topographical and Applied Human Anatomy. 2nd ed. Philadelphia, W. B. Saunders Company, 1964.

Pozniak M, Petasnick JP, Matalon TAS, et al: Computed tomography in the differential diagnosis of pelvic and extrapelvic disease. (RG), 5:587, 1985.

Wagner M, Lawson T: *Segmental Anatomy: Applications to Clinical Medicine.* New York, Macmillan, 1982.

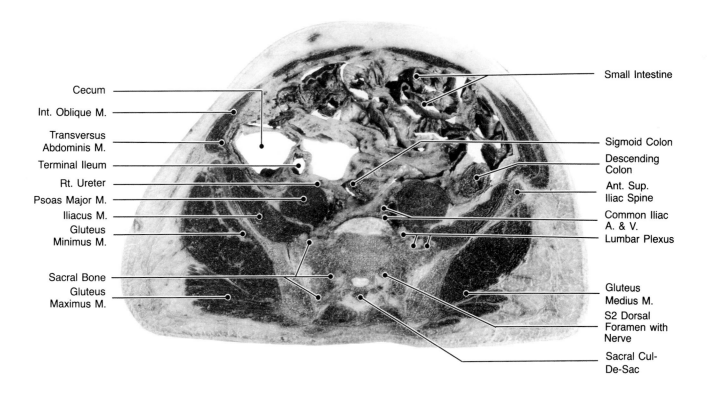

Small Intestine

Cecum

Int. Oblique M.

Transversus
Abdominis M.

Terminal Ileum

Rt. Ureter

Psoas Major M.

Iliacus M.

Gluteus
Minimus M.

Sacral Bone
Gluteus
Maximus M.

Sigmoid Colon

Descending
Colon

Ant. Sup.
Iliac Spine

Common Iliac
A. & V.

Lumbar Plexus

Gluteus
Medius M.

S2 Dorsal
Foramen with
Nerve

Sacral Cul-
De-Sac

1A Axial/Anatomic Plate/Common Iliac Level. This anatomic plate demonstrates all three gluteal muscles on the right. The large anterior root of the spinal nerve of L5 contributing to the lumbosacral plexus can be seen lying just posterior and medial to the iliopsoas muscle. The posterior sacral nerve root exit and SI nerve in it may be seen bilaterally as well. The thecal cul-de-sac at the S2 level is seen in the midline. Intraperitoneal structures consist chiefly of small bowel and its mesentery. A segment of midline sigmoid colon between the psoas muscles and common iliac muscles is identified.

1B Axial/CT/Common Iliac Level. The cadaver CT demonstrates calcifications in the common iliac arteries bilaterally. Fascial planes are seen to separate the three gluteal muscles. The right ureter is identified on the anterior surface of the right psoas muscle.

1C Axial/CT/Sacral Cul-de-Sac. The CT from the living patient demonstrates the L5-S1 intervertebral disc and the sacral cul-de-sac more posterior to it.

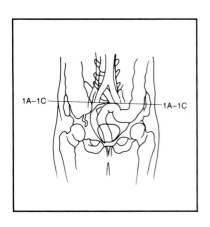

1A–1C

1A–1C

1B

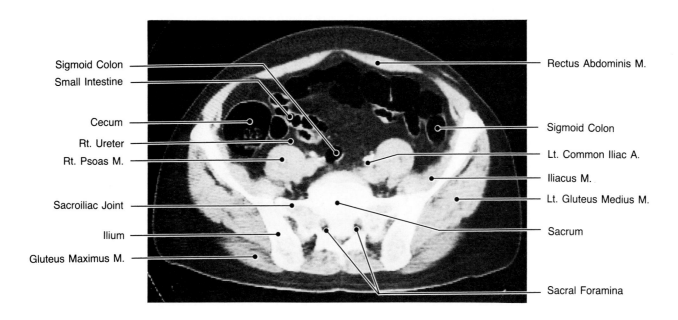

Sigmoid Colon

Small Intestine

Cecum

Rt. Ureter

Rt. Psoas M.

Sacroiliac Joint

Ilium

Gluteus Maximus M.

Rectus Abdominis M.

Sigmoid Colon

Lt. Common Iliac A.

Iliacus M.

Lt. Gluteus Medius M.

Sacrum

Sacral Foramina

1C

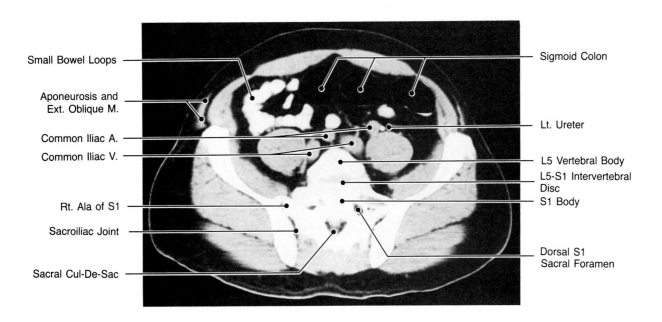

Small Bowel Loops

Aponeurosis and
Ext. Oblique M.

Common Iliac A.

Common Iliac V.

Rt. Ala of S1

Sacroiliac Joint

Sacral Cul-De-Sac

Sigmoid Colon

Lt. Ureter

L5 Vertebral Body

L5-S1 Intervertebral
Disc

S1 Body

Dorsal S1
Sacral Foramen

1D

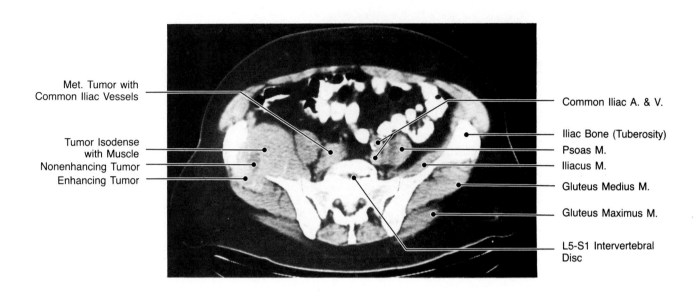

Met. Tumor with
Common Iliac Vessels

Tumor Isodense
with Muscle
Nonenhancing Tumor
Enhancing Tumor

Common Iliac A. & V.

Iliac Bone (Tuberosity)
Psoas M.
Iliacus M.
Gluteus Medius M.

Gluteus Maximus M.

L5-S1 Intervertebral
Disc

1D, 1E Axial/CT/Expanding Bone Tumor. Two CT images at the same iliac window as the anatomic plate clearly demonstrate pathology in the form of a large inhomogeneous tumor arising in the right iliac wing with elements that are isodense with muscle, hypodense with muscle, and clearly enhancing compared with muscle. This renal cell carcinoma metastasis may be seen at bone window (Fig. 1E) to destroy the iliac wing from its medial margin at the tuberosity to most lateral to the right sacroiliac joint.

1F Axial/MRI/Sacrum. This magnetic resonance image done with partial saturation technique demonstrates the sacrum and sacroiliac articulations and adjacent iliac bones because marrow fat in these bony structures returns a moderately bright signal. T1 weighted images show fat as a bright signal, and in this patient there is mesenteric and subcutaneous fat in abundance. Incidental note is made of the fibrous scars in the buttocks of this diabetic patient.

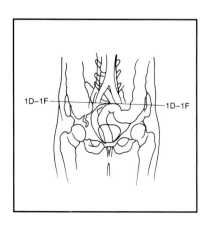

1D–1F   1D–1F

1E

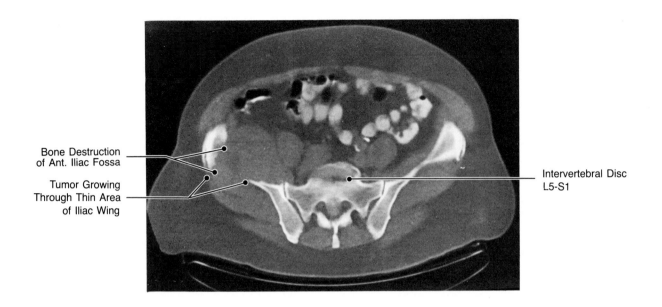

Bone Destruction
of Ant. Iliac Fossa

Tumor Growing
Through Thin Area
of Iliac Wing

Intervertebral Disc
L5-S1

1F

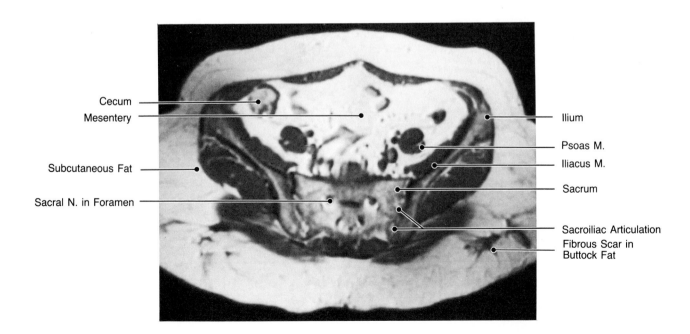

Cecum

Mesentery

Subcutaneous Fat

Sacral N. in Foramen

Ilium

Psoas M.

Iliacus M.

Sacrum

Sacroiliac Articulation

Fibrous Scar in
Buttock Fat

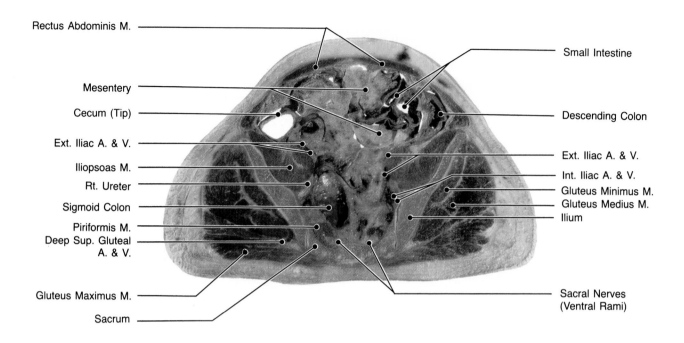

Rectus Abdominis M. —

Small Intestine

Mesentery —

Cecum (Tip) —

Descending Colon

Ext. Iliac A. & V. —

Iliopsoas M. —

Ext. Iliac A. & V.

Rt. Ureter —

Int. Iliac A. & V.

Sigmoid Colon —

Gluteus Minimus M.

Piriformis M. —

Gluteus Medius M.

Deep Sup. Gluteal A. & V. —

Ilium

Gluteus Maximus M. —

Sacrum —

Sacral Nerves (Ventral Rami)

2A Axial/Anatomic Plate/Sacral Fossa. The anatomic plate shows that the ventral concavity of the sacrum is much deeper at this level and contains the ventral rami of sacral nerves, forming the sacral plexus. On the right the piriformis muscle is well developed. Anterior to the piriformis the sigmoid colon and mesentery for large and small bowel fill the pelvis. On the right the ureter is seen to lie between the iliopsoas and the sigmoid colon, while on the left, internal and external iliac vessels are seen in tandem along the medial iliopsoas muscle.

2B Axial/CT/Sacral Fossa. The cadaver CT section demonstrates the same true and false pelvic structures as on the anatomic plate. Additionally, the sartorius muscle may be seen arising from the anterior superior iliac crest just behind the internal abdominal oblique and transverse abdominis muscles.

2C Axial/CT/Sacral Fossa. The CT section in the living patient with contrast enhancement demonstrates similar vascular, muscular, and bony relationships. However, the bowel that is present is chiefly sigmoid colon and the straight rectosigmoid.

2D Axial/MRI/Sacral Fossa. Plate 2D is the magnetic resonance image corresponding to the anatomic and CT plates described above. Piriformis muscles line the sacral fossa laterally, while sigmoid colon is seen in the midline. The superior gluteal artery may be seen on the anterior surface of the piriformis muscle and the continuation outside the pelvis of this vessel, the deep superior gluteal artery can be seen between the gluteus maximus and medius muscles. The sciatic nerve exits the pelvis with the superior gluteal vessels.

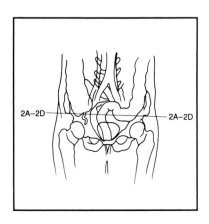

2A–2D   2A–2D

2B

Ext. Iliac A. & V.

Cecum
Sartorius M.
Tensor Fasciae
Latae M.
Iliopsoas M.

Ilium
Sigmoid Colon
Piriformis M.

Small Intestine

Descending Colon

Ext. Iliac A. & V.
Gluteus Minimus M.
Lt. Ureter

Gluteus Medius M.

Gluteus Maximus M.

Sacroiliac Joint

Sacral Vertebra

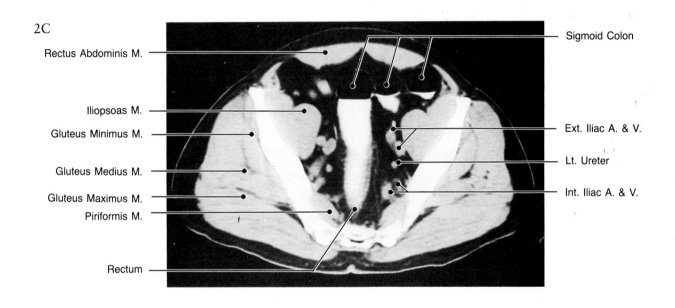

2C

Rectus Abdominis M.

Iliopsoas M.

Gluteus Minimus M.

Gluteus Medius M.

Gluteus Maximus M.
Piriformis M.

Rectum

Sigmoid Colon

Ext. Iliac A. & V.

Lt. Ureter

Int. Iliac A. & V.

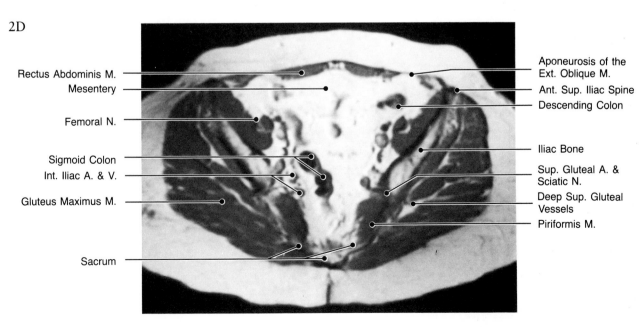

2D

Rectus Abdominis M.
Mesentery

Femoral N.

Sigmoid Colon
Int. Iliac A. & V.

Gluteus Maximus M.

Sacrum

Aponeurosis of the
Ext. Oblique M.
Ant. Sup. Iliac Spine
Descending Colon

Iliac Bone

Sup. Gluteal A. &
Sciatic N.
Deep Sup. Gluteal
Vessels
Piriformis M.

3A

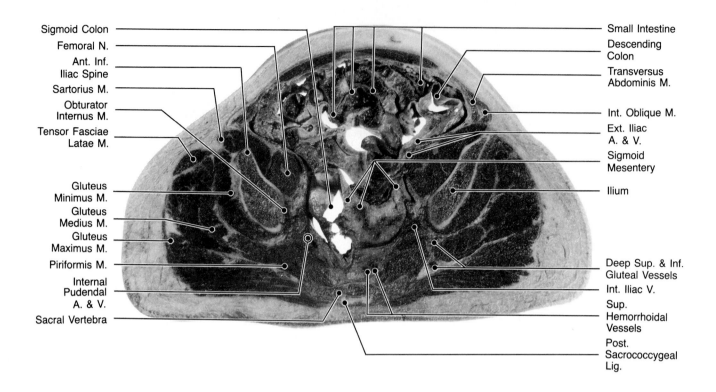

Sigmoid Colon — Small Intestine

Femoral N. — Descending Colon

Ant. Inf. Iliac Spine — Transversus Abdominis M.

Sartorius M. —

Obturator Internus M. — Int. Oblique M.

Tensor Fasciae Latae M. — Ext. Iliac A. & V.

— Sigmoid Mesentery

Gluteus Minimus M. — Ilium

Gluteus Medius M.

Gluteus Maximus M.

Piriformis M. — Deep Sup. & Inf. Gluteal Vessels

Internal Pudendal A. & V. — Int. Iliac V.

Sacral Vertebra — Sup. Hemorrhoidal Vessels

Post. Sacrococcygeal Lig.

3A Axial/Anatomic Male Plate/Greater Sciatic Notch. The anatomic plate at the level of the greater sciatic notch demonstrates the piriformis muscle nearly filling the notch, leaving space only for sciatic nerve and superior gluteal vessels anteriorly. Sigmoid colon and mesentery occupy most of the true pelvic cavity, while small intestine dominates the false pelvic cavity anteriorly. The anterior inferior iliac spine forms the anterior margin of the ilium on the right, and iliopsoas muscle at this level is draped over the spine.

3B Axial/CT/Sciatic Notch. The cadaver plate demonstrates the gas-containing loops of bowel and the fatty mesentery as on the anatomic plate. Further and progressive attenuation of the internal abdominal oblique muscle and nearly vanishingly small transversalis fascia are now present.

3C Axial/CT/Sciatic Notch. This CT study done in the living patient at the level of the greater sciatic notch shows contrast in the rectosigmoid and in a few small bowel loops in the anterior abdomen. Contrast enhancement visualizes the ureters. The piriformis muscle nearly fills the sciatic notch, and the sciatic nerve is seen on its anterior surface surrounded by fat.

3D Axial/MRI/Sciatic Notch. The partial saturation image from a magnetic resonance scan at the corresponding level of the supraacetabular pelvic bones demonstrates the piriformis muscle, internal and external iliac vessels, rectum and sigmoid colon, and an abundance of fat. Femoral nerves in the iliopsoas muscle and sciatic nerves are also identified.

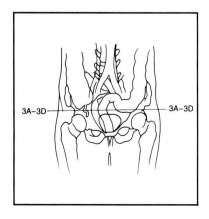

3A–3D   3A–3D

3B

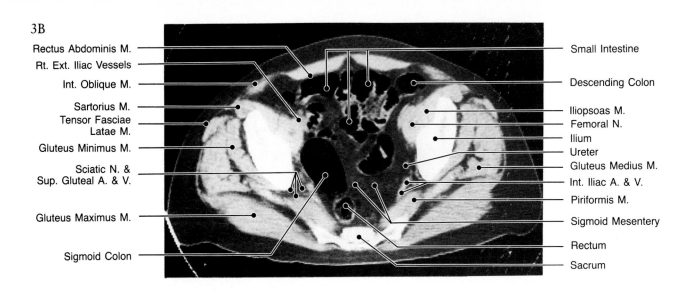

Rectus Abdominis M. — Small Intestine
Rt. Ext. Iliac Vessels
Int. Oblique M. — Descending Colon
Sartorius M.
Tensor Fasciae Latae M. — Iliopsoas M.
— Femoral N.
— Ilium
Gluteus Minimus M. — Ureter
— Gluteus Medius M.
Sciatic N. & Sup. Gluteal A. & V. — Int. Iliac A. & V.
— Piriformis M.
Gluteus Maximus M. — Sigmoid Mesentery
— Rectum
Sigmoid Colon — Sacrum

3C

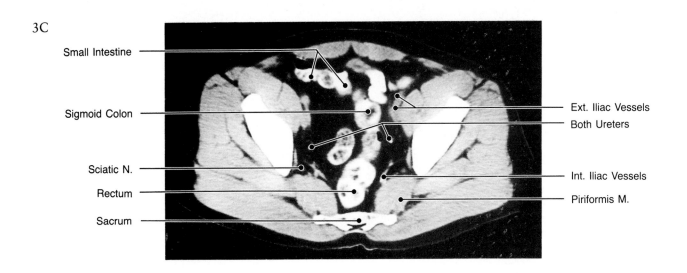

Small Intestine
Sigmoid Colon — Ext. Iliac Vessels
— Both Ureters
Sciatic N.
Rectum — Int. Iliac Vessels
Sacrum — Piriformis M.

3D

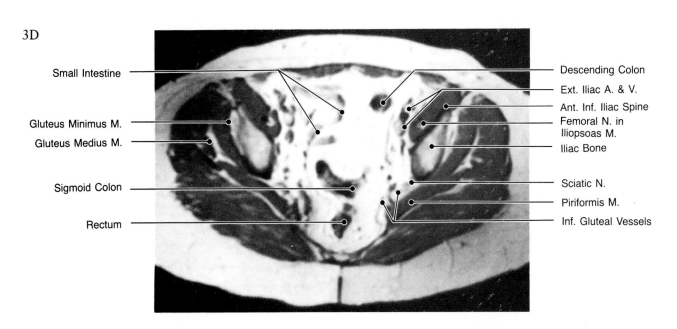

Small Intestine — Descending Colon
— Ext. Iliac A. & V.
— Ant. Inf. Iliac Spine
Gluteus Minimus M. — Femoral N. in Iliopsoas M.
Gluteus Medius M. — Iliac Bone
Sigmoid Colon — Sciatic N.
— Piriformis M.
Rectum — Inf. Gluteal Vessels

307

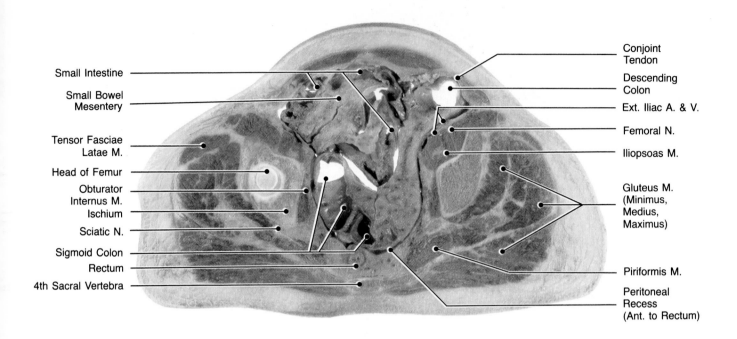

Small Intestine

Small Bowel Mesentery

Tensor Fasciae Latae M.

Head of Femur

Obturator Internus M.

Ischium

Sciatic N.

Sigmoid Colon

Rectum

4th Sacral Vertebra

Conjoint Tendon

Descending Colon

Ext. Iliac A. & V.

Femoral N.

Iliopsoas M.

Gluteus M. (Minimus, Medius, Maximus)

Piriformis M.

Peritoneal Recess (Ant. to Rectum)

4A Axial/Anatomic Plate/Acetabular Roof. The anatomic plate is cut obliquely through the roof of the acetabulum on the right (including a portion of the femoral head) and slightly higher on the left. Peritoneum covers the anterior surface of the rectum and the presacral fat and also the adjacent sigmoid loop seen in continuity with the descending colon. Small intestine and its mesentery fill the rest of the peritoneal cavity.

4B Axial/CT/Acetabular Roof. The cadaver CT demonstrates the widening in the ilium to form the acetabular roof. This CT window emphasizes the bony structures but does not clearly show the head of the femur. Sigmoid colon, rectum and descending colon, and a few small bowel loops are identified in the pelvis, but delineation of the peritoneal surfaces is not as easy as with the anatomic plate.

4C Axial/CT/Acetabular Roof. The CT in the living patient with contrast enhancement at the level of the acetabulum also demonstrates the anterior inferior iliac spine and internal and external iliac vessels. Rectum and sigmoid in this patient are only bowel loops clearly seen. The internal oblique abdominal muscle is thinned at this level to form the conjoint tendon.

4D Axial/MRI/Acetabular Roof. This partial saturation magnetic resonance image is at the level of the acetabular roof. The dark shadowing in the acetabulum is due to the ligamentous structures surrounding the top of the femoral heads. Rectum and a small amount of sigmoid are identified, and to the right of the midline and the sigmoid lies the urine containing bladder (a black signal at partial saturation).

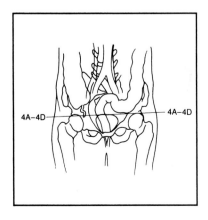

4A–4D

4A–4D

4B

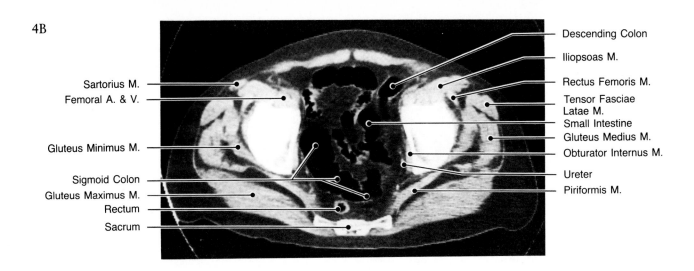

Sartorius M.

Femoral A. & V.

Gluteus Minimus M.

Sigmoid Colon

Gluteus Maximus M.

Rectum

Sacrum

Descending Colon

Iliopsoas M.

Rectus Femoris M.

Tensor Fasciae Latae M.

Small Intestine

Gluteus Medius M.

Obturator Internus M.

Ureter

Piriformis M.

4C

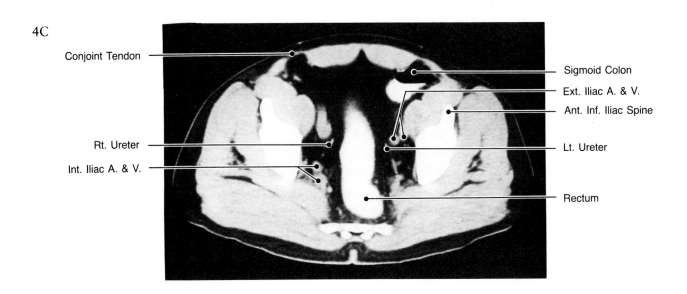

Conjoint Tendon

Rt. Ureter

Int. Iliac A. & V.

Sigmoid Colon

Ext. Iliac A. & V.

Ant. Inf. Iliac Spine

Lt. Ureter

Rectum

4D

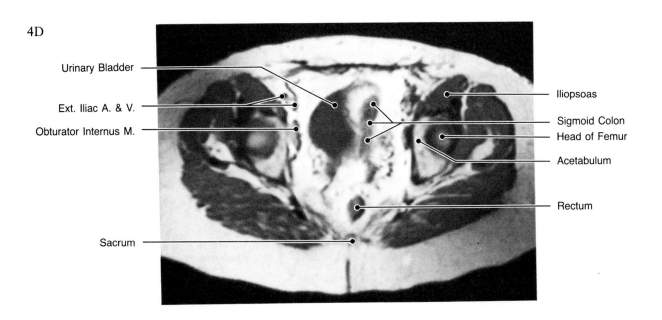

Urinary Bladder

Ext. Iliac A. & V.

Obturator Internus M.

Sacrum

Iliopsoas

Sigmoid Colon

Head of Femur

Acetabulum

Rectum

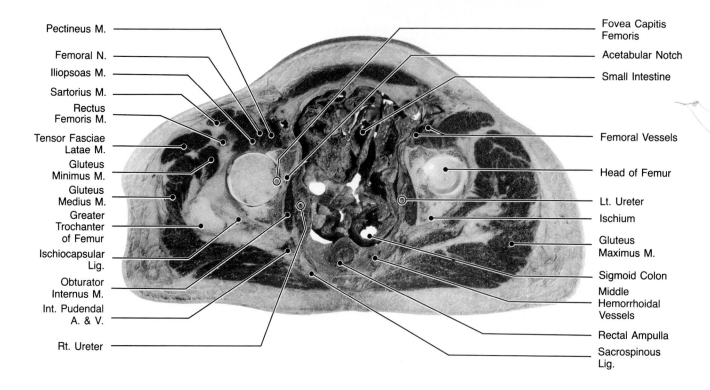

Pectineus M.
Femoral N.
Iliopsoas M.
Sartorius M.
Rectus Femoris M.
Tensor Fasciae Latae M.
Gluteus Minimus M.
Gluteus Medius M.
Greater Trochanter of Femur
Ischiocapsular Lig.
Obturator Internus M.
Int. Pudendal A. & V.
Rt. Ureter

Fovea Capitis Femoris
Acetabular Notch
Small Intestine
Femoral Vessels
Head of Femur
Lt. Ureter
Ischium
Gluteus Maximus M.
Sigmoid Colon
Middle Hemorrhoidal Vessels
Rectal Ampulla
Sacrospinous Lig.

5A Axial/Anatomic Plate/Femoral Head. This anatomic plate at the level of each femoral head identifies on the right the mid acetabular level because of the presence of the fovea capitis on the medial femoral head in direct apposition to the acetabular notch. The strong ischiocapsular ligament supporting the acetabulofemoral joint is also present posterior to the right femoral head. From near the spinous level of the ischium the sacrospinous ligament is identified, and in the presacral fat hemorrhoidal vessels are seen around the rectum. Both ureters are identified in the extraperitoneal fat just medial to the obturator internus muscles.

5B Axial/CT/Femoral Head. The CT cadaver plate clearly demonstrates the sacrospinous ligament, the rectum, and both acetabula. Nondescript intestinal loops are seen elsewhere in the pelvis.

5C Axial/CT/Femoral Head. CT demonstrates the pelvis at slightly lower level in this living patient with contrast in the bladder and outlining the edges of the rectum. The acetabular notch and both acetabula are seen with the obturator notch on the posterior aspect of each superior pubic ramus. Through this notch run the obturator artery, vein, and nerve.

5D Axial/MRI/Femoral Head. A magnetic resonance image through the femoral head at the level of the fovea capitis on the left shows a full bladder. Both prostate and seminal vesicles are also seen at this level between bladder and rectum. Obturator internus muscle is well developed as are the gluteal muscles.

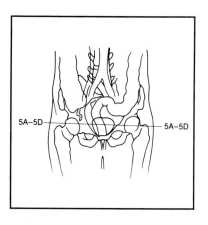

5A–5D   5A–5D

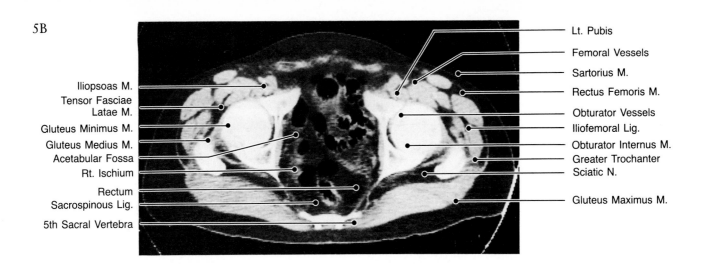

Iliopsoas M.
Tensor Fasciae Latae M.
Gluteus Minimus M.
Gluteus Medius M.
Acetabular Fossa
Rt. Ischium
Rectum
Sacrospinous Lig.
5th Sacral Vertebra

Lt. Pubis
Femoral Vessels
Sartorius M.
Rectus Femoris M.
Obturator Vessels
Iliofemoral Lig.
Obturator Internus M.
Greater Trochanter
Sciatic N.
Gluteus Maximus M.

5C

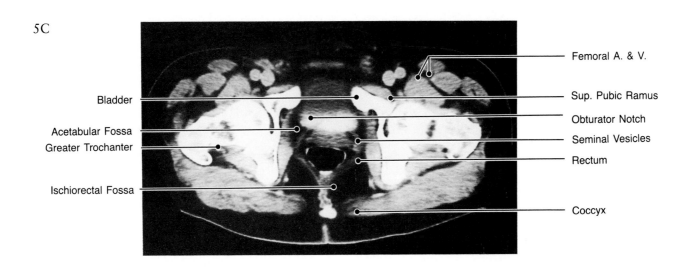

Bladder
Acetabular Fossa
Greater Trochanter
Ischiorectal Fossa

Femoral A. & V.
Sup. Pubic Ramus
Obturator Notch
Seminal Vesicles
Rectum
Coccyx

5D

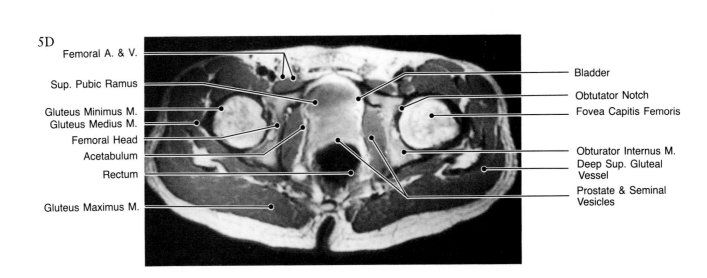

Femoral A. & V.
Sup. Pubic Ramus
Gluteus Minimus M.
Gluteus Medius M.
Femoral Head
Acetabulum
Rectum
Gluteus Maximus M.

Bladder
Obtutator Notch
Fovea Capitis Femoris
Obturator Internus M.
Deep Sup. Gluteal Vessel
Prostate & Seminal Vesicles

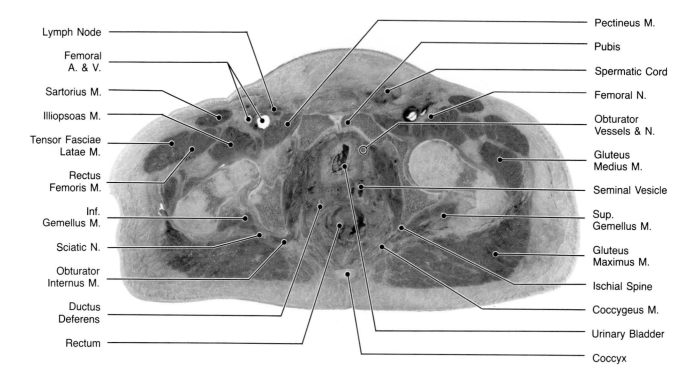

Lymph Node

Femoral
A. & V.

Sartorius M.

Illiopsoas M.

Tensor Fasciae
Latae M.

Rectus
Femoris M.

Inf.
Gemellus M.

Sciatic N.

Obturator
Internus M.

Ductus
Deferens

Rectum

Pectineus M.

Pubis

Spermatic Cord

Femoral N.

Obturator
Vessels & N.

Gluteus
Medius M.

Seminal Vesicle

Sup.
Gemellus M.

Gluteus
Maximus M.

Ischial Spine

Coccygeus M.

Urinary Bladder

Coccyx

6A Axial/Anatomic Plate/Ischial Spine. The anatomic plate demonstrates more of the superior pubic ramus and obturator notch and acetabulum with ischial spines identified posteriorly. On the right the obturator internus muscle is seen to turn lateralward behind the ischial spine. From anterior to posterior in the true pelvis the viscera are the urinary bladder, the seminal vesicles (and ductus deferens on the right), and the rectum. The femoral vessels and the spermatic cord are identified on the left. Muscles anterior to the femur principally for internal rotation and flexion consist of the pectineus, the iliopsoas, the sartorius, the rectus femoris, and the tensor fasciae latae. Three principal external rotators of the femur at this level are the obturator internus and superior and inferior gemellus muscles.

6B Axial/CT/Seminal Vesicles. The CT from the cadaver specimen demonstrates an air-fluid level in the bladder, behind which appears the rounded prostate and the "bow-tie" of the seminal vesicles. The spermatic cord, larger on left than right, is confirmed on this cadaver specimen.

6C Axial/CT/Seminal Vesicles. The contrast-enhanced CT in the living patient demonstrates prominent seminal vesicles between the bladder and the rectum with the ureters lying in the intervening fat between the seminal vesicles and bladder. The spermatic cord and vas deferens are seen in the inguinal canal.

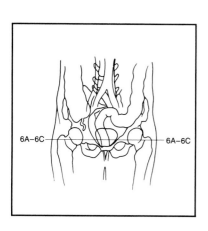

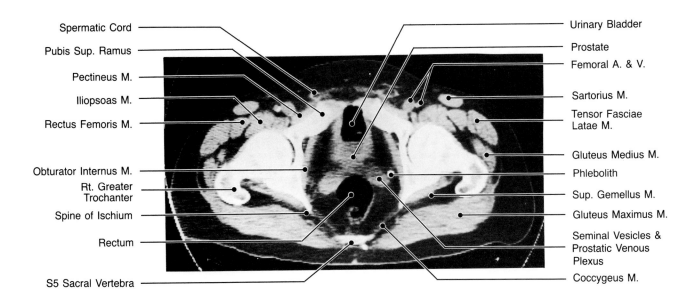

Spermatic Cord
Pubis Sup. Ramus
Pectineus M.
Iliopsoas M.
Rectus Femoris M.

Obturator Internus M.
Rt. Greater Trochanter
Spine of Ischium

Rectum

S5 Sacral Vertebra

Urinary Bladder
Prostate
Femoral A. & V.
Sartorius M.
Tensor Fasciae Latae M.
Gluteus Medius M.
Phlebolith
Sup. Gemellus M.
Gluteus Maximus M.
Seminal Vesicles & Prostatic Venous Plexus
Coccygeus M.

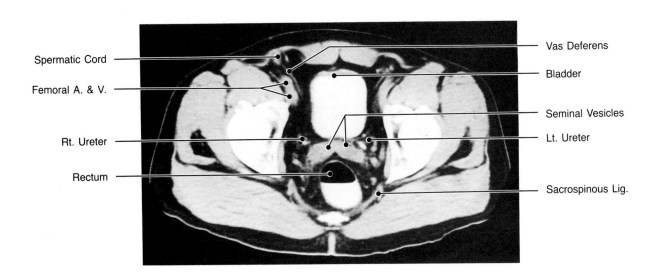

Spermatic Cord
Femoral A. & V.

Rt. Ureter

Rectum

Vas Deferens
Bladder
Seminal Vesicles
Lt. Ureter
Sacrospinous Lig.

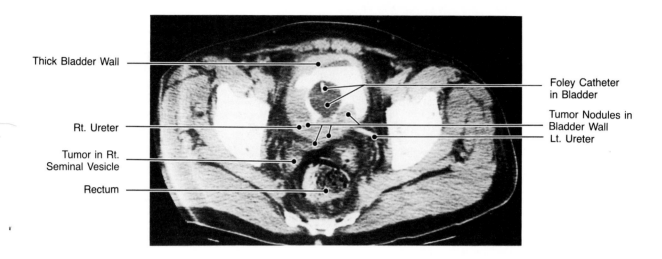

Thick Bladder Wall

Foley Catheter
in Bladder

Rt. Ureter

Tumor Nodules in
Bladder Wall
Lt. Ureter

Tumor in Rt.
Seminal Vesicle

Rectum

6D Axial/CT/Bladder Tumor. This CT demonstrates extensive infiltration, with tumor in the bladder wall and a Foley catheter in the bladder lumen. Some of the tumor extends into the right seminal vesicle and is seen at this level.

6E Axial/MRI/Ischial Spine. A partial saturation magnetic resonance image at this level of another patient shows marrow fat, which is more abundant in the femoral heads and hence projects a brighter signal there than in the pubis. In the obturator notch, fat surrounds the obturator vessels and nerve, and fat in the ischiorectal fossa highlights the levator ani muscles. The ileofemoral and ischiofemoral ligaments have low signal characteristics.

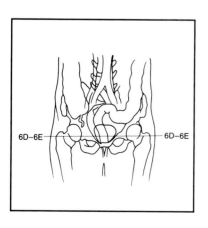

6D–6E

6D–6E

6E

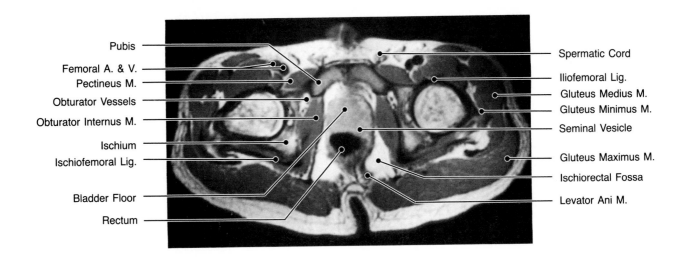

Pubis

Femoral A. & V.

Pectineus M.

Obturator Vessels

Obturator Internus M.

Ischium

Ischiofemoral Lig.

Bladder Floor

Rectum

Spermatic Cord

Iliofemoral Lig.

Gluteus Medius M.

Gluteus Minimus M.

Seminal Vesicle

Gluteus Maximus M.

Ischiorectal Fossa

Levator Ani M.

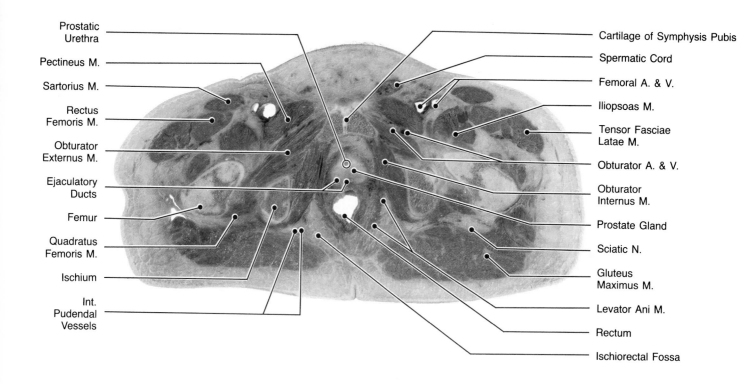

Prostatic Urethra — Cartilage of Symphysis Pubis

Pectineus M. — Spermatic Cord

Sartorius M. — Femoral A. & V.

Rectus Femoris M. — Iliopsoas M.

Obturator Externus M. — Tensor Fasciae Latae M.

Ejaculatory Ducts — Obturator A. & V.

Femur — Obturator Internus M.

Quadratus Femoris M. — Prostate Gland

Ischium — Sciatic N.

Int. Pudendal Vessels — Gluteus Maximus M.

Levator Ani M.

Rectum

Ischiorectal Fossa

7A Axial/Anatomic Plate/Pubic Symphysis. The anatomic plate demonstrates the bony pubis and fibrocartilaginous symphysis bridging the midline. Behind the symphysis is retropubic fat and the prostate gland. The levator ani extends from pubis to the coccyx. The quadratus femoris muscle and sciatic nerve are seen between the ischium and lesser trochanter.

7B Axial/CT/Pubic Symphysis. The CT cadaver plate demonstrates the midline pelvic structures surrounded by the obturator internus muscle. The obturator membrane separates the obturator internus from the obturator externus muscles.

7C Axial/CT/Pubic Symphysis. A CT from the living patient, at a corresponding level, demonstrates the levator sling containing the rectum and prostate and a small amount of contrast in the prostatic urethra. Spermatic cords are prominent structures medial to the femoral vessels and superficial to the pectineus muscle.

7D Axial/MRI/Pubic Symphysis. A partial saturation image of the pelvis at the level of the prostate gland and symphysis pubis demonstrates the corpora cavernosa of the penis just anterior to and between the spermatic cords. The muscle groups are demonstrated anteriorly and posteriorly to the femoral heads.

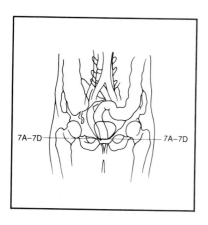

7A–7D   7A–7D

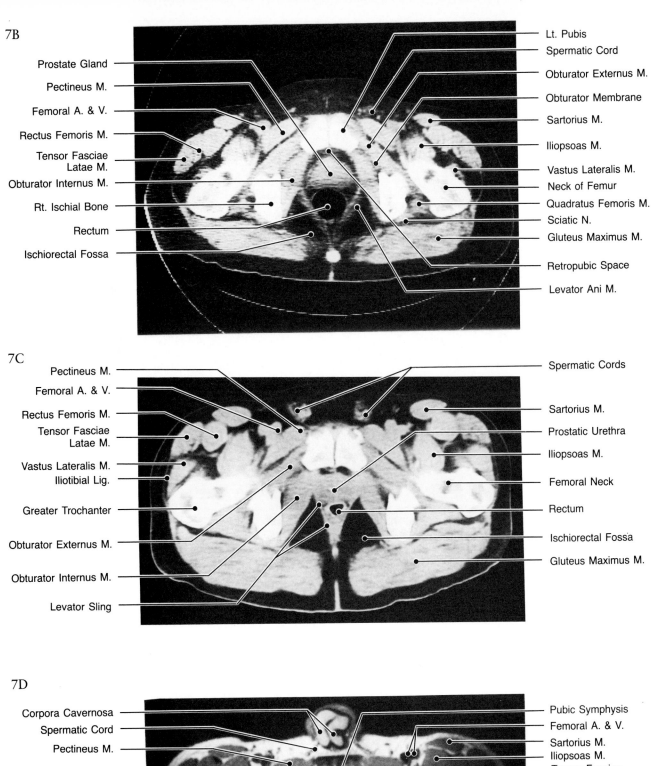

7B

Prostate Gland
Pectineus M.
Femoral A. & V.
Rectus Femoris M.
Tensor Fasciae Latae M.
Obturator Internus M.
Rt. Ischial Bone
Rectum
Ischiorectal Fossa

Lt. Pubis
Spermatic Cord
Obturator Externus M.
Obturator Membrane
Sartorius M.
Iliopsoas M.
Vastus Lateralis M.
Neck of Femur
Quadratus Femoris M.
Sciatic N.
Gluteus Maximus M.
Retropubic Space
Levator Ani M.

7C

Pectineus M.
Femoral A. & V.
Rectus Femoris M.
Tensor Fasciae Latae M.
Vastus Lateralis M.
Iliotibial Lig.
Greater Trochanter
Obturator Externus M.
Obturator Internus M.
Levator Sling

Spermatic Cords
Sartorius M.
Prostatic Urethra
Iliopsoas M.
Femoral Neck
Rectum
Ischiorectal Fossa
Gluteus Maximus M.

7D

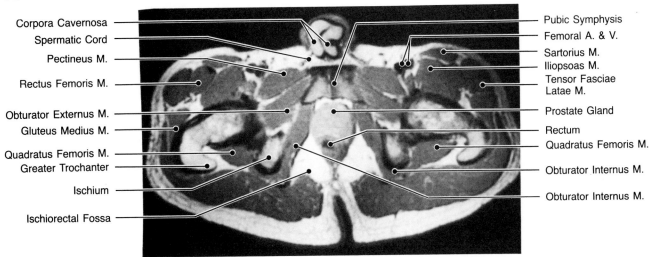

Corpora Cavernosa
Spermatic Cord
Pectineus M.
Rectus Femoris M.
Obturator Externus M.
Gluteus Medius M.
Quadratus Femoris M.
Greater Trochanter
Ischium
Ischiorectal Fossa

Pubic Symphysis
Femoral A. & V.
Sartorius M.
Iliopsoas M.
Tensor Fasciae Latae M.
Prostate Gland
Rectum
Quadratus Femoris M.
Obturator Internus M.
Obturator Internus M.

317

8A

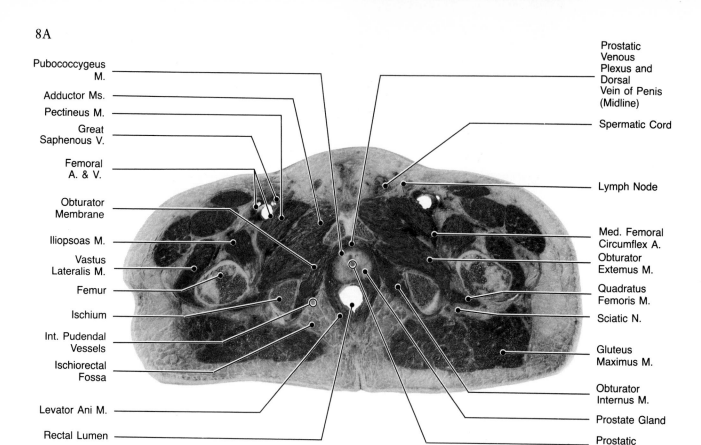

Pubococcygeus M.

Adductor Ms.

Pectineus M.

Great Saphenous V.

Femoral A. & V.

Obturator Membrane

Iliopsoas M.

Vastus Lateralis M.

Femur

Ischium

Int. Pudendal Vessels

Ischiorectal Fossa

Levator Ani M.

Rectal Lumen

Prostatic Venous Plexus and Dorsal Vein of Penis (Midline)

Spermatic Cord

Lymph Node

Med. Femoral Circumflex A.

Obturator Extemus M.

Quadratus Femoris M.

Sciatic N.

Gluteus Maximus M.

Obturator Internus M.

Prostate Gland

Prostatic Urethra

8A Axial/Anatomic Plate/Obturator Foramen. The anatomic plate at the lowest pubic symphysis level demonstrates the obturator membrane, which separates the origins of the medial posteriorly radiating fibers of the obturator internus muscle and the lateral radiating fibers of the obturator externus muscle. The levator sling contains the prostatic urethra and its periprostatic venous plexus and the rectum in the midline. Joining the strong flexor group of muscles to the thigh is the vastus lateralis portion of the quadratus femoris muscle. Internal pudendal vessels lie on the medial surfaces of the obturator muscle.

8B Axial/CT/Obturator Foramen. This plate, the cadaver CT without contrast, demonstrates thigh muscles, rectum, and prostate. From the posterolateral aspect of the ischial tuberosity the tendinous origin for the semitendinosus muscle is identified.

8C Axial/CT/Obturator Foramen. This plate is a contrast-enhanced CT from another living patient who demonstrates the medial femoral circumflex vessels running between the pectineus muscle and the iliopsoas muscle on the right. At this level muscles for flexion, extension, and rotation of the thigh predominate.

8D Axial/MRI/Obturator Foramen. This plate is a magnetic resonance partial saturation image centered at approximately the same level. The femoral artery and vein are demonstrated superficially between the pectineus and the iliopsoas muscles, and deeper between these muscles, the medial femoral circumflex vessels are seen.

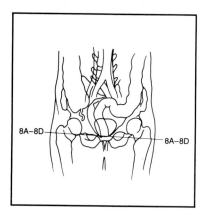

8A–8D   8A–8D

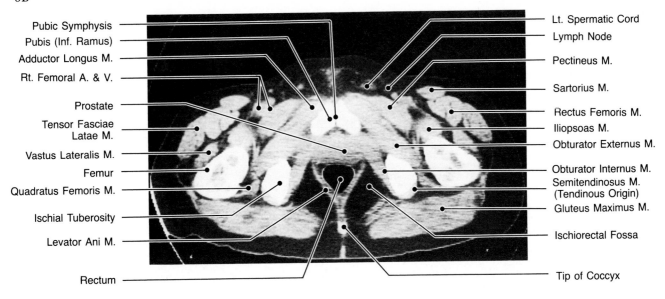

8B

Pubic Symphysis	Lt. Spermatic Cord
Pubis (Inf. Ramus)	Lymph Node
Adductor Longus M.	Pectineus M.
Rt. Femoral A. & V.	Sartorius M.
Prostate	Rectus Femoris M.
Tensor Fasciae Latae M.	Iliopsoas M.
Vastus Lateralis M.	Obturator Externus M.
Femur	Obturator Internus M.
Quadratus Femoris M.	Semitendinosus M. (Tendinous Origin)
Ischial Tuberosity	Gluteus Maximus M.
Levator Ani M.	Ischiorectal Fossa
Rectum	Tip of Coccyx

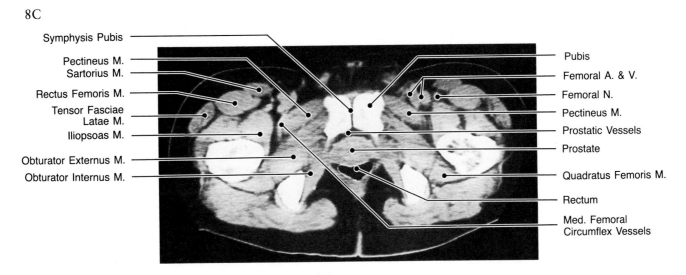

8C

Symphysis Pubis	Pubis
Pectineus M.	Femoral A. & V.
Sartorius M.	Femoral N.
Rectus Femoris M.	Pectineus M.
Tensor Fasciae Latae M.	Prostatic Vessels
Iliopsoas M.	Prostate
Obturator Externus M.	Quadratus Femoris M.
Obturator Internus M.	Rectum
	Med. Femoral Circumflex Vessels

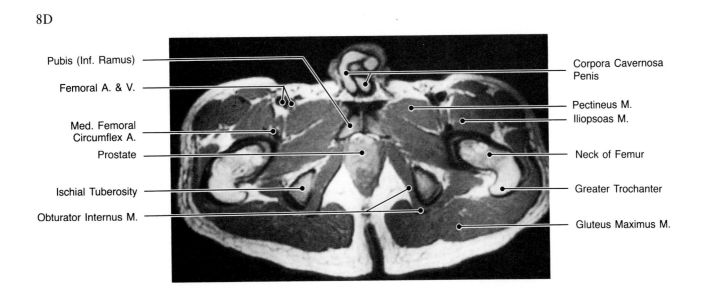

8D

Pubis (Inf. Ramus)	Corpora Cavernosa Penis
Femoral A. & V.	Pectineus M.
	Iliopsoas M.
Med. Femoral Circumflex A.	
Prostate	Neck of Femur
Ischial Tuberosity	Greater Trochanter
Obturator Internus M.	Gluteus Maximus M.

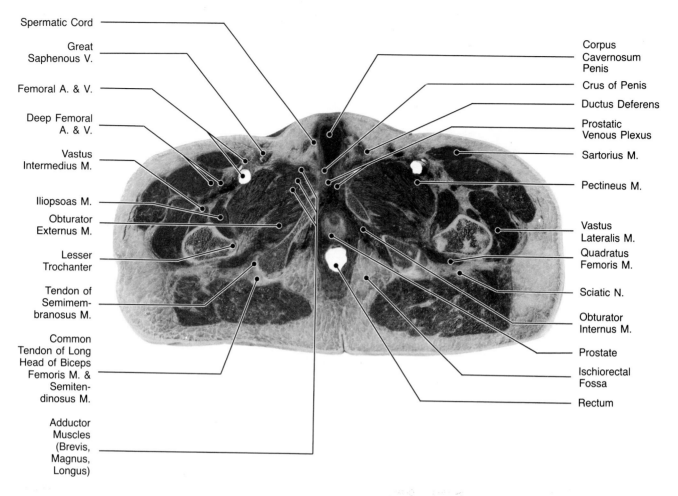

Spermatic Cord

Great
Saphenous V.

Femoral A. & V.

Deep Femoral
A. & V.

Vastus
Intermedius M.

Iliopsoas M.

Obturator
Externus M.

Lesser
Trochanter

Tendon of
Semimem-
branosus M.

Common
Tendon of Long
Head of Biceps
Femoris M. &
Semiten-
dinosus M.

Adductor
Muscles
(Brevis,
Magnus,
Longus)

Corpus
Cavernosum
Penis

Crus of Penis

Ductus Deferens

Prostatic
Venous Plexus

Sartorius M.

Pectineus M.

Vastus
Lateralis M.

Quadratus
Femoris M.

Sciatic N.

Obturator
Internus M.

Prostate

Ischiorectal
Fossa

Rectum

9A Axial/Anatomic Plate/Subpubic Arch. This anatomic plate demonstrates the midline corpora cavernosa of penis and just laterally the spermatic cords and ductus deferens. Between the inferior pubic rami are the prostatic venous plexus and the crura of the penis. At this level the adductor brevis, magnus, and longus muscles are seen as separate bundles lying medial to the pectineus muscle and anterior to the obturator externus muscle. The femur on its anterior surface is convex laterally for the vastus intermedius and vastus lateralis muscles and is concave medially where the iliopsoas muscle inserts on the lesser trochanter. From the posterior lateral aspects of the ischial tuberosities the tendinous origins for semimembranosus, semitendinosus, and long head of biceps femoris muscles originate. The sciatic nerve is just lateral to these.

9B Axial/CT/Subpubic Arch. The cadaver CT demonstrates the cavernous portion, body, and crus of the penis in the midline, and the spermatic cords just lateral. The adductor muscle group and characteristic shape of the intertrochanteric portion of the proximal femur, is repeated.

9C Axial/CT/Subpubic Arch. CT from a living patient demonstrates the corpora cavernosa and spermatic cords anterior to the subpubic arch and the prostate posterior to it. Muscle groups are outlined by intervening fatty layers.

9D Axial/MRI/Subpubic Arch. A partial saturation MR image of the pelvis demonstrates the corpora cavernosa penis and the cavernous urethra in corpus spongiosum and a testicle on the right. Adductor longus and brevis muscles are inseparable and lie anterior to the large adductor magnus muscle.

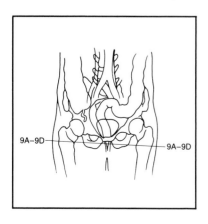

9A–9D   9A–9D

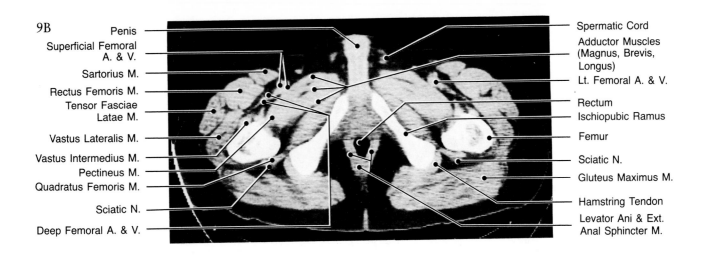

9B

Penis
Superficial Femoral A. & V.
Sartorius M.
Rectus Femoris M.
Tensor Fasciae Latae M.
Vastus Lateralis M.
Vastus Intermedius M.
Pectineus M.
Quadratus Femoris M.
Sciatic N.
Deep Femoral A. & V.

Spermatic Cord
Adductor Muscles (Magnus, Brevis, Longus)
Lt. Femoral A. & V.
Rectum
Ischiopubic Ramus
Femur
Sciatic N.
Gluteus Maximus M.
Hamstring Tendon
Levator Ani & Ext. Anal Sphincter M.

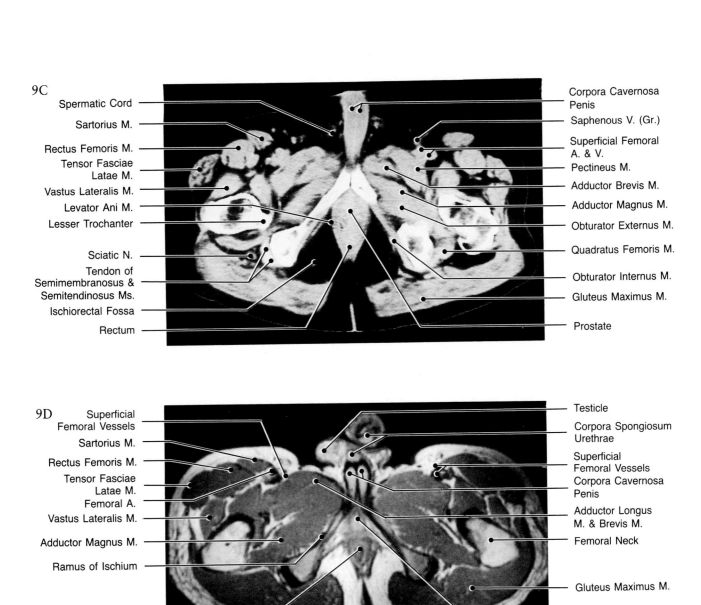

9C

Spermatic Cord
Sartorius M.
Rectus Femoris M.
Tensor Fasciae Latae M.
Vastus Lateralis M.
Levator Ani M.
Lesser Trochanter
Sciatic N.
Tendon of Semimembranosus & Semitendinosus Ms.
Ischiorectal Fossa
Rectum

Corpora Cavernosa Penis
Saphenous V. (Gr.)
Superficial Femoral A. & V.
Pectineus M.
Adductor Brevis M.
Adductor Magnus M.
Obturator Externus M.
Quadratus Femoris M.
Obturator Internus M.
Gluteus Maximus M.
Prostate

9D

Superficial Femoral Vessels
Sartorius M.
Rectus Femoris M.
Tensor Fasciae Latae M.
Femoral A.
Vastus Lateralis M.
Adductor Magnus M.
Ramus of Ischium
Rectum

Testicle
Corpora Spongiosum Urethrae
Superficial Femoral Vessels
Corpora Cavernosa Penis
Adductor Longus M. & Brevis M.
Femoral Neck
Gluteus Maximus M.
Bulb of Penis

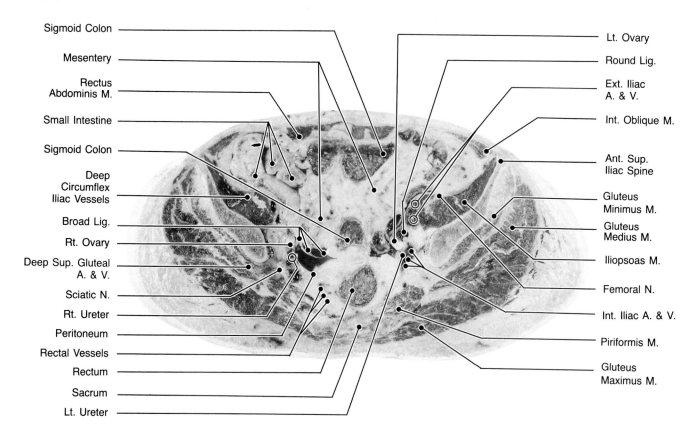

Sigmoid Colon

Mesentery

Rectus Abdominis M.

Small Intestine

Sigmoid Colon

Deep Circumflex Iliac Vessels

Broad Lig.

Rt. Ovary

Deep Sup. Gluteal A. & V.

Sciatic N.

Rt. Ureter

Peritoneum

Rectal Vessels

Rectum

Sacrum

Lt. Ureter

Lt. Ovary

Round Lig.

Ext. Iliac A. & V.

Int. Oblique M.

Ant. Sup. Iliac Spine

Gluteus Minimus M.

Gluteus Medius M.

Iliopsoas M.

Femoral N.

Int. Iliac A. & V.

Piriformis M.

Gluteus Maximus M.

10A Axial/Anatomic Plate/Ovarian Level. The anatomic plate demonstrates the whitish ovaries surrounded by pelvic fat just anterior to the internal iliac vessels and ureters on both sides. On the right, broad ligament is seen running from the medial aspect of the right ovary toward the midline. Abundant fat is seen in the mesentery to small bowel and colon and in the perirectal and subcutaneous areas. This is the level of the greater sciatic notch, and the sciatic nerve and superior gluteal vessels are also identified.

10B Axial/CT/Ovarian Level. A contrast-enhanced CT from a younger female is seen at the same level. In this patient the uterus is seen anterior to the rectum and behind small bowel loops at the level of the greater sciatic notch. Fullness in the left adnexa suggests an ovarian cyst.

10C Axial/CT/Ovarian Level. This CT, is on a different female patient with bilateral ovarian cysts, left larger than right.

10D Axial/MRI/Ovarian Level. This partial saturation magnetic resonance image is from a 29-year-old woman at the same anatomic level. Her cervix and uterus and a right ovarian cyst are all seen at this level.

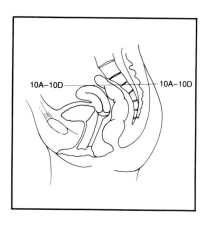

10A–10D   10A–10D

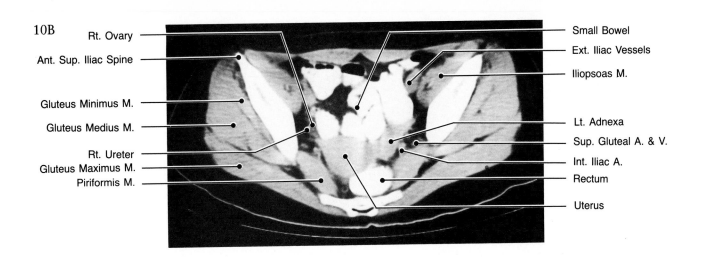

10B

Rt. Ovary
Ant. Sup. Iliac Spine
Gluteus Minimus M.
Gluteus Medius M.
Rt. Ureter
Gluteus Maximus M.
Piriformis M.

Small Bowel
Ext. Iliac Vessels
Iliopsoas M.
Lt. Adnexa
Sup. Gluteal A. & V.
Int. Iliac A.
Rectum
Uterus

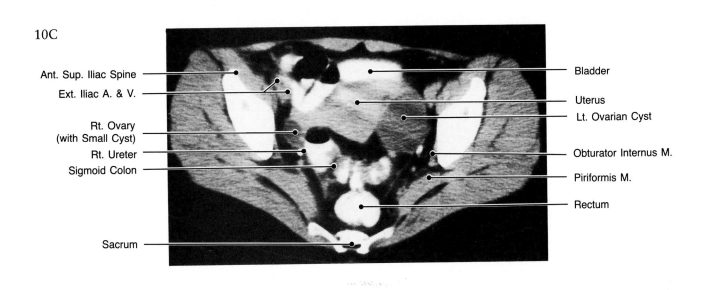

10C

Ant. Sup. Iliac Spine
Ext. Iliac A. & V.
Rt. Ovary
(with Small Cyst)
Rt. Ureter
Sigmoid Colon

Sacrum

Bladder
Uterus
Lt. Ovarian Cyst
Obturator Internus M.
Piriformis M.
Rectum

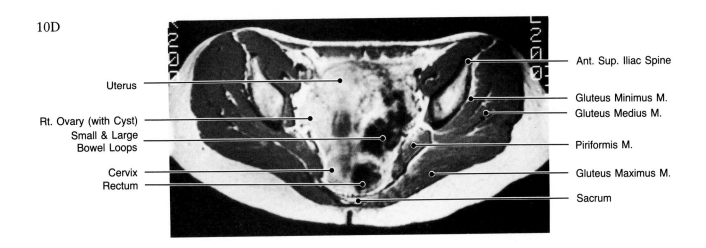

10D

Uterus
Rt. Ovary (with Cyst)
Small & Large
Bowel Loops
Cervix
Rectum

Ant. Sup. Iliac Spine
Gluteus Minimus M.
Gluteus Medius M.
Piriformis M.
Gluteus Maximus M.
Sacrum

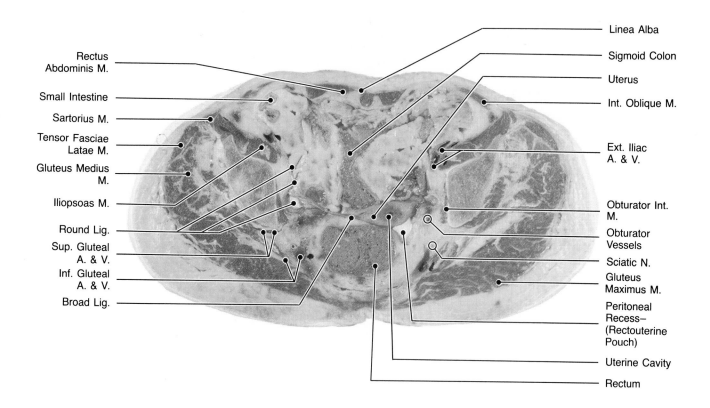

Rectus Abdominis M.
Small Intestine
Sartorius M.
Tensor Fasciae Latae M.
Gluteus Medius M.
Iliopsoas M.
Round Lig.
Sup. Gluteal A. & V.
Inf. Gluteal A. & V.
Broad Lig.

Linea Alba
Sigmoid Colon
Uterus
Int. Oblique M.
Ext. Iliac A. & V.
Obturator Int. M.
Obturator Vessels
Sciatic N.
Gluteus Maximus M.
Peritoneal Recess— (Rectouterine Pouch)
Uterine Cavity
Rectum

11A Axial/Anatomic Plate/Body of Uterus. An anatomic plate 2 cm below the previous plate demonstrates the uterine cavity and body of the uterus in this 85-year-old patient. Posteriorly the rectouterine recess separates the uterus from the rectum, which is skewed slightly to the left of midline. The broad ligament spans from the right wall of the uterus to the adnexal region and continues anteriorly as the round ligament on the right lateral pelvic wall. Sciatic nerve and superior and inferior gluteal vessels are seen in the sciatic notch.

11B Axial/CT/Uterus. The CT in the young living patient with contrast in the gastrointestinal tract and intravenously demonstrates the left broad ligament, the uterus in the midline, and the uterine vascular plexus in the region of the right broad ligament. Circling laterally from the adnexa on the right and coursing anteriorly is the round ligament.

11C Axial/CT/Uterus. This is a CT of another patient at the same level with cervical cancer enlarging the cervix and extending into the uterus.

11D Axial/MRI/Uterus and Ovary. A partial saturation MRI image on a different young female patient demonstrates a left ovary, the uterine wall and cavity, and the uterine vascular plexus in the adnexa region.

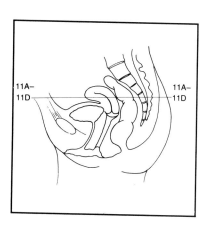

11A—
11D

11A—
11D

11B

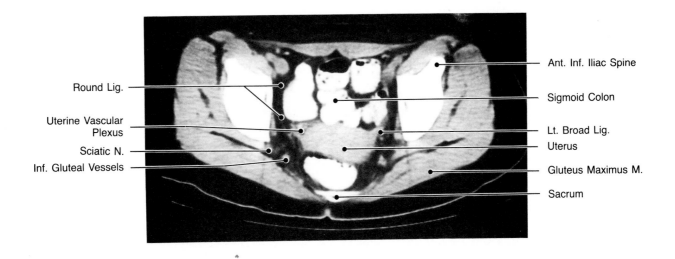

Round Lig.

Uterine Vascular Plexus

Sciatic N.

Inf. Gluteal Vessels

Ant. Inf. Iliac Spine

Sigmoid Colon

Lt. Broad Lig.

Uterus

Gluteus Maximus M.

Sacrum

11C

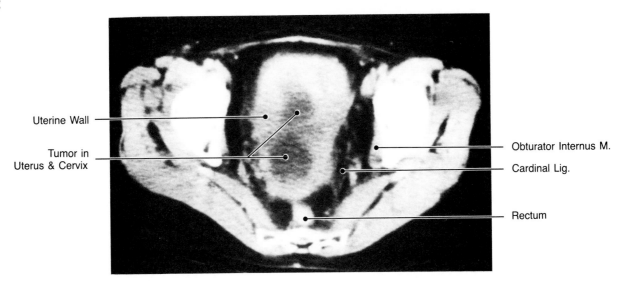

Uterine Wall

Tumor in Uterus & Cervix

Obturator Internus M.

Cardinal Lig.

Rectum

11D

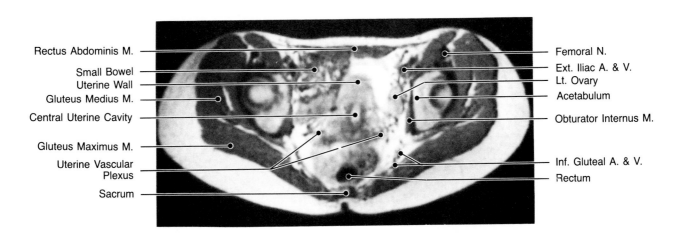

Rectus Abdominis M.

Small Bowel

Uterine Wall

Gluteus Medius M.

Central Uterine Cavity

Gluteus Maximus M.

Uterine Vascular Plexus

Sacrum

Femoral N.

Ext. Iliac A. & V.

Lt. Ovary

Acetabulum

Obturator Internus M.

Inf. Gluteal A. & V.

Rectum

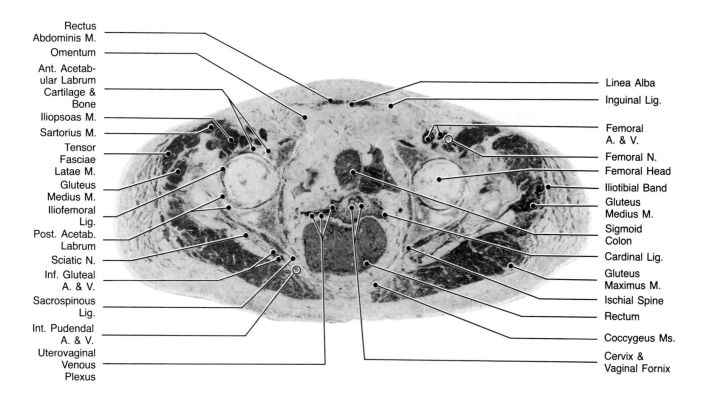

Rectus Abdominis M.
Omentum
Ant. Acetabular Labrum Cartilage & Bone
Iliopsoas M.
Sartorius M.
Tensor Fasciae Latae M.
Gluteus Medius M.
Iliofemoral Lig.
Post. Acetab. Labrum
Sciatic N.
Inf. Gluteal A. & V.
Sacrospinous Lig.
Int. Pudendal A. & V.
Uterovaginal Venous Plexus

Linea Alba
Inguinal Lig.
Femoral A. & V.
Femoral N.
Femoral Head
Iliotibial Band
Gluteus Medius M.
Sigmoid Colon
Cardinal Lig.
Gluteus Maximus M.
Ischial Spine
Rectum
Coccygeus Ms.
Cervix & Vaginal Fornix

12A Axial/Anatomic Plate/Cervix Level. An anatomic plate in an elderly female at the level of the uterine cervix demonstrates also the uterovaginal venous plexus laterally in the cardinal ligament. This level is also the level of the femoral head acetabulum and the ischial spine with elements of the sacrospinous ligament and coccygeus muscle demonstrated.

12B Axial/CT/Cervix. The contrast enhanced CT plate in the living patient at the level of the cardinal ligament shows the right ureter entering the adnexal fat and the dome of the bladder identified on the right. The cervix is seen at the level of the acetabulum; its lateral extension is the cardinal ligament.

12C Axial/MRI/Cervix. The magnetic resonance image done with partial saturation technique at the acetabular level demonstrates the rectum as a signal void to the right of midline and the cervix adjacent to it just to the left of the midline. The uterine body is seen anterior to both of these structures. Ligaments which return poor signal and hence appear black are seen around the capsule of the acetabulofemoral joint.

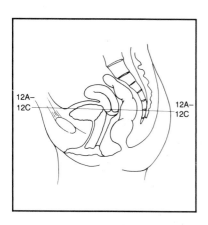

12A–12C

12A–12C

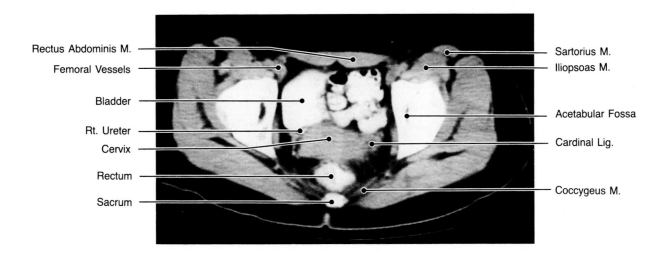

Rectus Abdominis M.
Femoral Vessels
Bladder
Rt. Ureter
Cervix
Rectum
Sacrum

Sartorius M.
Iliopsoas M.
Acetabular Fossa
Cardinal Lig.
Coccygeus M.

12C

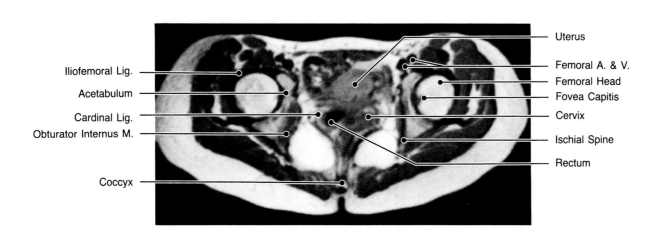

Iliofemoral Lig.
Acetabulum
Cardinal Lig.
Obturator Internus M.
Coccyx

Uterus
Femoral A. & V.
Femoral Head
Fovea Capitis
Cervix
Ischial Spine
Rectum

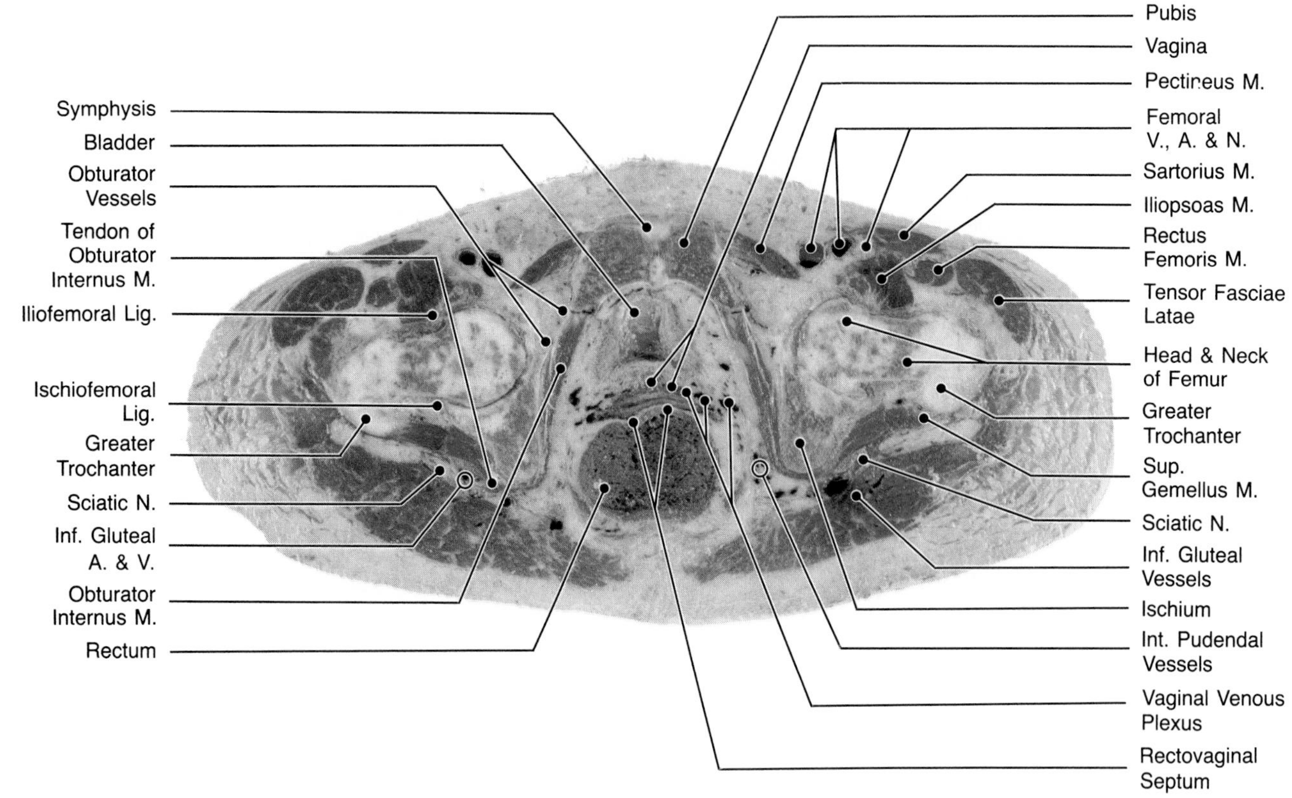
Pubis
Vagina
Pectineus M.
Femoral V., A. & N.
Sartorius M.
Iliopsoas M.
Rectus Femoris M.
Tensor Fasciae Latae
Head & Neck of Femur
Greater Trochanter
Sup. Gemellus M.
Sciatic N.
Inf. Gluteal Vessels
Ischium
Int. Pudendal Vessels
Vaginal Venous Plexus
Rectovaginal Septum
Symphysis
Bladder
Obturator Vessels
Tendon of Obturator Internus M.
Iliofemoral Lig.
Ischiofemoral Lig.
Greater Trochanter
Sciatic N.
Inf. Gluteal A. & V.
Obturator Internus M.
Rectum

13A Axial/Anatomic Plate/Top of Symphysis pubis. The anatomic plate in this patient demonstrates, from anterior to posterior, the symphysis pubis, the urinary bladder, the vagina, the rectovaginal septum and rectum, and levator ani and gluteus maximus muscles. This is also the level of the femoral neck and greater trochanter, and more of the capsular ligaments about the acetabulum are demonstrated.

13B Axial/CT/Symphysis Pubis. Contrast-enhanced CT in the living female patient demonstrates the bladder distended in the midline and the vagina and rectum sequentially posterior to that. Ischiorectal fossa fat is prominent in this patient.

13C Axial/MRI/Symphysis Pubis. Magnetic resonance imaging using partial saturation technique at this level demonstrates the bladder, vagina, and rectum in tandem as on the anatomic and CT images. The muscle groups of the anterior thigh and the obturator muscles predominate.

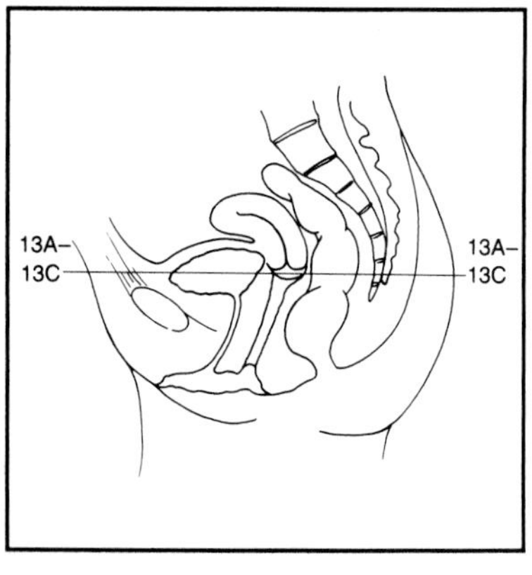
13A–
13C
13A–
13C

13B

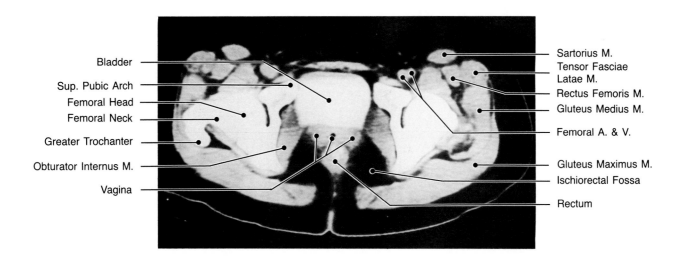

Bladder

Sup. Pubic Arch

Femoral Head

Femoral Neck

Greater Trochanter

Obturator Internus M.

Vagina

Sartorius M.

Tensor Fasciae Latae M.

Rectus Femoris M.

Gluteus Medius M.

Femoral A. & V.

Gluteus Maximus M.

Ischiorectal Fossa

Rectum

13C

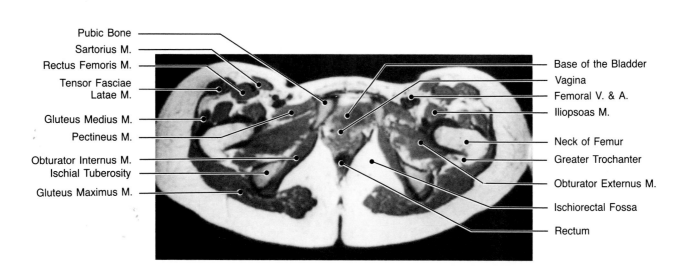

Pubic Bone

Sartorius M.

Rectus Femoris M.

Tensor Fasciae Latae M.

Gluteus Medius M.

Pectineus M.

Obturator Internus M.

Ischial Tuberosity

Gluteus Maximus M.

Base of the Bladder

Vagina

Femoral V. & A.

Iliopsoas M.

Neck of Femur

Greater Trochanter

Obturator Externus M.

Ischiorectal Fossa

Rectum

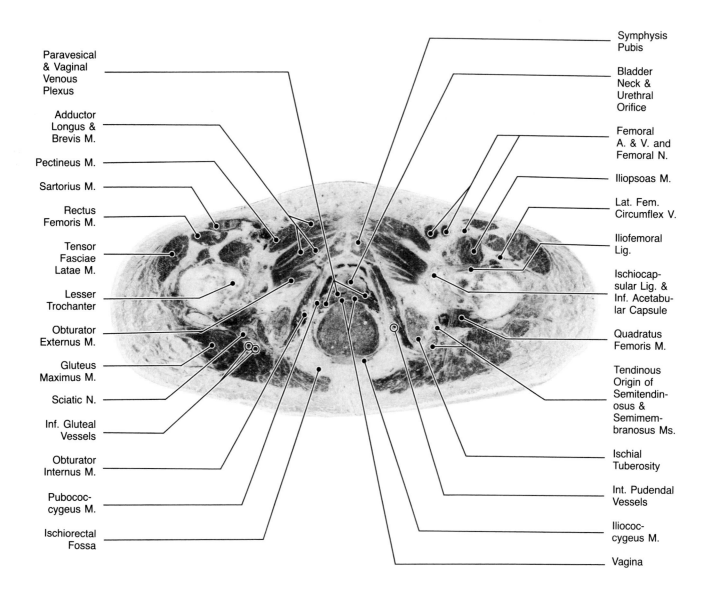

Paravesical & Vaginal Venous Plexus

Adductor Longus & Brevis M.

Pectineus M.

Sartorius M.

Rectus Femoris M.

Tensor Fasciae Latae M.

Lesser Trochanter

Obturator Externus M.

Gluteus Maximus M.

Sciatic N.

Inf. Gluteal Vessels

Obturator Internus M.

Pubococcygeus M.

Ischiorectal Fossa

Symphysis Pubis

Bladder Neck & Urethral Orifice

Femoral A. & V. and Femoral N.

Iliopsoas M.

Lat. Fem. Circumflex V.

Iliofemoral Lig.

Ischiocapsular Lig. & Inf. Acetabular Capsule

Quadratus Femoris M.

Tendinous Origin of Semitendinosus & Semimembranosus Ms.

Ischial Tuberosity

Int. Pudendal Vessels

Iliococcygeus M.

Vagina

14A Axial/Anatomic Plate/Bladder Neck. The anatomic plate at the lower end of the symphysis pubis demonstrates the bladder neck and urethral orifice in the midline surrounded by

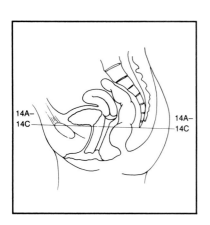

paravesical and paravaginal venous plexuses. Anterior pubococcygeus and posterior iliococcygeus muscles are seen to contribute to the levator sling. This is also the level of the ischial tuberosity, and the tendinous origins of semitendinosus and semimembranosus muscles are seen to arise from the posterolateral aspects of these tuberosities.

14B Axial/CT/Bladder Neck. The CT plate in the living patient shows only the inferior portion of the bladder in the retropubic location. At this level the structures of the peritoneum are not clearly discerned because of an absence of fascial planes and contrast.

14C Axial/MRI/Bladder Neck. The magnetic resonance image at the lower end of the symphysis demonstrates the retropubic space filled with fat and shows the bladder neck region. The anus is represented by a small dot of low density surrounded by fat.

Pectineus M.

Iliotibial Fascia

Obturator Externus M.

Obturator Internus M.

Symphysis Pubis

Bladder

Vagina

Rectum (collapsed)

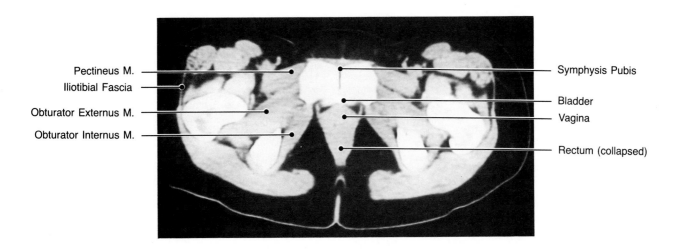

Pectineus M.

Sartorius M.

Rectus Femoris M.

Tensor Fasciae
Latae M.

Iliopsoas M.

Adductor Ms.

Quadratus Femoris M.

Gluteus Maximus M.

Symphysis Pubis

Retropubic Fat

Bladder

Femoral Neck

Anus

Ischial Tuberosity

Ischiorectal Fat

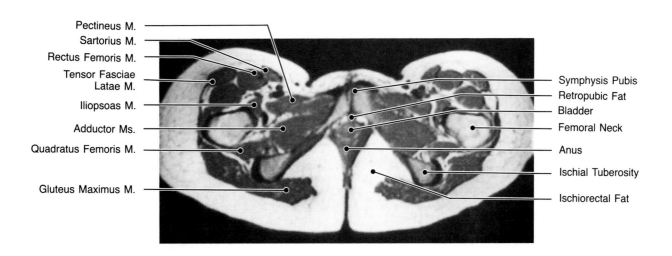

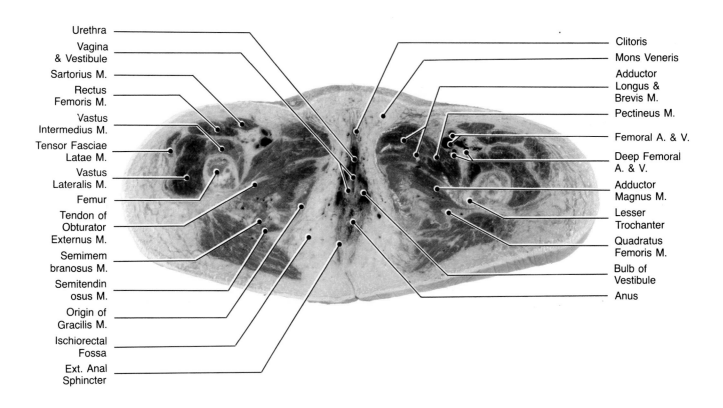

15A Axial/Anatomic Plate/Clitoris and Vestibule. This anatomic plate is at the level of the clitoris in the midline, which lies just anterior to the urethra; this in turn lies just anterior to the vaginal vestibule and the anus. Around these midline structures an abundant venous plexus is seen.

15B Axial/CT/Clitoris-Vagina. This CT demonstrates the thickened vaginal wall due to cervical cancer that had spread to the lower third of it. Other perineal structures, as on the anatomic plate, are shown.

15C Axial/MRI/Clitoris. The magnetic resonance image at the level of the subpubic arch demonstrates the clitoris in the midline anteriorly between the fat-filled mons veneris. The anal canal is flanked by ischiorectal fossa fat, and anterior to it the vagina and urethra open on the peritoneal surface.

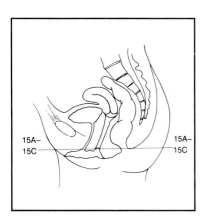

15B

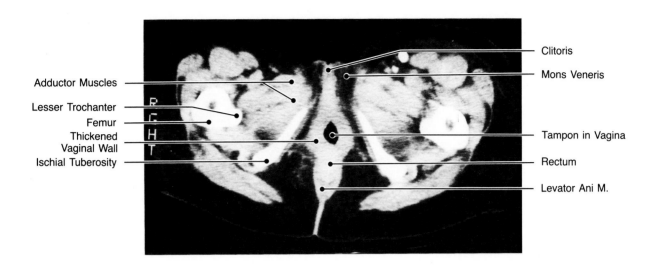

Adductor Muscles

Lesser Trochanter

Femur

Thickened
Vaginal Wall

Ischial Tuberosity

Clitoris

Mons Veneris

Tampon in Vagina

Rectum

Levator Ani M.

15C

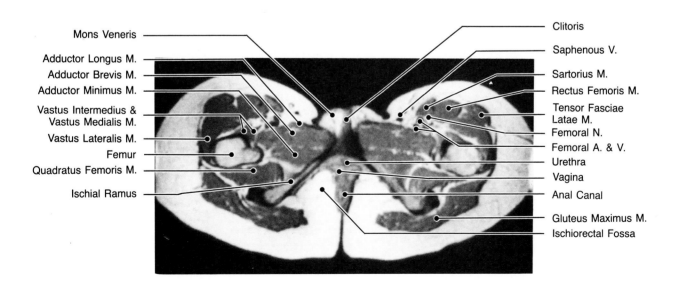

Mons Veneris

Adductor Longus M.

Adductor Brevis M.

Adductor Minimus M.

Vastus Intermedius &
Vastus Medialis M.

Vastus Lateralis M.

Femur

Quadratus Femoris M.

Ischial Ramus

Clitoris

Saphenous V.

Sartorius M.

Rectus Femoris M.

Tensor Fasciae
Latae M.

Femoral N.

Femoral A. & V.

Urethra

Vagina

Anal Canal

Gluteus Maximus M.

Ischiorectal Fossa

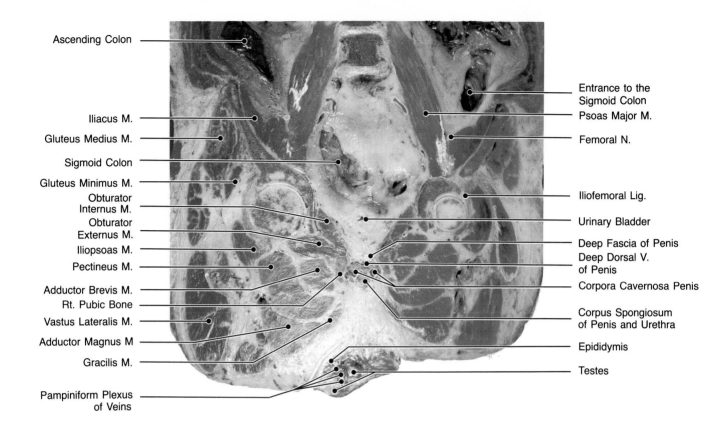

Ascending Colon

Iliacus M.

Gluteus Medius M.

Sigmoid Colon

Gluteus Minimus M.

Obturator
Internus M.

Obturator
Externus M.

Iliopsoas M.

Pectineus M.

Adductor Brevis M.

Rt. Pubic Bone

Vastus Lateralis M.

Adductor Magnus M

Gracilis M.

Pampiniform Plexus
of Veins

Entrance to the
Sigmoid Colon

Psoas Major M.

Femoral N.

Iliofemoral Lig.

Urinary Bladder

Deep Fascia of Penis

Deep Dorsal V.
of Penis

Corpora Cavernosa Penis

Corpus Spongiosum
of Penis and Urethra

Epididymis

Testes

16A Coronal/Anatomic/Anterior Pelvic Area. This male pelvis was sectioned just posterior to the pubic symphysis, and the male genital organs that are seen include the scrotum containing testes, epididymis, and pampiniform plexus and the root of the penis, consisting of copora cavernosa and corpus cavernosum of the urethra and penis. All are contained within the deep fascia of the penis. On the superficial superior aspect is the deep dorsal vein of the penis. Above the suprapubic fat lies the contracted urinary bladder flanked by the obturator internus muscles. The obturator membrane bridges from pubis to ischium and serves as a common origin for both obturator internus and externus muscles. Adductor muscles are demonstrated together with pectineus, gracilis, and sartorius muscles. Ascending and descending colon lie in the fat filling the groove between iliacus and psoas muscles on the right and on the left, respectively.

16B Coronal/MRI/Anterior Pelvic Area. This partial saturation magnetic resonance image is slightly anterior to the anatomic plate and demonstrates the corpora cavernosa of the penis and the corpus cavernosum of the penile urethra. All are contained within the deep fascia of the penis and surrounded by perineal fat. The relationship of the iliopsoas muscle coursing over the anterior pelvic brim is demonstrated bilaterally. Just above the iliac crest the inferior vena cava is seen to be receiving right and left common iliac tributaries.

16C Coronal/MRI/Anterior Pelvic Area. The same male patient as demonstrated in *B*, 1.2 cm posteriorly, demonstrates the epididymis of the right testis receding from the region of the scrotum into the perineal fat region. Femoral heads are clearly seen seated in acetabular fossae, and the iliacus and psoas muscles are much more prominent.

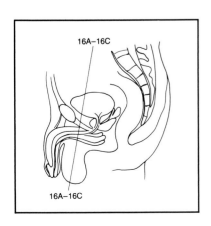

16A–16C

16A–16C

16B

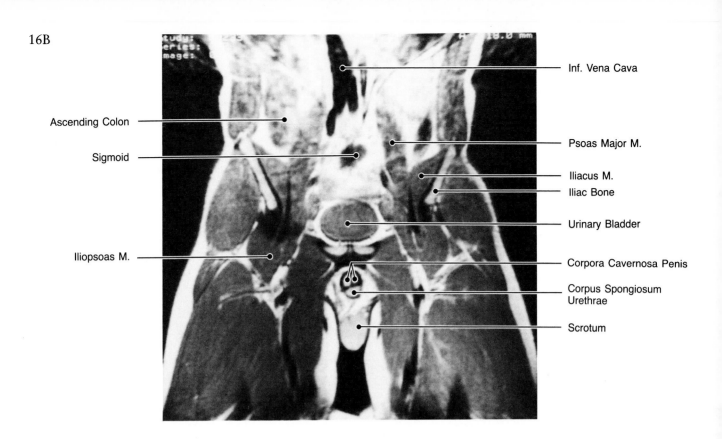

Inf. Vena Cava

Ascending Colon

Sigmoid

Psoas Major M.

Iliacus M.
Iliac Bone

Urinary Bladder

Iliopsoas M.

Corpora Cavernosa Penis

Corpus Spongiosum
Urethrae

Scrotum

16C

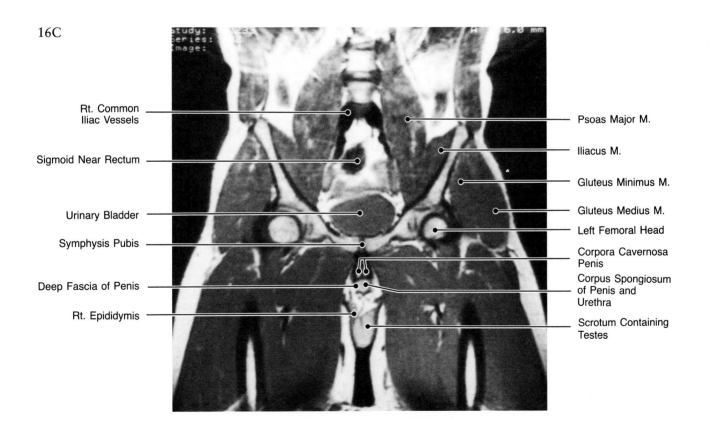

Rt. Common
Iliac Vessels

Sigmoid Near Rectum

Urinary Bladder

Symphysis Pubis

Deep Fascia of Penis

Rt. Epididymis

Psoas Major M.

Iliacus M.

Gluteus Minimus M.

Gluteus Medius M.

Left Femoral Head

Corpora Cavernosa
Penis

Corpus Spongiosum
of Penis and
Urethra

Scrotum Containing
Testes

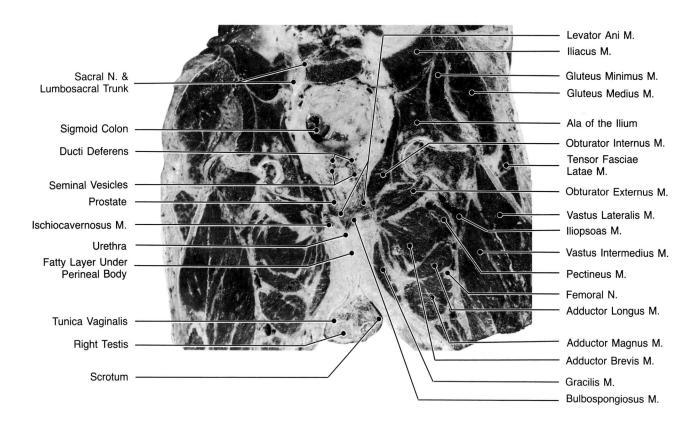

Sacral N. & Lumbosacral Trunk

Sigmoid Colon

Ducti Deferens

Seminal Vesicles

Prostate

Ischiocavernosus M.

Urethra

Fatty Layer Under Perineal Body

Tunica Vaginalis

Right Testis

Scrotum

Levator Ani M.

Iliacus M.

Gluteus Minimus M.

Gluteus Medius M.

Ala of the Ilium

Obturator Internus M.

Tensor Fasciae Latae M.

Obturator Externus M.

Vastus Lateralis M.

Iliopsoas M.

Vastus Intermedius M.

Pectineus M.

Femoral N.

Adductor Longus M.

Adductor Magnus M.

Adductor Brevis M.

Gracilis M.

Bulbospongiosus M.

17A Coronal/Anatomic Plate/Seminal Vesicles and Prostate. This anatomic section from a level about 2 cm posterior to the previous anatomic plate demonstrates the prostate lobes and just above the prostate on each side the seminal vesicles. Just above the seminal vesicles the ducti deferens are identified. These structures are surrounded by retrovesicle, perineal, and pelvic fat contained chiefly within the sling of the levator ani muscles. Below the levator sling but not entirely separate from it are the midline bulbospongiosus muscles and the farther lateral ischiocavernosus muscles. The fatty layer under the perineum separates the scrotum and testes from these pelvic viscera and muscles. The strong adductors of the thigh are demonstrated as are the vastus lateralis and intermedius heads of the quadriceps femoris. To the right of S1 and S2 is the confluence of sacral and lumbosacral trunk of nerves.

17B Coronal/MRI/Seminal Vesicles and Prostate. This magnetic resonance image is at the level seen on the anatomic specimen and demonstrates the prostate, which in this more elderly male elevates the bladder floor slightly.

17C Coronal/MRI/Seminal Vesicles and Prostate. This MRI partial saturation image is obtained on the same patient as seen in Figure 16 and demonstrates the bladder with seminal vesicles as bilateral crescentic structures indenting the inferior bladder wall. Lateral to the right seminal vesicle surface is the distal portion of the vas deferens just proximal to ductus deferens. The corpus spongiosum of the penis and corpora cavernosa of the penis are also identified. The obturator externus muscle is seen diving posterior to each femoral neck.

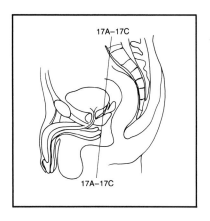

17A–17C

17A–17C

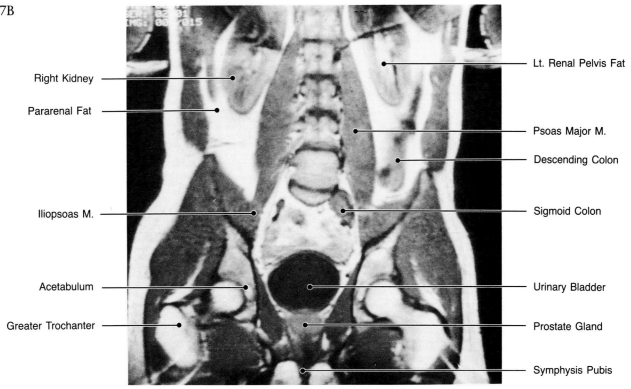

17B

Right Kidney

Pararenal Fat

Iliopsoas M.

Acetabulum

Greater Trochanter

Lt. Renal Pelvis Fat

Psoas Major M.

Descending Colon

Sigmoid Colon

Urinary Bladder

Prostate Gland

Symphysis Pubis

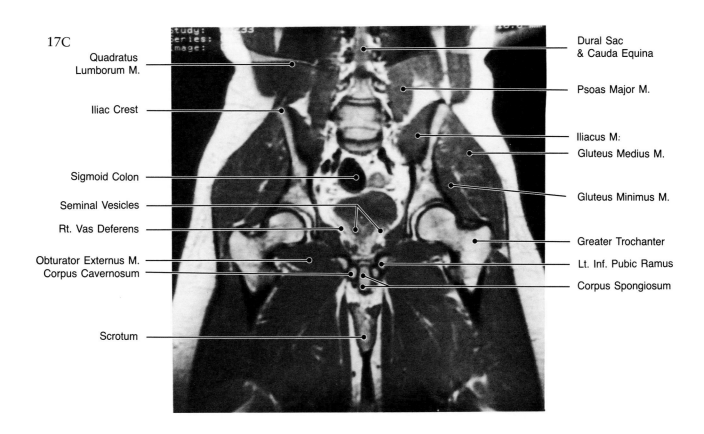

17C

Quadratus
Lumborum M.

Iliac Crest

Sigmoid Colon

Seminal Vesicles

Rt. Vas Deferens

Obturator Externus M.

Corpus Cavernosum

Scrotum

Dural Sac
& Cauda Equina

Psoas Major M.

Iliacus M.

Gluteus Medius M.

Gluteus Minimus M.

Greater Trochanter

Lt. Inf. Pubic Ramus

Corpus Spongiosum

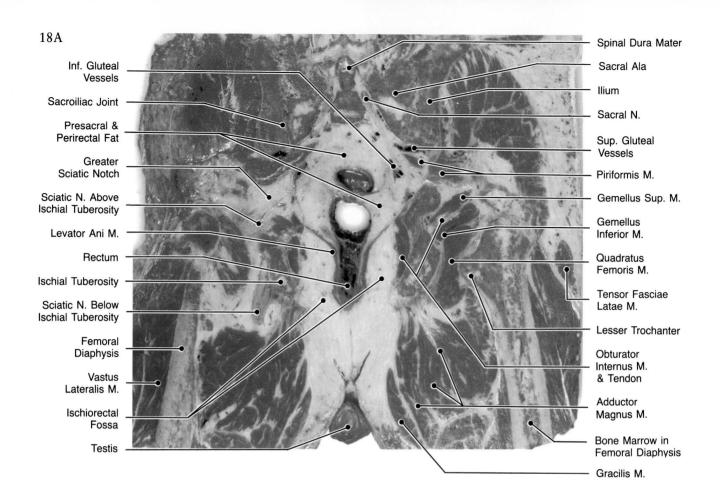

18A

Inf. Gluteal Vessels

Sacroiliac Joint

Presacral & Perirectal Fat

Greater Sciatic Notch

Sciatic N. Above Ischial Tuberosity

Levator Ani M.

Rectum

Ischial Tuberosity

Sciatic N. Below Ischial Tuberosity

Femoral Diaphysis

Vastus Lateralis M.

Ischiorectal Fossa

Testis

Spinal Dura Mater

Sacral Ala

Ilium

Sacral N.

Sup. Gluteal Vessels

Piriformis M.

Gemellus Sup. M.

Gemellus Inferior M.

Quadratus Femoris M.

Tensor Fasciae Latae M.

Lesser Trochanter

Obturator Internus M. & Tendon

Adductor Magnus M.

Bone Marrow in Femoral Diaphysis

Gracilis M.

18A Coronal/Anatomic Plate/Rectal Level. The rectum is in the midline and is surrounded by the levator sling and ischiorectal fossa fat laterally and perirectal fat superiorly. Just lateral to the ischial tuberosity on the right is the sciatic nerve and farther cephalad about 5 cm is the sciatic nerve just exiting the greater sciatic notch. Below the iliac and sacroilliac joint on the right is the piriformis muscle with the superior gluteal artery and vein above it and inferior gluteal artery and vein beneath it; both are about to exit the pelvis. On the left just lateral to the ischium and between the ischium and the femoral neck are the superior and inferior gemellus muscles separated by the tendon of the obturator internus muscle. Arising from a tendinous origin from the ischial tuberosity are fibers of the adductor magnus muscle.

18B Coronal/MRI/Rectal Level. This is a partial saturation magnetic resonance image on a young man at about the same level as A demonstrating some prostate and a testicle in the midline separated by perineal fat. Above the prostate is the rectum and a plexus of perirectal vessels. The greater sciatic notch is seen bilaterally with a slip of piriformis muscle and internal iliac vessels near the origin of the gluteals.

18C Coronal/MRI/Rectal Level. This is a magnetic resonance image 1.5 cm more posterior on the same patient. There is a small portion of one testicle. Gluteal and adductor muscles predominate with a slip of gracillis muscle lying medial on each adductor magnus. Above the rectum lies a portion of the sigmoid just at the rectosigmoid reflection.

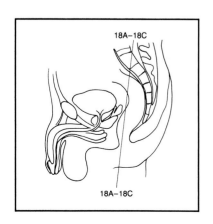

18A–18C

18A–18C

18B

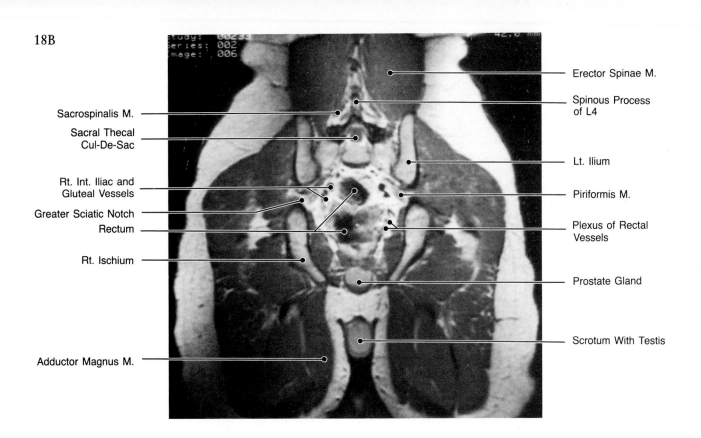

Sacrospinalis M.

Sacral Thecal
Cul-De-Sac

Rt. Int. Iliac and
Gluteal Vessels

Greater Sciatic Notch

Rectum

Rt. Ischium

Adductor Magnus M.

Erector Spinae M.

Spinous Process
of L4

Lt. Ilium

Piriformis M.

Plexus of Rectal
Vessels

Prostate Gland

Scrotum With Testis

18C

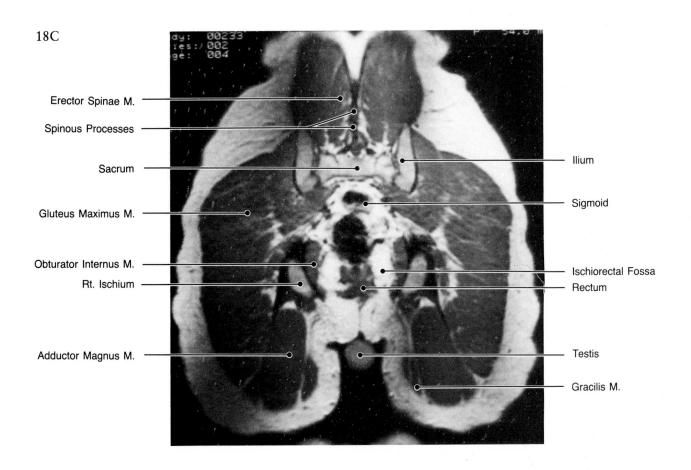

Erector Spinae M.

Spinous Processes

Sacrum

Gluteus Maximus M.

Obturator Internus M.

Rt. Ischium

Adductor Magnus M.

Ilium

Sigmoid

Ischiorectal Fossa

Rectum

Testis

Gracilis M.

19A

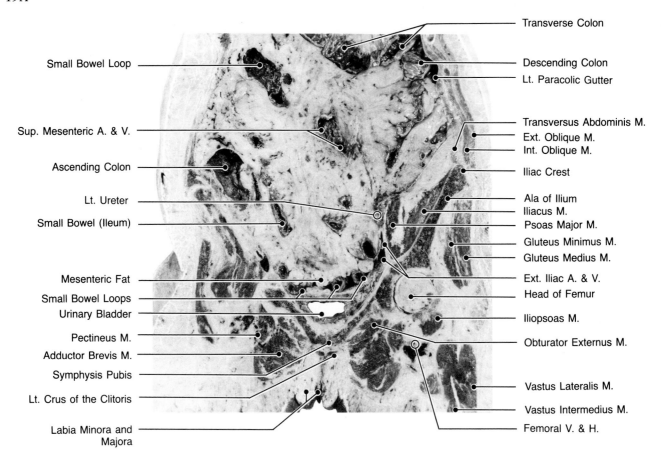

Labels (left side, top to bottom):
- Small Bowel Loop
- Sup. Mesenteric A. & V.
- Ascending Colon
- Lt. Ureter
- Small Bowel (Ileum)
- Mesenteric Fat
- Small Bowel Loops
- Urinary Bladder
- Pectineus M.
- Adductor Brevis M.
- Symphysis Pubis
- Lt. Crus of the Clitoris
- Labia Minora and Majora

Labels (right side, top to bottom):
- Transverse Colon
- Descending Colon
- Lt. Paracolic Gutter
- Transversus Abdominis M.
- Ext. Oblique M.
- Int. Oblique M.
- Iliac Crest
- Ala of Ilium
- Iliacus M.
- Psoas Major M.
- Gluteus Minimus M.
- Gluteus Medius M.
- Ext. Iliac A. & V.
- Head of Femur
- Iliopsoas M.
- Obturator Externus M.
- Vastus Lateralis M.
- Vastus Intermedius M.
- Femoral V. & H.

19A Coronal/Anatomic Plate/Anterior Pelvic Area. The coronal anatomic plate through the anterior pelvis and symphysis pubis demonstrates the labia and the crura of the clitoris. A portion of the bladder is seen just above the suprapubic fat. Ascending, transverse, and proximal descending colon are identified in the peripheral peritoneal space with small bowel loops identified more centrally surrounded by mesenteric fat. The small bowel loops in the pelvis lie on the dome of the bladder. The pelvis is farther posteriorly seen on the left, and along the medial aspect of the psoas muscle the external iliac artery and vein are identified with ureter just above them. The lateral and anterior abdominal wall musculature is particularly well seen on the left.

19B Coronal/MRI/Anterior Pelvic Area. This and the next magnetic resonance image are from a far thinner female patient who is in her early 20's. Her bladder is more distended, and the cecum lies against the right lateral wall of the bladder dome. The pubic symphysis and superior pubic rami are outlined by marrow fat, and the labia may be seen in the subpubic area.

19C Coronal/MRI/Anterior Pelvic Area. This MRI partial saturation image demonstrates superior mesenteric artery and vein lying above transverse colon near the midline. Ascending the descending colon are identified with greater omentum between. Small bowel loops are not identified as discreet structures because of intrinsic motion. Flow voids identify the left external iliac artery and vein crossing the pubis to become left femoral vessels in the thigh.

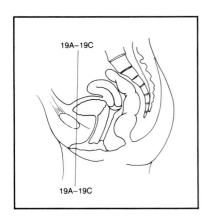

19A–19C

19A–19C

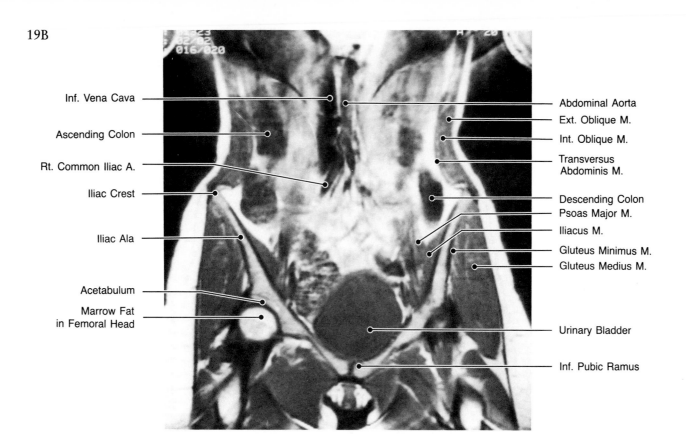

Inf. Vena Cava

Ascending Colon

Rt. Common Iliac A.

Iliac Crest

Iliac Ala

Acetabulum

Marrow Fat
in Femoral Head

Abdominal Aorta

Ext. Oblique M.

Int. Oblique M.

Transversus
Abdominis M.

Descending Colon

Psoas Major M.

Iliacus M.

Gluteus Minimus M.

Gluteus Medius M.

Urinary Bladder

Inf. Pubic Ramus

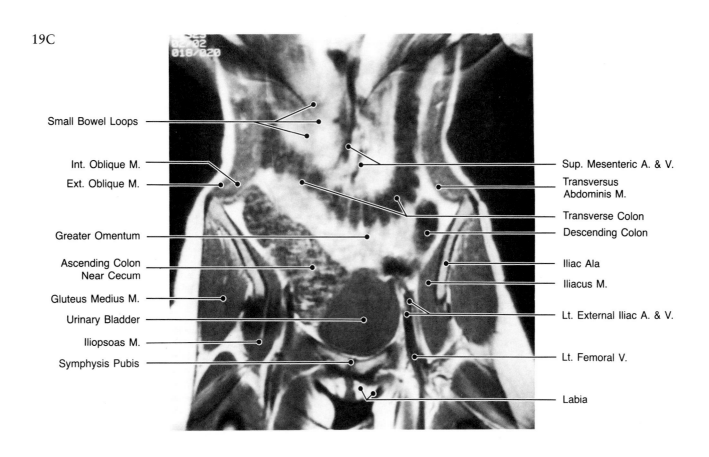

Small Bowel Loops

Int. Oblique M.

Ext. Oblique M.

Greater Omentum

Ascending Colon
Near Cecum

Gluteus Medius M.

Urinary Bladder

Iliopsoas M.

Symphysis Pubis

Sup. Mesenteric A. & V.

Transversus
Abdominis M.

Transverse Colon

Descending Colon

Iliac Ala

Iliacus M.

Lt. External Iliac A. & V.

Lt. Femoral V.

Labia

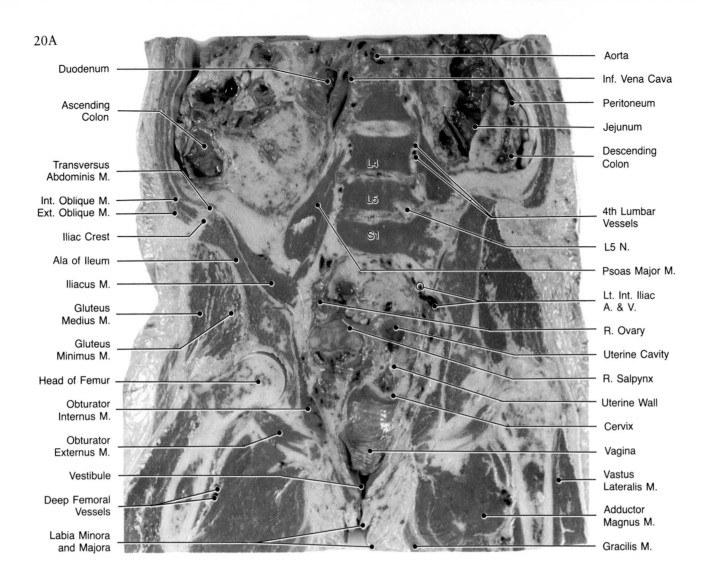

20A

Duodenum

Ascending Colon

Transversus Abdominis M.

Int. Oblique M.
Ext. Oblique M.

Iliac Crest

Ala of Ileum

Iliacus M.

Gluteus Medius M.

Gluteus Minimus M.

Head of Femur

Obturator Internus M.

Obturator Externus M.

Vestibule

Deep Femoral Vessels

Labia Minora and Majora

Aorta

Inf. Vena Cava

Peritoneum

Jejunum

Descending Colon

L4

L5

S1

4th Lumbar Vessels

L5 N.

Psoas Major M.

Lt. Int. Iliac A. & V.

R. Ovary

Uterine Cavity

R. Salpynx

Uterine Wall

Cervix

Vagina

Vastus Lateralis M.

Adductor Magnus M.

Gracilis M.

20A Coronal/Anatomic Plate/Uterus and Cervix. This anatomic section anterior to the rectum demonstrates the posterior vaginal wall, vestibule, and labia. In intimate contact with the vaginal fornix is the uterine cervix and body, within which a small uterine cavity is identified. From the right uterine wall the right salpinx is identified, ending laterally in the right ovary. A venous plexus is identified to the left of the uterus principally supplied by internal iliac vessels. This patient's uterus is neither anteflexed nor retroflexed but is oriented vertically in the sagittal and coronal planes. On the right, which is at a slightly more anterior plane than the left, iliacus and psoas muscles with the femoral nerve in the groove between them are

identified. On the right paraspinous level above L4 the collapsed inferior vena cava is identified with the transected aorta in the midline above it.

20B Coronal/MRI/Uterus and Cervix. This partial saturation magnetic resonance image from a young female patient demonstrates the posterior wall of the vaginal canal in the midline with ischiorectal fossa fat bilaterally. The uterus in this patient is anteflexed and lies over the dome of the bladder in the midline. Adnexal structures are seen bilaterally. Retroperitoneal fat outlines the psoas major muscles and lower poles of each kidney, and more of descending colon is identified just within the properitoneal fat on the left.

20C Coronal/MRI/Uterus and Cervix. This magnetic resonance image on the same patient is several centimeters anterior to the image above. It demonstrates lumbar vertebral bodies 1 through 5, the right kidney, and both psoas major muscles. The retroperitoneal fat highlights the V-like groove of the junction of iliacus and psoas muscles. The bladder, which is distended in the midline, elevates the uterus anteflexed over it. Small bowel loops are seen to the right of the bladder dome, and the left ovary is on the contralateral side. The bladder neck points toward the urethra, and the vagina is in the midline below the bladder. Obturator internus and externus muscles with their common origin from the obturator membrane are demonstrated bilaterally.

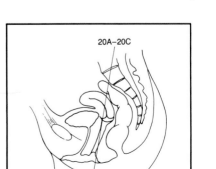

20A–20C

20A–20C

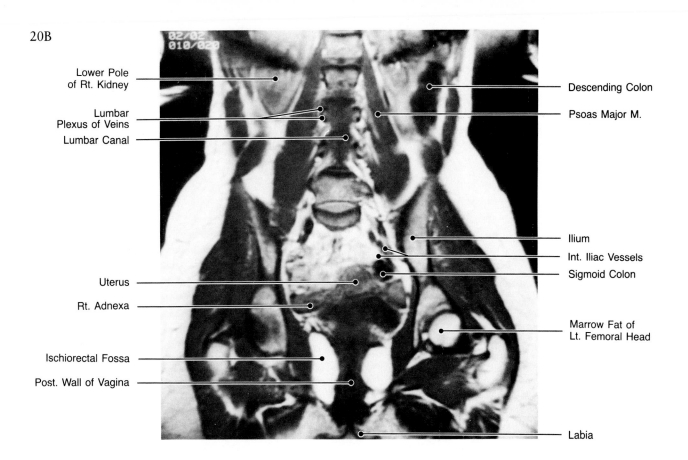

20B

Lower Pole of Rt. Kidney

Lumbar Plexus of Veins

Lumbar Canal

Uterus

Rt. Adnexa

Ischiorectal Fossa

Post. Wall of Vagina

Descending Colon

Psoas Major M.

Ilium

Int. Iliac Vessels

Sigmoid Colon

Marrow Fat of Lt. Femoral Head

Labia

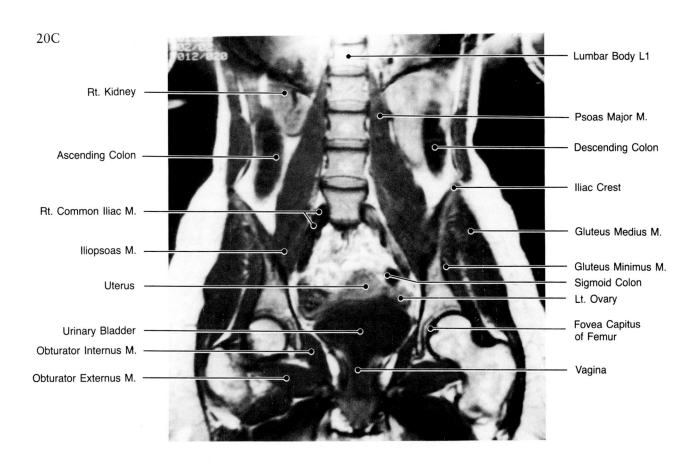

20C

Rt. Kidney

Ascending Colon

Rt. Common Iliac M.

Iliopsoas M.

Uterus

Urinary Bladder

Obturator Internus M.

Obturator Externus M.

Lumbar Body L1

Psoas Major M.

Descending Colon

Iliac Crest

Gluteus Medius M.

Gluteus Minimus M.

Sigmoid Colon

Lt. Ovary

Fovea Capitus of Femur

Vagina

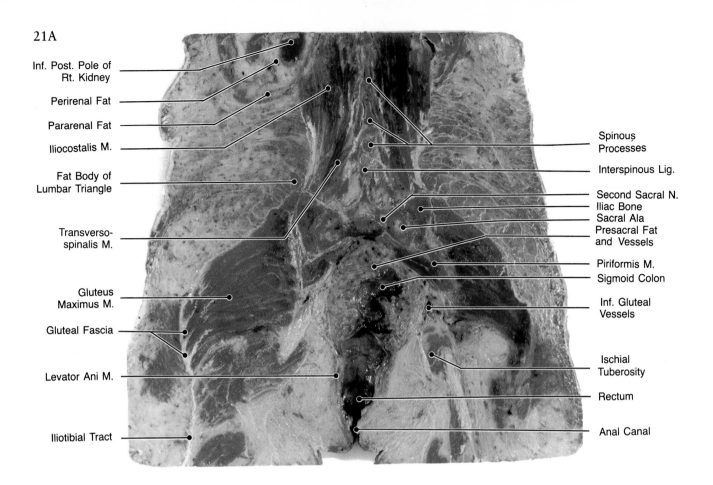

21A

Inf. Post. Pole of
Rt. Kidney

Perirenal Fat

Pararenal Fat

Iliocostalis M.

Fat Body of
Lumbar Triangle

Transverso-
spinalis M.

Gluteus
Maximus M.

Gluteal Fascia

Levator Ani M.

Iliotibial Tract

Spinous
Processes

Interspinous Lig.

Second Sacral N.
Iliac Bone
Sacral Ala
Presacral Fat
and Vessels

Piriformis M.
Sigmoid Colon

Inf. Gluteal
Vessels

Ischial
Tuberosity

Rectum

Anal Canal

21A Coronal/Anatomic Plate/Sacral Level. This anatomic plate on an obese patient demonstrates abundant fat, particularly the fat body of the lumbar triangle bilaterally, lateral to iliocostalis, longissimus, and transversus spinalis muscles. In addition to the sacral ala one can identify the left S2 nerve and a small portion of the sacroiliac joint on the left. The piriformis muscle exiting the greater sciatic notch has superior gluteal vessels above it and inferior gluteal vessels below it. In the midline is the anal canal and above it confined by levator ani muscle is the rectum and above that sigmoid colon surrounded by pelvic presacral and pericolonic fat. On the right the iliotibial tract marks the lateral margin of the gluteus maximus and continues around the gluteus as the gluteal fascia.

21B Coronal /MRI/Sacral Level. This magnetic resonance image from a thinner female demonstrates, as does the anatomic specimen, the spinal muscles, including longissimus, transversus spinalis, and iliocostalis. Spinous processes and the sacral canal of S1-2 are demonstrated above the sacral body. The greater sciatic notch demonstrates as on the anatomic specimen the gluteal vessels and the piriformis muscle.

21C Coronal/MRI/Sacral Level. Compared to the first magnetic resonance image this is slightly anterior and demonstrates more of transverse processes and laminae of the lower lumbar vertebra. Larger segments of sacral alae are also appreciated. Ischial tuberosities are now identified bilaterally, and a better developed obturator internus muscle can be seen with fibers on the left curving around the ischial tuberosity laterally.

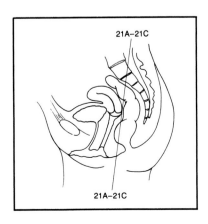

21A–21C

21A–21C

21B

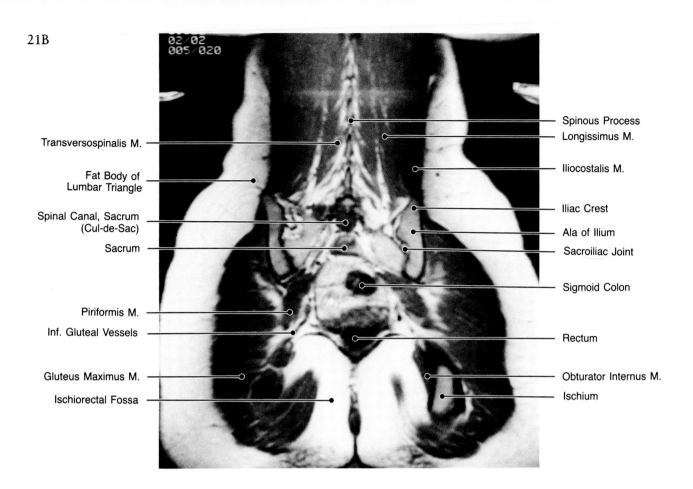

Transversospinalis M.

Fat Body of
Lumbar Triangle

Spinal Canal, Sacrum
(Cul-de-Sac)

Sacrum

Piriformis M.

Inf. Gluteal Vessels

Gluteus Maximus M.

Ischiorectal Fossa

Spinous Process

Longissimus M.

Iliocostalis M.

Iliac Crest

Ala of Ilium

Sacroiliac Joint

Sigmoid Colon

Rectum

Obturator Internus M.

Ischium

21C

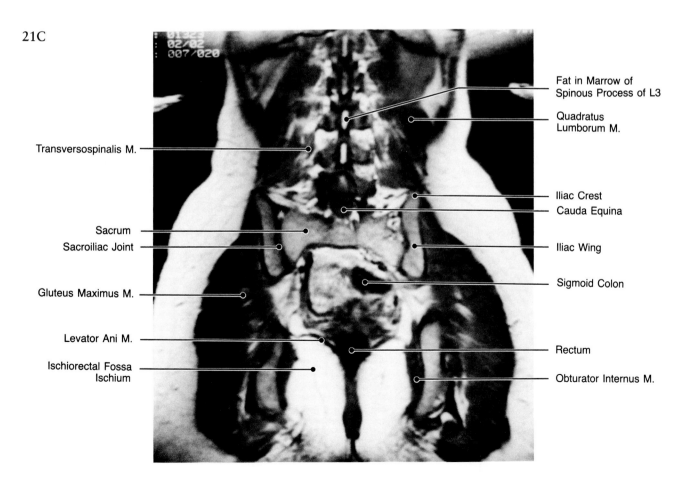

Transversospinalis M.

Sacrum

Sacroiliac Joint

Gluteus Maximus M.

Levator Ani M.

Ischiorectal Fossa
Ischium

Fat in Marrow of
Spinous Process of L3

Quadratus
Lumborum M.

Iliac Crest

Cauda Equina

Iliac Wing

Sigmoid Colon

Rectum

Obturator Internus M.

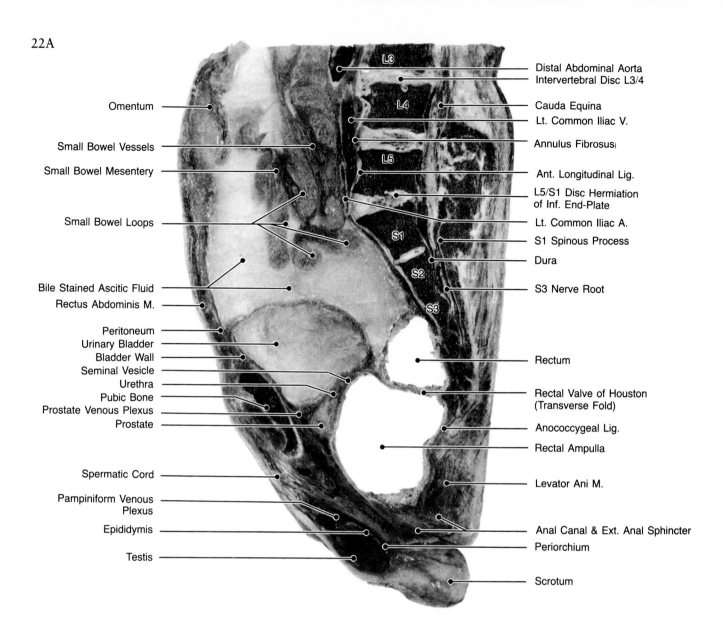

Omentum

Small Bowel Vessels

Small Bowel Mesentery

Small Bowel Loops

Bile Stained Ascitic Fluid
Rectus Abdominis M.

Peritoneum
Urinary Bladder
Bladder Wall
Seminal Vesicle
Urethra
Pubic Bone
Prostate Venous Plexus
Prostate

Spermatic Cord

Pampiniform Venous
Plexus

Epididymis

Testis

Distal Abdominal Aorta
Intervertebral Disc L3/4

Cauda Equina
Lt. Common Iliac V.

Annulus Fibrosus

Ant. Longitudinal Lig.

L5/S1 Disc Hermiation
of Inf. End-Plate

Lt. Common Iliac A.
S1 Spinous Process
Dura

S3 Nerve Root

Rectum

Rectal Valve of Houston
(Transverse Fold)

Anococcygeal Lig.

Rectal Ampulla

Levator Ani M.

Anal Canal & Ext. Anal Sphincter
Periorchium

Scrotum

22A Sagittal/Anatomic Plate/Midline Level. The anatomic plate is a midline sagittal cut in a *male* with bile-stained ascites and a number of bowel loops floating with it. In the pelvis the distended rectum lies posterior against the sacrum and coccyx, ending in the anal canal. Anteriorly the bladder rests against the rectum except where separated by the wedge of seminal vesicle above and prostate below. Anterior to the pubic bone one spermatic cord is seen arising out of the region of the scrotum and cut section of one testis, epididymis, and pampin-

iform plexus. From this midsagittal cut, the cauda equina and portions of L3 through S3 are clearly identified with their intervertebral discs, anulus fibrosus, and prevertebral vessels.

22B Sagittal/MRI/Midline. A magnetic resonance image utilizing parital saturation on a young male volunteer demonstrates prominent bulb of the penis, corpora cavernosum and corpus spongiosum, and parts of both testicles. The urethra may be seen traversing prostate and bulb of penis. The bladder rests on the prostate and pubic bone. Midline lumbar spine from L3 through the cauda equina is seen. The colon from sigmoid through rectum to the anal canal is also demonstrated.

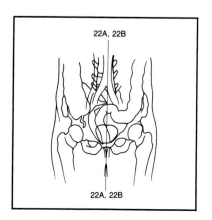

22A, 22B

22A, 22B

22B

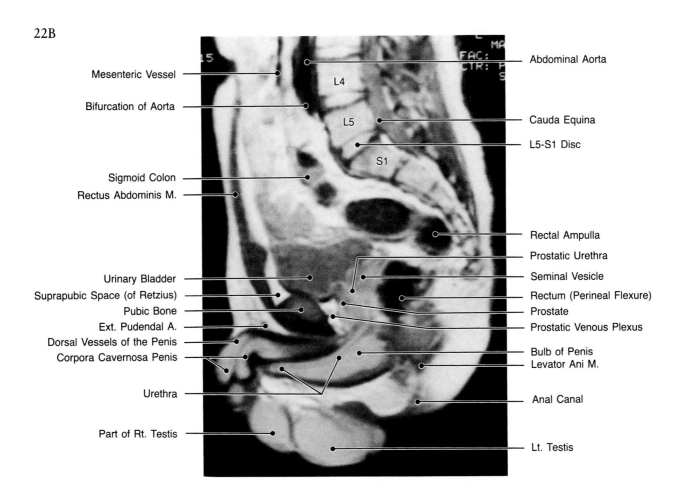

Mesenteric Vessel

Bifurcation of Aorta

Sigmoid Colon

Rectus Abdominis M.

Urinary Bladder

Suprapubic Space (of Retzius)

Pubic Bone

Ext. Pudendal A.

Dorsal Vessels of the Penis

Corpora Cavernosa Penis

Urethra

Part of Rt. Testis

Abdominal Aorta

Cauda Equina

L5-S1 Disc

Rectal Ampulla

Prostatic Urethra

Seminal Vesicle

Rectum (Perineal Flexure)

Prostate

Prostatic Venous Plexus

Bulb of Penis

Levator Ani M.

Anal Canal

Lt. Testis

L4

L5

S1

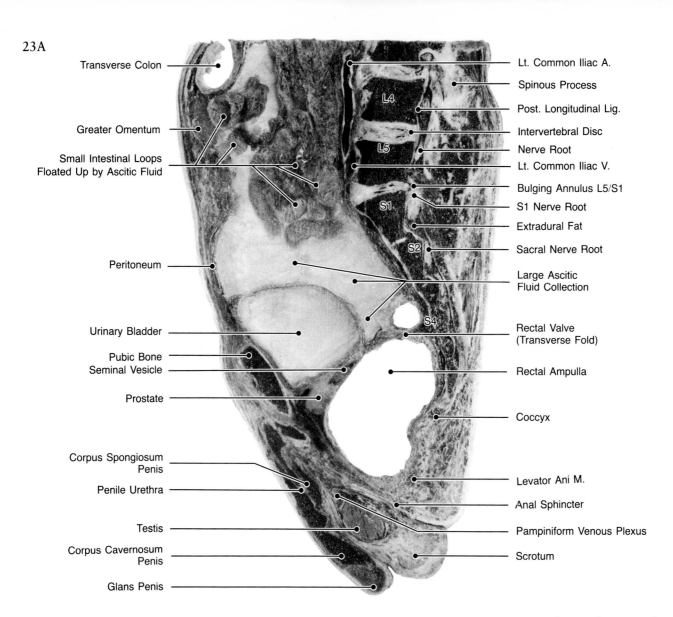

23A

Transverse Colon

Greater Omentum

Small Intestinal Loops
Floated Up by Ascitic Fluid

Peritoneum

Urinary Bladder

Pubic Bone
Seminal Vesicle

Prostate

Corpus Spongiosum
Penis

Penile Urethra

Testis

Corpus Cavernosum
Penis

Glans Penis

Lt. Common Iliac A.

Spinous Process

Post. Longitudinal Lig.

Intervertebral Disc

Nerve Root

Lt. Common Iliac V.

Bulging Annulus L5/S1

S1 Nerve Root

Extradural Fat

Sacral Nerve Root

Large Ascitic
Fluid Collection

Rectal Valve
(Transverse Fold)

Rectal Ampulla

Coccyx

Levator Ani M.

Anal Sphincter

Pampiniform Venous Plexus

Scrotum

L4

L5

S1

S2

S4

23A Sagittal/Anatomic Plate/0.5 Cm Left of Midline. The anatomic plate demonstrates more of the male external genitalia, including corpus spongiosum, penile urethra, and corpora cavernosa ending in glans penis. Behind the pubis and below the bladder, portions of the prostate and seminal vesicles are seen. Small bowel loops float in the bile-stained ascites. The left common iliac artery is seen through a long section, and posterior to it at the level of L5 a short transected segment of left common iliac vein is seen. Exiting nerve roots, L4 through S2, are identified.

23B Sagittal/MRI/0.5 Cm Left of Midline. This magnetic resonance image with partial saturation technique is a centimeter to the left of the midline. The external genitalia are little altered, and the fasciae of the penis are demonstrated separating the corpora cavernosa and corpus spongiosum and on both sides of them. The left common iliac vessels are seen anterior to L4 and L5, and epidural fat indicates a location lateral to the cauda equina behind the vertebral bodies.

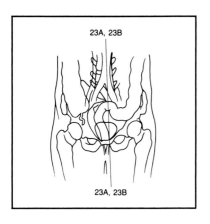

23A, 23B

23A, 23B

23B

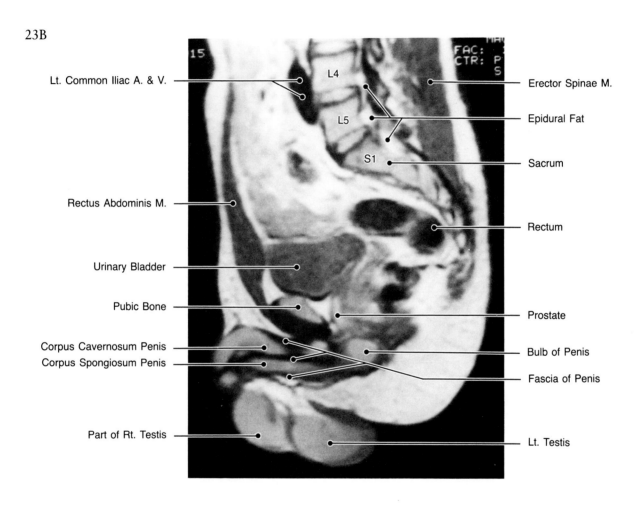

Lt. Common Iliac A. & V. —

Rectus Abdominis M. —

Urinary Bladder —

Pubic Bone —

Corpus Cavernosum Penis —
Corpus Spongiosum Penis —

Part of Rt. Testis —

L4

L5

S1

— Erector Spinae M.

— Epidural Fat

— Sacrum

— Rectum

— Prostate

— Bulb of Penis

— Fascia of Penis

— Lt. Testis

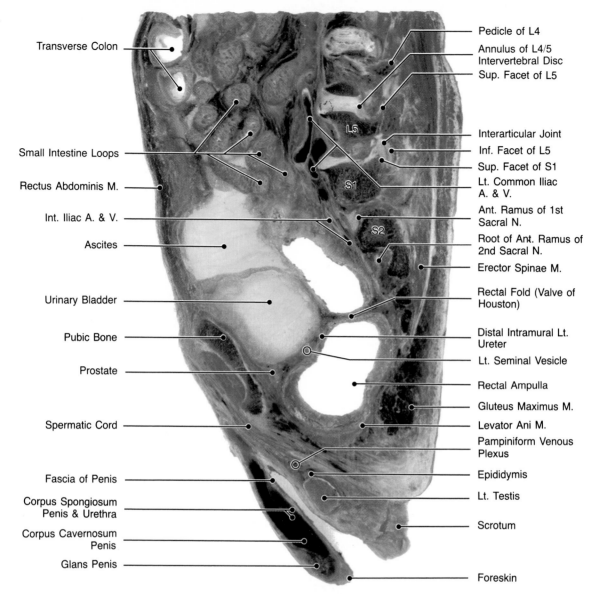

Transverse Colon

Small Intestine Loops

Rectus Abdominis M.

Int. Iliac A. & V.

Ascites

Urinary Bladder

Pubic Bone

Prostate

Spermatic Cord

Fascia of Penis

Corpus Spongiosum Penis & Urethra

Corpus Cavernosum Penis

Glans Penis

Pedicle of L4

Annulus of L4/5 Intervertebral Disc

Sup. Facet of L5

Interarticular Joint

Inf. Facet of L5

Sup. Facet of S1

Lt. Common Iliac A. & V.

Ant. Ramus of 1st Sacral N.

Root of Ant. Ramus of 2nd Sacral N.

Erector Spinae M.

Rectal Fold (Valve of Houston)

Distal Intramural Lt. Ureter

Lt. Seminal Vesicle

Rectal Ampulla

Gluteus Maximus M.

Levator Ani M.

Pampiniform Venous Plexus

Epididymis

Lt. Testis

Scrotum

Foreskin

24A Sagittal/Anatomic Plate/2 Cm Left of Midline. The anatomic plate demonstrates the left spermatic cord and left testicle with epididymis and pampiniform venous plexus cranially. Just above the left seminal vesicle is the intramural portion of the left ureter. Anterior to the spine, portions of the common iliac artery and vein at L5 and S1 and internal iliac artery and vein at S1-S2 are seen.

24B Sagittal/MRI/2 Cm Left of Midline. The partial saturation magnetic resonance image at the corresponding level demonstrates the external iliac artery and vein anterior to L4 and L5. Just below S1 is the internal iliac artery and vein lying on the superior surface of piriformis muscle.

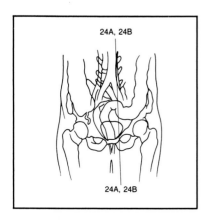

24A, 24B

24A, 24B

24B

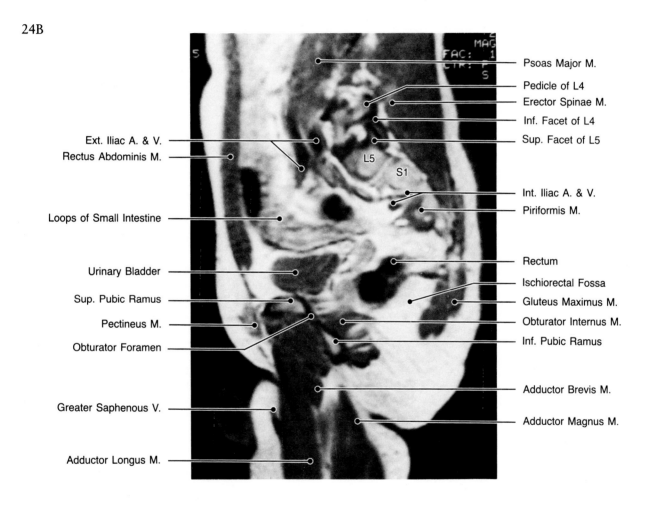

Ext. Iliac A. & V.

Rectus Abdominis M.

Loops of Small Intestine

Urinary Bladder

Sup. Pubic Ramus

Pectineus M.

Obturator Foramen

Greater Saphenous V.

Adductor Longus M.

Psoas Major M.

Pedicle of L4

Erector Spinae M.

Inf. Facet of L4

Sup. Facet of L5

L5

S1

Int. Iliac A. & V.

Piriformis M.

Rectum

Ischiorectal Fossa

Gluteus Maximus M.

Obturator Internus M.

Inf. Pubic Ramus

Adductor Brevis M.

Adductor Magnus M.

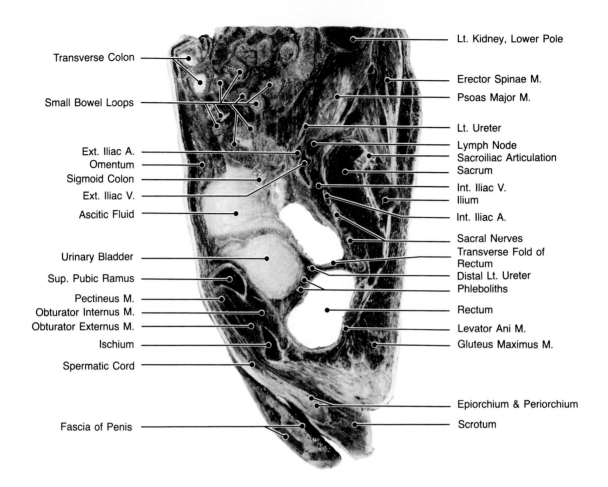

25A

Transverse Colon

Small Bowel Loops

Ext. Iliac A.
Omentum
Sigmoid Colon
Ext. Iliac V.

Ascitic Fluid

Urinary Bladder

Sup. Pubic Ramus

Pectineus M.
Obturator Internus M.
Obturator Externus M.

Ischium

Spermatic Cord

Fascia of Penis

Lt. Kidney, Lower Pole

Erector Spinae M.
Psoas Major M.

Lt. Ureter
Lymph Node
Sacroiliac Articulation
Sacrum
Int. Iliac V.
Ilium
Int. Iliac A.

Sacral Nerves
Transverse Fold of Rectum
Distal Lt. Ureter
Phleboliths

Rectum

Levator Ani M.
Gluteus Maximus M.

Epiorchium & Periorchium
Scrotum

25A Sagittal/Anatomic Plate/3 Cm Left of Midplane. The anatomic plate lies 3 cm to the left of the midplane and demonstrates some phleboliths in the periprostatic region between bladder and rectum. Just anterior to the transverse rectal fold lies the distal left ureter. The more proximal portion of the left ureter is seen above the external iliac artery and vein on the surface of the psoas muscle. Presacrally, internal iliac vessels and sacral nerves are demonstrated. In the external genitalia only epiorchium and periorchium of the left testis remain together with remnants of the spermatic cord.

25B Sagittal/MRI/3 Cm Left of Midline. This is a magnetic resonance image about 3 cm left of the midline in a young adult male with well-developed psoas muscles and posterior paraspinal muscles. The sacrum with its fatty marrow lies comma-like above the sciatic nerve and branches of the internal iliac vessels and piriformis muscle. Anterior to the sacrum are the external iliac vessels. The lateral aspect of the bladder is identified superior to the superior pubic ramus and fibers of the obturator internus muscle. An abundance of fat is seen in the ischial rectal fossa.

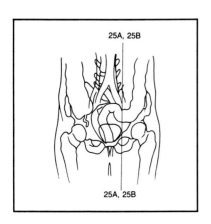

25A, 25B

25A, 25B

25B

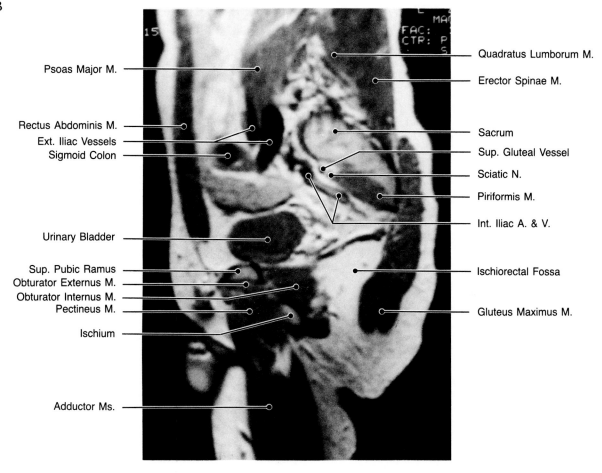

Psoas Major M. ——————————

Rectus Abdominis M. ——————
Ext. Iliac Vessels ——————
Sigmoid Colon ——————

Urinary Bladder ——————

Sup. Pubic Ramus ——————
Obturator Externus M. ——————
Obturator Internus M. ——————
Pectineus M. ——————

Ischium ——————

Adductor Ms. ——————

Quadratus Lumborum M.
Erector Spinae M.

Sacrum
Sup. Gluteal Vessel
Sciatic N.
Piriformis M.
Int. Iliac A. & V.

Ischiorectal Fossa

Gluteus Maximus M.

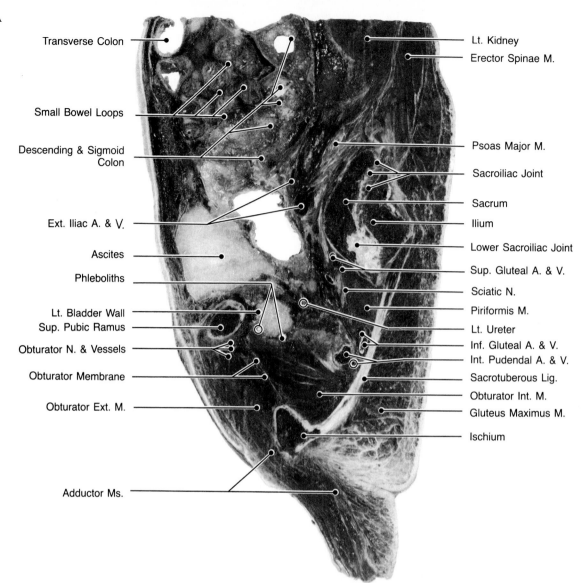

26A

Transverse Colon

Small Bowel Loops

Descending & Sigmoid Colon

Ext. Iliac A. & V.

Ascites

Phleboliths

Lt. Bladder Wall
Sup. Pubic Ramus

Obturator N. & Vessels

Obturator Membrane

Obturator Ext. M.

Adductor Ms.

Lt. Kidney
Erector Spinae M.

Psoas Major M.

Sacroiliac Joint

Sacrum
Ilium

Lower Sacroiliac Joint

Sup. Gluteal A. & V.

Sciatic N.
Piriformis M.
Lt. Ureter
Inf. Gluteal A. & V.
Int. Pudendal A. & V.
Sacrotuberous Lig.
Obturator Int. M.
Gluteus Maximus M.

Ischium

26B Sagittal/MRI/Lateral Pelvic Side Wall. This partial saturation magnetic resonance image from a young male who volunteered demonstrates external iliac vessels in the mid abdomen below the psoas major muscle. The sacrum is anterior to the ilium, and above them both are the prominent erector spinae muscles. At this level the other prominent muscles include the piriformis in the concavity of the sacrum and the gluteus maximus behind it; obturator internus and externus muscles bridge the obturator fossa, and the muscles of the thigh are the adductor longus and adductor magnus.

26B Sagittal/MRI/Lateral Pelvic Side Wall. This partial saturation magnetic resonance image from a young male who volunteered demonstrates external iliac vessels in the mid abdomen below the psoas major muscle. The sacrum is anterior to the ilium, and above them both are the prominent erector spinae muscles. At this level the other prominent muscles include the piriformis in the concavity of the sacrum and the gluteus maximus behind it; obturator internus and externus muscles bridge the obturator fossa, and the muscles of the thigh are the adductor longus and adductor magnus.

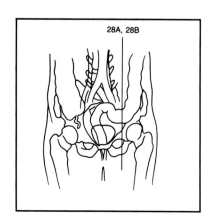

28A, 28B

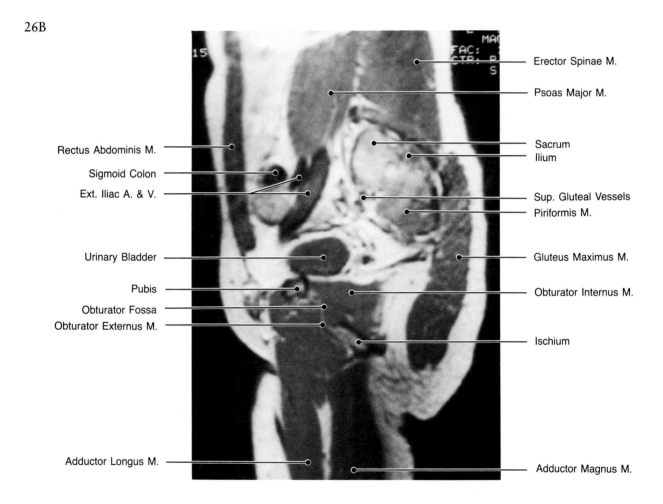

Erector Spinae M.

Psoas Major M.

Rectus Abdominis M.

Sigmoid Colon

Ext. Iliac A. & V.

Sacrum

Ilium

Sup. Gluteal Vessels

Piriformis M.

Urinary Bladder

Gluteus Maximus M.

Pubis

Obturator Internus M.

Obturator Fossa

Obturator Externus M.

Ischium

Adductor Longus M.

Adductor Magnus M.

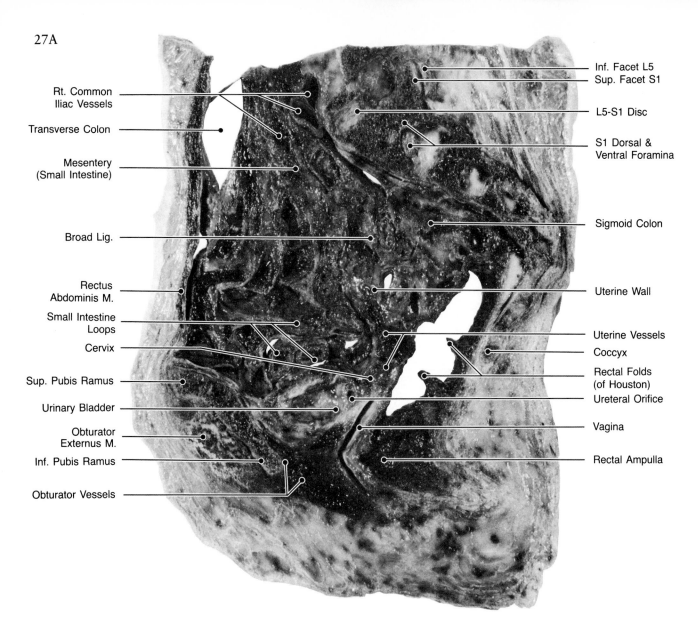

27A

Rt. Common Iliac Vessels

Transverse Colon

Mesentery (Small Intestine)

Broad Lig.

Rectus Abdominis M.

Small Intestine Loops

Cervix

Sup. Pubis Ramus

Urinary Bladder

Obturator Externus M.

Inf. Pubis Ramus

Obturator Vessels

Inf. Facet L5
Sup. Facet S1

L5-S1 Disc

S1 Dorsal & Ventral Foramina

Sigmoid Colon

Uterine Wall

Uterine Vessels

Coccyx

Rectal Folds (of Houston)

Ureteral Orifice

Vagina

Rectal Ampulla

27A Sagittal/Anatomical Plate/2 Cm Right of Midline. This anatomic plate of the female pelvis in sagittal section about 1 inch to the right of the midline demonstrates rectum and anterior to it the vagina and anterior to that the urethral orifice in the right lateral wall of the bladder. Above the vagina is the cervix and after an intervening section, where uterine vessels predominate, the lateral uterine wall ending cephalically to the broad ligament. Above the bladder are small bowel loops and mesentery.

27B Sagittal/MRI/2 Cm Right of Midline. The magnetic resonance image of a normal young female done with partial saturation technique about 2 cm to right of the midline demonstrates an anteverted uterus lying over the dome of the bladder and separating the bladder from the higher small bowel loops. Behind the uterus is the rectum and below the rectum is the levator ani muscle. Note the five lumbar paramedian segments, the sacral segments, the cauda equina, the thecal sac, and the intervertebral discs.

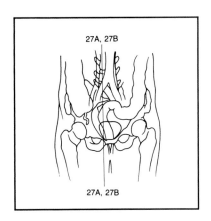

27A, 27B

27A, 27B

27B

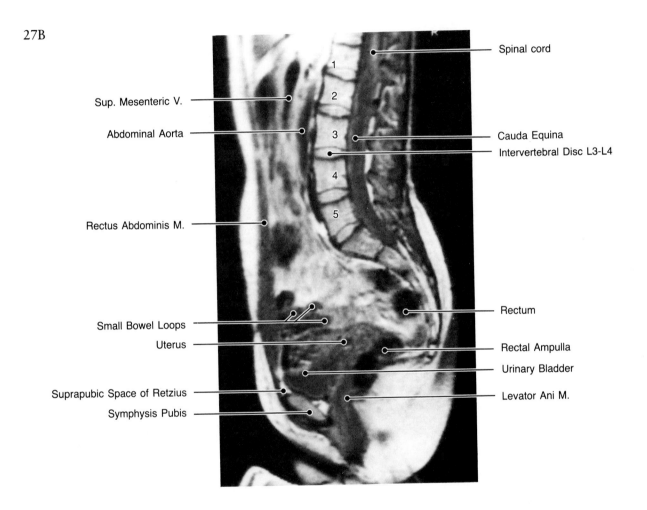

Spinal cord

Sup. Mesenteric V.

Abdominal Aorta

Cauda Equina

Intervertebral Disc L3-L4

Rectus Abdominis M.

Small Bowel Loops

Rectum

Uterus

Rectal Ampulla

Urinary Bladder

Suprapubic Space of Retzius

Levator Ani M.

Symphysis Pubis

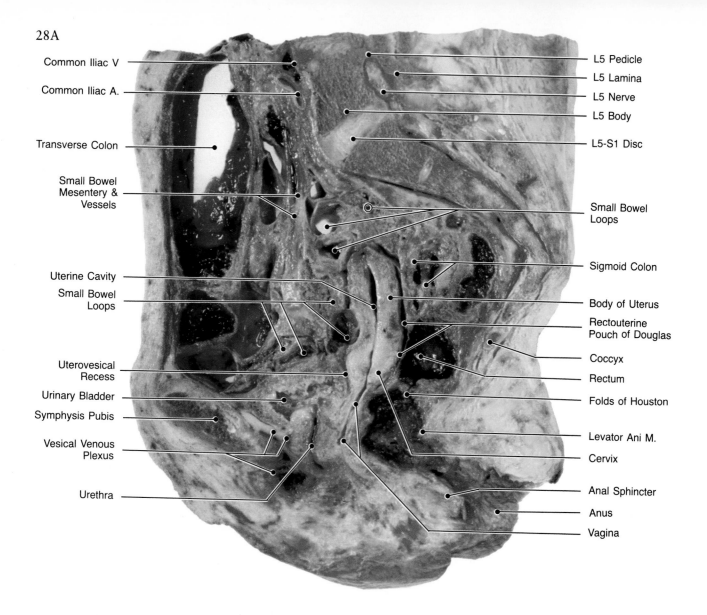

Common Iliac V

Common Iliac A.

Transverse Colon

Small Bowel Mesentery & Vessels

Uterine Cavity

Small Bowel Loops

Uterovesical Recess

Urinary Bladder

Symphysis Pubis

Vesical Venous Plexus

Urethra

L5 Pedicle

L5 Lamina

L5 Nerve

L5 Body

L5-S1 Disc

Small Bowel Loops

Sigmoid Colon

Body of Uterus

Rectouterine Pouch of Douglas

Coccyx

Rectum

Folds of Houston

Levator Ani M.

Cervix

Anal Sphincter

Anus

Vagina

28A Sagittal/Anatomical Female Plate/Midline Level. The midline is seen on the anatomic plate. The retroverted uterus lies vertical as though a continuation of the vagina in this elderly patient supported anteriorly by small bowel loops and posteriorly by sigmoid colon and rectum. The uterine cavity is clearly demonstrated, and the cervix is seen at the lowest extent of the uterine body. The urethra is seen exiting from the bladder.

28B Sagittal/Anatomic Female Plate/Midline Level. The MRI image in the same patient as seen in Figure 27B but at the midline is shown. In the abdomen the abdominal aorta and superior mesenteric vein are identified; anterior to L5 and S1 are the common iliac artery and vein. The vagina is faintly seen behind the bladder and anterior to the rectum. The uterus, which is anteflexed, has its cervix on the anterior rectal wall and body on the dome of the bladder. Intrinsic motion in small bowel loops has caused them to be blurred in this patient.

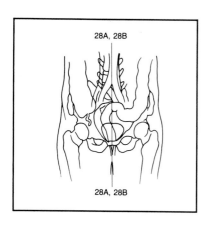

28A, 28B

28A, 28B

28B

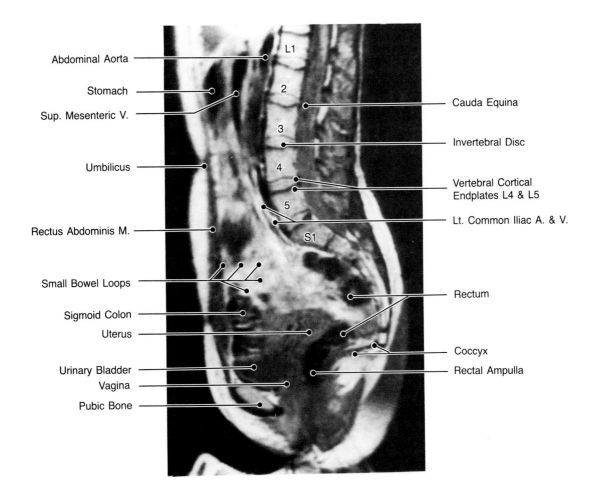

Abdominal Aorta

Stomach

Sup. Mesenteric V.

Umbilicus

Rectus Abdominis M.

Small Bowel Loops

Sigmoid Colon

Uterus

Urinary Bladder

Vagina

Pubic Bone

L1

2

3

4

5

S1

Cauda Equina

Invertebral Disc

Vertebral Cortical
Endplates L4 & L5

Lt. Common Iliac A. & V.

Rectum

Coccyx

Rectal Ampulla

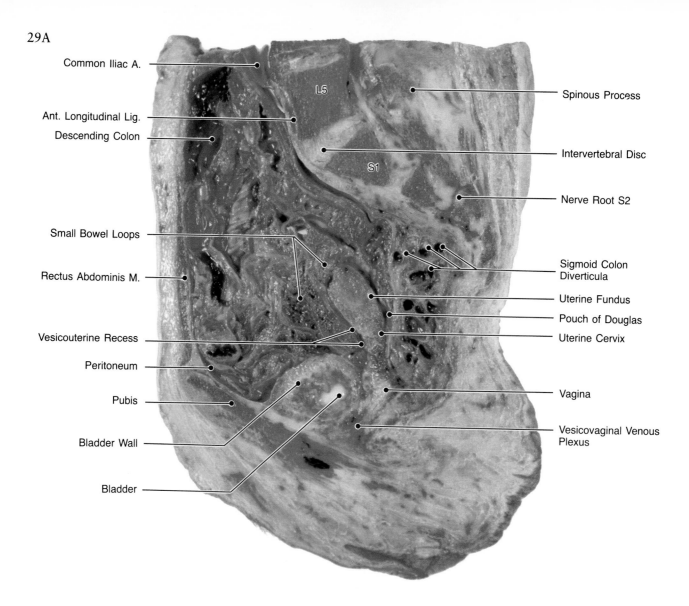

29A

Common Iliac A.

Ant. Longitudinal Lig.

Descending Colon

Small Bowel Loops

Rectus Abdominis M.

Vesicouterine Recess

Peritoneum

Pubis

Bladder Wall

Bladder

L5

S1

Spinous Process

Intervertebral Disc

Nerve Root S2

Sigmoid Colon Diverticula

Uterine Fundus

Pouch of Douglas

Uterine Cervix

Vagina

Vesicovaginal Venous Plexus

29A Sagittal/Anatomic Plate/2 Cm to the Left of the Midline. This anatomic plate of the female pelvis 2 cm to the left of the midline demonstrates the vagina at the cervix and cardinal ligament level. Posterior to the uterus is the pouch of Douglas, the peritoneal reflexion between the uterus and the colon at the rectosigmoid level. Small bowel loops fill the vesicouterine recess.

29B Sagittal Female/MRI/2 Cm to Left of Midline. The magnetic resonance image at partial saturation technique displays the bladder, vagina, and rectum as on Figure 28*B*. In the lower abdomen identification of the abdominal aorta and mesenteric vessels as flow voids, including the large superior mesenteric vein, is made. Epidural fat surrounds the nerves in the exit foramina of the lower lumbar and sacral levels.

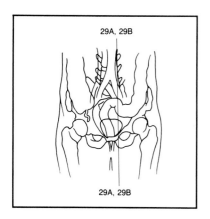

29A, 29B

29A, 29B

29B

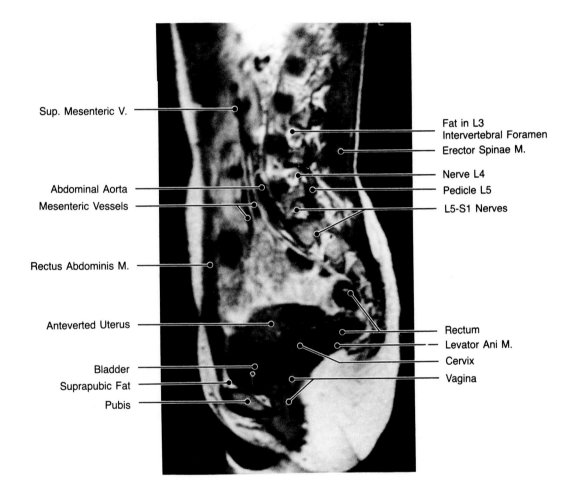

Sup. Mesenteric V.

Fat in L3
Intervertebral Foramen

Erector Spinae M.

Nerve L4

Abdominal Aorta

Pedicle L5

Mesenteric Vessels

L5-S1 Nerves

Rectus Abdominis M.

Anteverted Uterus

Rectum

Levator Ani M.

Cervix

Bladder

Vagina

Suprapubic Fat

Pubis

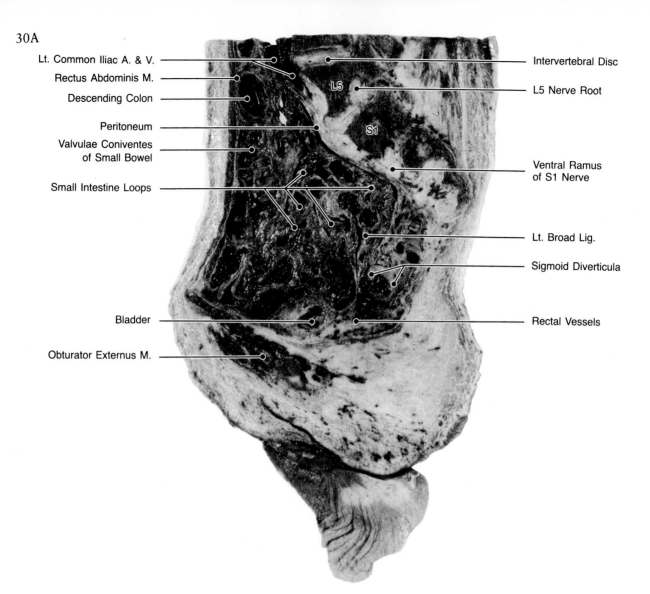

30A

Lt. Common Iliac A. & V. —

Rectus Abdominis M. —

Descending Colon —

Peritoneum —

Valvulae Coniventes of Small Bowel —

Small Intestine Loops —

Bladder —

Obturator Externus M. —

L5

S1

— Intervertebral Disc

— L5 Nerve Root

— Ventral Ramus of S1 Nerve

— Lt. Broad Lig.

— Sigmoid Diverticula

— Rectal Vessels

30A Sagittal/Anatomic Plate/Left Broad Ligament. This anatomic plate, about 1 cm lateral to the previous anatomic plate, demonstrates the lateral wall of the bladder and the lateral rectouterine recess behind it. In the same vertical orientation as the uterus at this level lies the left broad ligament. All around the broad ligament are small intestinal loops; their valvulae conniventes are demonstrated. Anterior to the L5 vertebral body is the left common iliac artery and vein, and anterior to those vessels is the descending colon.

30B Sagittal/MRI/. A slightly more lateral plate than Figure 29B in this magnetic resonance partial saturation image demonstrates the left common iliac artery anterior to L5 and the L5-S1 interspace. The configuration of the sacrum is similar to that seen on A. The bladder is still a prominent structure in this plate.

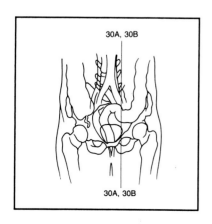

30A, 30B

30A, 30B

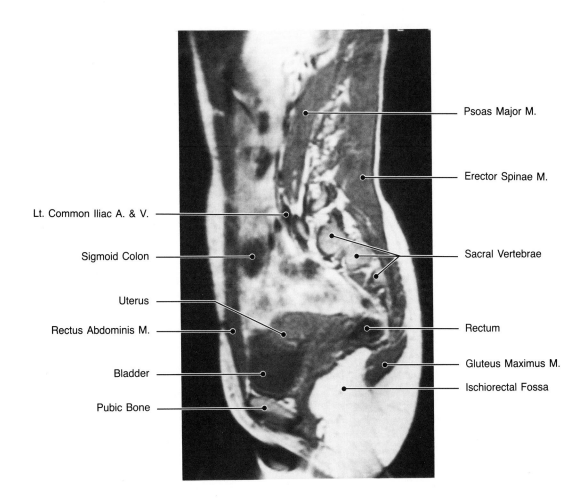

Psoas Major M.

Erector Spinae M.

Lt. Common Iliac A. & V.

Sigmoid Colon

Sacral Vertebrae

Uterus

Rectus Abdominis M.

Rectum

Bladder

Gluteus Maximus M.

Ischiorectal Fossa

Pubic Bone

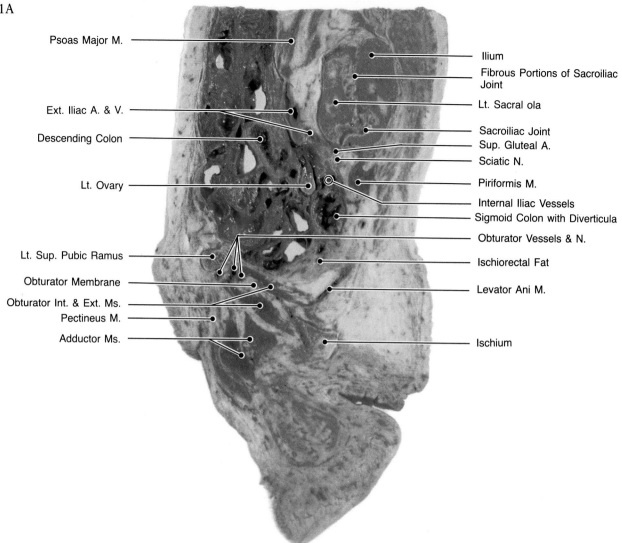

Psoas Major M.

Ext. Iliac A. & V.

Descending Colon

Lt. Ovary

Lt. Sup. Pubic Ramus

Obturator Membrane

Obturator Int. & Ext. Ms.

Pectineus M.

Adductor Ms.

Ilium

Fibrous Portions of Sacroiliac Joint

Lt. Sacral ola

Sacroiliac Joint

Sup. Gluteal A.

Sciatic N.

Piriformis M.

Internal Iliac Vessels

Sigmoid Colon with Diverticula

Obturator Vessels & N.

Ischiorectal Fat

Levator Ani M.

Ischium

31A Sagittal/Anatomic Female Plate/Left Ovary. This anatomic plate is through the left ovary, a whitish structure lying anterior to the sigmoid colon and internal iliac vessels, posterior to the descending colon and just inferior to external iliac artery and vein. The lateral aspect of the left sacrum and sacroiliac joint are seen above the ovary. This level also demonstrates the obturator fossa with the obturator membrane separating obturator internus from obturator externus muscles. In the anterior portion, the obturator notch and the obturator vessels and nerve are identified.

31B Sagittal Female/MRI/Sacroiliac Joint. This magnetic resonance image is at the level of the sacrum and sacroiliac joint and demonstrates some vessels of the broad ligament, but no ovarian tissue or abnormality is seen. External iliac vessels are seen at the inferior aspect of the psoas major muscle.

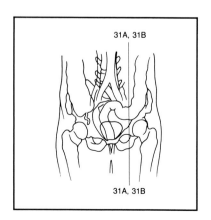

31A, 31B

31A, 31B

31B

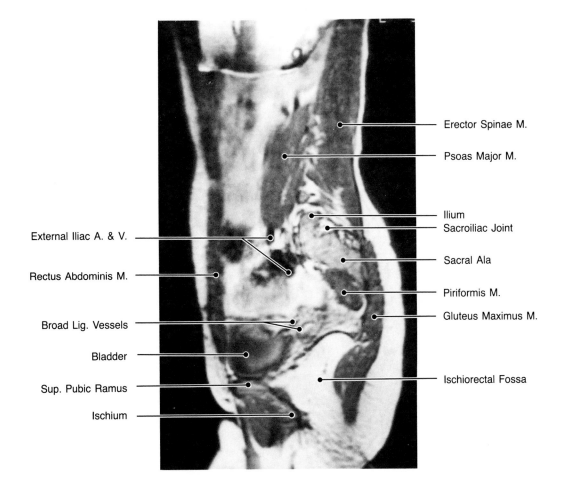

Erector Spinae M.

Psoas Major M.

External Iliac A. & V.

Ilium

Sacroiliac Joint

Sacral Ala

Rectus Abdominis M.

Piriformis M.

Gluteus Maximus M.

Broad Lig. Vessels

Bladder

Sup. Pubic Ramus

Ischiorectal Fossa

Ischium

SECTION 6

Extremities

Javier Beltran, M.D.
Delmas Allen, Ph. D.
A. J. Christoforidis, M.D., Ph. D.

In this section the normal anatomy of the upper and lower extremities is presented. Axial planes of sections are used as a baseline, but in several areas sagittal and coronal planes are added to illustrate specific anatomic details. These areas include the shoulder, hand, wrist, hip, knee, and ankle. Several of the CT images of the shoulder were obtained following intra-articular injection of positive contrast material and air to better demonstrate the internal anatomy of the joint space.

Cadaver sections were obtained with the same technique used in other areas. After freezing, the region to be studied was sectioned with a band saw in the selected plane. The section thickness was approximately 2 cm. Following sectioning, the surface was treated and prepared as indicated in the general introduction and then it was photographed. The extremities of five different cadavers were used in order to provide enough sections to correlate with the CT and MR images. However, some of the muscular groups identified on the anatomic section do not correspond exactly to the same groups of muscles included on the MR and CT sections. Nevertheless, the clinically important information regarding the anatomy of the joint space and surrounding areas is well correlated. The sagittal sections of the hip joint and some of the axial sections of the knee were obtained from a cadaver of a patient who died of obstructive jaundice. In these sections, the tissues are stained yellowish by bilirubin. The axial section at the level of the suprapatellar bursa was distended by a joint effusion. This was preserved and photographed to illustrate the superior extension of the joint space.

All the CT images were obtained using fourth generation scanners (Technicare 2060Q or 1440). Window settings were adjusted to demonstrate mainly the soft tissue components of the sections.

The MR images were obtained with a 1.5 Tesla unit (Signa, General Electric). Normal anatomic structures were best demonstrated by T1 weighted images. Spin echo sequences used throughout the studies were repetition times (TR) varying from 600 to 1000 msec with echo times (TE) of 25 to 50 msec. The body coil was used for the axial images obtained at the level of the thigh. The rest of the images were obtained using surface coils of different designs. The shoulder and wrist images were obtained using a transmit-receive circular surface coil measuring 5 inches in diameter. The images from the elbow, forearm, hip, knee, leg, and foot and some of the hand were obtained using a receive-only rectangle-shaped surface coil. In most studies two averages were obtained using a matrix size of 256 × 256. The field of view changed depending on the size of the region to be imaged. For small parts such as the wrist or foot a field of view of 16 cm was used. For larger parts, correspondingly larger fields of view, such as 20 cm, were used. The slice thickness used was 3 mm for all the images. Interslice gaps of 0.6 mm were used when the joints were imaged. In larger areas such as the thigh or leg, larger interslice gaps were used (5 to 10 mm).

On the CT images, the muscles offer an intermediate density between the low attenuation density of the fat and the higher attenuation density of the tendons. Blood vessels and nerves cannot be distinguished on CT without the use of intravenous contrast material. On MRI, the cortical bone, calcifications, and tendons are represented as dark structures due to their very prolonged T1 relaxation time. Meniscus cartilage (fibrocartilage) is also represented as a dark, low intensity signal, which contrasts significantly with the high intensity signal of the hyaline articular cartilage. These differences in signal are due to the low water content of the fibrocartilage against the high water content of the hyaline cartilage. Ligaments have a signal intensity similar to that of the tendons. Muscles offer an intermediate signal, and fat has a very high intensity signal (short T1). Nerves are of intermediate signal intensity, being slightly higher than tendons. Blood vessels are depicted on MR images differently depending on the velocity of the flow. Veins, with slow flowing blood, in general offer a high intensity signal when imaged axially, and arteries have low intensity signal in the first echo image (flow void phenomenon). Some arteries show a high intensity signal due to the paradoxical enhancement and second echo rephasing artifacts.

In selected areas, like the joints, examples of common pathology have been added to enhance the clinical significant of a given technique or section.

In summary, efforts have been made to provide the reader with images that demonstrate the CT and MRI depiction of the normal anatomy not only on the axial plane but also on the sagittal and coronal planes, especially at the level of the joints, where their clinical usefulness is more significant.

Bibliography

Anderson JE: *Grant's Atlas of Anatomy.* 7th ed. Baltimore, Williams and Wilkins, 1978.

Beltran J, Noto AM, Herman LJ, Mosure JC, Burk JM, Christoforidis AJ: Joint effusions: MR imaging. *Radiology* 158:133–137, 1986.

Bo WJ, Meschan I, Krueger WA: *Basic Atlas of Cross-Sectional Anatomy.* Philadelphia, W. B. Saunders Co., 1980.

Bretzke CA, Crass JR, Craig EV, Feinberg SB: Ultrasongraphy of the rotator cuff: Normal and pathological anatomy. *Invest Radiol* 20:311–315, 1958.

Ghelman B: Meniscal tears of the knee: Evaluation by high resolution CT combined with arthrography. *Radiology* 157:23–28, 1985.

Goldman AB, Ghelman R: The double-contrast shoulder arthrogram: A review of 159 studies. *Radiology* 127:655–663, 1978.

Huber DJ, Sauter R, Mueller E, Requardt H, Weber H: MR imaging of the normal shoulder. *Radiology* 158:405–408, 1986.

Kieft GL, Obermann WR, Bloem JL, Rozing PM, Verbout AJ: MR Imaging of the shoulder in normal and pathologic conditions. Scientific Exhibit. Presented at the Annual Meeting of the Radiological Society of North America, Chicago, 1985.

Li DKB, Mayo J, Fache JS, et al: MRI of the knee with cruciate ligament injuries. Presented at the Annual Meeting of the Radiological Society of North America, Washington, D.C., 1984.

Li KC, Henkelman RM, Poon PY, Rubenstein J: MR imaging of the normal knee. *J Comput Assist Tomogr* 6:1147–1154, 1984.

Passariello R, Trecoo F, dePaulis F, Masciocchi C, Bonanni G, Beomont E, Zobel B: Meniscal lesions of the knee joint: CT diagnosis. *Radiology* 157:29–34, 1985.

Reicher MA, Bassett LW, Gold RH: High resolution magnetic resonance imaging of the knee joint: Pathologic correlations. *AJR* 145:903–909, 1985.

Reicher MA, Rauschning W, Gold RH, et al: High resolution magnetic resonance imaging of the knee joint: Pathologic correlations. *AJR* 145:895–902, 1985.

Schuman ET, Kilcoyne RF, Matsen FA, Rogers JB, Mack LA: Double-contrast computed tomography of the glenoid labrum. *AJR* 141:581–584, 1983.

Totty WG, Murphy WA, Ganz WI, Kumar B, Daum WJ, Siegel BA: Magnetic resonance imaging of the normal and ischemic femoral head. *AJR* 143:1273–1281, 1984.

Turner DA, Prodomos CC, Clark JW: MRI in detecting acute injury of ligaments of the knee. Presented at the Annual Meeting of the Radiological Society of North America, Washington D.C., 1984.

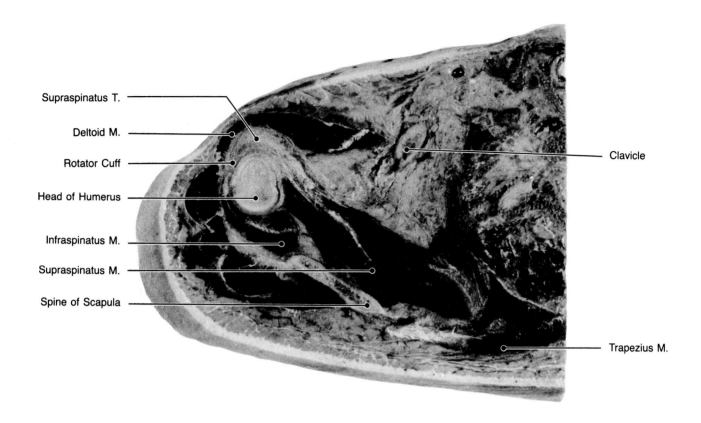

Supraspinatus T.

Deltoid M.

Rotator Cuff

Head of Humerus

Infraspinatus M.

Supraspinatus M.

Spine of Scapula

Clavicle

Trapezius M.

1A–1C Axial/Anatomic Plate/CT/MRI/Glenohumeral Joint.
Axial section through the glenohumeral joint, superior aspect
demonstrating the supraspinatus muscle in almost its entire
extension from the supraspinatus fossa of the scapula to the
humeral head.

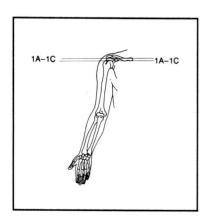

1A–1C

1A–1C

1B

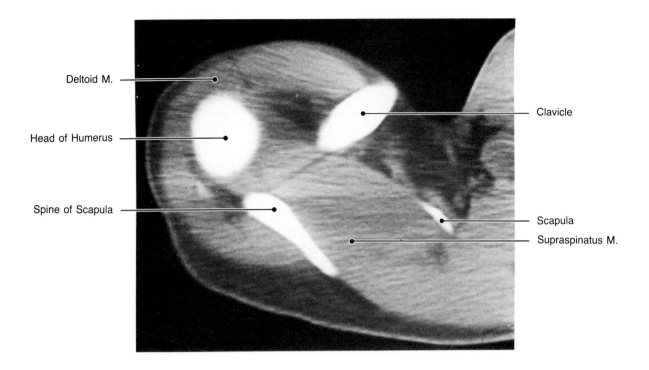

Deltoid M.

Head of Humerus

Spine of Scapula

Clavicle

Scapula

Supraspinatus M.

1C

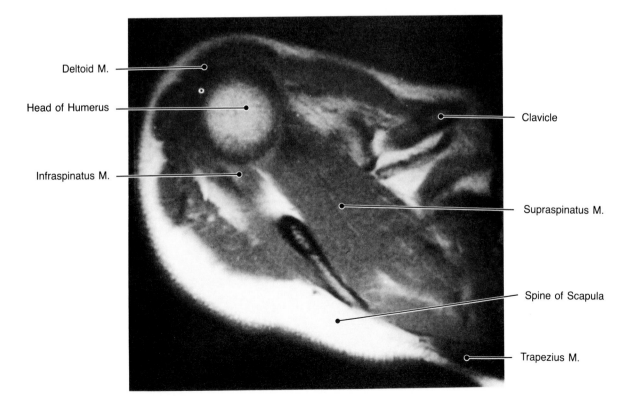

Deltoid M.

Head of Humerus

Infraspinatus M.

Clavicle

Supraspinatus M.

Spine of Scapula

Trapezius M.

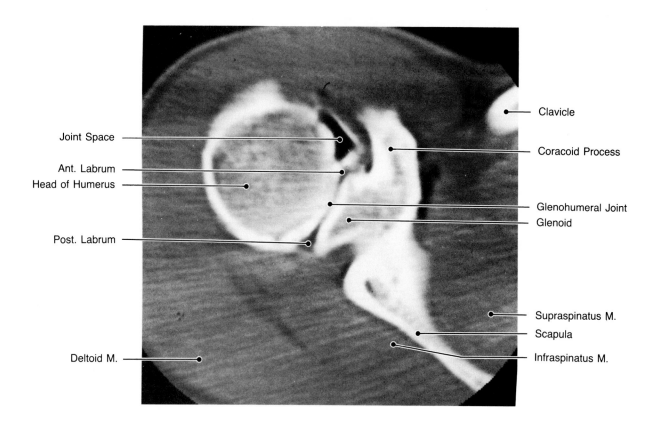

Joint Space

Ant. Labrum

Head of Humerus

Post. Labrum

Deltoid M.

Clavicle

Coracoid Process

Glenohumeral Joint

Glenoid

Supraspinatus M.

Scapula

Infraspinatus M.

1D–1E Axial/CT/MRI/Glenohumeral Joint. Axial images at the level of the coracoid process. The CT image was obtained following intrarticular administration of air and positive contrast. The anterior and posterior aspects of the labrum are demonstrated with both imaging techniques.

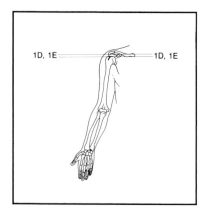

1D, 1E ———— ———— 1D, 1E

Deltoid M.

Coracoid Process of
Scapula

Clavicle

Head of Humerus

Glenohumeral Joint
with Articular Cartilage

Teres Minor M.

Serratus Ant. M.

Spine of Scapula

Infraspinatus M.

Supraspinatus M.

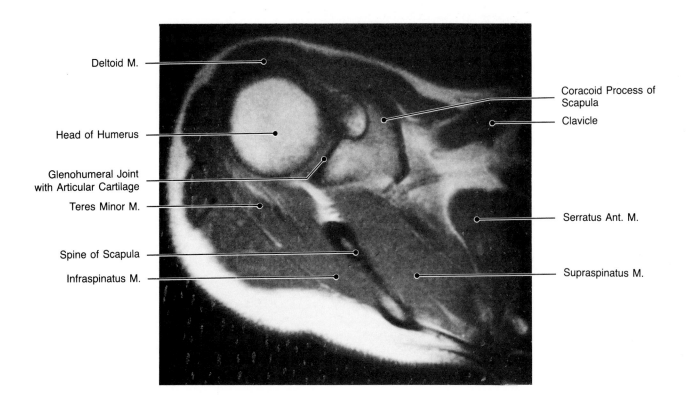

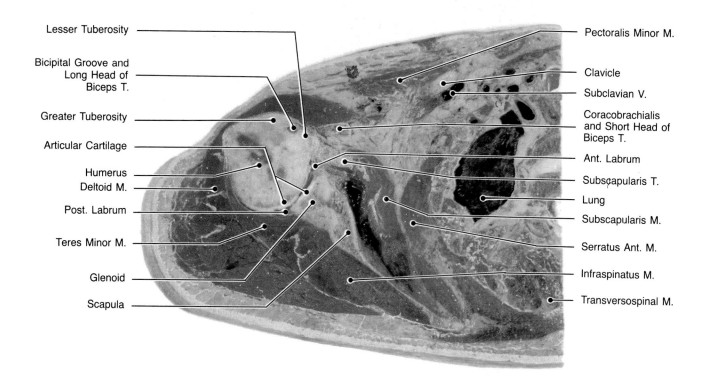

Lesser Tuberosity

Bicipital Groove and Long Head of Biceps T.

Greater Tuberosity

Articular Cartilage

Humerus

Deltoid M.

Post. Labrum

Teres Minor M.

Glenoid

Scapula

Pectoralis Minor M.

Clavicle

Subclavian V.

Coracobrachialis and Short Head of Biceps T.

Ant. Labrum

Subscapularis T.

Lung

Subscapularis M.

Serratus Ant. M.

Infraspinatus M.

Transversospinal M.

2A, 2B, 2C Axial/Anatomic Plate/CT/MRI/Mid Glenohumeral Joint. Axial sections through the mid glenohumeral joint. The long head of the biceps brachii tendon is seen within the bicipital groove in both CT (2B) and MRI (2C) sections. The labrum and articular cartilage of the glenoid can be assessed by both techniques. The subscapularis tendon can be seen surrounded by contrast material in the CT section.

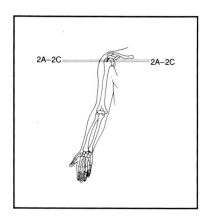

2A–2C   2A–2C

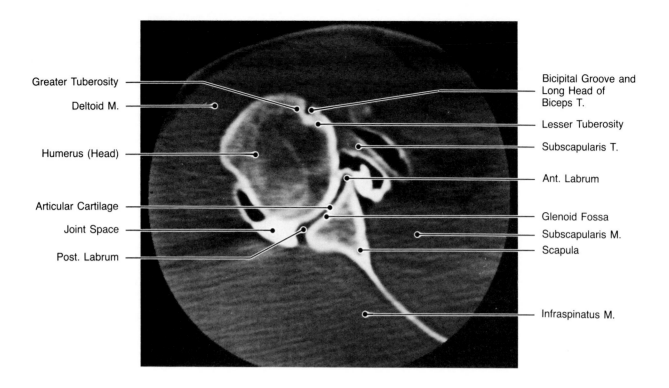

Greater Tuberosity

Deltoid M.

Humerus (Head)

Articular Cartilage

Joint Space

Post. Labrum

Bicipital Groove and
Long Head of
Biceps T.

Lesser Tuberosity

Subscapularis T.

Ant. Labrum

Glenoid Fossa

Subscapularis M.

Scapula

Infraspinatus M.

2C

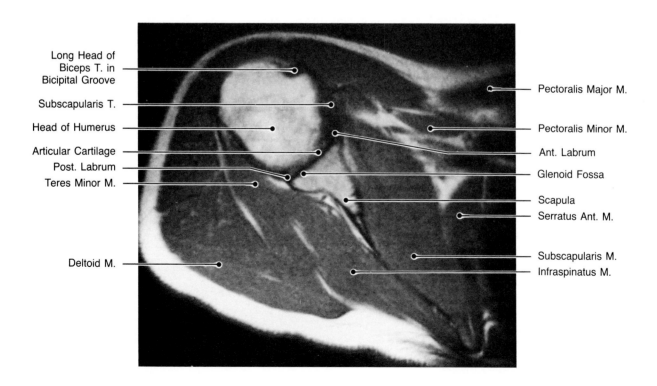

Long Head of
Biceps T. in
Bicipital Groove

Subscapularis T.

Head of Humerus

Articular Cartilage

Post. Labrum

Teres Minor M.

Deltoid M.

Pectoralis Major M.

Pectoralis Minor M.

Ant. Labrum

Glenoid Fossa

Scapula

Serratus Ant. M.

Subscapularis M.

Infraspinatus M.

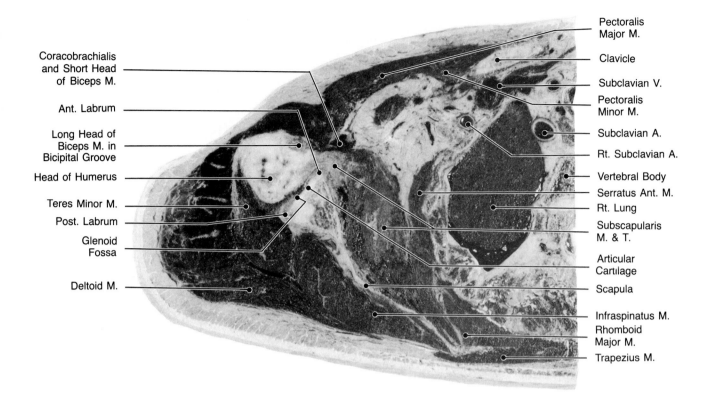

Pectoralis Major M.

Coracobrachialis and Short Head of Biceps M.

Clavicle

Subclavian V.

Ant. Labrum

Pectoralis Minor M.

Long Head of Biceps M. in Bicipital Groove

Subclavian A.

Rt. Subclavian A.

Head of Humerus

Vertebral Body

Serratus Ant. M.

Teres Minor M.

Rt. Lung

Post. Labrum

Subscapularis M. & T.

Glenoid Fossa

Articular Cartilage

Deltoid M.

Scapula

Infraspinatus M.

Rhomboid Major M.

Trapezius M.

3A, 3B, 3C Axial/Anatomic Plate/CT/MRI/Low Glenohumeral Joint. Axial sections through the lower aspect of the glenohumeral joint about 1 cm below the previous section. Again the posterior and anterior aspects of the labrum can be assessed by MR (2C). The tendon of the long head of the biceps brachii muscle is seen surrounded by contrast material in the CT section (2C). Note the teres minor muscle surrounding the head of the humerus posteriorly.

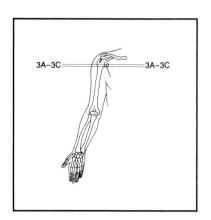

3A–3C   3A–3C

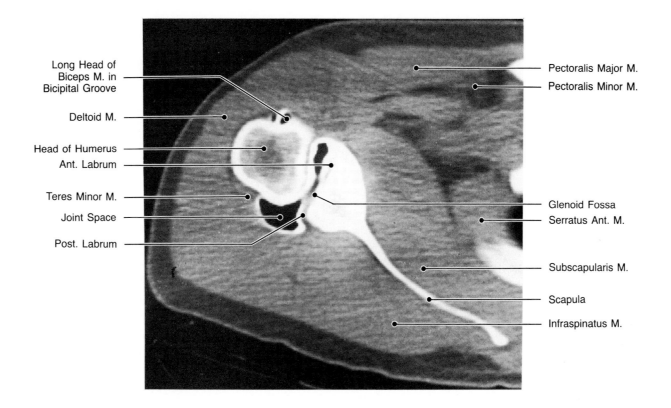

Long Head of
Biceps M. in
Bicipital Groove

Deltoid M.

Head of Humerus

Ant. Labrum

Teres Minor M.

Joint Space

Post. Labrum

Pectoralis Major M.

Pectoralis Minor M.

Glenoid Fossa

Serratus Ant. M.

Subscapularis M.

Scapula

Infraspinatus M.

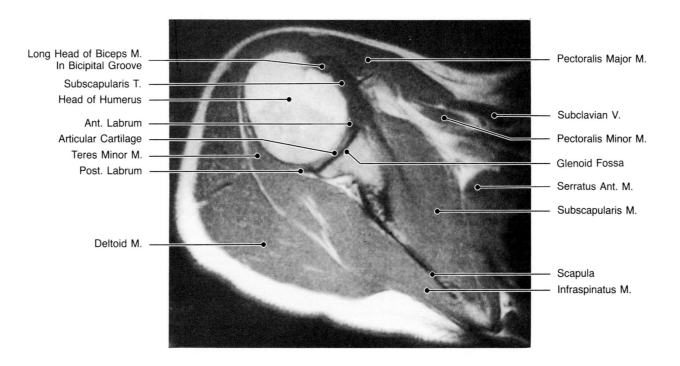

Long Head of Biceps M.
In Bicipital Groove

Subscapularis T.

Head of Humerus

Ant. Labrum

Articular Cartilage

Teres Minor M.

Post. Labrum

Deltoid M.

Pectoralis Major M.

Subclavian V.

Pectoralis Minor M.

Glenoid Fossa

Serratus Ant. M.

Subscapularis M.

Scapula

Infraspinatus M.

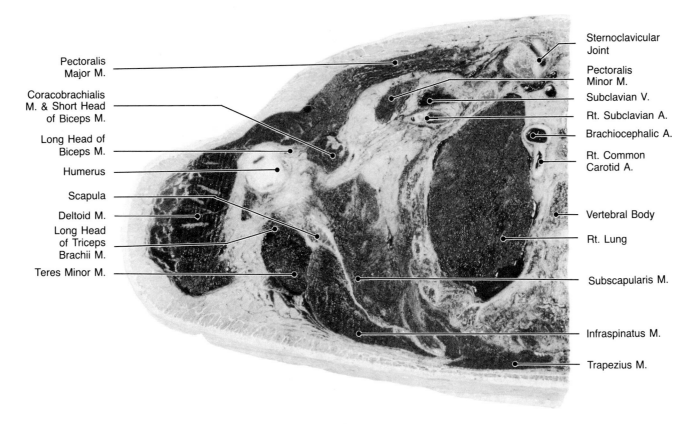

Pectoralis Major M.

Coracobrachialis M. & Short Head of Biceps M.

Long Head of Biceps M.

Humerus

Scapula

Deltoid M.

Long Head of Triceps Brachii M.

Teres Minor M.

Sternoclavicular Joint

Pectoralis Minor M.

Subclavian V.

Rt. Subclavian A.

Brachiocephalic A.

Rt. Common Carotid A.

Vertebral Body

Rt. Lung

Subscapularis M.

Infraspinatus M.

Trapezius M.

4A, 4B, 4C Axial/Anatomic Plate/CT/MRI/Low Glenohumeral Joint. Axial sections through the most inferior aspect of the glenohumeral joint, about 1 cm below the previous section. The coracobrachialis muscle and short head of the biceps brachii muscle are seen medially and anteriorly to the humerus.

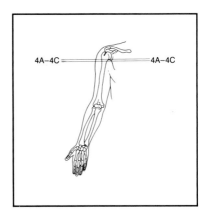

4A–4C   4A–4C

4B

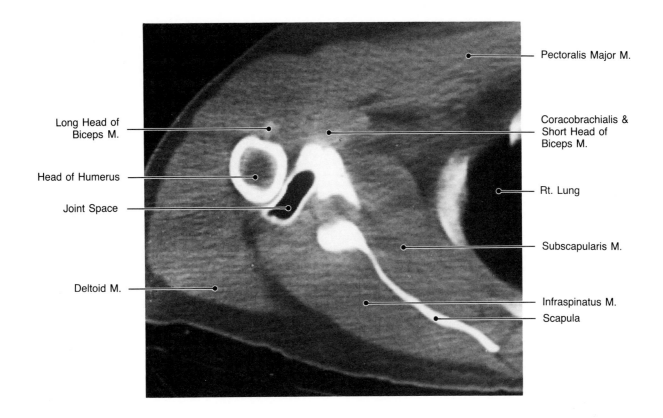

Long Head of
Biceps M.

Head of Humerus

Joint Space

Deltoid M.

Pectoralis Major M.

Coracobrachialis &
Short Head of
Biceps M.

Rt. Lung

Subscapularis M.

Infraspinatus M.
Scapula

4C

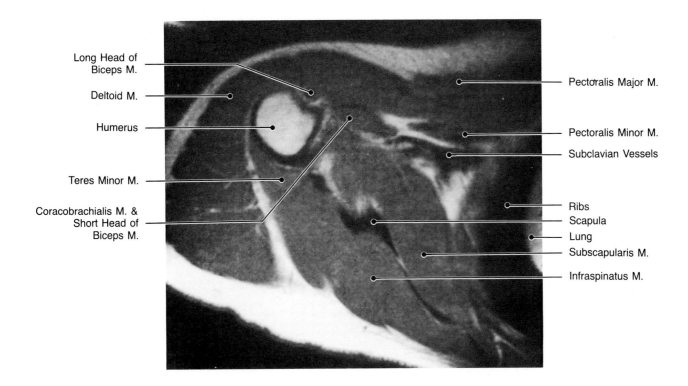

Long Head of
Biceps M.

Deltoid M.

Humerus

Teres Minor M.

Coracobrachialis M. &
Short Head of
Biceps M.

Pectoralis Major M.

Pectoralis Minor M.
Subclavian Vessels

Ribs
Scapula
Lung
Subscapularis M.
Infraspinatus M.

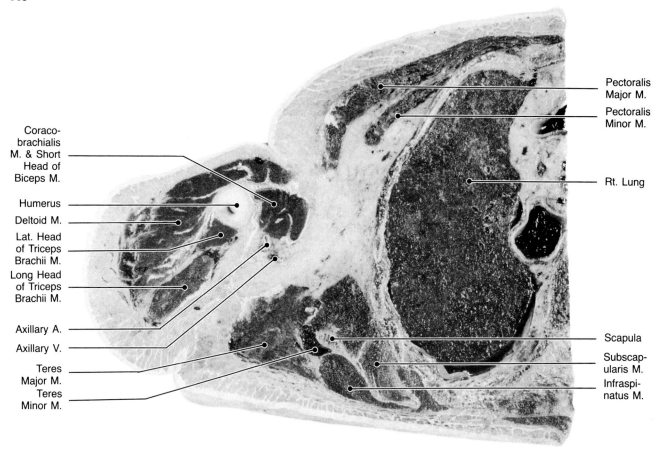

Pectoralis Major M.

Pectoralis Minor M.

Coraco-brachialis M. & Short Head of Biceps M.

Rt. Lung

Humerus

Deltoid M.

Lat. Head of Triceps Brachii M.

Long Head of Triceps Brachii M.

Axillary A.

Axillary V.

Scapula

Teres Major M.

Subscapularis M.

Teres Minor M.

Infraspinatus M.

5A, 5B, 5C Axial/Anatomic Plate/CT/MRI/Proximal Humerus. Axial section through the proximal humerus, just above the axilla. Note the axillary vessels in the anatomic (5A) and MRI sections (5C). The teres minor muscle is seen at the level of its origin in the lateral aspect of the scapula. The lateral head of the triceps brachii muscle is seen immediately posterior to the proximal humerus.

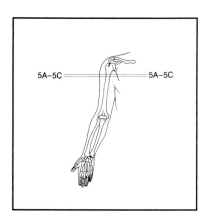

5A–5C   5A–5C

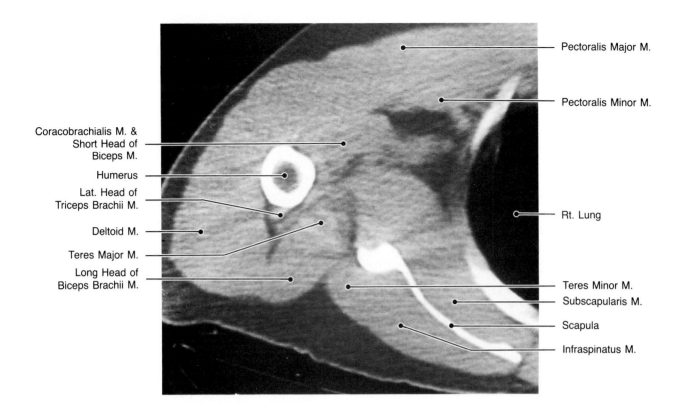

Pectoralis Major M.

Pectoralis Minor M.

Coracobrachialis M. &
Short Head of
Biceps M.

Humerus

Lat. Head of
Triceps Brachii M.

Deltoid M.

Teres Major M.

Long Head of
Biceps Brachii M.

Rt. Lung

Teres Minor M.

Subscapularis M.

Scapula

Infraspinatus M.

5C

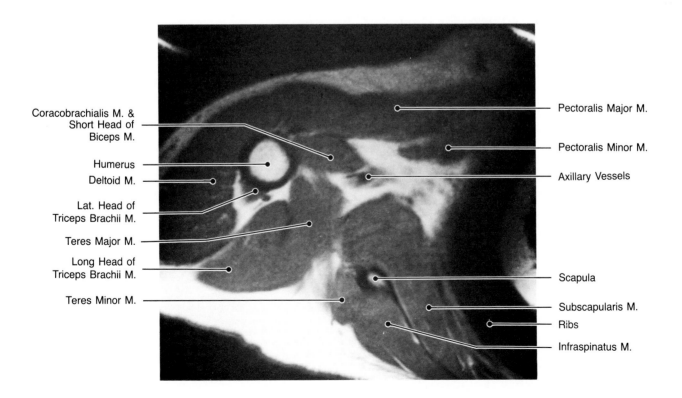

Coracobrachialis M. &
Short Head of
Biceps M.

Humerus

Deltoid M.

Lat. Head of
Triceps Brachii M.

Teres Major M.

Long Head of
Triceps Brachii M.

Teres Minor M.

Pectoralis Major M.

Pectoralis Minor M.

Axillary Vessels

Scapula

Subscapularis M.

Ribs

Infraspinatus M.

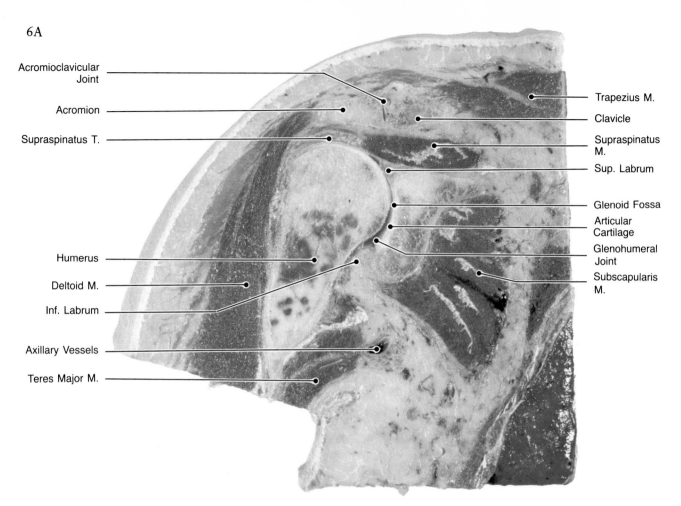

Acromioclavicular Joint

Acromion

Supraspinatus T.

Humerus

Deltoid M.

Inf. Labrum

Axillary Vessels

Teres Major M.

Trapezius M.

Clavicle

Supraspinatus M.

Sup. Labrum

Glenoid Fossa

Articular Cartilage

Glenohumeral Joint

Subscapularis M.

6A, 6B Coronal/Anatomic Plate/MRI/Glenohumeral Joint. Coronal sections through the glenohumeral joint. Note the acromioclavicular joint. The supraspinatus tendon is seen in both anatomic (6A) and MRI sections (6B), between the acromion process and humeral head.

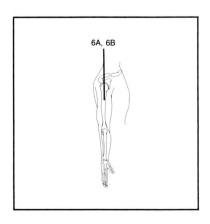

6A, 6B

381

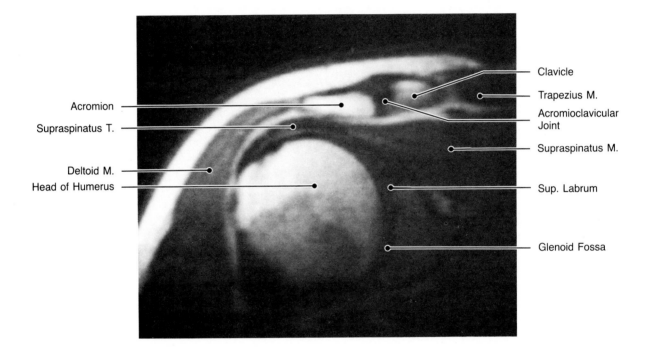

Acromion

Supraspinatus T.

Deltoid M.

Head of Humerus

Clavicle

Trapezius M.

Acromioclavicular
Joint

Supraspinatus M.

Sup. Labrum

Glenoid Fossa

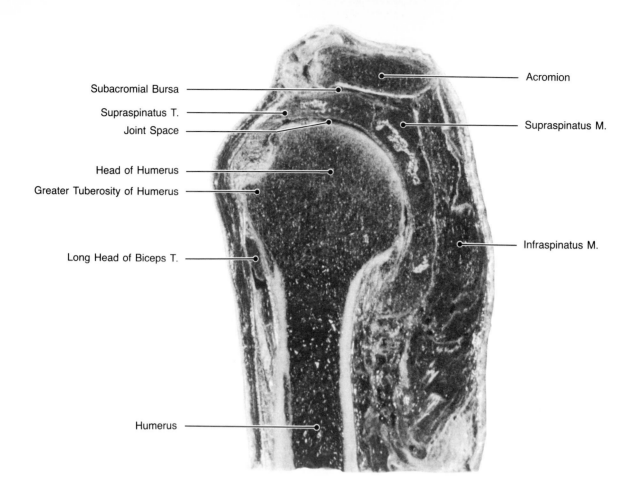

Subacromial Bursa —
Supraspinatus T. —
Joint Space —
Head of Humerus —
Greater Tuberosity of Humerus —
Long Head of Biceps T. —
Humerus —

Acromion
Supraspinatus M.
Infraspinatus M.

7A–7C Sagittal/Anatomic Plate/CT/Glenohumeral Joint. Sagittal sections through the glenohumeral joint. The anatomic section (7A) was obtained through the acromion and proximal humerus. The supraspinatus muscle and tendon are seen to the level of their insertion into the greater tuberosity of the humerus, forming part of the rotator cuff. Sections 7B and 7C are direct sagittal CT sections through the glenohumeral joint after double-contrast arthrography with the arm in abduction. Window settings are adjusted to demonstrate the structures within the joint. Note the intracapsular portion of the long head of the bicipital tendon from its insertion into the glenoid labrum superiorly and surrounding the humeral head. The thickness of the rotator cuff can be evaluated using this projection.

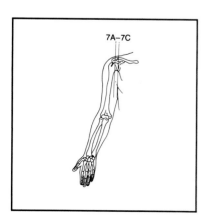

7A–7C

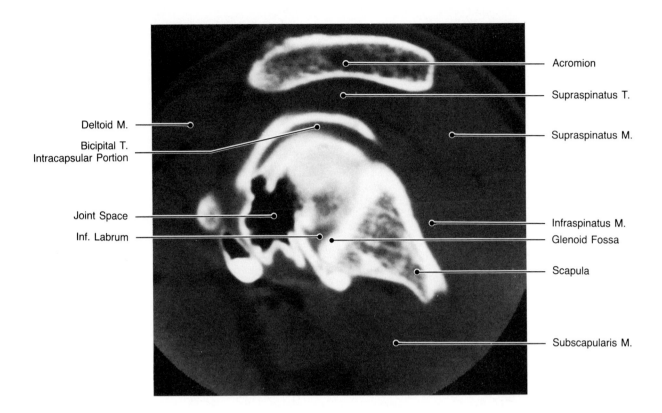

Acromion

Supraspinatus T.

Deltoid M.

Supraspinatus M.

Bicipital T.
Intracapsular Portion

Joint Space

Infraspinatus M.

Inf. Labrum

Glenoid Fossa

Scapula

Subscapularis M.

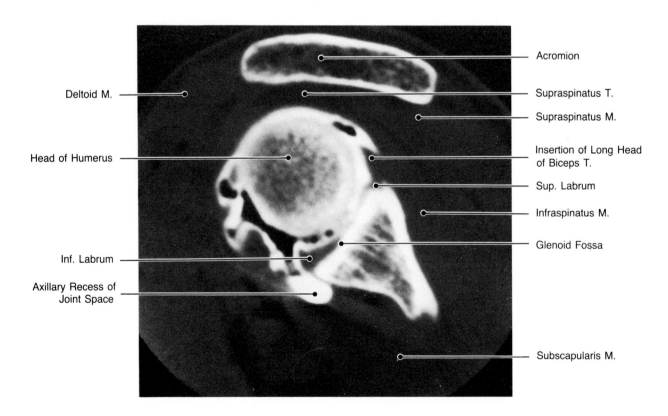

Acromion

Deltoid M.

Supraspinatus T.

Supraspinatus M.

Head of Humerus

Insertion of Long Head
of Biceps T.

Sup. Labrum

Infraspinatus M.

Inf. Labrum

Glenoid Fossa

Axillary Recess of
Joint Space

Subscapularis M.

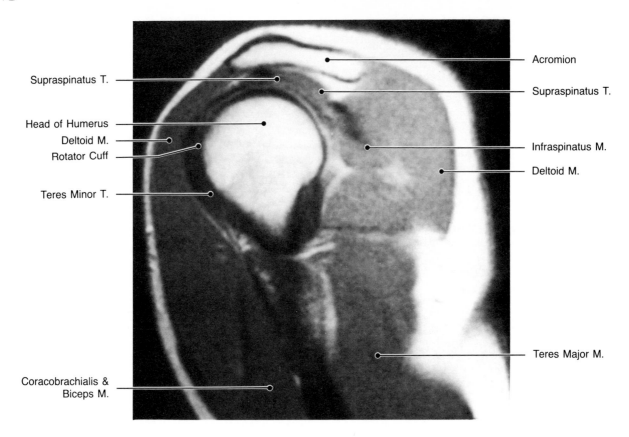

Supraspinatus T.

Head of Humerus

Deltoid M.

Rotator Cuff

Teres Minor T.

Coracobrachialis &
Biceps M.

Acromion

Supraspinatus T.

Infraspinatus M.

Deltoid M.

Teres Major M.

7D, 7E Saggital/MRI/Glenohumeral Joint. Sections 7D and 7E are saggital MRI sections at a plane similar to that of the anatomic and CT sections. The supraspinatus tendon can also be assessed in the subacromial region.

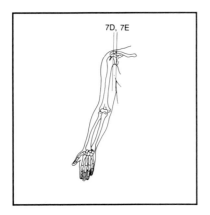

7D, 7E

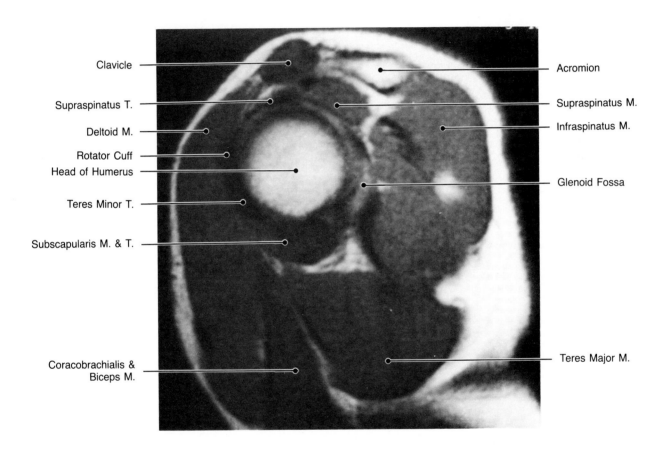

Clavicle

Supraspinatus T.

Deltoid M.

Rotator Cuff

Head of Humerus

Teres Minor T.

Subscapularis M. & T.

Coracobrachialis & Biceps M.

Acromion

Supraspinatus M.

Infraspinatus M.

Glenoid Fossa

Teres Major M.

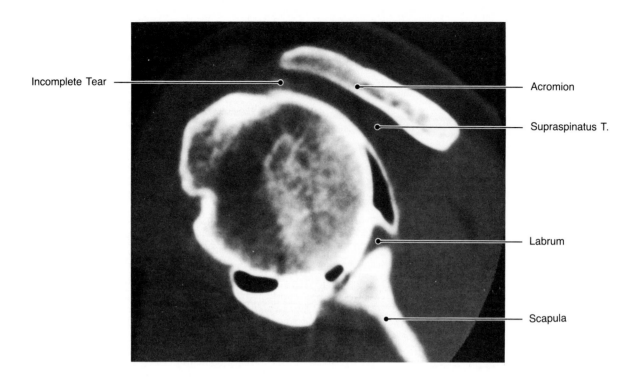

Incomplete Tear —————————— Acromion

————————— Supraspinatus T.

————— Labrum

————— Scapula

8A Sagittal/CT/Glenohumeral Joint. CT arthrogram of a sagittal section is shown. There is an incomplete tear of the rotator cuff at the level of the supraspinatus tendon. Contrast material is seen within the substance of the tendon.

8B Sagittal/CT/Glenohumeral Joint. Complete rotator cuff tear. Sagittal CT section post arthrogram. Air and contrast material are seen in subacromial-subdeltoid bursa. Contrast material within the joint capsule outlines the intracapsular portion of the long head of the bicipital tendon.

8C Sagittal/CT/Glenohumeral Joint. Chronic rotator cuff tear. Sagittal CT section post arthrogram. Air and contrast material are seen within the subacromial bursa. Note the thinning of the supraspinatus tendon and the decreased distance between the acromion and humeral head (impingement of the supraspinatus tendon).

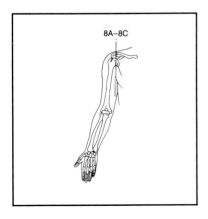

8A–8C

8B

Air and Contrast
Material in
Subacromial-Subdeltoid
Bursa

Intracapsular Bicipital T.

Acromion

Joint Capsule

Scapula

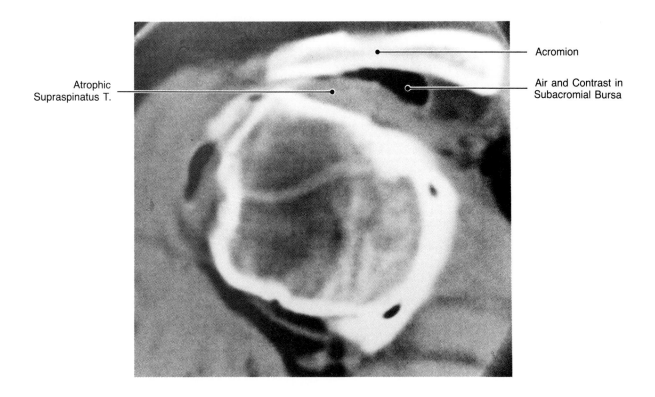

8C

Atrophic
Supraspinatus T.

Acromion

Air and Contrast in
Subacromial Bursa

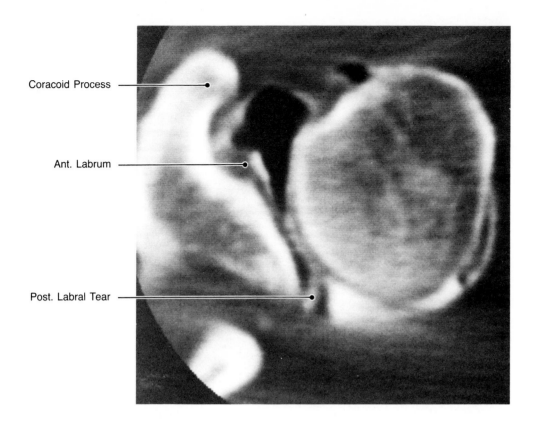

Coracoid Process

Ant. Labrum

Post. Labral Tear

8D, 8E Axial/CT/MRI/Glenohumeral Joint. Torn posterior labrum. Axial CT (8D) and MRI (8E) images of the same patient. The CT was obtained after intra-articular injection of contrast material and air. Contrast material is seen filling the defect in the posterior labrum. The MRI shows posterior and lateral displacement of the torn posterior labrum. Compare with the normal location of the anterior labrum.

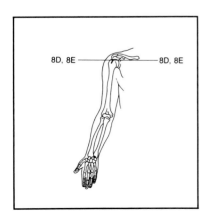

8D, 8E   8D, 8E

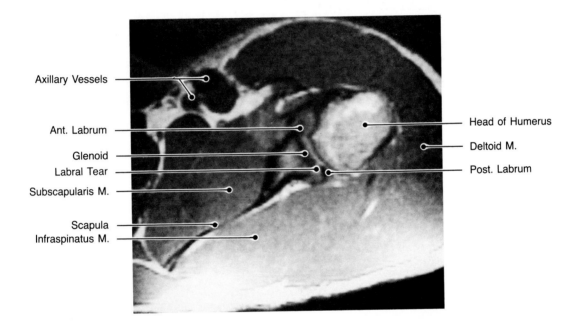

Axillary Vessels

Ant. Labrum

Glenoid

Labral Tear

Subscapularis M.

Scapula
Infraspinatus M.

Head of Humerus

Deltoid M.

Post. Labrum

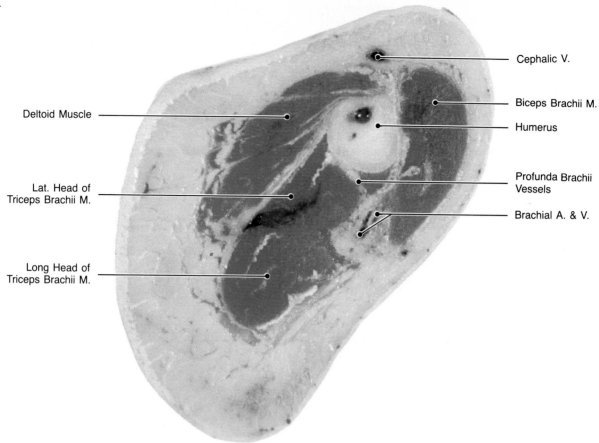

Cephalic V.

Biceps Brachii M.

Deltoid Muscle

Humerus

Lat. Head of
Triceps Brachii M.

Profunda Brachii
Vessels

Brachial A. & V.

Long Head of
Triceps Brachii M.

9A, 9B, 9C Axial/Anatomic Plate/CT/MRI/Upper Arm.
Axial sections through the upper arm, immediately below the
axilla. The brachial vessels and median nerve are seen in the
medial aspect of the arm, posterior to the biceps brachii muscle. The lateral and long heads of the triceps brachii can be
distinguished only on the MRI section (9C).

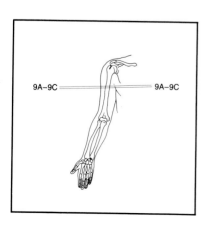

9A–9C   9A–9C

9B

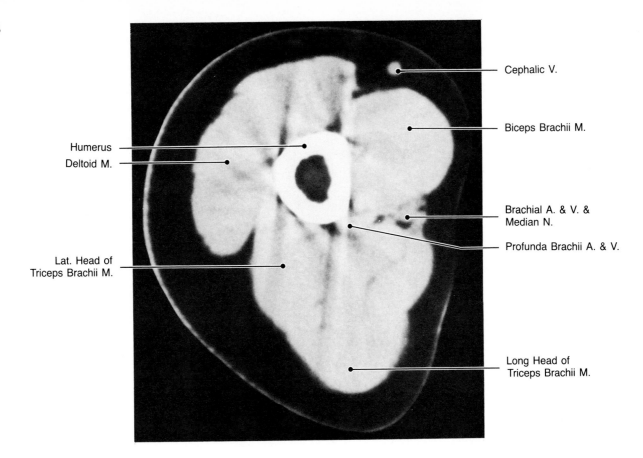

Cephalic V.

Biceps Brachii M.

Humerus

Deltoid M.

Brachial A. & V. &
Median N.

Profunda Brachii A. & V.

Lat. Head of
Triceps Brachii M.

Long Head of
Triceps Brachii M.

9C

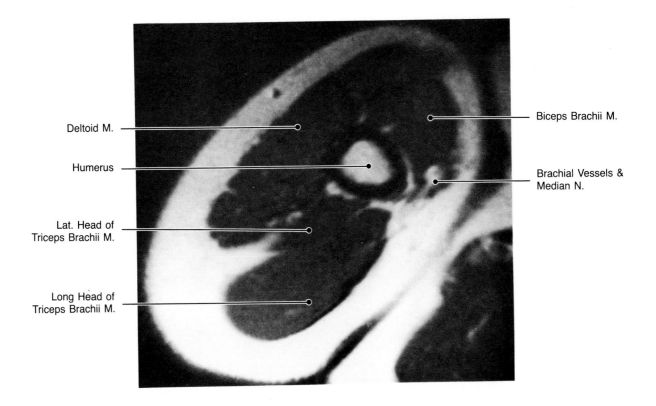

Biceps Brachii M.

Deltoid M.

Humerus

Brachial Vessels &
Median N.

Lat. Head of
Triceps Brachii M.

Long Head of
Triceps Brachii M.

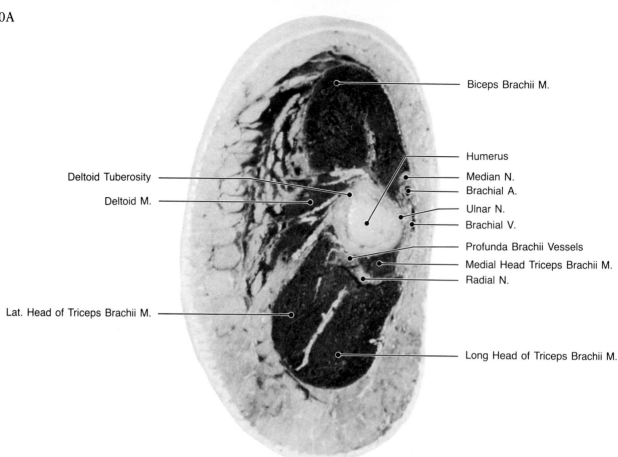

Biceps Brachii M.

Humerus

Deltoid Tuberosity

Median N.

Brachial A.

Deltoid M.

Ulnar N.

Brachial V.

Profunda Brachii Vessels

Medial Head Triceps Brachii M.

Radial N.

Lat. Head of Triceps Brachii M.

Long Head of Triceps Brachii M.

10A, 10B, 10C Axial/Anatomic Plate/CT/MRI/Upper Arm.
Axial sections through the upper arm, about 3 cm below previous section. The anterior compartment of the arm is occupied by the biceps brachii muscle. The posterior compartment is occupied by the triceps muscle with its lateral, long, and medial heads. The median and ulnar nerves are identified in the most medial aspect of the arm near the brachial vessels. The MRI section was obtained 3 cm below the CT section, at the level of the deltoid tuberosity.

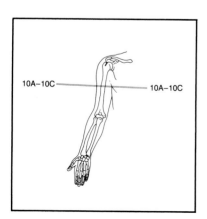

10A–10C

10A–10C

10B

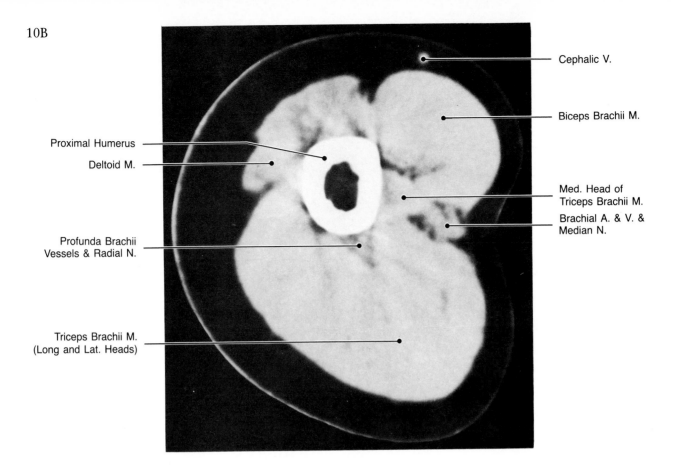

Cephalic V.

Biceps Brachii M.

Proximal Humerus

Deltoid M.

Med. Head of
Triceps Brachii M.

Brachial A. & V. &
Median N.

Profunda Brachii
Vessels & Radial N.

Triceps Brachii M.
(Long and Lat. Heads)

10C

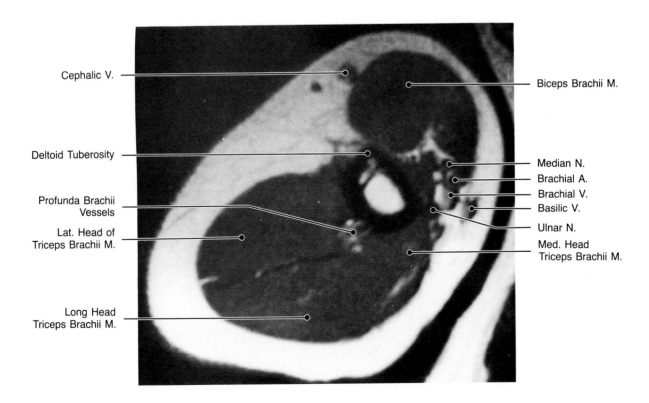

Cephalic V.

Biceps Brachii M.

Deltoid Tuberosity

Median N.
Brachial A.
Brachial V.
Basilic V.
Ulnar N.

Profunda Brachii
Vessels

Lat. Head of
Triceps Brachii M.

Med. Head
Triceps Brachii M.

Long Head
Triceps Brachii M.

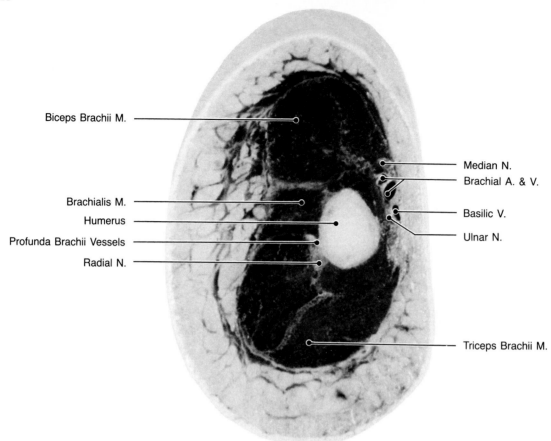

Biceps Brachii M.

Median N.

Brachial A. & V.

Brachialis M.

Basilic V.

Humerus

Ulnar N.

Profunda Brachii Vessels

Radial N.

Triceps Brachii M.

11A, 11B, 11C Axial/Anatomic Plate/CT/MRI/Mid-Arm.
Axial sections through the mid third of the arm. The brachialis muscle is identified between the biceps and triceps muscles, in the lateral aspect of the arm. Medially, the brachial vessels and median nerve are seen in both CT (11B) and MRI sections (11C), immediately behind the biceps brachii muscle. The profunda brachii vessels and radial nerve are identified in the lateral aspect of the arm, between the triceps and brachialis muscles.

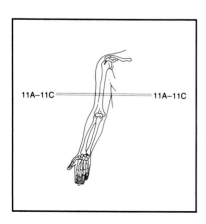

11A–11C

11A–11C

11B

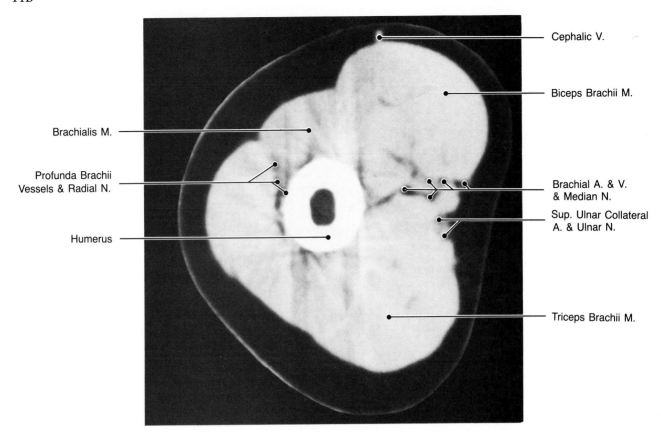

Cephalic V.

Biceps Brachii M.

Brachialis M.

Profunda Brachii
Vessels & Radial N.

Humerus

Brachial A. & V.
& Median N.

Sup. Ulnar Collateral
A. & Ulnar N.

Triceps Brachii M.

11C

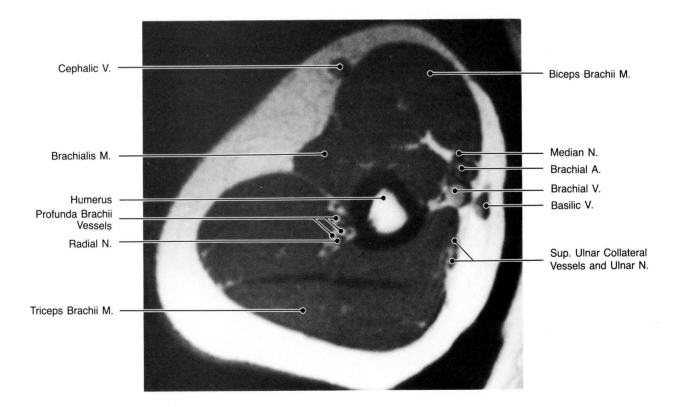

Cephalic V.

Biceps Brachii M.

Brachialis M.

Median N.

Brachial A.

Brachial V.

Humerus

Basilic V.

Profunda Brachii
Vessels

Radial N.

Sup. Ulnar Collateral
Vessels and Ulnar N.

Triceps Brachii M.

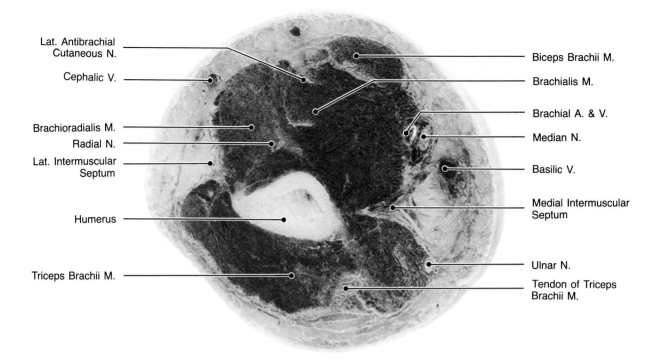

Lat. Antibrachial Cutaneous N.

Cephalic V.

Brachioradialis M.

Radial N.

Lat. Intermuscular Septum

Humerus

Triceps Brachii M.

Biceps Brachii M.

Brachialis M.

Brachial A. & V.

Median N.

Basilic V.

Medial Intermuscular Septum

Ulnar N.

Tendon of Triceps Brachii M.

12A, 12B, 12C Axial/Anatomic Plate/CT/MRI/Distal Arm.
Axial section through the supracondylar area of the humerus. The distal humerus has an elongated appearance on these sections. The radial nerve is identified in the MRI section (12C) between the brachioradialis and brachialis muscles. The ulnar nerve is located medial to the medial epicondyle and is better seen on the CT section (12B). The triceps brachii muscle and tendon are seen posterior to the distal humerus.

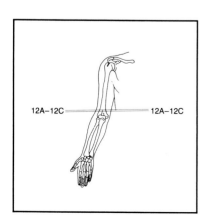

12A–12C   12A–12C

12B

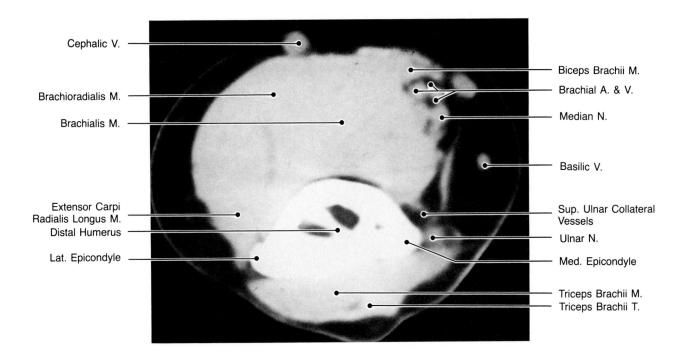

Cephalic V.

Brachioradialis M.

Brachialis M.

Extensor Carpi
Radialis Longus M.

Distal Humerus

Lat. Epicondyle

Biceps Brachii M.

Brachial A. & V.

Median N.

Basilic V.

Sup. Ulnar Collateral
Vessels

Ulnar N.

Med. Epicondyle

Triceps Brachii M.

Triceps Brachii T.

12C

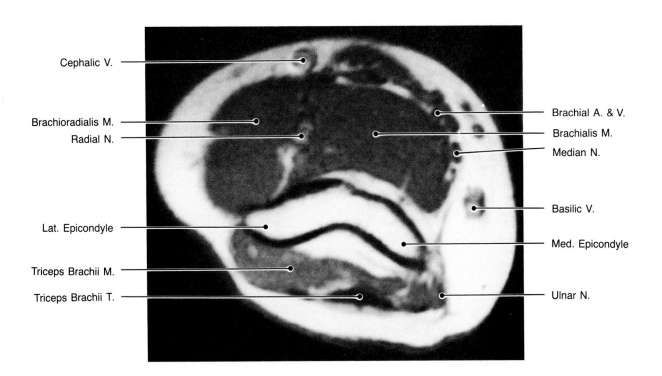

Cephalic V.

Brachioradialis M.

Radial N.

Lat. Epicondyle

Triceps Brachii M.

Triceps Brachii T.

Brachial A. & V.

Brachialis M.

Median N.

Basilic V.

Med. Epicondyle

Ulnar N.

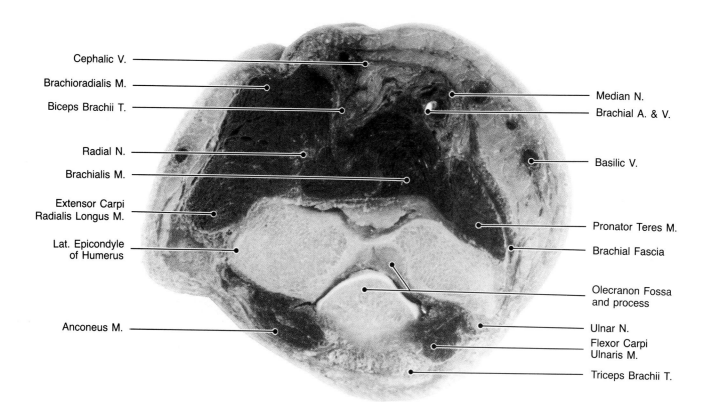

Cephalic V.

Brachioradialis M.

Biceps Brachii T.

Radial N.

Brachialis M.

Extensor Carpi
Radialis Longus M.

Lat. Epicondyle
of Humerus

Anconeus M.

Median N.

Brachial A. & V.

Basilic V.

Pronator Teres M.

Brachial Fascia

Olecranon Fossa
and process

Ulnar N.

Flexor Carpi
Ulnaris M.

Triceps Brachii T.

13A, 13B, 13C Axial/Anatomic Plate/CT/MRI/Proximal Elbow. Axial sections through the distal humerus, including the olecranon fossa. Anteriorly, from lateral to medial, the following are identified: extensor carpi radialis longus muscle, brachioradialis muscle, biceps brachii tendon, and brachialis and pronator teres muscles. Posteriorly, lateral to the olecranon, the anconeus muscle is seen. The ulnar nerve is identified in both CT (13B) and MRI (13C) sections, immediately posterior to the medial epicondyle. Posterior to the ulnar nerve the flexor carpi radialis muscle is seen adjacent to the olecranon.

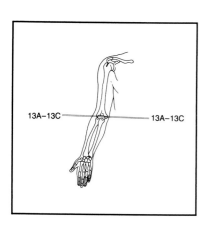

13A–13C   13A–13C

13B

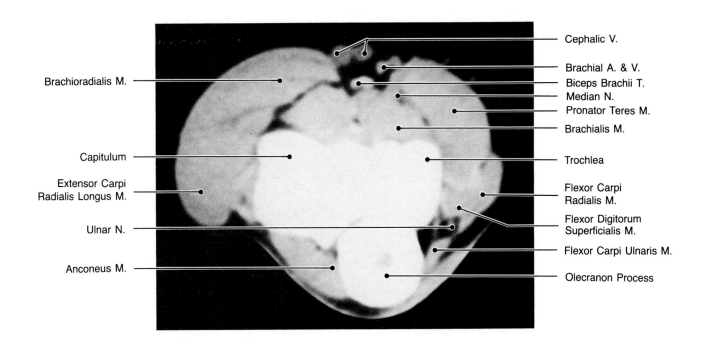

Brachioradialis M.

Capitulum

Extensor Carpi
Radialis Longus M.

Ulnar N.

Anconeus M.

Cephalic V.

Brachial A. & V.

Biceps Brachii T.

Median N.

Pronator Teres M.

Brachialis M.

Trochlea

Flexor Carpi
Radialis M.

Flexor Digitorum
Superficialis M.

Flexor Carpi Ulnaris M.

Olecranon Process

13C

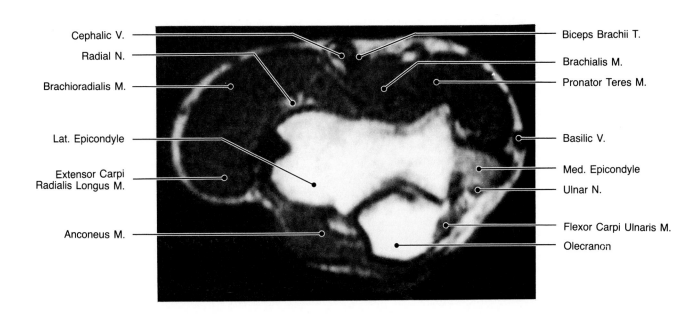

Cephalic V.

Radial N.

Brachioradialis M.

Lat. Epicondyle

Extensor Carpi
Radialis Longus M.

Anconeus M.

Biceps Brachii T.

Brachialis M.

Pronator Teres M.

Basilic V.

Med. Epicondyle

Ulnar N.

Flexor Carpi Ulnaris M.

Olecranon

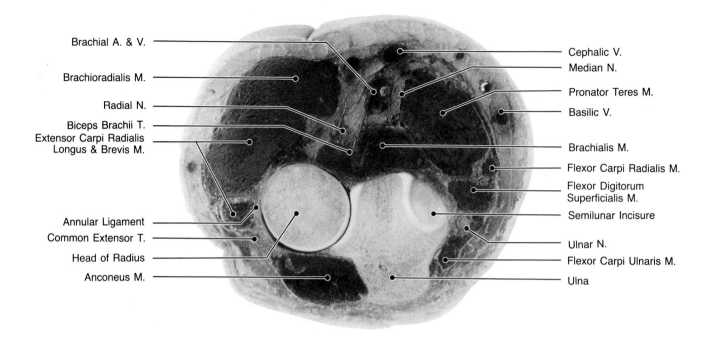

Brachial A. & V.

Brachioradialis M.

Radial N.

Biceps Brachii T.
Extensor Carpi Radialis
Longus & Brevis M.

Annular Ligament

Common Extensor T.

Head of Radius

Anconeus M.

Cephalic V.

Median N.

Pronator Teres M.

Basilic V.

Brachialis M.

Flexor Carpi Radialis M.

Flexor Digitorum
Superficialis M.

Semilunar Incisure

Ulnar N.

Flexor Carpi Ulnaris M.

Ulna

14A, 14B, 14C Axial/Anatomic Plate/CT/MRI/Proximal Radioulnar Joint. Axial section through the radial head and proximal ulna. The proximal radioulnar joint is seen. The radial nerve is identified between the brachioradialis muscle and the brachialis muscle medially. The space between the brachioradialis muscle laterally and the pronator teres muscle medially is occupied by the brachial vessels, biceps brachii tendon, median nerve, and cephalic vein.

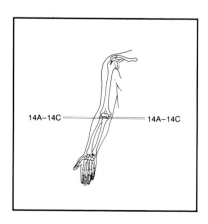

14A–14C

14A–14C

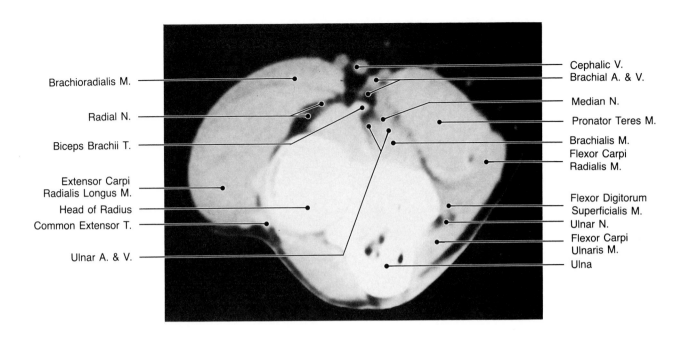

Brachioradialis M. — Cephalic V.
Brachial A. & V.
Radial N. — Median N.
Pronator Teres M.
Biceps Brachii T. — Brachialis M.
Flexor Carpi
Radialis M.
Extensor Carpi
Radialis Longus M. — Flexor Digitorum
Head of Radius — Superficialis M.
Common Extensor T. — Ulnar N.
Flexor Carpi
Ulnaris M.
Ulnar A. & V. — Ulna

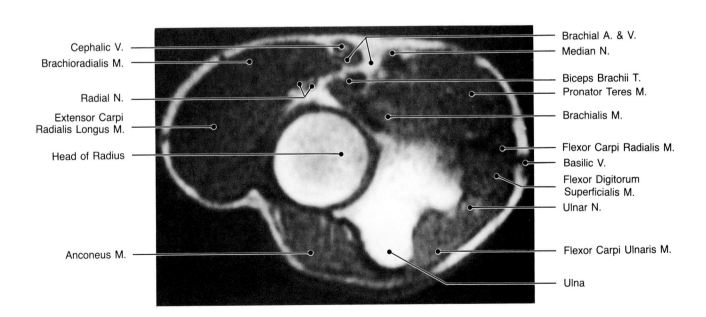

Cephalic V. — Brachial A. & V.
Brachioradialis M. — Median N.
Radial N. — Biceps Brachii T.
Pronator Teres M.
Extensor Carpi
Radialis Longus M. — Brachialis M.
Head of Radius — Flexor Carpi Radialis M.
Basilic V.
Flexor Digitorum
Superficialis M.
Ulnar N.
Anconeus M. — Flexor Carpi Ulnaris M.
Ulna

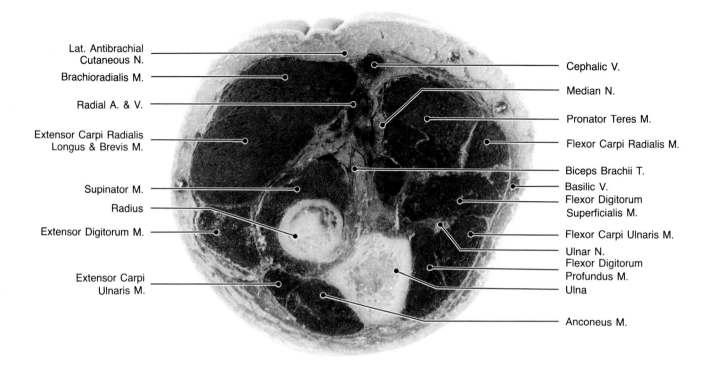

Lat. Antibrachial Cutaneous N.

Brachioradialis M.

Radial A. & V.

Extensor Carpi Radialis Longus & Brevis M.

Supinator M.

Radius

Extensor Digitorum M.

Extensor Carpi Ulnaris M.

Cephalic V.

Median N.

Pronator Teres M.

Flexor Carpi Radialis M.

Biceps Brachii T.

Basilic V.

Flexor Digitorum Superficialis M.

Flexor Carpi Ulnaris M.

Ulnar N.

Flexor Digitorum Profundus M.

Ulna

Anconeus M.

15A, 15B, 15C Axial/Anatomic Plate/CT/MRI/Proximal Forearm. Axial sections through the proximal forearm, 1 cm below previous section. The head of the radius is seen slightly more anterior than the ulna. Surrounding the radial head, the supinator muscle lies between the radius and the extensor carpi radialis longus and brevis muscles. Posterolaterally the extensor digitorum and extensor carpi ulnaris muscles are seen adjacent to the anconeous muscle. Medially the following muscles are identified better on the MRI section (15C) pronator teres, flexor carpi radialis, flexor digitorum superficalis, flexor carpi ulnaris, and flexor digitorum profundus.

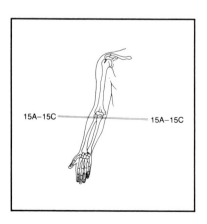

15A–15C

15A–15C

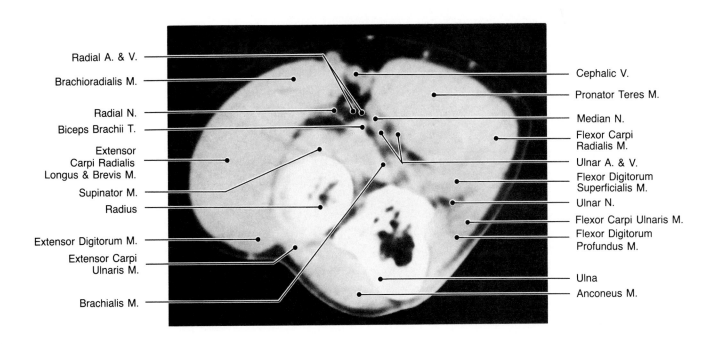

Radial A. & V.
Brachioradialis M.
Radial N.
Biceps Brachii T.
Extensor Carpi Radialis Longus & Brevis M.
Supinator M.
Radius
Extensor Digitorum M.
Extensor Carpi Ulnaris M.
Brachialis M.

Cephalic V.
Pronator Teres M.
Median N.
Flexor Carpi Radialis M.
Ulnar A. & V.
Flexor Digitorum Superficialis M.
Ulnar N.
Flexor Carpi Ulnaris M.
Flexor Digitorum Profundus M.
Ulna
Anconeus M.

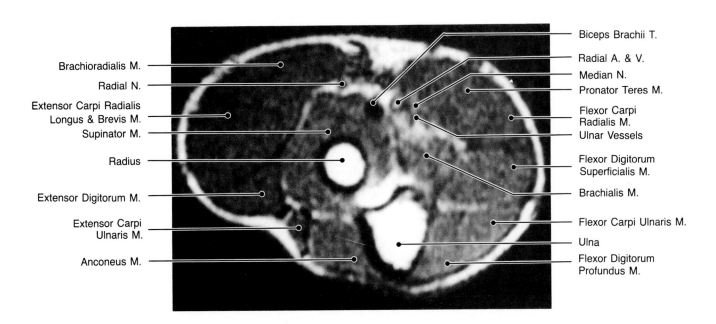

Brachioradialis M.
Radial N.
Extensor Carpi Radialis Longus & Brevis M.
Supinator M.
Radius
Extensor Digitorum M.
Extensor Carpi Ulnaris M.
Anconeus M.

Biceps Brachii T.
Radial A. & V.
Median N.
Pronator Teres M.
Flexor Carpi Radialis M.
Ulnar Vessels
Flexor Digitorum Superficialis M.
Brachialis M.
Flexor Carpi Ulnaris M.
Ulna
Flexor Digitorum Profundus M.

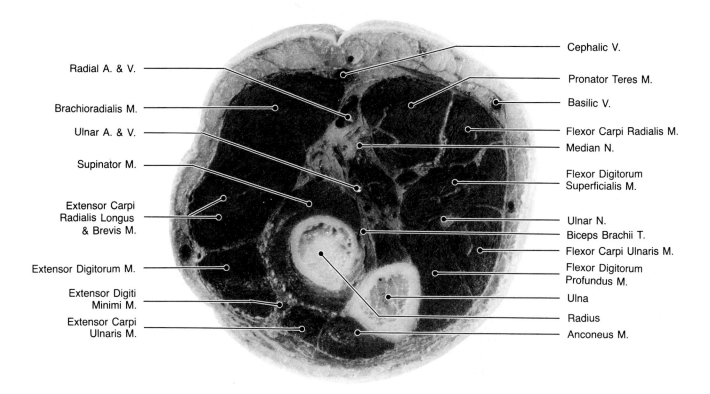

Radial A. & V.

Brachioradialis M.

Ulnar A. & V.

Supinator M.

Extensor Carpi
Radialis Longus
& Brevis M.

Extensor Digitorum M.

Extensor Digiti
Minimi M.

Extensor Carpi
Ulnaris M.

Cephalic V.

Pronator Teres M.

Basilic V.

Flexor Carpi Radialis M.

Median N.

Flexor Digitorum
Superficialis M.

Ulnar N.

Biceps Brachii T.

Flexor Carpi Ulnaris M.

Flexor Digitorum
Profundus M.

Ulna

Radius

Anconeus M.

16A, 16B, 16C Axial/Anatomic Plate/CT/MRI/Proximal Forearm. Axial sections through the proximal third of the forearm, about 2 cm below previous section. The deep space between the pronator teres muscle medially and the brachioradialis and supinator muscles laterally is occupied by the following vessels and nerves, from anterior to posterior: cephalic vein, radial nerve, radial vessels, median nerve, and ulnar vessels and slightly more medial by the biceps brachii tendon.

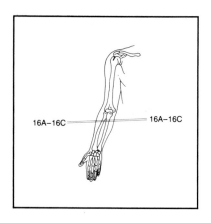

16A–16C   16A–16C

16B

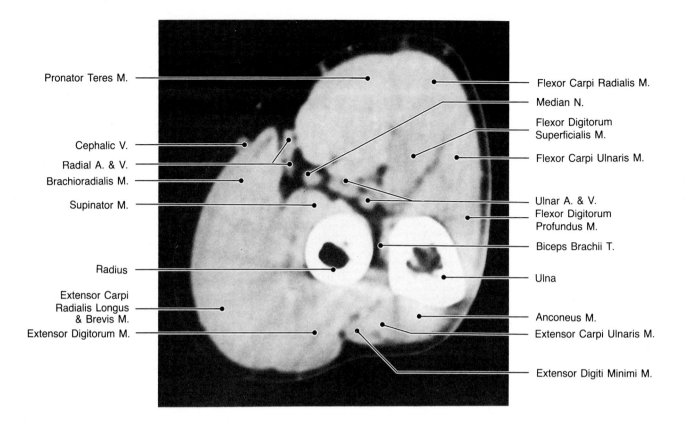

Pronator Teres M.

Cephalic V.

Radial A. & V.

Brachioradialis M.

Supinator M.

Radius

Extensor Carpi
Radialis Longus
& Brevis M.

Extensor Digitorum M.

Flexor Carpi Radialis M.

Median N.

Flexor Digitorum
Superficialis M.

Flexor Carpi Ulnaris M.

Ulnar A. & V.

Flexor Digitorum
Profundus M.

Biceps Brachii T.

Ulna

Anconeus M.

Extensor Carpi Ulnaris M.

Extensor Digiti Minimi M.

16C

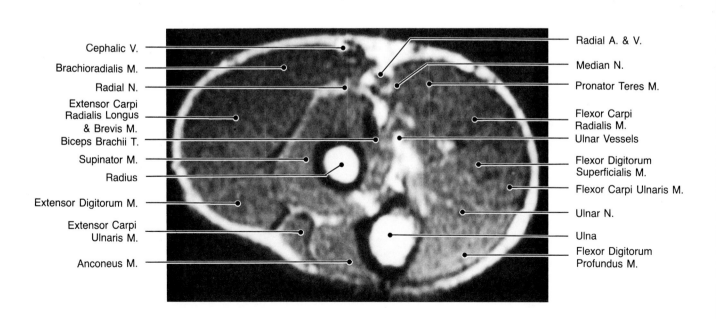

Cephalic V.

Brachioradialis M.

Radial N.

Extensor Carpi
Radialis Longus
& Brevis M.

Biceps Brachii T.

Supinator M.

Radius

Extensor Digitorum M.

Extensor Carpi
Ulnaris M.

Anconeus M.

Radial A. & V.

Median N.

Pronator Teres M.

Flexor Carpi
Radialis M.

Ulnar Vessels

Flexor Digitorum
Superficialis M.

Flexor Carpi Ulnaris M.

Ulnar N.

Ulna

Flexor Digitorum
Profundus M.

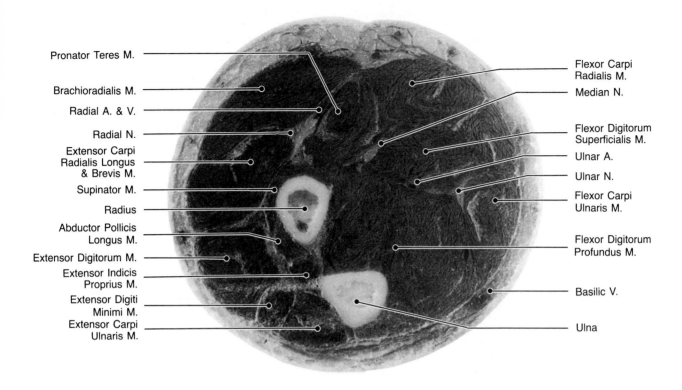

Pronator Teres M.

Brachioradialis M.

Radial A. & V.

Radial N.

Extensor Carpi
Radialis Longus
& Brevis M.

Supinator M.

Radius

Abductor Pollicis
Longus M.

Extensor Digitorum M.

Extensor Indicis
Proprius M.

Extensor Digiti
Minimi M.

Extensor Carpi
Ulnaris M.

Flexor Carpi
Radialis M.

Median N.

Flexor Digitorum
Superficialis M.

Ulnar A.

Ulnar N.

Flexor Carpi
Ulnaris M.

Flexor Digitorum
Profundus M.

Basilic V.

Ulna

17A, 17B, 17C Axial/Anatomic Plate/CT/MRI/Proximal Forearm. Axial sections through the proximal third of the forearm, about 3 cm below previous section. The anterior compartment is occupied by the following muscles, from lateral to medial: extensor carpii radialis longus and brevis, brachioradialis, pronator teres, flexor carpi radialis, flexor digitorum superficialis, flexor carpi ulnaris, and flexor digitorum profundus. The posterior compartment is occupied by the abductor pollicis longus, extensor digitorum, extensor digiti minimi, and extensor carpi radialis muscles. Owing to the large amount of muscular tissue in this section, the vessels are seen only on the MRI section (17C) because of the flow void and paradoxic enhancement phenomena.

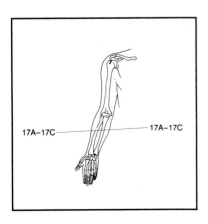

17A–17C   17A–17C

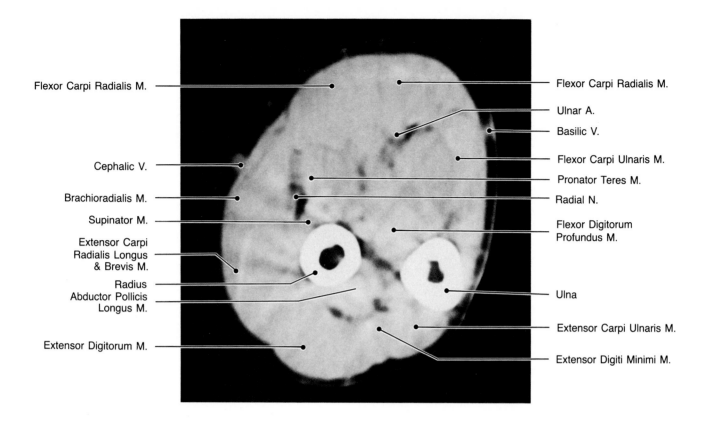

Flexor Carpi Radialis M.

Flexor Carpi Radialis M.

Ulnar A.

Basilic V.

Cephalic V.

Flexor Carpi Ulnaris M.

Pronator Teres M.

Brachioradialis M.

Radial N.

Supinator M.

Flexor Digitorum
Profundus M.

Extensor Carpi
Radialis Longus
& Brevis M.

Radius

Abductor Pollicis
Longus M.

Ulna

Extensor Carpi Ulnaris M.

Extensor Digitorum M.

Extensor Digiti Minimi M.

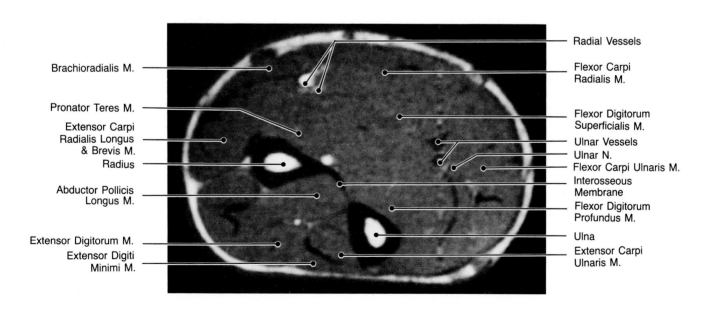

Radial Vessels

Brachioradialis M.

Flexor Carpi
Radialis M.

Pronator Teres M.

Flexor Digitorum
Superficialis M.

Extensor Carpi
Radialis Longus
& Brevis M.

Ulnar Vessels

Ulnar N.

Radius

Flexor Carpi Ulnaris M.

Abductor Pollicis
Longus M.

Interosseous
Membrane

Flexor Digitorum
Profundus M.

Extensor Digitorum M.

Ulna

Extensor Digiti
Minimi M.

Extensor Carpi
Ulnaris M.

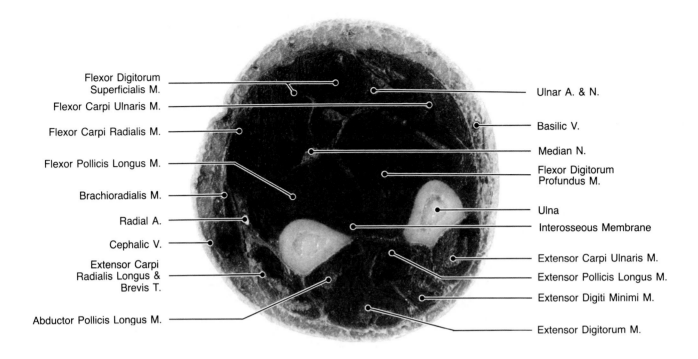

Flexor Digitorum Superficialis M.

Flexor Carpi Ulnaris M.

Flexor Carpi Radialis M.

Flexor Pollicis Longus M.

Brachioradialis M.

Radial A.

Cephalic V.

Extensor Carpi Radialis Longus & Brevis T.

Abductor Pollicis Longus M.

Ulnar A. & N.

Basilic V.

Median N.

Flexor Digitorum Profundus M.

Ulna

Interosseous Membrane

Extensor Carpi Ulnaris M.

Extensor Pollicis Longus M.

Extensor Digiti Minimi M.

Extensor Digitorum M.

18A, 18B, 18C Axial/Anatomic Plate/CT/MRI/Mid Forearm. Axial section through the mid third of the forearm, about 3 cm below the previous section. The tendons of the muscles, both flexor and extensor groups, can be identified on CT (18B) as high density linear structures, and on the MRI sections (18C) as low signal intensity structures.

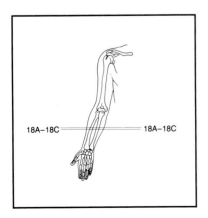

18A–18C 18A–18C

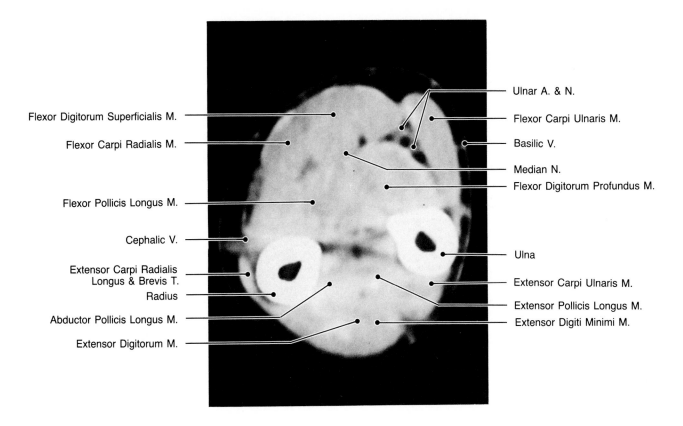

Flexor Digitorum Superficialis M.

Flexor Carpi Radialis M.

Flexor Pollicis Longus M.

Cephalic V.

Extensor Carpi Radialis
Longus & Brevis T.

Radius

Abductor Pollicis Longus M.

Extensor Digitorum M.

Ulnar A. & N.

Flexor Carpi Ulnaris M.

Basilic V.

Median N.

Flexor Digitorum Profundus M.

Ulna

Extensor Carpi Ulnaris M.

Extensor Pollicis Longus M.

Extensor Digiti Minimi M.

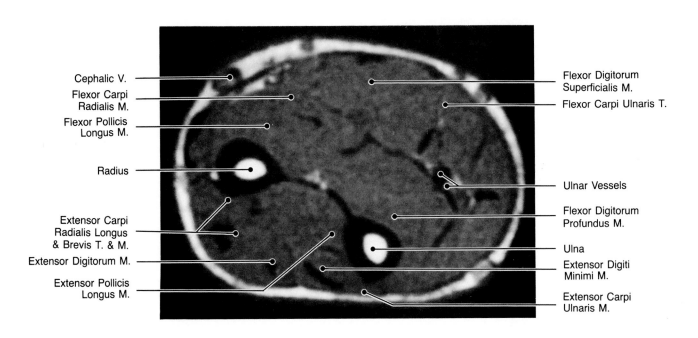

Cephalic V.

Flexor Carpi
Radialis M.

Flexor Pollicis
Longus M.

Radius

Extensor Carpi
Radialis Longus
& Brevis T. & M.

Extensor Digitorum M.

Extensor Pollicis
Longus M.

Flexor Digitorum
Superficialis M.

Flexor Carpi Ulnaris T.

Ulnar Vessels

Flexor Digitorum
Profundus M.

Ulna

Extensor Digiti
Minimi M.

Extensor Carpi
Ulnaris M.

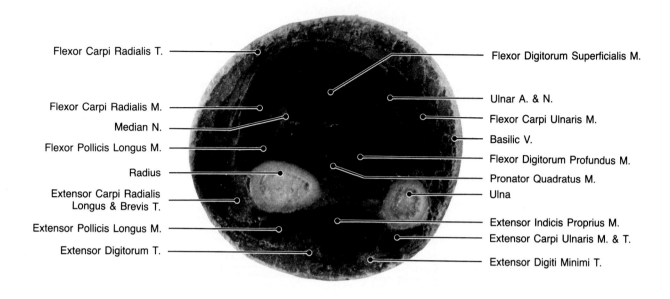

Flexor Carpi Radialis T.

Flexor Carpi Radialis M.

Median N.

Flexor Pollicis Longus M.

Radius

Extensor Carpi Radialis Longus & Brevis T.

Extensor Pollicis Longus M.

Extensor Digitorum T.

Flexor Digitorum Superficialis M.

Ulnar A. & N.

Flexor Carpi Ulnaris M.

Basilic V.

Flexor Digitorum Profundus M.

Pronator Quadratus M.

Ulna

Extensor Indicis Proprius M.

Extensor Carpi Ulnaris M. & T.

Extensor Digiti Minimi T.

19A, 19B, 19C Axial/Anatomic Plate/CT/MRI/Distal Forearm. Axial sections through the distal third of the forearm, about 3 cm below the previous section. Most of the tendons of the extensor and flexor muscles are now identified. The ulnar vessels and nerves are seen between the flexor carpi ulnaris posterolaterally and the flexor digitorum superficialis and profundus muscles laterally. The radial artery is very superficial, lateral and anterior to the flexor pollicis longus and pronator quadratus muscles.

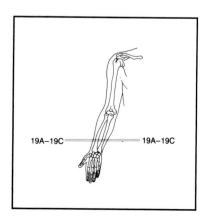

19A–19C   19A–19C

19B

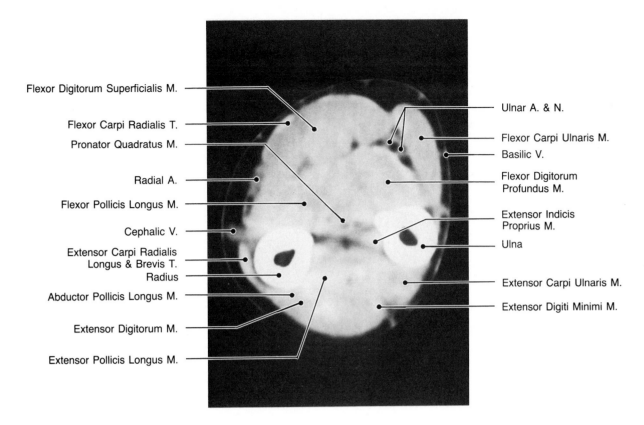

Flexor Digitorum Superficialis M.

Flexor Carpi Radialis T.

Pronator Quadratus M.

Radial A.

Flexor Pollicis Longus M.

Cephalic V.

Extensor Carpi Radialis
Longus & Brevis T.

Radius

Abductor Pollicis Longus M.

Extensor Digitorum M.

Extensor Pollicis Longus M.

Ulnar A. & N.

Flexor Carpi Ulnaris M.

Basilic V.

Flexor Digitorum
Profundus M.

Extensor Indicis
Proprius M.

Ulna

Extensor Carpi Ulnaris M.

Extensor Digiti Minimi M.

19C

Radial A.

Flexor Pollicis
Longus M.

Cephalic V.

Extensor Carpi
Radialis & Brevis T.

Radius

Pronator Quadratus M.

Abductor Pollicis
Longus M.

Extensor Digitorum T.

Extensor Pollicis
Longus M.

Extensor
Digiti Minimi T.

Flexor Carpi
Radialis M. & T.

Flexor Digitorum
Superficialis M.

Flexor Digitorum
Profundus M.

Ulnar Vessels

Ulnar N.

Flexor Carpi Ulnaris M.

Extensor Indicis
Proprius M.

Ulna

Extensor Carpi
Ulnaris M.

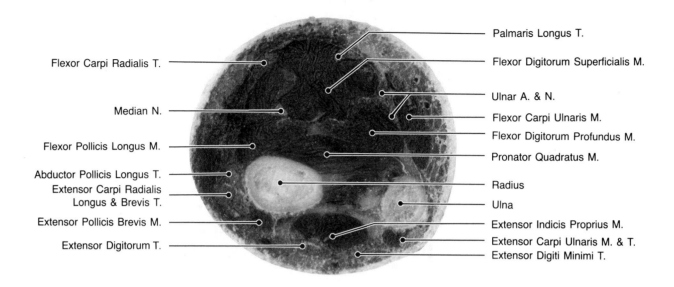

Flexor Carpi Radialis T.

Median N.

Flexor Pollicis Longus M.

Abductor Pollicis Longus T.

Extensor Carpi Radialis Longus & Brevis T.

Extensor Pollicis Brevis M.

Extensor Digitorum T.

Palmaris Longus T.

Flexor Digitorum Superficialis M.

Ulnar A. & N.

Flexor Carpi Ulnaris M.

Flexor Digitorum Profundus M.

Pronator Quadratus M.

Radius

Ulna

Extensor Indicis Proprius M.

Extensor Carpi Ulnaris M. & T.

Extensor Digiti Minimi T.

20A, 20B, 20C Axial/Anatomic Plate/CT/MRI/Distal Forearm. Axial sections through the distal forearm, about 3 cm below the previous section. The muscular tissue is thinner, and multiple tendons of both flexor and extensor groups can be identified. At this level the anterior compartment of the forearm includes the brachioradialis tendon, flexor pollicis longus muscle, flexor carpi radialis muscle and tendon, pronator quadratus muscle, and flexor digitorum superficialis and flexor carpi ulnaris muscles. The posterior compartment includes the extensor pollicis brevis and longus muscles and extensor digitorum, extensor indicis proprius, and extensor digiti minimi muscles and tendons. The interosseous membrane with interosseous vessels is identified on the MRI section (20C) and is surrounded by fat.

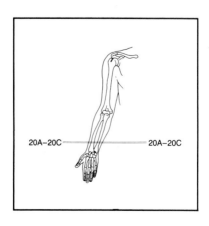

20A–20C   20A–20C

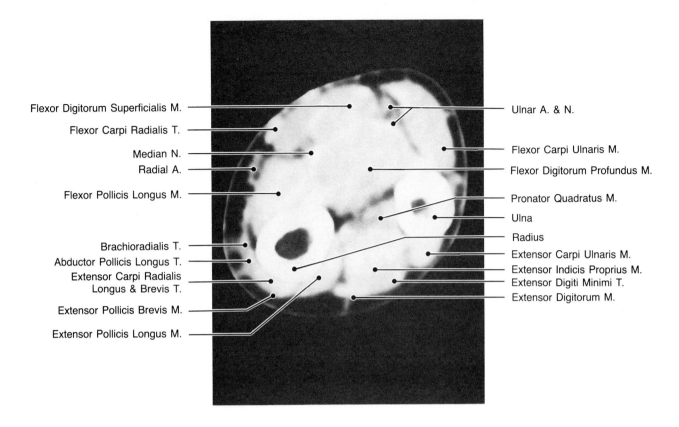

Flexor Digitorum Superficialis M.

Flexor Carpi Radialis T.

Median N.

Radial A.

Flexor Pollicis Longus M.

Brachioradialis T.

Abductor Pollicis Longus T.

Extensor Carpi Radialis
Longus & Brevis T.

Extensor Pollicis Brevis M.

Extensor Pollicis Longus M.

Ulnar A. & N.

Flexor Carpi Ulnaris M.

Flexor Digitorum Profundus M.

Pronator Quadratus M.

Ulna

Radius

Extensor Carpi Ulnaris M.

Extensor Indicis Proprius M.

Extensor Digiti Minimi T.

Extensor Digitorum M.

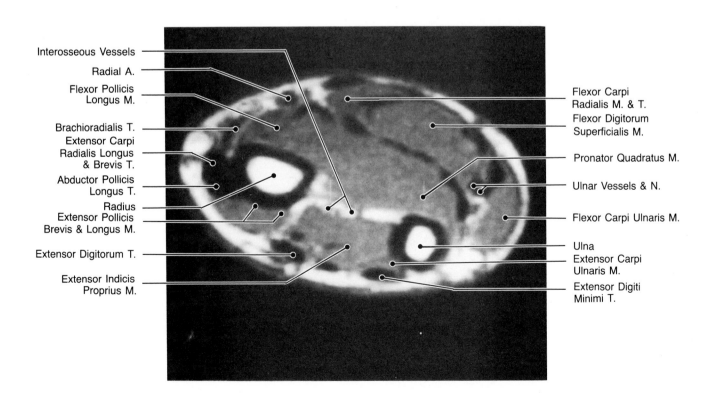

Interosseous Vessels

Radial A.

Flexor Pollicis
Longus M.

Brachioradialis T.

Extensor Carpi
Radialis Longus
& Brevis T.

Abductor Pollicis
Longus T.

Radius

Extensor Pollicis
Brevis & Longus M.

Extensor Digitorum T.

Extensor Indicis
Proprius M.

Flexor Carpi
Radialis M. & T.

Flexor Digitorum
Superficialis M.

Pronator Quadratus M.

Ulnar Vessels & N.

Flexor Carpi Ulnaris M.

Ulna

Extensor Carpi
Ulnaris M.

Extensor Digiti
Minimi T.

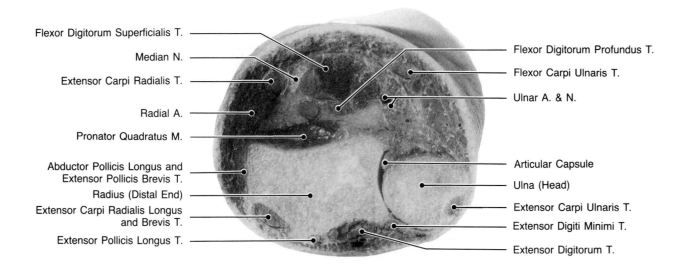

Flexor Digitorum Superficialis T.

Median N.

Extensor Carpi Radialis T.

Radial A.

Pronator Quadratus M.

Abductor Pollicis Longus and
Extensor Pollicis Brevis T.

Radius (Distal End)

Extensor Carpi Radialis Longus
and Brevis T.

Extensor Pollicis Longus T.

Flexor Digitorum Profundus T.

Flexor Carpi Ulnaris T.

Ulnar A. & N.

Articular Capsule

Ulna (Head)

Extensor Carpi Ulnaris T.

Extensor Digiti Minimi T.

Extensor Digitorum T.

21A, 21B, 21C Axial/Anatomic Plate/CT/MRI/Distal Forearm. Axial section through the distal third of the forearm just above the wrist, at the level of the distal radioulnar joint. Tendons are identified in both anatomic section (21A) and CT (21B). The MRI section (21C) corresponds to a plane about 2 cm above the anatomic section. The anterior aspect is occupied by the superficial and profundus flexor tendons. The lateral aspect is occupied by the abductor pollicis longus muscle and extensor pollicis brevis tendons and more posteriorly by the extensor carpi radialis longus and brevis tendons. The space between the posterior ulna and radius is occupied, from lateral to medial, by the extensor pollicis longus tendon, extensor digitorum longus tendons, extensor digiti minimi tendon, and extensor carpi ulnaris tendon.

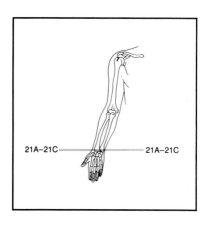

21A–21C   21A–21C

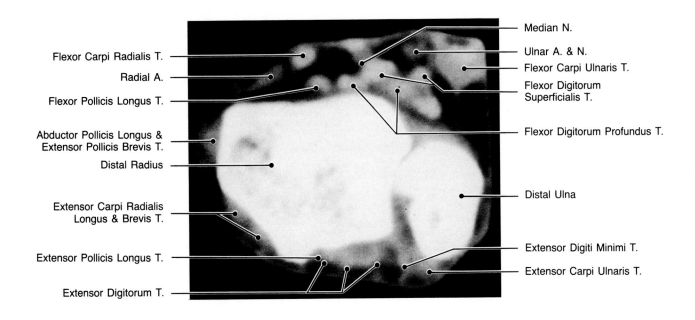

Flexor Carpi Radialis T.

Radial A.

Flexor Pollicis Longus T.

Abductor Pollicis Longus &
Extensor Pollicis Brevis T.

Distal Radius

Extensor Carpi Radialis
Longus & Brevis T.

Extensor Pollicis Longus T.

Extensor Digitorum T.

Median N.

Ulnar A. & N.

Flexor Carpi Ulnaris T.

Flexor Digitorum
Superficialis T.

Flexor Digitorum Profundus T.

Distal Ulna

Extensor Digiti Minimi T.

Extensor Carpi Ulnaris T.

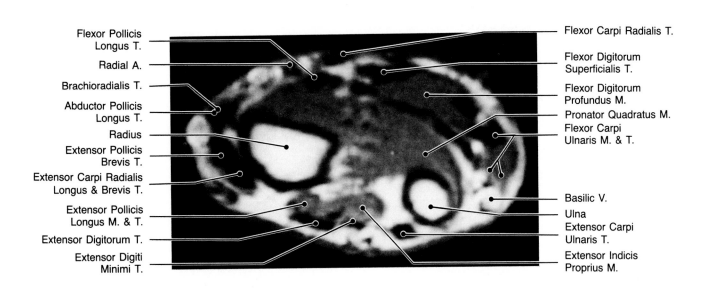

Flexor Pollicis
Longus T.

Radial A.

Brachioradialis T.

Abductor Pollicis
Longus T.

Radius

Extensor Pollicis
Brevis T.

Extensor Carpi Radialis
Longus & Brevis T.

Extensor Pollicis
Longus M. & T.

Extensor Digitorum T.

Extensor Digiti
Minimi T.

Flexor Carpi Radialis T.

Flexor Digitorum
Superficialis T.

Flexor Digitorum
Profundus M.

Pronator Quadratus M.

Flexor Carpi
Ulnaris M. & T.

Basilic V.

Ulna

Extensor Carpi
Ulnaris T.

Extensor Indicis
Proprius M.

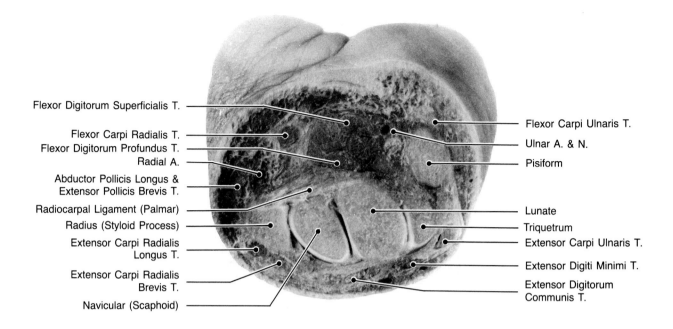

Flexor Digitorum Superficialis T.

Flexor Carpi Radialis T.

Flexor Digitorum Profundus T.

Radial A.

Abductor Pollicis Longus & Extensor Pollicis Brevis T.

Radiocarpal Ligament (Palmar)

Radius (Styloid Process)

Extensor Carpi Radialis Longus T.

Extensor Carpi Radialis Brevis T.

Navicular (Scaphoid)

Flexor Carpi Ulnaris T.

Ulnar A. & N.

Pisiform

Lunate

Triquetrum

Extensor Carpi Ulnaris T.

Extensor Digiti Minimi T.

Extensor Digitorum Communis T.

22A, 22B, 22C Axial/Anatomic Plate/CT/MRI/Proximal Carpus. Axial section through the proximal carpal row. The joint between the pisiform and triquetrum bone is seen in the lateral aspect of the wrist. The carpal tunnel is occupied anteriorly by the flexor digitorum superficialis and more posteriorly by the flexor digitorum profundus tendons. The median nerve is identified on the MRI section (22C) as a higher intensity structure than the surrounding tendons. Lateral to the carpal tunnel the tendons of the abductor pollicis brevis and abductor pollicis longus muscles can be identified. The dorsal aspect of the wrist is occupied by the extensor tendons.

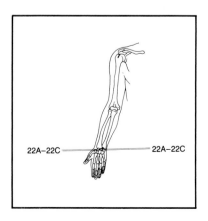

22A–22C   22A–22C

22B

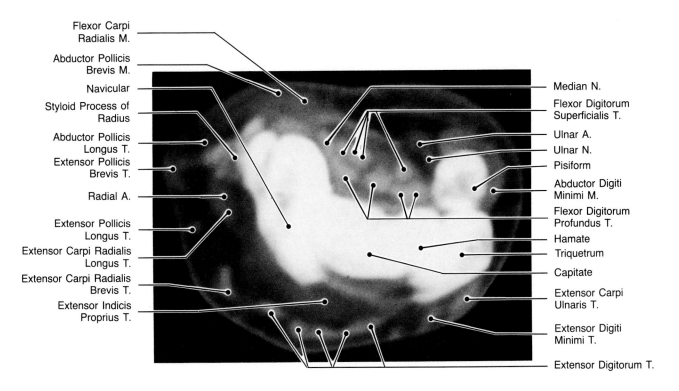

Flexor Carpi Radialis M.

Abductor Pollicis Brevis M.

Navicular

Styloid Process of Radius

Abductor Pollicis Longus T.

Extensor Pollicis Brevis T.

Radial A.

Extensor Pollicis Longus T.

Extensor Carpi Radialis Longus T.

Extensor Carpi Radialis Brevis T.

Extensor Indicis Proprius T.

Median N.

Flexor Digitorum Superficialis T.

Ulnar A.

Ulnar N.

Pisiform

Abductor Digiti Minimi M.

Flexor Digitorum Profundus T.

Hamate

Triquetrum

Capitate

Extensor Carpi Ulnaris T.

Extensor Digiti Minimi T.

Extensor Digitorum T.

22C

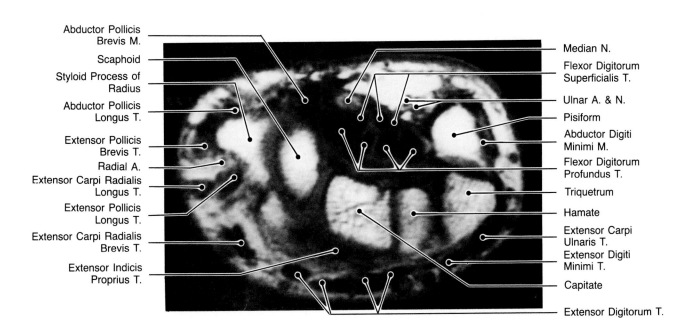

Abductor Pollicis Brevis M.

Scaphoid

Styloid Process of Radius

Abductor Pollicis Longus T.

Extensor Pollicis Brevis T.

Radial A.

Extensor Carpi Radialis Longus T.

Extensor Pollicis Longus T.

Extensor Carpi Radialis Brevis T.

Extensor Indicis Proprius T.

Median N.

Flexor Digitorum Superficialis T.

Ulnar A. & N.

Pisiform

Abductor Digiti Minimi M.

Flexor Digitorum Profundus T.

Triquetrum

Hamate

Extensor Carpi Ulnaris T.

Extensor Digiti Minimi T.

Capitate

Extensor Digitorum T.

Flexor Digitorum Profundus T.

Abductor Pollicis Brevis M.

Opponens Pollicis M.

Extensor Pollicis Brevis T.

Trapezium (Greater Multangular)

Radial A.

Trapezoid (Lesser Multangular)

Flexor Digitorum Superficialis T.

Ulnar A. & N.

Abductor Digiti Minimi (Quinti) M.

Hook of Hamate

Capitate

Extensor Carpi Ulnaris T.

Hamate

Extensor Digiti Minimi T.

Extensor Digitorum T.

23A, 23B, 23C Axial/Anatomic Plate/CT/MRI/Distal Carpus. Axial section through the distal carpal row. The hook of the hamate is clearly identified medially. The carpal tunnel is well delineated anteriorly by the surrounding flexor retinaculum. The flexor tendons and median nerve are again noted within the carpal tunnel.

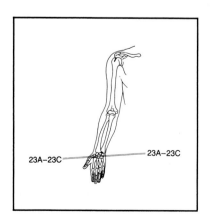

23A–23C

23A–23C

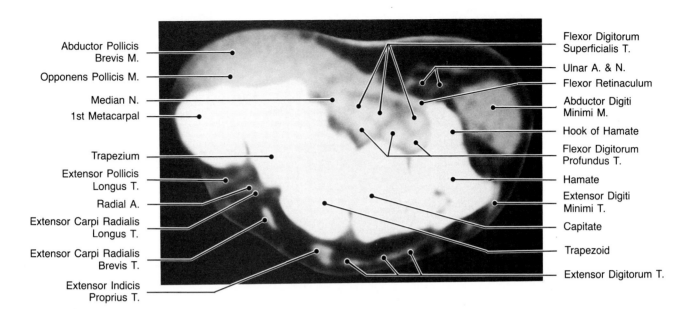

Abductor Pollicis
Brevis M.

Opponens Pollicis M.

Median N.

1st Metacarpal

Trapezium

Extensor Pollicis
Longus T.

Radial A.

Extensor Carpi Radialis
Longus T.

Extensor Carpi Radialis
Brevis T.

Extensor Indicis
Proprius T.

Flexor Digitorum
Superficialis T.

Ulnar A. & N.

Flexor Retinaculum

Abductor Digiti
Minimi M.

Hook of Hamate

Flexor Digitorum
Profundus T.

Hamate

Extensor Digiti
Minimi T.

Capitate

Trapezoid

Extensor Digitorum T.

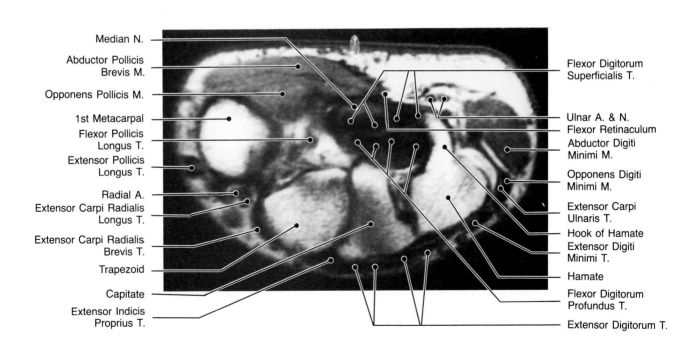

Median N.

Abductor Pollicis
Brevis M.

Opponens Pollicis M.

1st Metacarpal

Flexor Pollicis
Longus T.

Extensor Pollicis
Longus T.

Radial A.
Extensor Carpi Radialis
Longus T.

Extensor Carpi Radialis
Brevis T.

Trapezoid

Capitate

Extensor Indicis
Proprius T.

Flexor Digitorum
Superficialis T.

Ulnar A. & N.

Flexor Retinaculum

Abductor Digiti
Minimi M.

Opponens Digiti
Minimi M.

Extensor Carpi
Ulnaris T.

Hook of Hamate

Extensor Digiti
Minimi T.

Hamate

Flexor Digitorum
Profundus T.

Extensor Digitorum T.

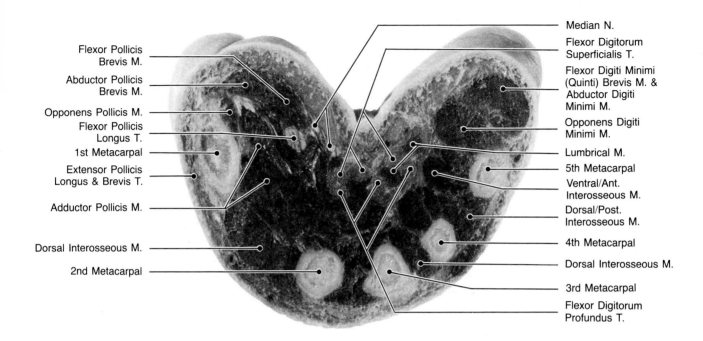

Flexor Pollicis Brevis M.

Abductor Pollicis Brevis M.

Opponens Pollicis M.

Flexor Pollicis Longus T.

1st Metacarpal

Extensor Pollicis Longus & Brevis T.

Adductor Pollicis M.

Dorsal Interosseous M.

2nd Metacarpal

Median N.

Flexor Digitorum Superficialis T.

Flexor Digiti Minimi (Quinti) Brevis M. & Abductor Digiti Minimi M.

Opponens Digiti Minimi M.

Lumbrical M.

5th Metacarpal

Ventral/Ant. Interosseous M.

Dorsal/Post. Interosseous M.

4th Metacarpal

Dorsal Interosseous M.

3rd Metacarpal

Flexor Digitorum Profundus T.

24A, 24B, 24C Axial/Anatomic Plate/CT/MRI/Proximal Head. Axial sections through the proximal hand, at the level of the proximal metacarpal bones. The thenar eminence is formed by the flexor pollicis brevis, adductor pollicis brevis, and opponens pollicis muscles. The hypothenar eminence is formed by the abductor digiti minimi, flexor digiti minimi and opponens digiti minimi muscles.

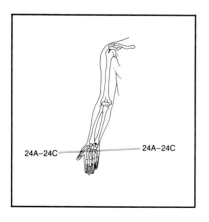

24A–24C   24A–24C

Flexor Pollicis Brevis M.

Abductor Pollicis Brevis M.

Opponens Pollicis M.

Extensor Pollicis Brevis T.

Extensor Pollicis Longus T.

1st Metacarpal

Flexor Pollicis Longus T.

Abductor Pollicis M.

Dorsal Interosseous M.

2nd Metacarpal

Superficial Palmar Arterial Arch

Flexor & Abductor Digiti Minimi M.

Opponens Digiti Minimi M.

Ant. Interosseous M.

5th Metacarpal

Dorsal Interosseous M.

Flexor Digitorum Profundus T.

4th Metacarpal

Extensor Digitorum T.

Dorsal Interosseous M.

3rd Metacarpal

Flexor Digitorum Superficialis T.

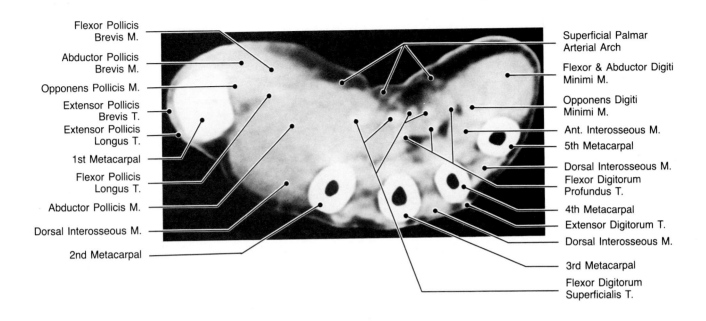

Flexor Pollicis Brevis M.

Abductor Pollicis Brevis M.

Opponens Pollicis M.

1st Metacarpal

Extensor Pollicis Brevis T.

Extensor Pollicis Longus T.

Dorsal Interosseous M.

Radial A.

Extensor Carpi Radialis Longus T.

2nd Metacarpal

Extensor Carpi Radialis Brevis T.

Extensor Indicis Proprius T.

Abductor Pollicis M.

Flexor Pollicis Longus T.

Flexor Digitorum Superficialis T.

Ulnar A.

Flexor Digiti Minimi M.

Abductor Digiti Minimi M.

Flexor Digitorum Profundus T.

5th Metacarpal

Extensor Digiti Minimi T.

4th Metacarpal

3rd Metacarpal

Extensor Digitorum T.

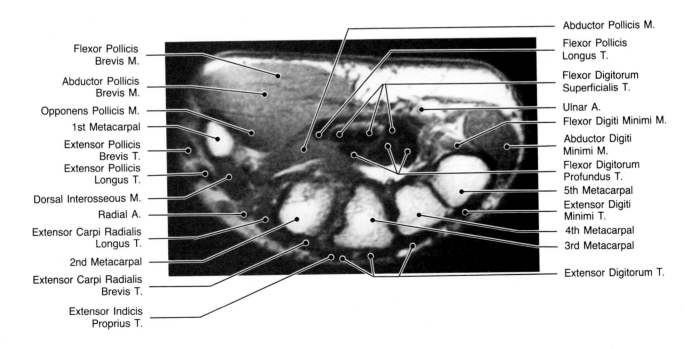

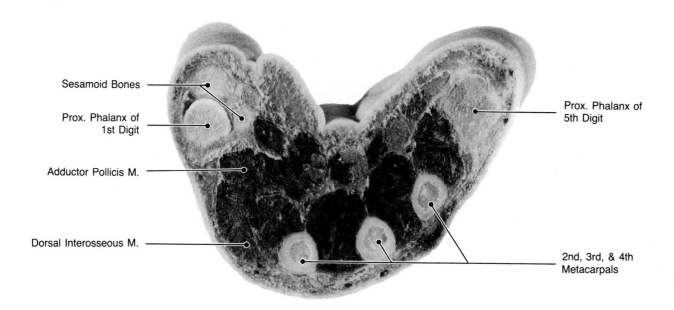

Sesamoid Bones

Prox. Phalanx of
1st Digit

Adductor Pollicis M.

Dorsal Interosseous M.

Prox. Phalanx of
5th Digit

2nd, 3rd, & 4th
Metacarpals

25A, 25B, Axial/Anatomic Plate/MRI/Mid Metacarpals.
Axial sections through the mid portions of the metacarpal
bones. The flexor digitorum superficialis and profundus tendons can be seen in the palmar aspect of the hand, anterior to
the ventral interosseii muscles. The dorsal interosseii muscles
are seen between the metacarpal bones.

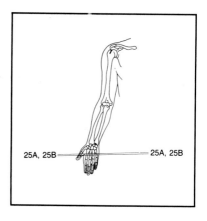

25A, 25B

25A, 25B

Flexor Digitorum
Superficialis &
Profundus T.

Flexor Pollicis
Longus T.

Prox. Phalanx of
1st Digit

Adductor Pollicis M.

2nd, 3rd, & 4th
Metacarpals

Dorsal Interosseii M.

Ventral Interosseii M.

Prox. Phalanx of
5th Digit

Extensor
Digitorum T.

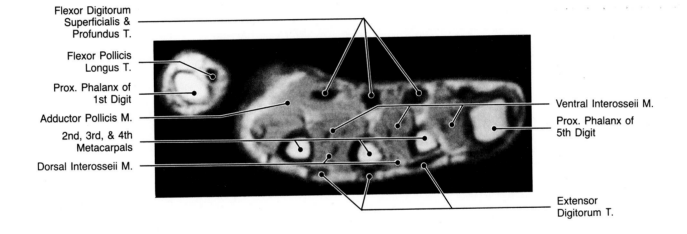

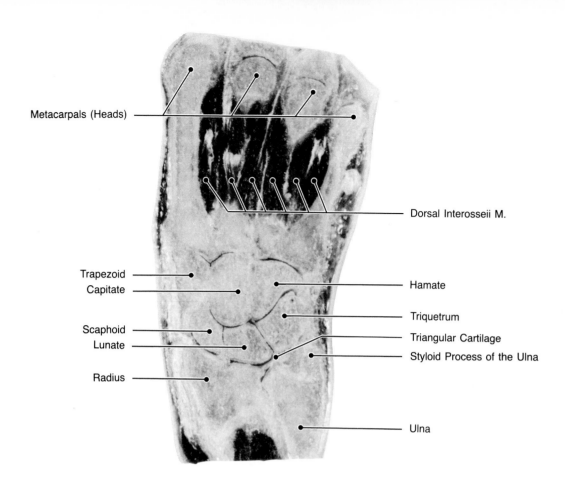

Metacarpals (Heads)

Dorsal Interosseii M.

Trapezoid
Capitate

Hamate

Triquetrum

Scaphoid
Lunate

Triangular Cartilage
Styloid Process of the Ulna

Radius

Ulna

26A, 26B Coronal/Anatomic Plate/MRI/Hand and Wrist. Coronal section through the hand and wrist. The dorsal interosseii muscles are seen between the metacarpals. The carpal bones are well seen in both anatomic and MRI sections. Note the low intensity signal of the triangular cartilage at the level of the ulnar-carpal joint.

26C Coronal/MRI/Carpus. Coronal MRI through the carpal bones in a patient with rheumatoid arthritis. Note the multiple carpal bone cysts and erosions as well as the irregularity of the triangular cartilage.

26D Coronal/MRI/Carpus. Coronal MRI through carpal bones. Note the low signal intensity of the proximal half of the navicular bone. This patient sustained a nonunited fracture of the navicular. The lack of signal from the normal fat within the bone marrow indicates avascular necrosis.

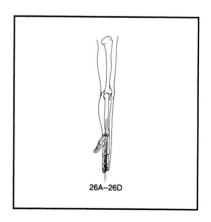

26A–26D

26B

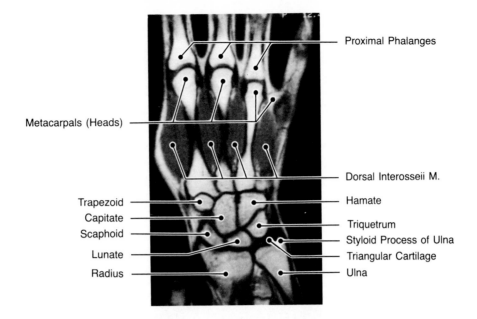

Proximal Phalanges

Metacarpals (Heads)

Dorsal Interosseii M.

Trapezoid — Hamate
Capitate — Triquetrum
Scaphoid — Styloid Process of Ulna
Lunate — Triangular Cartilage
Radius — Ulna

26C

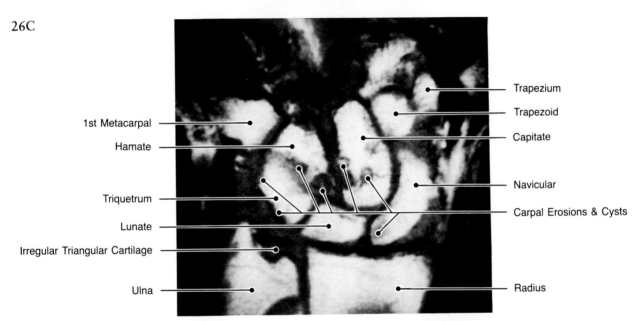

Trapezium
Trapezoid
1st Metacarpal — Capitate
Hamate

Navicular

Triquetrum
Lunate — Carpal Erosions & Cysts

Irregular Triangular Cartilage

Ulna — Radius

26D

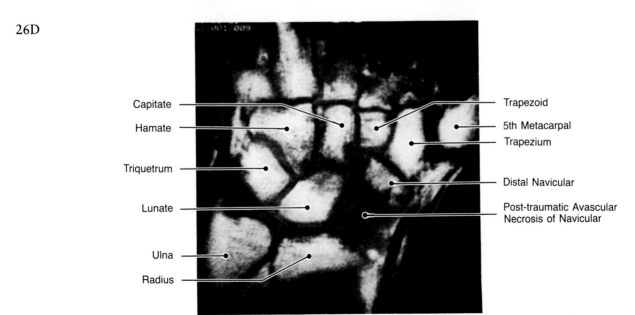

Capitate — Trapezoid
Hamate — 5th Metacarpal
Trapezium
Triquetrum
Distal Navicular
Lunate
Post-traumatic Avascular
Necrosis of Navicular
Ulna
Radius

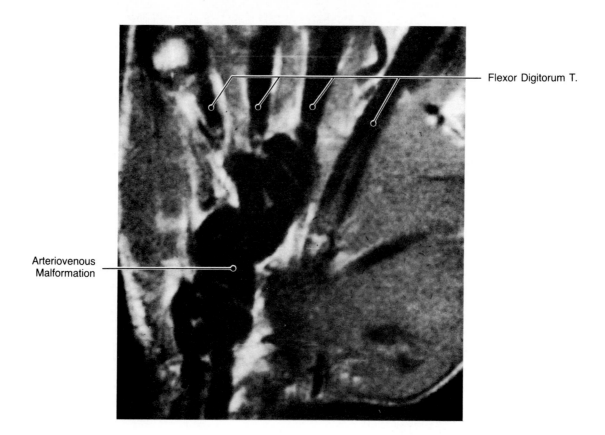

Flexor Digitorum T.

Arteriovenous
Malformation

26E Coronal/MRI/Hand. Coronal MRI section through the hand. Note the large vascular structures (flow void phenomenon) in the volar aspect of the hand. This corresponds to the most distal extension of an arteriovenous malformation.

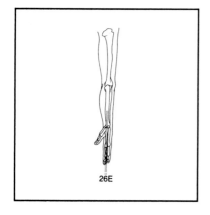

26E

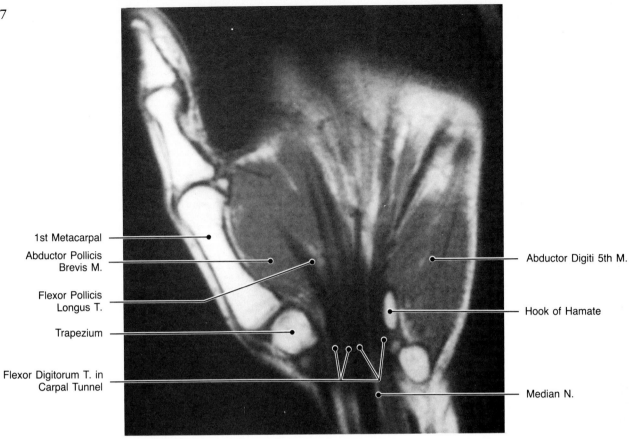

1st Metacarpal

Abductor Pollicis
Brevis M.

Flexor Pollicis
Longus T.

Trapezium

Flexor Digitorum T. in
Carpal Tunnel

Abductor Digiti 5th M.

Hook of Hamate

Median N.

27 Coronal/MRI/Carpal Tunnel. Coronal MRI section through the carpal tunnel. The flexor digitorum tendons are seen within the carpal tunnel. The median nerve can be identified owing to its signal intensity, which is slightly higher than that of the surrounding tendons.

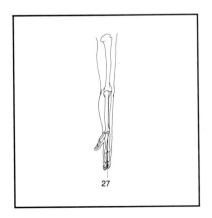

27

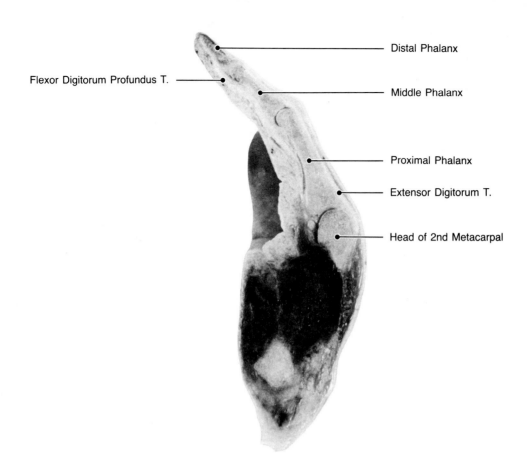

Distal Phalanx

Flexor Digitorum Profundus T.

Middle Phalanx

Proximal Phalanx

Extensor Digitorum T.

Head of 2nd Metacarpal

28A, 28B, 28C Sagittal/Anatomic Plate/MRI/Index Finger.
Sagittal section through the index finger. Note the insertion of
the flexor digitorum profundus at the level of the distal phalanx (28B) and the insertion of the flexor digitorum superficialis at the level of the middle phalanx (28C).

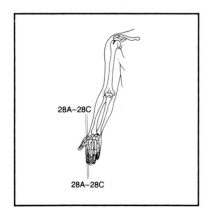

28A–28C

28A–28C

28B

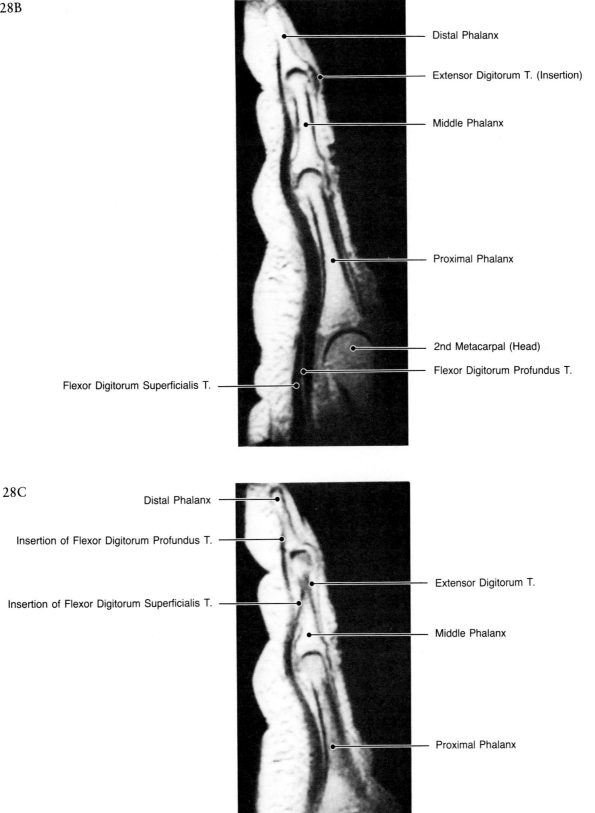

Distal Phalanx

Extensor Digitorum T. (Insertion)

Middle Phalanx

Proximal Phalanx

2nd Metacarpal (Head)

Flexor Digitorum Profundus T.

Flexor Digitorum Superficialis T.

28C

Distal Phalanx

Insertion of Flexor Digitorum Profundus T.

Insertion of Flexor Digitorum Superficialis T.

Extensor Digitorum T.

Middle Phalanx

Proximal Phalanx

2nd Metacarpal (Head)

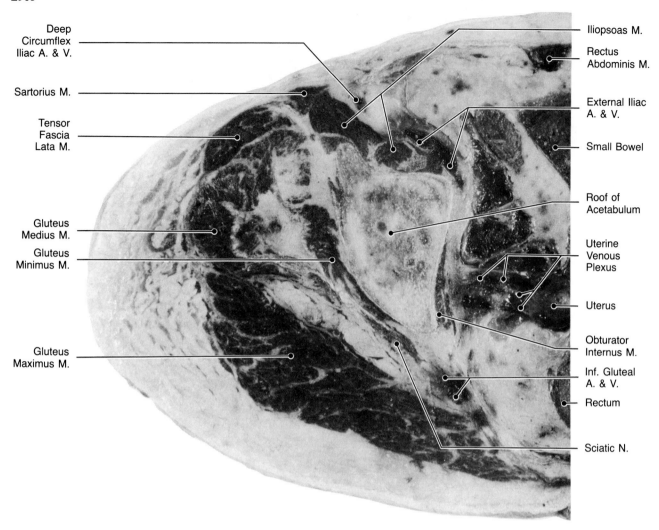

Deep
Circumflex
Iliac A. & V.

Sartorius M.

Tensor
Fascia
Lata M.

Gluteus
Medius M.

Gluteus
Minimus M.

Gluteus
Maximus M.

Iliopsoas M.

Rectus
Abdominis M.

External Iliac
A. & V.

Small Bowel

Roof of
Acetabulum

Uterine
Venous
Plexus

Uterus

Obturator
Internus M.

Inf. Gluteal
A. & V.

Rectum

Sciatic N.

29A–29D Axial/Anatomic Plate/CT/MRI/Acetabular Roof.
Axial section through the roof of the acetabulum. The hip
joint at this level is surrounded by the gluteal muscles posteriorly and laterally and the iliopsoas muscle anteriorly. The different muscles can be better delineated on the MRI sections
(28C and 28D). The sciatic nerve is located immediately posterior to the posterior column of the acetabulum.

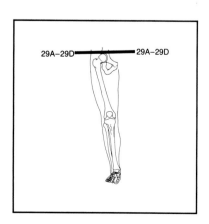

29A–29D   29A–29D

29B

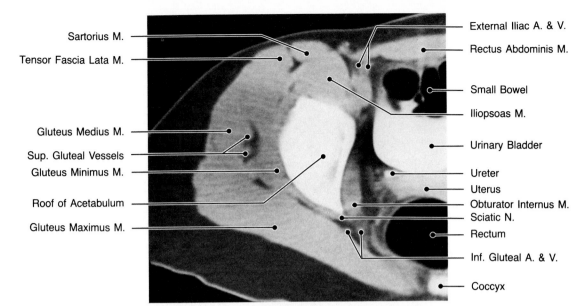

Sartorius M.
Tensor Fascia Lata M.
Gluteus Medius M.
Sup. Gluteal Vessels
Gluteus Minimus M.
Roof of Acetabulum
Gluteus Maximus M.

External Iliac A. & V.
Rectus Abdominis M.
Small Bowel
Iliopsoas M.
Urinary Bladder
Ureter
Uterus
Obturator Internus M.
Sciatic N.
Rectum
Inf. Gluteal A. & V.
Coccyx

29C

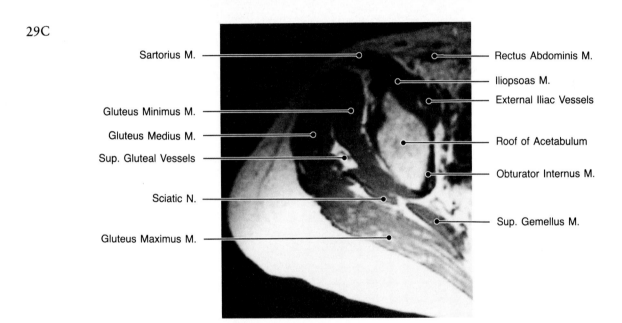

Sartorius M.
Gluteus Minimus M.
Gluteus Medius M.
Sup. Gluteal Vessels
Sciatic N.
Gluteus Maximus M.

Rectus Abdominis M.
Iliopsoas M.
External Iliac Vessels
Roof of Acetabulum
Obturator Internus M.
Sup. Gemellus M.

29D

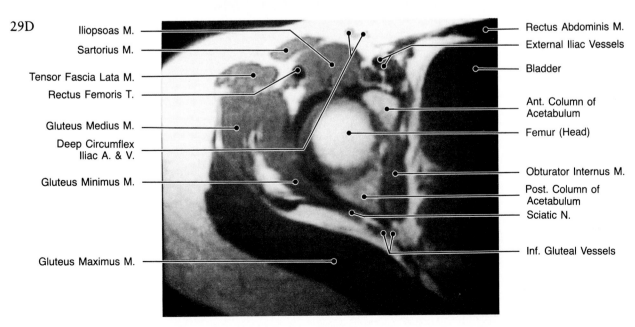

Iliopsoas M.
Sartorius M.
Tensor Fascia Lata M.
Rectus Femoris T.
Gluteus Medius M.
Deep Circumflex
Iliac A. & V.
Gluteus Minimus M.
Gluteus Maximus M.

Rectus Abdominis M.
External Iliac Vessels
Bladder
Ant. Column of
Acetabulum
Femur (Head)
Obturator Internus M.
Post. Column of
Acetabulum
Sciatic N.
Inf. Gluteal Vessels

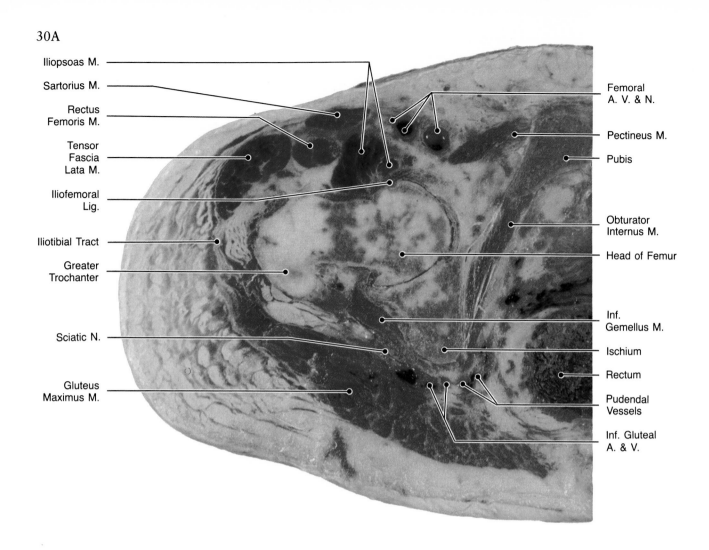

Iliopsoas M.

Sartorius M.

Rectus Femoris M.

Tensor Fascia Lata M.

Iliofemoral Lig.

Iliotibial Tract

Greater Trochanter

Sciatic N.

Gluteus Maximus M.

Femoral A. V. & N.

Pectineus M.

Pubis

Obturator Internus M.

Head of Femur

Inf. Gemellus M.

Ischium

Rectum

Pudendal Vessels

Inf. Gluteal A. & V.

30A–30C Axial/Anatomic Plate/CT/MRI/Femoral Head.
Axial sections through the hip joint, at the level of the upper half of the femoral head. The hip joint is well seen in this projection. The anterior and posterior columns of the acetabulum are also shown.

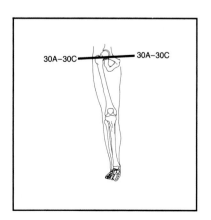

30A–30C   30A–30C

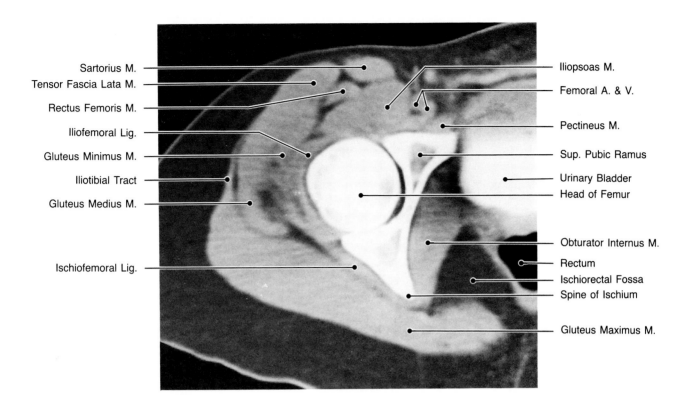

Sartorius M.
Tensor Fascia Lata M.
Rectus Femoris M.
Iliofemoral Lig.
Gluteus Minimus M.
Iliotibial Tract
Gluteus Medius M.

Ischiofemoral Lig.

Iliopsoas M.
Femoral A. & V.
Pectineus M.
Sup. Pubic Ramus
Urinary Bladder
Head of Femur
Obturator Internus M.
Rectum
Ischiorectal Fossa
Spine of Ischium
Gluteus Maximus M.

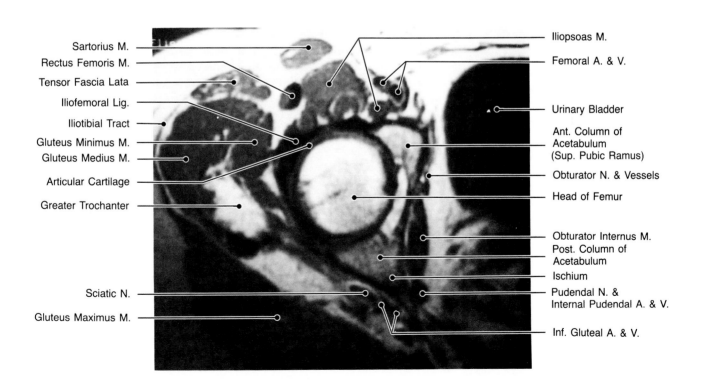

Sartorius M.
Rectus Femoris M.
Tensor Fascia Lata
Iliofemoral Lig.
Iliotibial Tract
Gluteus Minimus M.
Gluteus Medius M.
Articular Cartilage
Greater Trochanter

Sciatic N.
Gluteus Maximus M.

Iliopsoas M.
Femoral A. & V.
Urinary Bladder
Ant. Column of
Acetabulum
(Sup. Pubic Ramus)
Obturator N. & Vessels
Head of Femur
Obturator Internus M.
Post. Column of
Acetabulum
Ischium
Pudendal N. &
Internal Pudendal A. & V.
Inf. Gluteal A. & V.

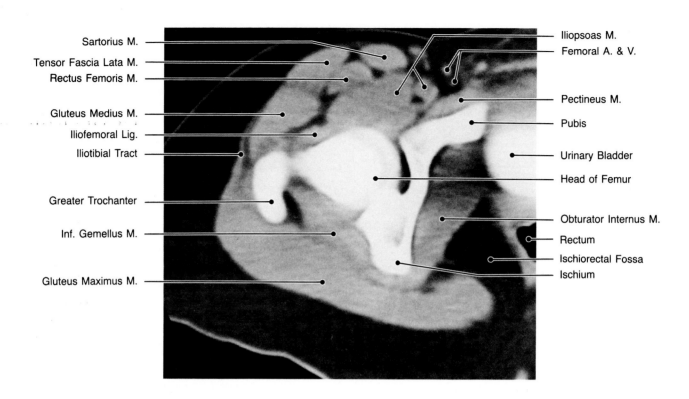

Sartorius M. ———

Tensor Fascia Lata M. ———

Rectus Femoris M. ———

Gluteus Medius M. ———

Iliofemoral Lig. ———

Iliotibial Tract ———

Greater Trochanter ———

Inf. Gemellus M. ———

Gluteus Maximus M. ———

Iliopsoas M.

Femoral A. & V.

Pectineus M.

Pubis

Urinary Bladder

Head of Femur

Obturator Internus M.

Rectum

Ischiorectal Fossa

Ischium

30D, 30E Axial/CT/MRI/Femoral Head. These sections were obtained about 1 cm below previous sections. The greater trochanter is seen laterally. Note the sciatic nerve posterior to the ischium on the MRI (30E).

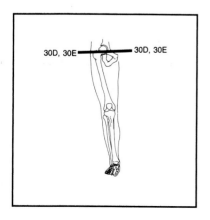

30D, 30E ——— 30D, 30E

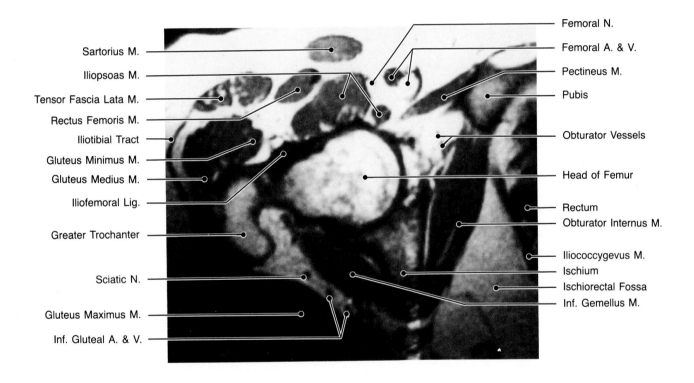

Sartorius M.

Iliopsoas M.

Tensor Fascia Lata M.

Rectus Femoris M.

Iliotibial Tract

Gluteus Minimus M.

Gluteus Medius M.

Iliofemoral Lig.

Greater Trochanter

Sciatic N.

Gluteus Maximus M.

Inf. Gluteal A. & V.

Femoral N.

Femoral A. & V.

Pectineus M.

Pubis

Obturator Vessels

Head of Femur

Rectum

Obturator Internus M.

Iliococcygevus M.

Ischium

Ischiorectal Fossa

Inf. Gemellus M.

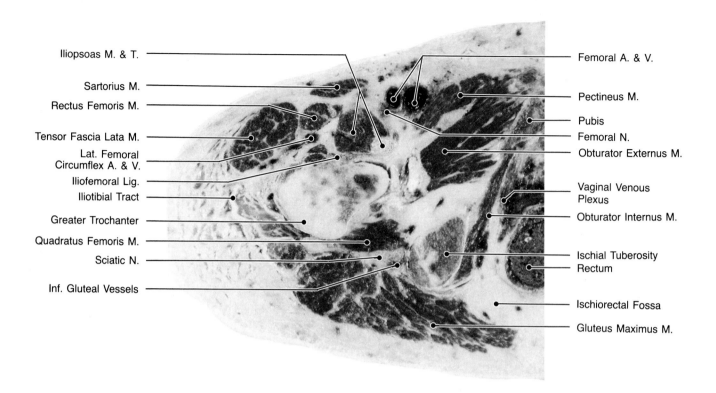

Iliopsoas M. & T.

Sartorius M.

Rectus Femoris M.

Tensor Fascia Lata M.

Lat. Femoral
Circumflex A. & V.

Iliofemoral Lig.

Iliotibial Tract

Greater Trochanter

Quadratus Femoris M.

Sciatic N.

Inf. Gluteal Vessels

Femoral A. & V.

Pectineus M.

Pubis

Femoral N.

Obturator Externus M.

Vaginal Venous
Plexus

Obturator Internus M.

Ischial Tuberosity
Rectum

Ischiorectal Fossa

Gluteus Maximus M.

31A, 31B, 31C Axial/Anatomic Plate/CT/MRI/Femoral Neck. Axial section through the femoral neck. Due to the oblique orientation of the femoral neck, this structure can be seen in several planes. These sections cross the obturator foramen. The sciatic nerve can be identified posterior to the quadratus femoris muscle.

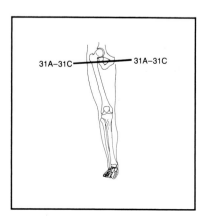

31A–31C   31A–31C

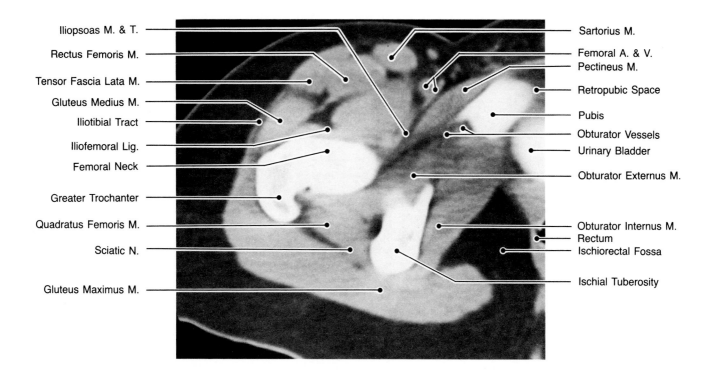

Iliopsoas M. & T. — Sartorius M.
Rectus Femoris M. — Femoral A. & V.
— Pectineus M.
Tensor Fascia Lata M. — Retropubic Space
Gluteus Medius M. — Pubis
Iliotibial Tract — Obturator Vessels
Iliofemoral Lig. — Urinary Bladder
Femoral Neck — Obturator Externus M.
Greater Trochanter
Quadratus Femoris M. — Obturator Internus M.
— Rectum
Sciatic N. — Ischiorectal Fossa
Gluteus Maximus M. — Ischial Tuberosity

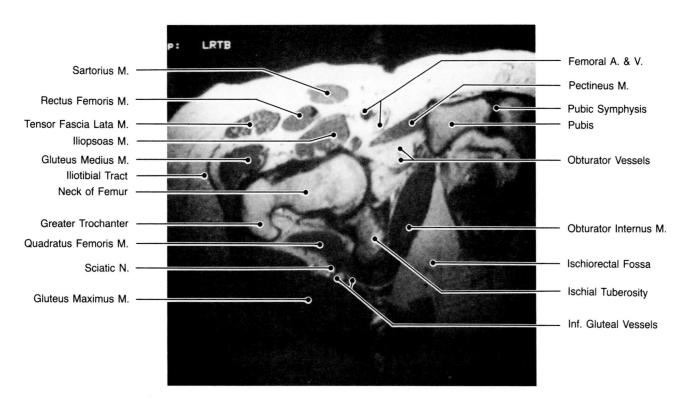

Sartorius M. — Femoral A. & V.
— Pectineus M.
Rectus Femoris M. — Pubic Symphysis
Tensor Fascia Lata M. — Pubis
Iliopsoas M.
Gluteus Medius M. — Obturator Vessels
Iliotibial Tract
Neck of Femur
Greater Trochanter — Obturator Internus M.
Quadratus Femoris M.
Sciatic N. — Ischiorectal Fossa
— Ischial Tuberosity
Gluteus Maximus M.
— Inf. Gluteal Vessels

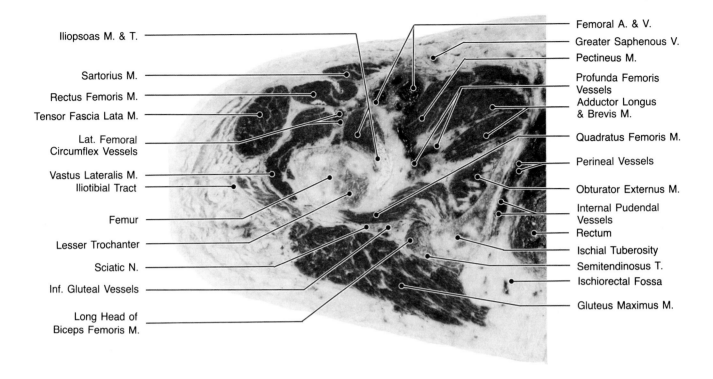

Iliopsoas M. & T.

Sartorius M.

Rectus Femoris M.

Tensor Fascia Lata M.

Lat. Femoral
Circumflex Vessels

Vastus Lateralis M.
Iliotibial Tract

Femur

Lesser Trochanter

Sciatic N.

Inf. Gluteal Vessels

Long Head of
Biceps Femoris M.

Femoral A. & V.

Greater Saphenous V.

Pectineus M.

Profunda Femoris
Vessels

Adductor Longus
& Brevis M.

Quadratus Femoris M.

Perineal Vessels

Obturator Externus M.

Internal Pudendal
Vessels

Rectum

Ischial Tuberosity

Semitendinosus T.

Ischiorectal Fossa

Gluteus Maximus M.

32A, 32B, 32C Axial/Anatomic Plate/CT/MRI/Proximal Femur. Axial sections through the intertrochanteric region and ischial tuberosity. In these sections the femoral vessels can be identified between the sartorius muscle and pectineus muscle medially. At the level of the ischial tuberosity, the long head of the biceps femoris muscle and semitendinosus tendon is well seen.

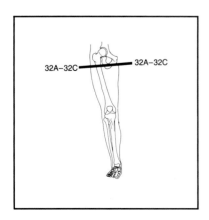

32A–32C   32A–32C

32B

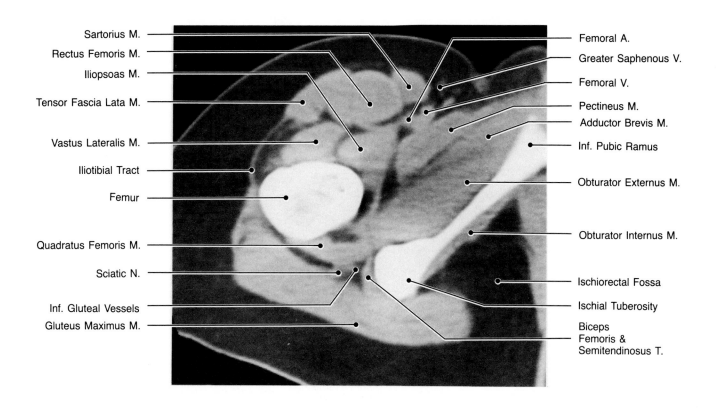

Sartorius M.

Rectus Femoris M.

Iliopsoas M.

Tensor Fascia Lata M.

Vastus Lateralis M.

Iliotibial Tract

Femur

Quadratus Femoris M.

Sciatic N.

Inf. Gluteal Vessels

Gluteus Maximus M.

Femoral A.

Greater Saphenous V.

Femoral V.

Pectineus M.

Adductor Brevis M.

Inf. Pubic Ramus

Obturator Externus M.

Obturator Internus M.

Ischiorectal Fossa

Ischial Tuberosity

Biceps Femoris & Semitendinosus T.

32C

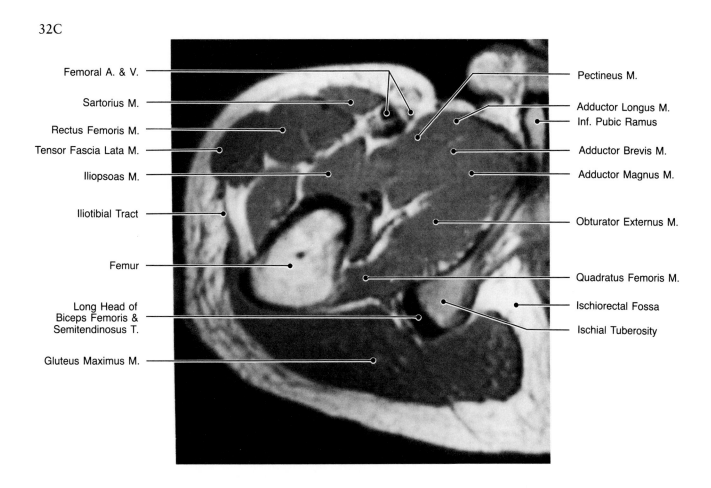

Femoral A. & V.

Sartorius M.

Rectus Femoris M.

Tensor Fascia Lata M.

Iliopsoas M.

Iliotibial Tract

Femur

Long Head of Biceps Femoris & Semitendinosus T.

Gluteus Maximus M.

Pectineus M.

Adductor Longus M.

Inf. Pubic Ramus

Adductor Brevis M.

Adductor Magnus M.

Obturator Externus M.

Quadratus Femoris M.

Ischiorectal Fossa

Ischial Tuberosity

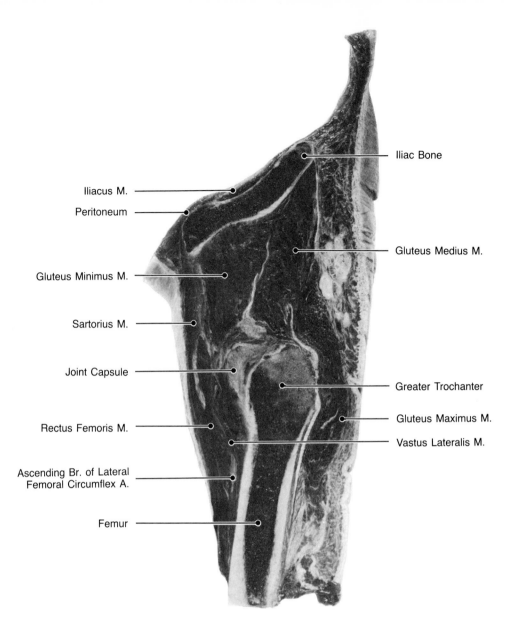

Iliac Bone

Iliacus M.

Peritoneum

Gluteus Medius M.

Gluteus Minimus M.

Sartorius M.

Joint Capsule

Greater Trochanter

Rectus Femoris M.

Gluteus Maximus M.

Vastus Lateralis M.

Ascending Br. of Lateral
Femoral Circumflex A.

Femur

33A, 33B Sagittal/Anatomic Plate/MRI/Lateral Hip Joint.
Sagittal sections through the lateral aspect of the hip. The femoral head, the greater trochanter, and the proximal femoral shaft are seen in these sections. The joint capsule is seen tangentially.

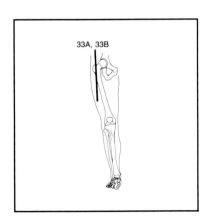

33A, 33B

441

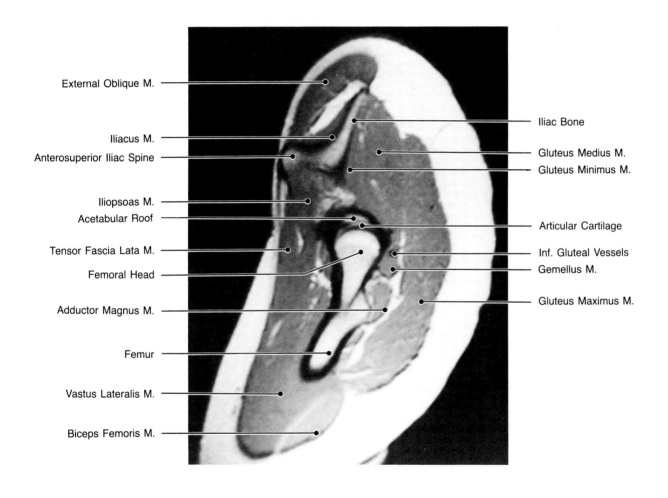

External Oblique M.

Iliacus M.

Anterosuperior Iliac Spine

Iliopsoas M.

Acetabular Roof

Tensor Fascia Lata M.

Femoral Head

Adductor Magnus M.

Femur

Vastus Lateralis M.

Biceps Femoris M.

Iliac Bone

Gluteus Medius M.

Gluteus Minimus M.

Articular Cartilage

Inf. Gluteal Vessels

Gemellus M.

Gluteus Maximus M.

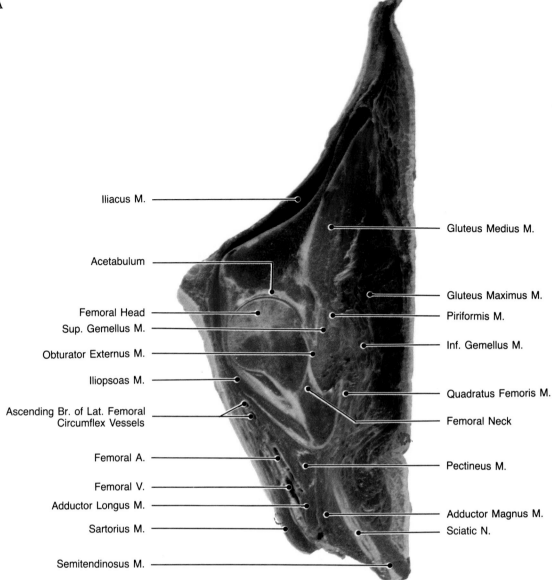

Iliacus M.

Gluteus Medius M.

Acetabulum

Gluteus Maximus M.

Femoral Head

Piriformis M.

Sup. Gemellus M.

Obturator Externus M.

Inf. Gemellus M.

Iliopsoas M.

Ascending Br. of Lat. Femoral
Circumflex Vessels

Quadratus Femoris M.

Femoral Neck

Femoral A.

Pectineus M.

Femoral V.

Adductor Longus M.

Adductor Magnus M.

Sartorius M.

Sciatic N.

Semitendinosus M.

34A, 34B Sagittal/Anatomic Plate/MRI/Hip Joint. Sagittal sections through the hip joint. The joint space can be identified. The MRI (34B) shows the articular cartilage as a high intensity signal between the subchondral cortex of the femoral head and acetabulum. The iliopsoas muscle is located immediately anterior to the femoral head.

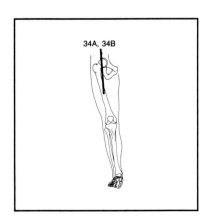

34A, 34B

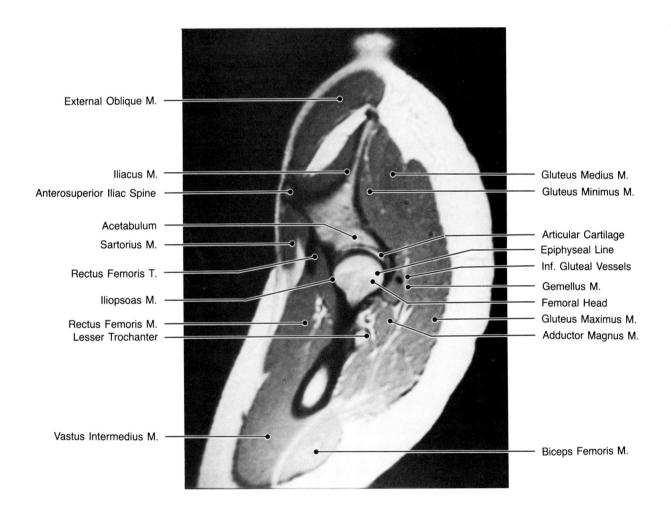

External Oblique M.

Iliacus M.

Anterosuperior Iliac Spine

Acetabulum

Sartorius M.

Rectus Femoris T.

Iliopsoas M.

Rectus Femoris M.

Lesser Trochanter

Vastus Intermedius M.

Gluteus Medius M.

Gluteus Minimus M.

Articular Cartilage

Epiphyseal Line

Inf. Gluteal Vessels

Gemellus M.

Femoral Head

Gluteus Maximus M.

Adductor Magnus M.

Biceps Femoris M.

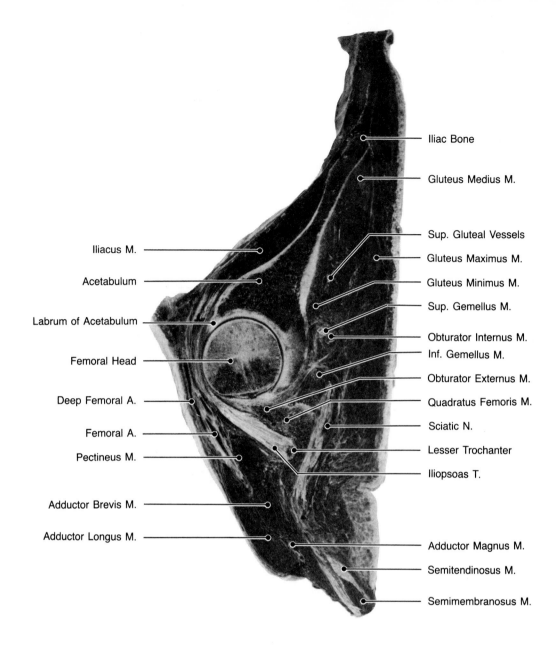

Iliac Bone

Gluteus Medius M.

Sup. Gluteal Vessels

Gluteus Maximus M.

Gluteus Minimus M.

Sup. Gemellus M.

Obturator Internus M.

Inf. Gemellus M.

Obturator Externus M.

Quadratus Femoris M.

Sciatic N.

Lesser Trochanter

Iliopsoas T.

Iliacus M.

Acetabulum

Labrum of Acetabulum

Femoral Head

Deep Femoral A.

Femoral A.

Pectineus M.

Adductor Brevis M.

Adductor Longus M.

Adductor Magnus M.

Semitendinosus M.

Semimembranosus M.

35A, 35B Sagittal/Anatomic Plate/MRI/Femoral Head. Sagittal sections through the femoral head. The sciatic nerve is identified between the gluteus maximus posteriorly and quadratus femoris anteriorly. The posterior column of the acetabulum is seen in both anatomic and MRI sections. The anterior aspect of the femoral head is in close contact with the iliopsoas muscle and tendon. The obturator internus, inferior gemellus, and obturator externus muscles are identified between the posterior column of the acetabulum and gluteus maximus.

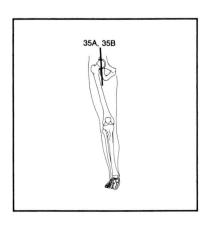

35A, 35B

External Oblique M. —————————————

—————————————— Iliac Crest

Iliacus M. —————————————

—————————————— Gluteus Medius M.

—————————————— Gluteus Minimus M.
—————————————— Acetabulum

—————————————— Articular Cartilage

Epiphyseal Line —————————————

—————————————— Obturator Internus M.

Sartorius M. —————————————

—————————————— Inf. Gemellus M.

Iliopsoas M. —————————————

—————————————— Gluteus Maximus M.
—————————————— Obturator Externus M.

Pectineus M. —————————————

—————————————— Quadratus Femoris M.

Rectus Femoris M. —————————————

Adductor Magnus M. —————————————

—————————————— Sciatic N.

Vastus Intermedius M. —————————————

Vastus Lateralis M. —————————————

—————————————— Biceps Femoris M.

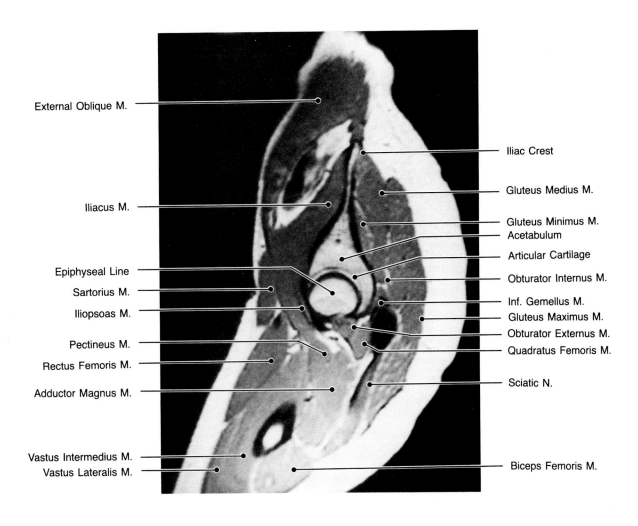

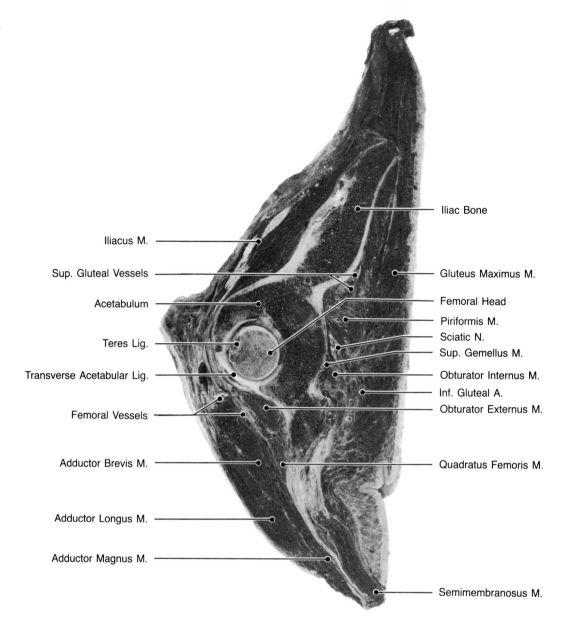

Iliac Bone

Iliacus M.

Sup. Gluteal Vessels

Acetabulum

Teres Lig.

Transverse Acetabular Lig.

Femoral Vessels

Adductor Brevis M.

Adductor Longus M.

Adductor Magnus M.

Gluteus Maximus M.

Femoral Head

Piriformis M.

Sciatic N.

Sup. Gemellus M.

Obturator Internus M.

Inf. Gluteal A.

Obturator Externus M.

Quadratus Femoris M.

Semimembranosus M.

36A, 36B Sagittal/Anatomic Plate/MRI/Medial Hip Joint.
Sagittal section through the most medial aspect of the hip
joint. The femoral head is almost totally surrounded by the
acetabulum except for its most anterior-inferior aspect, where
the transverse acetabular ligament bridges the gap.

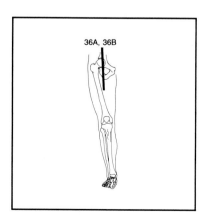

36A, 36B

External Oblique M.

Colon

Iliacus M.

Iliopsoas M.

Femoral Head

Sartorius M.

Rectus Femoris M.

Vastus Medialis M.

Femur

Vastus Intermedius M.

Gluteus Medius M.

Sup. Gluteal Vessels

Piriformis M.

Gluteus Maximus M.

Inf. Gluteal A.

Obturator Externus M.

Ischial Tuberosity

Adductor Magnus M.

Biceps Femoris M.

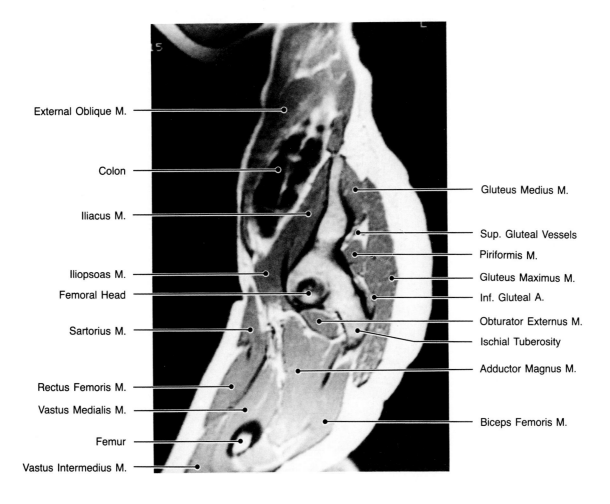

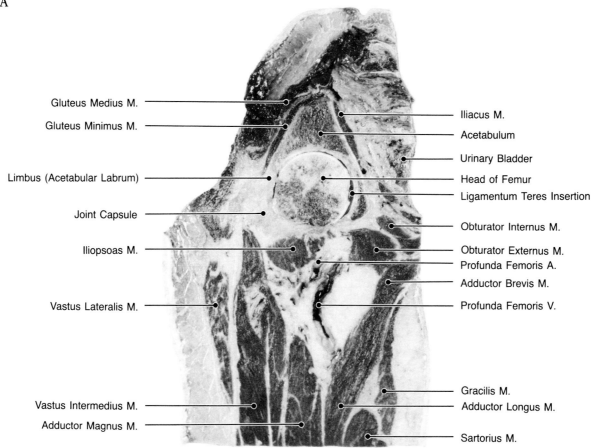

Gluteus Medius M.	Iliacus M.
Gluteus Minimus M.	Acetabulum
	Urinary Bladder
Limbus (Acetabular Labrum)	Head of Femur
	Ligamentum Teres Insertion
Joint Capsule	Obturator Internus M.
Iliopsoas M.	Obturator Externus M.
	Profunda Femoris A.
	Adductor Brevis M.
Vastus Lateralis M.	Profunda Femoris V.
	Gracilis M.
Vastus Intermedius M.	Adductor Longus M.
Adductor Magnus M.	Sartorius M.

37A, 37B Coronal/Anatomic Plate/MRI/Anterior Hip Joint.
Coronal sections through the anterior aspect of the hip joints.
The cadaver sections were obtained in a slightly more than
oblique plane the MRI. The low intensity, dark structure surrounding the femoral neck on the MRI section corresponds to
the joint capsule. When the joint is distended by fluid, this area
can be seen as a high intensity signal on a T2 weighted image.
The limbus (acetabular labrum) is seen on the MRI as a dark
triangle between the lateral aspect of the acetabulum and the
femoral head. The articular cartilage can be separated from
the dark, low signal intensity of the subchondral cortex of the
femoral head and acetabulum.

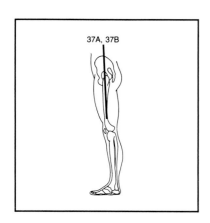

37A, 37B

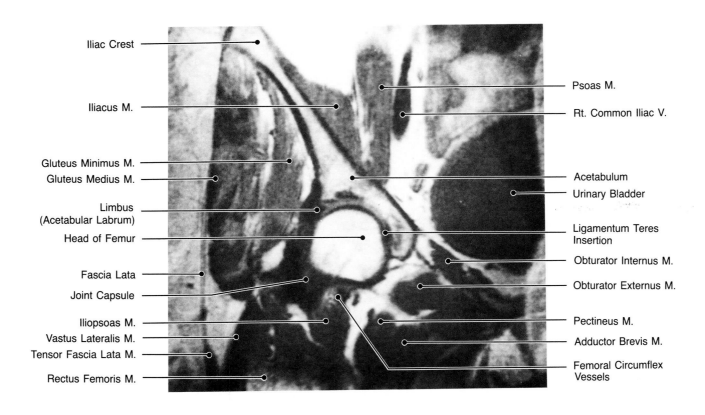

Iliac Crest

Iliacus M.

Gluteus Minimus M.
Gluteus Medius M.

Limbus
(Acetabular Labrum)

Head of Femur

Fascia Lata

Joint Capsule

Iliopsoas M.
Vastus Lateralis M.
Tensor Fascia Lata M.

Rectus Femoris M.

Psoas M.

Rt. Common Iliac V.

Acetabulum
Urinary Bladder

Ligamentum Teres
Insertion

Obturator Internus M.

Obturator Externus M.

Pectineus M.

Adductor Brevis M.

Femoral Circumflex
Vessels

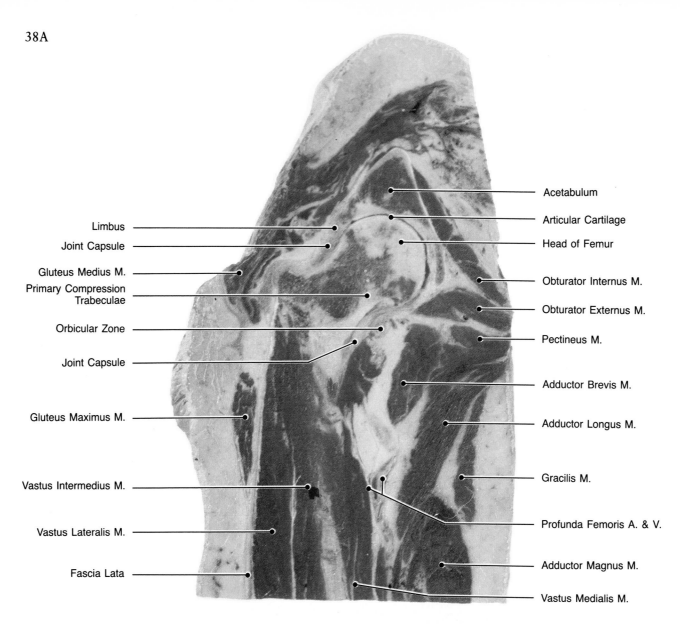

Acetabulum

Articular Cartilage

Head of Femur

Obturator Internus M.

Obturator Externus M.

Pectineus M.

Adductor Brevis M.

Adductor Longus M.

Gracilis M.

Profunda Femoris A. & V.

Adductor Magnus M.

Vastus Medialis M.

Limbus

Joint Capsule

Gluteus Medius M.

Primary Compression Trabeculae

Orbicular Zone

Joint Capsule

Gluteus Maximus M.

Vastus Intermedius M.

Vastus Lateralis M.

Fascia Lata

38A, 38B, 38C Coronal/Anatomic Plate/MRI/Mid Hip Joint. Coronal sections through the hip joint, including femoral head and neck. In this section, the entire hip joint is seen in the coronal plane, including the insertion of the ligamentum teres. The superior aspect of the femoral neck is surrounded by a low intensity signal structure corresponding to the joint capsule. A hip joint effusion would displace the joint capsule lat-erally. The high intensity signal between the femoral head and acetabulum does not correspond to fluid, but it represents articular, hyaline cartilage.

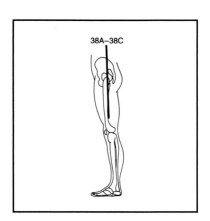

38A–38C

38B

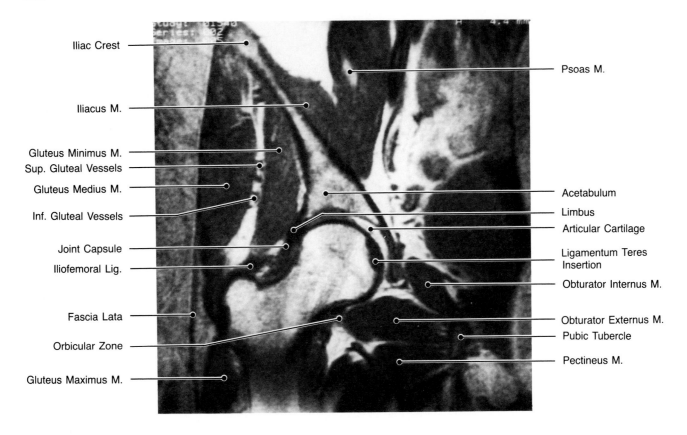

Iliac Crest

Iliacus M.

Gluteus Minimus M.
Sup. Gluteal Vessels
Gluteus Medius M.
Inf. Gluteal Vessels

Joint Capsule
Iliofemoral Lig.

Fascia Lata

Orbicular Zone

Gluteus Maximus M.

Psoas M.

Acetabulum
Limbus
Articular Cartilage
Ligamentum Teres
Insertion
Obturator Internus M.

Obturator Externus M.
Pubic Tubercle
Pectineus M.

38C

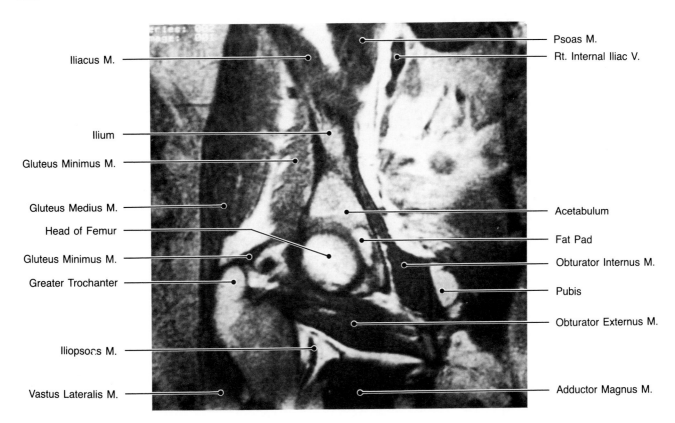

Iliacus M.

Ilium

Gluteus Minimus M.

Gluteus Medius M.
Head of Femur
Gluteus Minimus M.
Greater Trochanter

Iliopsoas M.

Vastus Lateralis M.

Psoas M.
Rt. Internal Iliac V.

Acetabulum
Fat Pad
Obturator Internus M.
Pubis
Obturator Externus M.

Adductor Magnus M.

Gluteus Medius M. ————

Greater Trochanter of Femur ————

Vastus Lateralis M. ————

Femoral Diaphysis ————

Fascia Lata (Iliotibial Tract) ————

———— Acetabular Roof
———— Obturator Internus M.

———— Obturator Externus M.

———— Adductor Magnus M.

39A, 39B Coronal/Anatomic Plate/MRI/Posterior Hip Joint. Coronal section through the posterior aspect of the hip joint. In this plane the obturator externus muscle crosses the femoral neck posteriorly. The hip joint is not as well identified as on previous sections. The fat pad within the acetabular cavity is easily identified on the MRI section.

39C Coronal/MRI/Hip Joints. T2 weighted MRI of a patient sustaining right hip pain. A right joint effusion is demonstrated distending the capsule. Compare with the normal left side.

39D Coronal/MRI/Hip Joint. T1 weighted image. Patient complaining of left hip pain post renal transplant. Avascular necrosis is seen as an area of decreased signal intensity in the superior half of the left femoral head.

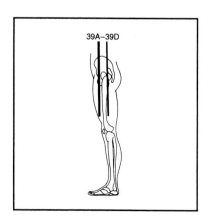

39A–39D

39B

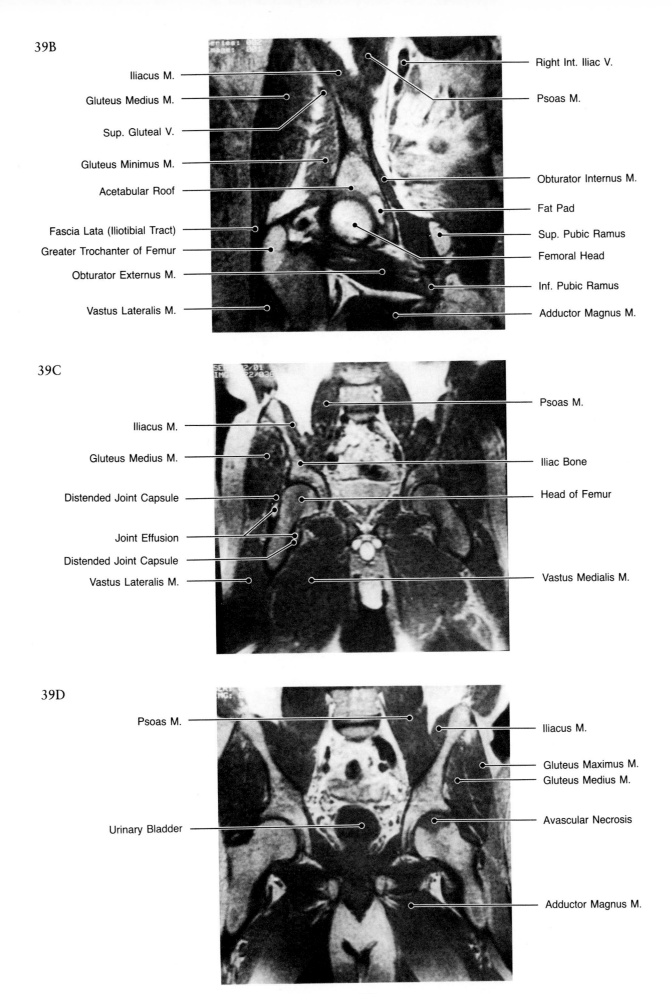

Iliacus M.

Gluteus Medius M.

Sup. Gluteal V.

Gluteus Minimus M.

Acetabular Roof

Fascia Lata (Iliotibial Tract)

Greater Trochanter of Femur

Obturator Externus M.

Vastus Lateralis M.

Right Int. Iliac V.

Psoas M.

Obturator Internus M.

Fat Pad

Sup. Pubic Ramus

Femoral Head

Inf. Pubic Ramus

Adductor Magnus M.

39C

Iliacus M.

Gluteus Medius M.

Distended Joint Capsule

Joint Effusion

Distended Joint Capsule

Vastus Lateralis M.

Psoas M.

Iliac Bone

Head of Femur

Vastus Medialis M.

39D

Psoas M.

Urinary Bladder

Iliacus M.

Gluteus Maximus M.

Gluteus Medius M.

Avascular Necrosis

Adductor Magnus M.

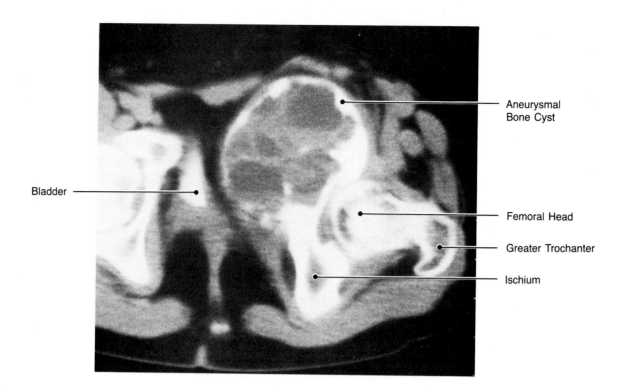

Bladder

Aneurysmal
Bone Cyst

Femoral Head

Greater Trochanter

Ischium

39E, 39F, 39G, 39H Axial/CT/Coronal MRI/Sagittal MRI/Pelvis. Aneurysmal bone cyst arising from the anterior column of the left acetabulum. The CT (39E) demonstrates the expanded cortex, with multiple septations inside the tumor, and fluid levels, which are thought to be characteristic of this tumor. MR images in the coronal (39F, 39G) and sagittal (39H) planes demonstrate the presence of multiple cysts with fluid levels. The cysts have different signal intensities, probably due to different ages of its hemorrhagic contents.

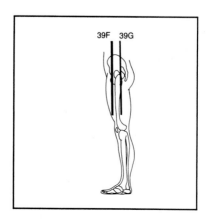

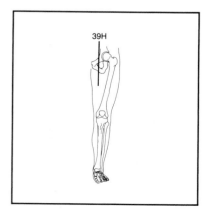

39F

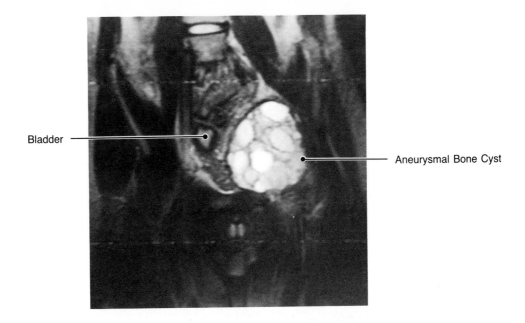

Bladder —

— Aneurysmal Bone Cyst

39G

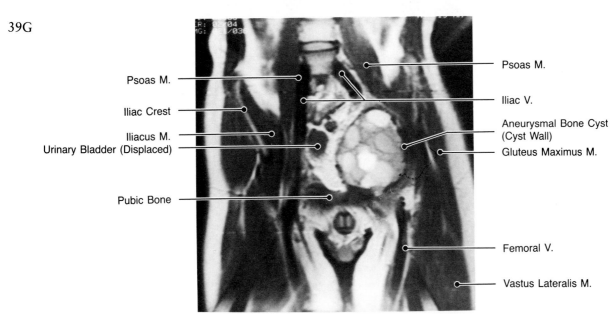

Psoas M. —

Iliac Crest —

Iliacus M. —
Urinary Bladder (Displaced) —

Pubic Bone —

— Psoas M.

— Iliac V.

— Aneurysmal Bone Cyst
(Cyst Wall)
— Gluteus Maximus M.

— Femoral V.

— Vastus Lateralis M.

39H

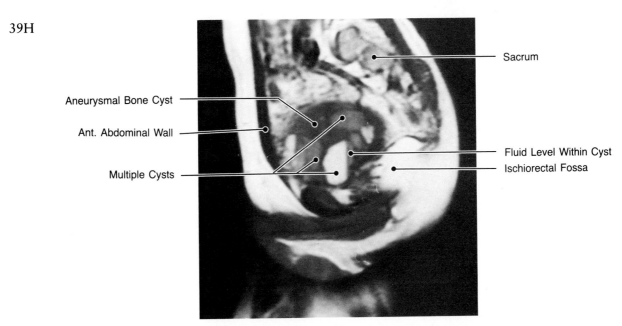

— Sacrum

Aneurysmal Bone Cyst —

Ant. Abdominal Wall —

Multiple Cysts —

— Fluid Level Within Cyst
— Ischiorectal Fossa

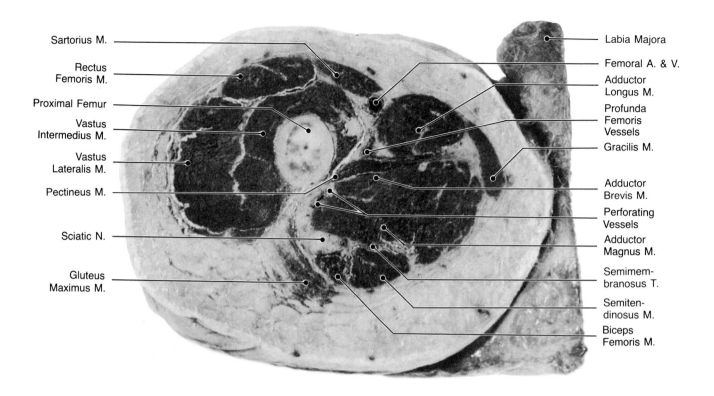

Sartorius M.

Rectus Femoris M.

Proximal Femur

Vastus Intermedius M.

Vastus Lateralis M.

Pectineus M.

Sciatic N.

Gluteus Maximus M.

Labia Majora

Femoral A. & V.

Adductor Longus M.

Profunda Femoris Vessels

Gracilis M.

Adductor Brevis M.

Perforating Vessels

Adductor Magnus M.

Semimembranosus T.

Semitendinosus M.

Biceps Femoris M.

40A, 40B, 40C Axial/Anatomic Plate/CT/MRI/Upper Thigh.
Axial sections through the inguinal region of the upper thigh. The most inferior aspect of the gluteus maximus muscle is seen posteriorly. The femoral artery and veins are seen immediately posterior to the sartorius muscle. The proximal femoral shaft is surrounded anteriorly by the vastus muscles. The sciatic nerve can be seen immediately lateral to the semimembranosus and semitendinosus muscles.

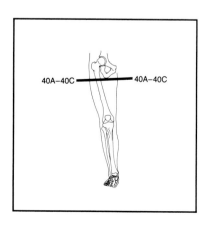

40A–40C   40A–40C

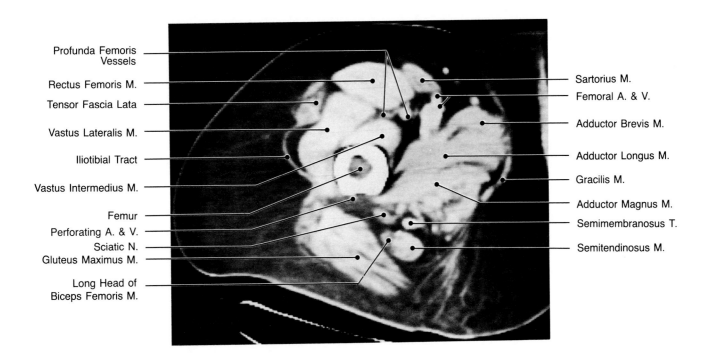

Profunda Femoris Vessels

Rectus Femoris M.

Tensor Fascia Lata

Vastus Lateralis M.

Iliotibial Tract

Vastus Intermedius M.

Femur

Perforating A. & V.

Sciatic N.

Gluteus Maximus M.

Long Head of Biceps Femoris M.

Sartorius M.

Femoral A. & V.

Adductor Brevis M.

Adductor Longus M.

Gracilis M.

Adductor Magnus M.

Semimembranosus T.

Semitendinosus M.

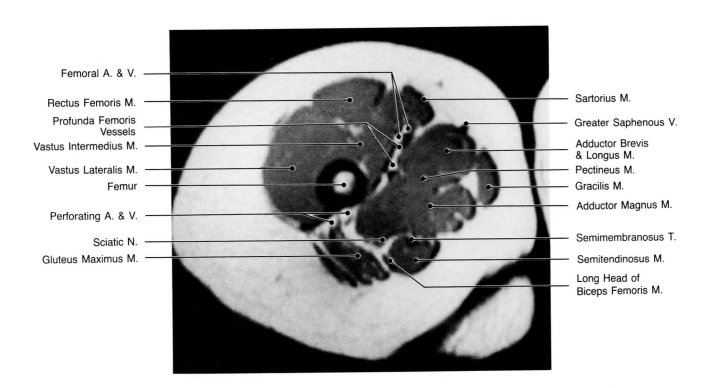

Femoral A. & V.

Rectus Femoris M.

Profunda Femoris Vessels

Vastus Intermedius M.

Vastus Lateralis M.

Femur

Perforating A. & V.

Sciatic N.

Gluteus Maximus M.

Sartorius M.

Greater Saphenous V.

Adductor Brevis & Longus M.

Pectineus M.

Gracilis M.

Adductor Magnus M.

Semimembranosus T.

Semitendinosus M.

Long Head of Biceps Femoris M.

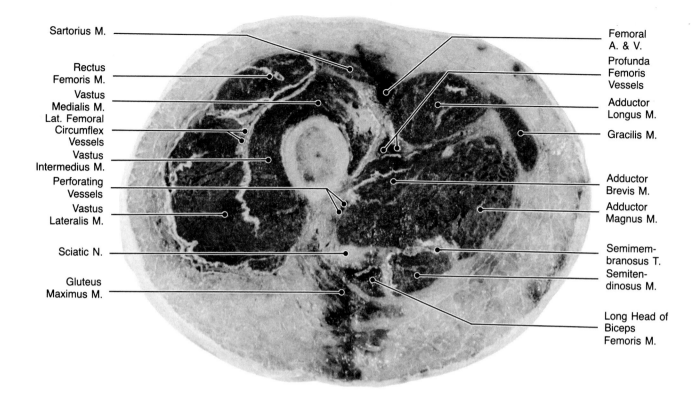

Sartorius M.

Rectus Femoris M.

Vastus Medialis M.

Lat. Femoral Circumflex Vessels

Vastus Intermedius M.

Perforating Vessels

Vastus Lateralis M.

Sciatic N.

Gluteus Maximus M.

Femoral A. & V.

Profunda Femoris Vessels

Adductor Longus M.

Gracilis M.

Adductor Brevis M.

Adductor Magnus M.

Semimembranosus T.

Semitendinosus M.

Long Head of Biceps Femoris M.

41A, 41B, 41C Axial/Anatomic Plate/CT/MRI/Upper Thigh.
Axial sections through the upper thigh, about 3 centimeters below the previous section. The sciatic nerve is better seen on the CT section than on the MRI probably because of the similar signal intensity of the nerve and muscles on MRI.

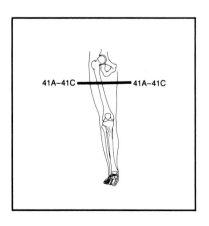

41A–41C   41A–41C

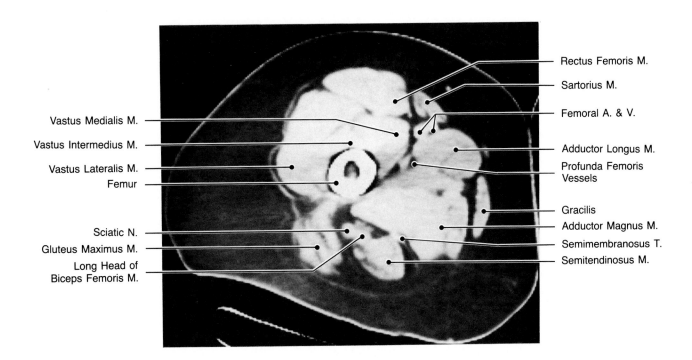

Rectus Femoris M.

Sartorius M.

Femoral A. & V.

Vastus Medialis M.

Vastus Intermedius M.

Adductor Longus M.

Vastus Lateralis M.

Profunda Femoris
Vessels

Femur

Gracilis

Adductor Magnus M.

Sciatic N.

Semimembranosus T.

Gluteus Maximus M.

Semitendinosus M.

Long Head of
Biceps Femoris M.

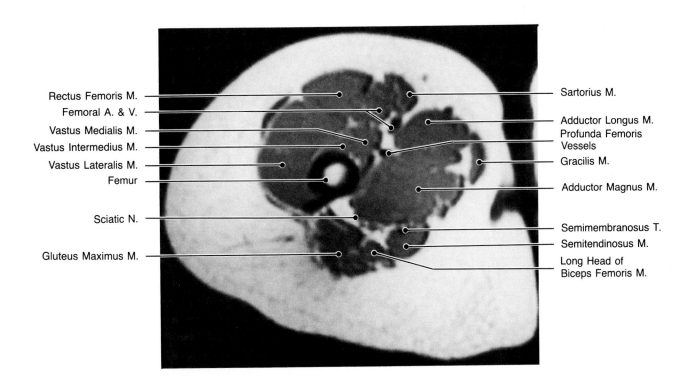

Rectus Femoris M.

Sartorius M.

Femoral A. & V.

Adductor Longus M.

Vastus Medialis M.

Profunda Femoris
Vessels

Vastus Intermedius M.

Vastus Lateralis M.

Gracilis M.

Femur

Adductor Magnus M.

Sciatic N.

Semimembranosus T.

Semitendinosus M.

Gluteus Maximus M.

Long Head of
Biceps Femoris M.

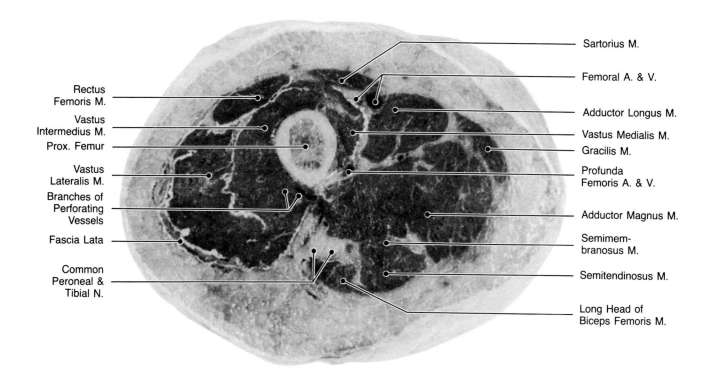

Rectus Femoris M.

Vastus Intermedius M.

Prox. Femur

Vastus Lateralis M.

Branches of Perforating Vessels

Fascia Lata

Common Peroneal & Tibial N.

Sartorius M.

Femoral A. & V.

Adductor Longus M.

Vastus Medialis M.

Gracilis M.

Profunda Femoris A. & V.

Adductor Magnus M.

Semimembranosus M.

Semitendinosus M.

Long Head of Biceps Femoris M.

42A, 42B, 42C Axial/Anatomic Plate/CT/MRI/Upper Thigh.
Axial sections through the upper third of the thigh. The profunda femoris vessels are seen in both CT and MRI located immediately posterior to the femur. The anterior compartment of the thigh is occupied by the vastus and rectus femoris muscles. The posterior compartment is occupied by the biceps femoris, semitendinosus, semimembranosus, gracilis, and sartorius muscles.

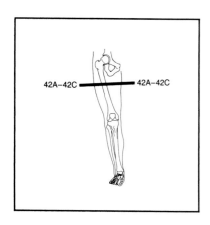

42A–42C   42A–42C

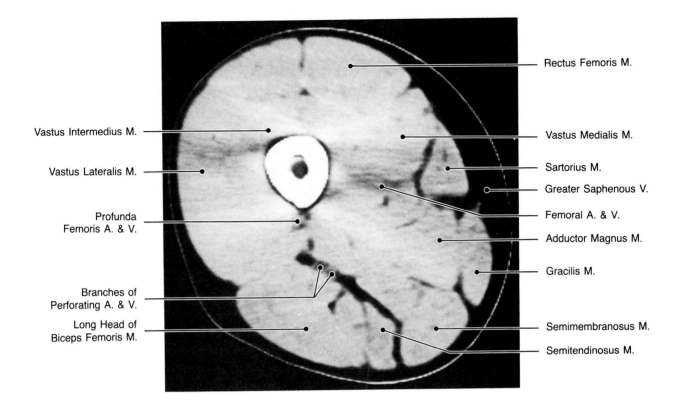

Rectus Femoris M.

Vastus Intermedius M.

Vastus Medialis M.

Vastus Lateralis M.

Sartorius M.

Greater Saphenous V.

Profunda Femoris A. & V.

Femoral A. & V.

Adductor Magnus M.

Gracilis M.

Branches of Perforating A. & V.

Long Head of Biceps Femoris M.

Semimembranosus M.

Semitendinosus M.

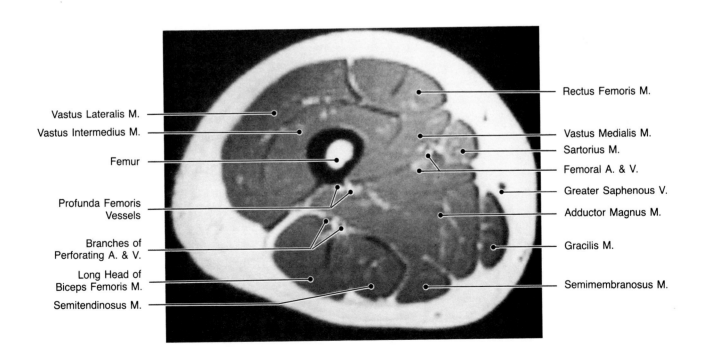

Rectus Femoris M.

Vastus Lateralis M.

Vastus Intermedius M.

Vastus Medialis M.

Sartorius M.

Femur

Femoral A. & V.

Greater Saphenous V.

Profunda Femoris Vessels

Adductor Magnus M.

Branches of Perforating A. & V.

Gracilis M.

Long Head of Biceps Femoris M.

Semimembranosus M.

Semitendinosus M.

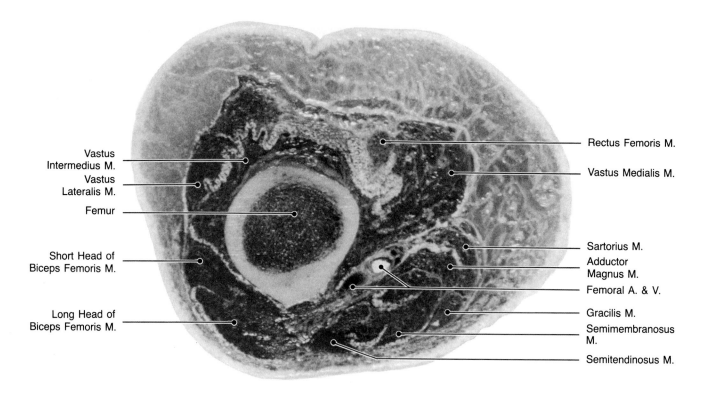

Vastus Intermedius M.

Vastus Lateralis M.

Femur

Short Head of Biceps Femoris M.

Long Head of Biceps Femoris M.

Rectus Femoris M.

Vastus Medialis M.

Sartorius M.

Adductor Magnus M.

Femoral A. & V.

Gracilis M.

Semimembranosus M.

Semitendinosus M.

43A, 43B, 43C Axial/Anatomic Plate/CT/MRI/Mid Thigh.
Axial section through the mid thigh. The femoral vessels as well as the sartorius and gracilis muscles are located in a more posterior position.

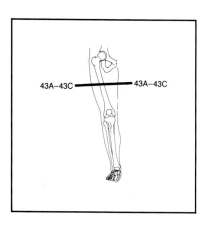

43A–43C   43A–43C

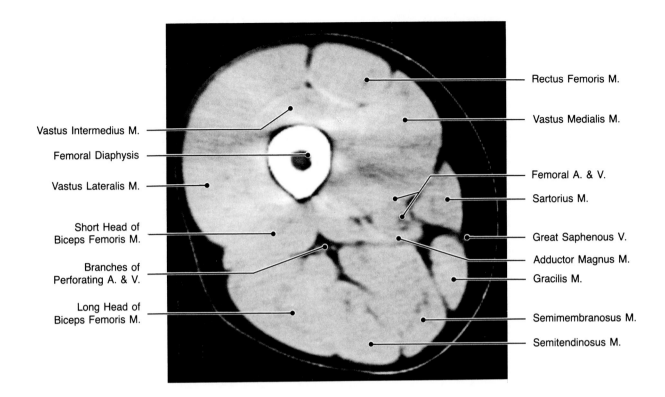

Vastus Intermedius M.

Femoral Diaphysis

Vastus Lateralis M.

Short Head of
Biceps Femoris M.

Branches of
Perforating A. & V.

Long Head of
Biceps Femoris M.

Rectus Femoris M.

Vastus Medialis M.

Femoral A. & V.

Sartorius M.

Great Saphenous V.

Adductor Magnus M.

Gracilis M.

Semimembranosus M.

Semitendinosus M.

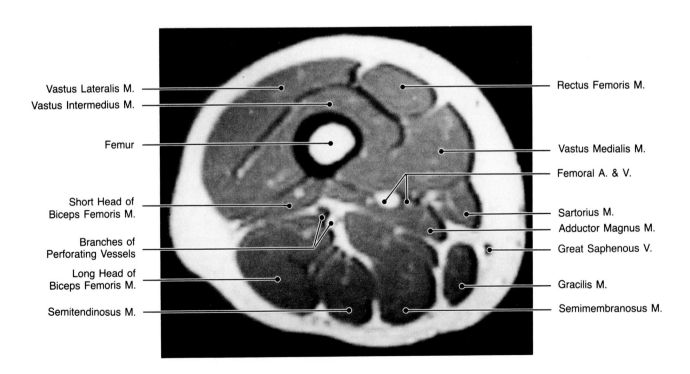

Vastus Lateralis M.

Vastus Intermedius M.

Femur

Short Head of
Biceps Femoris M.

Branches of
Perforating Vessels

Long Head of
Biceps Femoris M.

Semitendinosus M.

Rectus Femoris M.

Vastus Medialis M.

Femoral A. & V.

Sartorius M.

Adductor Magnus M.

Great Saphenous V.

Gracilis M.

Semimembranosus M.

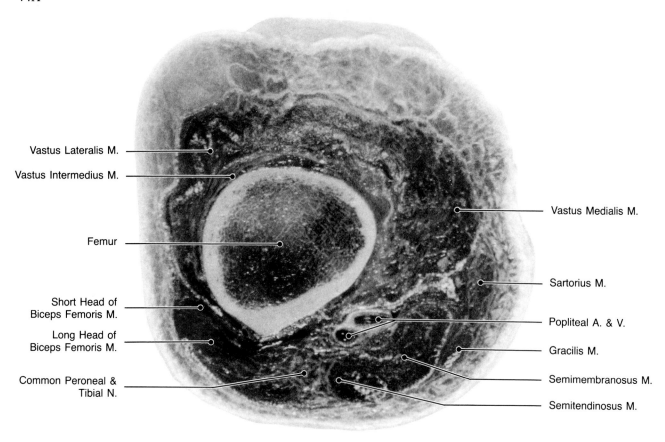

Vastus Lateralis M.

Vastus Intermedius M.

Femur

Short Head of
Biceps Femoris M.

Long Head of
Biceps Femoris M.

Common Peroneal &
Tibial N.

Vastus Medialis M.

Sartorius M.

Popliteal A. & V.

Gracilis M.

Semimembranosus M.

Semitendinosus M.

44A, 44B, 44C Axial/Anatomic Plate/CT/MRI/Lower Thigh.
Section through the lower third of the thigh. In this section the
common peroneal and tibial nerves can be identified between
the short head of the biceps femoris and the semimembranosus
muscles. The femoral vessels are located immediately posterior
to the body of the vastus medialis muscle. The rectus femoris
has become thinner.

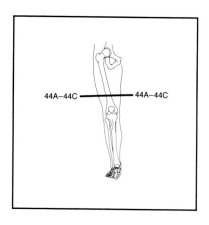

44A–44C   44A–44C

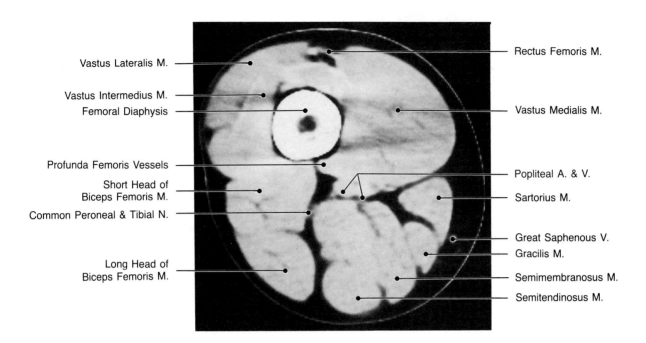

Vastus Lateralis M.

Vastus Intermedius M.
Femoral Diaphysis

Profunda Femoris Vessels
Short Head of
Biceps Femoris M.
Common Peroneal & Tibial N.

Long Head of
Biceps Femoris M.

Rectus Femoris M.

Vastus Medialis M.

Popliteal A. & V.
Sartorius M.

Great Saphenous V.
Gracilis M.
Semimembranosus M.
Semitendinosus M.

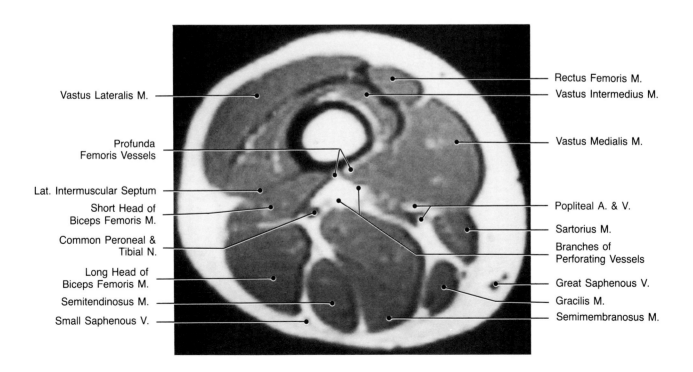

Vastus Lateralis M.

Profunda
Femoris Vessels

Lat. Intermuscular Septum
Short Head of
Biceps Femoris M.

Common Peroneal &
Tibial N.

Long Head of
Biceps Femoris M.
Semitendinosus M.
Small Saphenous V.

Rectus Femoris M.
Vastus Intermedius M.

Vastus Medialis M.

Popliteal A. & V.

Sartorius M.

Branches of
Perforating Vessels

Great Saphenous V.
Gracilis M.
Semimembranosus M.

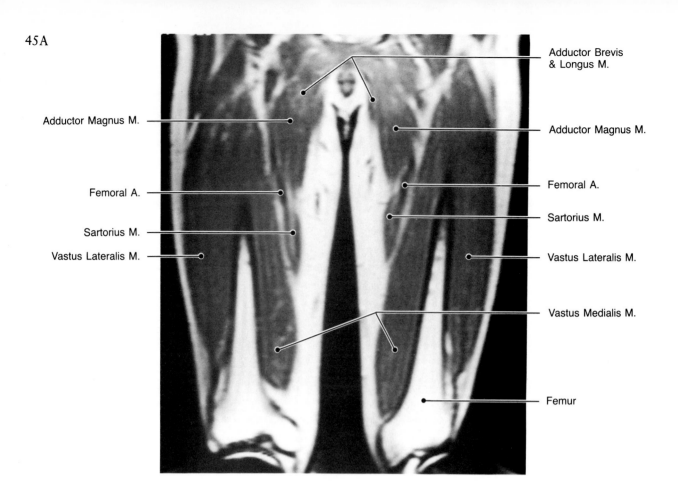

Adductor Brevis
& Longus M.

Adductor Magnus M.

Adductor Magnus M.

Femoral A.

Femoral A.

Sartorius M.

Sartorius M.

Vastus Lateralis M.

Vastus Lateralis M.

Vastus Medialis M.

Femur

45A, 45B, 45C Coronal/MRI/Thigh. Coronal sections through the anterior (45A), mid (45B) and posterior (45C) thigh. These sections illustrate the relationship between the adductor and the vastus medialis muscles. The sciatic nerve can be identified as a darker, low signal intensity structure in the upper third of the thigh in the most posterior coronal section (45C).

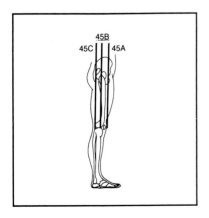

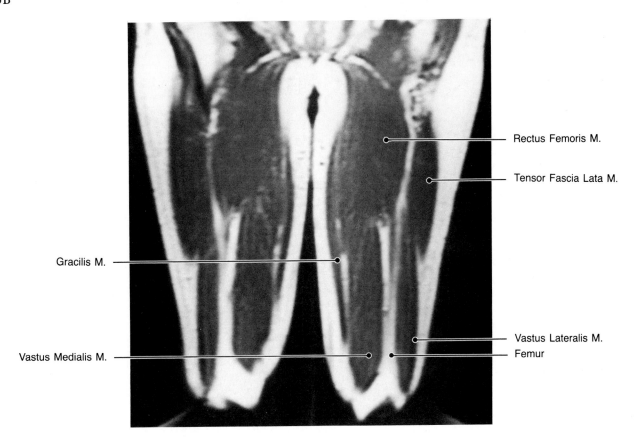

Rectus Femoris M.

Tensor Fascia Lata M.

Gracilis M.

Vastus Lateralis M.

Vastus Medialis M.

Femur

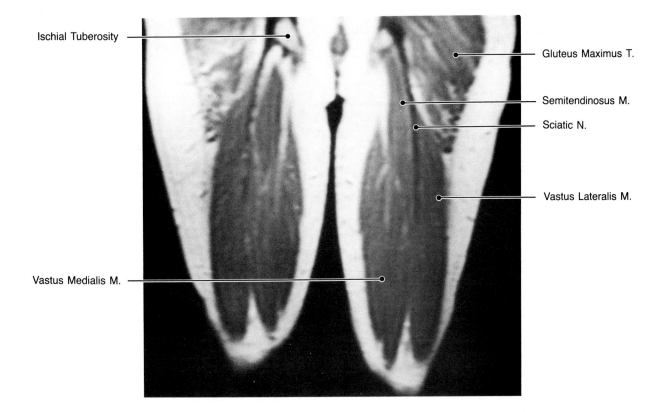

Ischial Tuberosity

Gluteus Maximus T.

Semitendinosus M.

Sciatic N.

Vastus Lateralis M.

Vastus Medialis M.

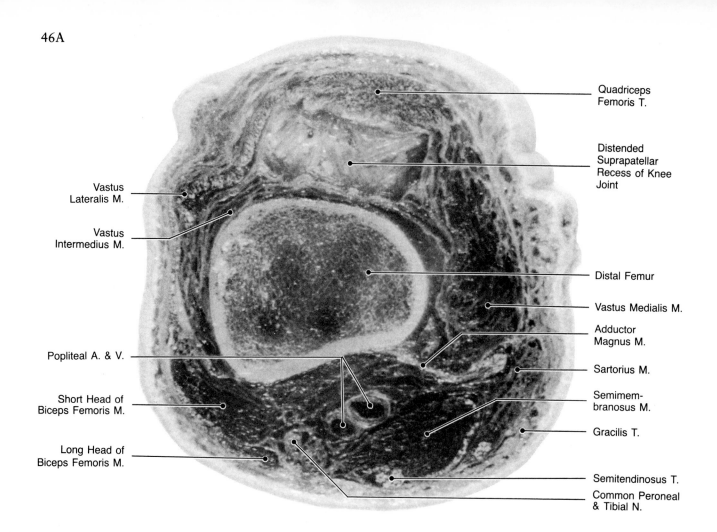

Quadriceps
Femoris T.

Distended
Suprapatellar
Recess of Knee
Joint

Vastus
Lateralis M.

Vastus
Intermedius M.

Distal Femur

Vastus Medialis M.

Adductor
Magnus M.

Popliteal A. & V.

Sartorius M.

Semimem-
branosus M.

Short Head of
Biceps Femoris M.

Gracilis T.

Long Head of
Biceps Femoris M.

Semitendinosus T.

Common Peroneal
& Tibial N.

46A, 46B, 46C Axial/Anatomic/CT/MRI/Distal Thigh.
Axial sections through the distal thigh just above the patella.
On the anatomic specimen the quadriceps femoris tendon is
displaced anteriorly by the distended suprapatellar recess of the
knee joint filled with fluid. The femoral artery and vein are located between the semimembranosus muscle posteriorly and the
femur anteriorly. The sciatic nerve has divided into common
peroneal and tibial nerves, which are seen between the semimembranosus muscle medially and biceps femoris laterally.

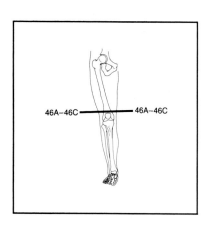

46A–46C   46A–46C

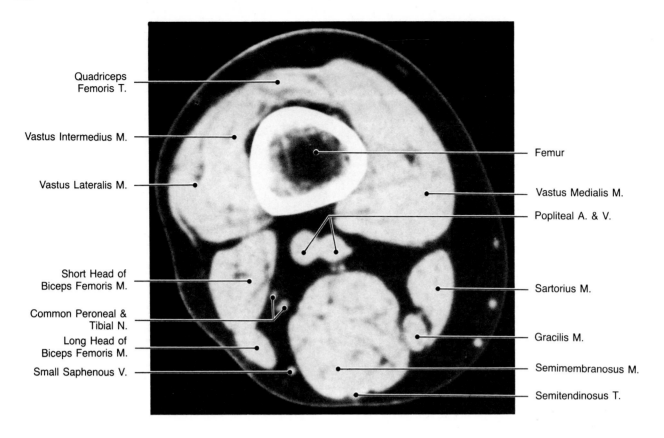

Quadriceps Femoris T.

Vastus Intermedius M.

Vastus Lateralis M.

Femur

Vastus Medialis M.

Popliteal A. & V.

Short Head of Biceps Femoris M.

Sartorius M.

Common Peroneal & Tibial N.

Long Head of Biceps Femoris M.

Gracilis M.

Small Saphenous V.

Semimembranosus M.

Semitendinosus T.

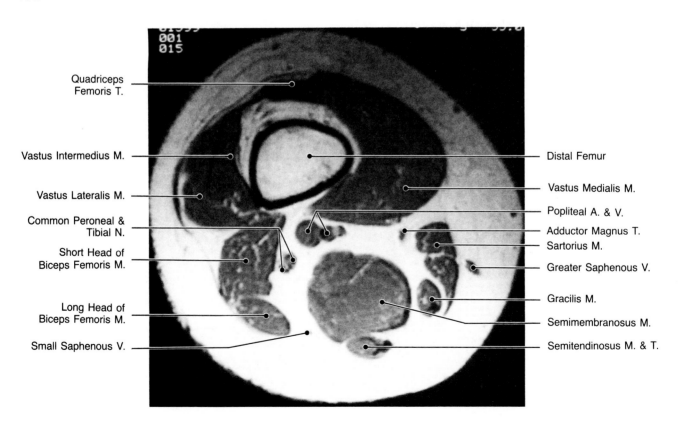

Quadriceps Femoris T.

Vastus Intermedius M.

Distal Femur

Vastus Lateralis M.

Vastus Medialis M.

Common Peroneal & Tibial N.

Popliteal A. & V.

Adductor Magnus T.

Sartorius M.

Short Head of Biceps Femoris M.

Greater Saphenous V.

Long Head of Biceps Femoris M.

Gracilis M.

Semimembranosus M.

Small Saphenous V.

Semitendinosus M. & T.

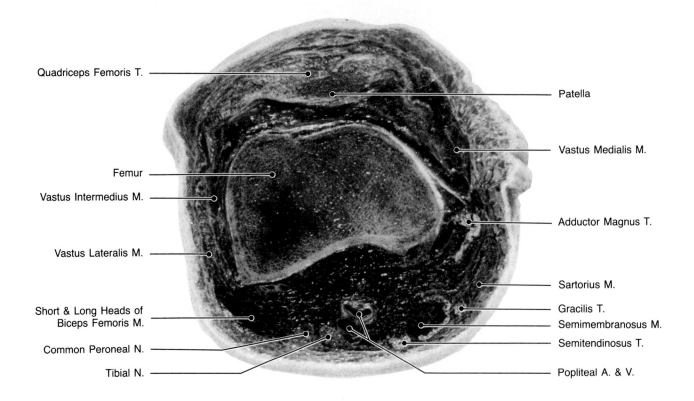

Quadriceps Femoris T.

Patella

Vastus Medialis M.

Femur

Vastus Intermedius M.

Adductor Magnus T.

Vastus Lateralis M.

Sartorius M.

Gracilis T.

Short & Long Heads of
Biceps Femoris M.

Semimembranosus M.

Semitendinosus T.

Common Peroneal N.

Tibial N.

Popliteal A. & V.

47A, 47B, 47C Axial/Anatomic Plate/CT/MRI/Distal Thigh.
Axial sections through the distal thigh, at the insertion of the
quadriceps femoris tendon in the superior aspect of the patella.
The vastus lateralis and vastus intermedius muscles have become thinner. At this level the amount of fat in the posterior
aspect of the femur has increased and surrounds completely
the popliteal artery and vein and the common peroneal and
tibial nerves. These structures are clearly demonstrated on
both the CT and MRI images.

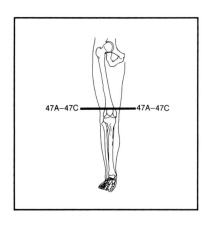

47A–47C   47A–47C

47B

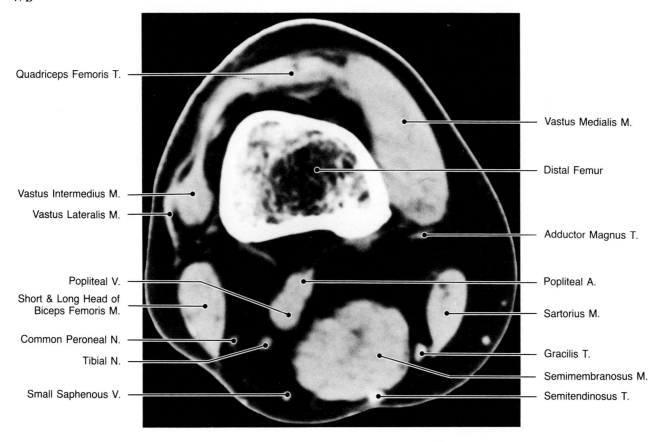

Quadriceps Femoris T.

Vastus Medialis M.

Distal Femur

Vastus Intermedius M.

Vastus Lateralis M.

Adductor Magnus T.

Popliteal V.

Popliteal A.

Short & Long Head of
Biceps Femoris M.

Sartorius M.

Common Peroneal N.

Tibial N.

Gracilis T.

Semimembranosus M.

Small Saphenous V.

Semitendinosus T.

47C

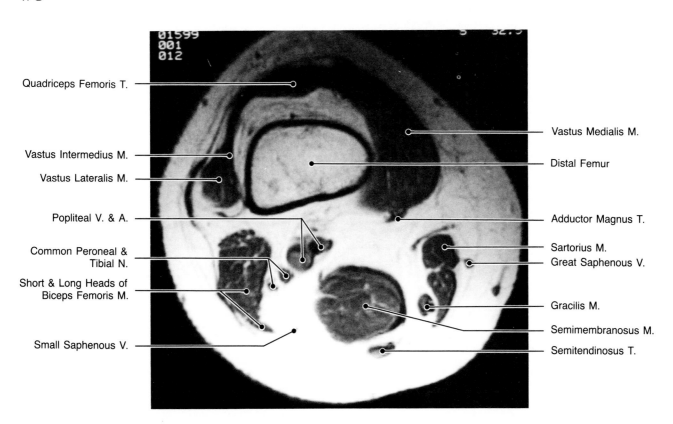

Quadriceps Femoris T.

Vastus Medialis M.

Distal Femur

Vastus Intermedius M.

Vastus Lateralis M.

Adductor Magnus T.

Popliteal V. & A.

Sartorius M.

Common Peroneal &
Tibial N.

Great Saphenous V.

Short & Long Heads of
Biceps Femoris M.

Gracilis M.

Semimembranosus M.

Small Saphenous V.

Semitendinosus T.

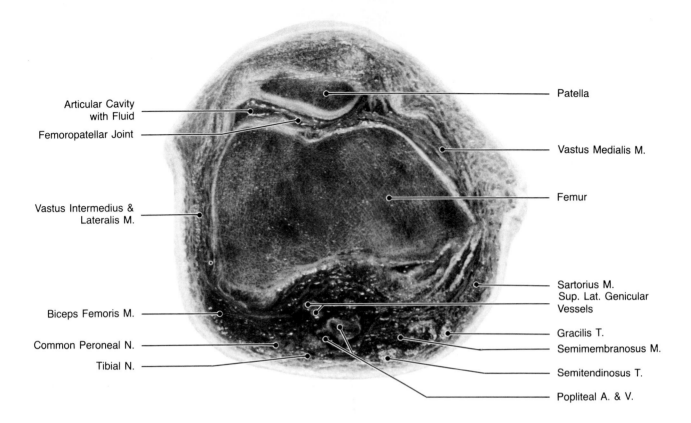

Articular Cavity with Fluid

Femoropatellar Joint

Vastus Intermedius & Lateralis M.

Biceps Femoris M.

Common Peroneal N.

Tibial N.

Patella

Vastus Medialis M.

Femur

Sartorius M.
Sup. Lat. Genicular Vessels

Gracilis T.

Semimembranosus M.

Semitendinosus T.

Popliteal A. & V.

48A, 48B, 48C Axial/Anatomic Plate/CT/MRI/Upper Knee Joint.   Axial sections through the upper half of the patella. The articular joint space is filled with fluid in the anatomic section. At this level the medial and lateral heads of the gastrocnemius muscles are seen at their origins in the posterosuperior aspect of the femoral condyles.

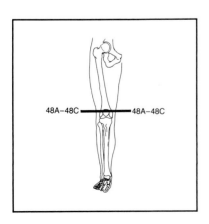

48A–48C ——— 48A–48C

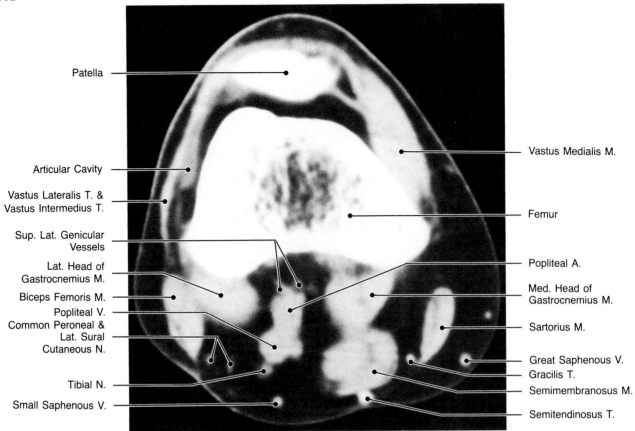

Patella

Articular Cavity

Vastus Lateralis T. &
Vastus Intermedius T.

Sup. Lat. Genicular
Vessels

Lat. Head of
Gastrocnemius M.

Biceps Femoris M.

Popliteal V.
Common Peroneal &
Lat. Sural
Cutaneous N.

Tibial N.

Small Saphenous V.

Vastus Medialis M.

Femur

Popliteal A.

Med. Head of
Gastrocnemius M.

Sartorius M.

Great Saphenous V.
Gracilis T.
Semimembranosus M.

Semitendinosus T.

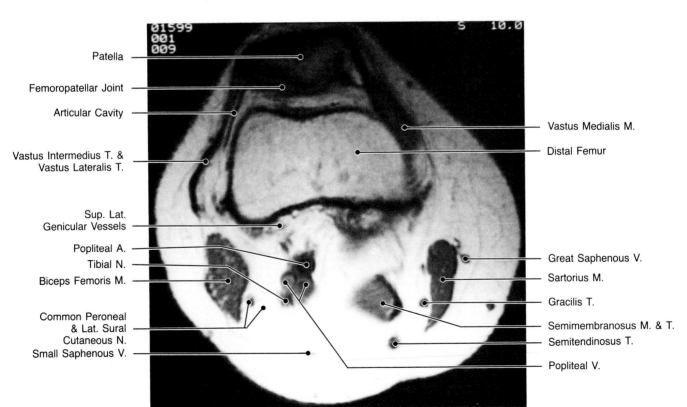

Patella

Femoropatellar Joint

Articular Cavity

Vastus Intermedius T. &
Vastus Lateralis T.

Sup. Lat.
Genicular Vessels

Popliteal A.
Tibial N.
Biceps Femoris M.

Common Peroneal
& Lat. Sural
Cutaneous N.
Small Saphenous V.

Vastus Medialis M.

Distal Femur

Great Saphenous V.

Sartorius M.

Gracilis T.

Semimembranosus M. & T.

Semitendinosus T.

Popliteal V.

01599
001
009

S   10.0

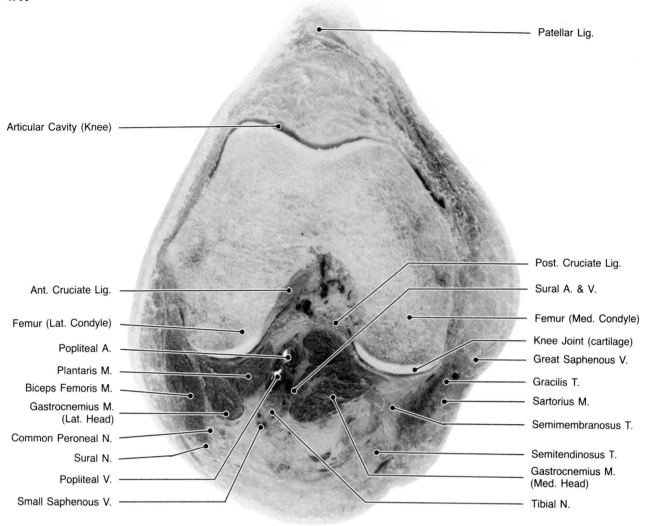

Patellar Lig.

Articular Cavity (Knee)

Ant. Cruciate Lig.

Femur (Lat. Condyle)

Popliteal A.

Plantaris M.

Biceps Femoris M.

Gastrocnemius M. (Lat. Head)

Common Peroneal N.

Sural N.

Popliteal V.

Small Saphenous V.

Post. Cruciate Lig.

Sural A. & V.

Femur (Med. Condyle)

Knee Joint (cartilage)

Great Saphenous V.

Gracilis T.

Sartorius M.

Semimembranosus T.

Semitendinosus T.

Gastrocnemius M. (Med. Head)

Tibial N.

49A, 49B, 49C Axial/Anatomic Plate/CT/MRI/Upper Knee Joint. Axial sections through the midpatellar region. Note the articular cartilage covering the posterior aspect of the patella as well as the posterior aspect of the femoral condyles seen on the MR image (49C). The origins of the anterior cruciate ligament in the medial aspect of the lateral femoral condyle and the posterior cruciate ligament in the lateral aspect of the medial femoral condyle are also demonstrated. In this section the tibial nerve remains close to the midline, whereas the common peroneal nerve is more lateral in position, just posterior to the biceps femoris muscle.

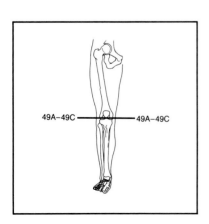

49A–49C   49A–49C

49B

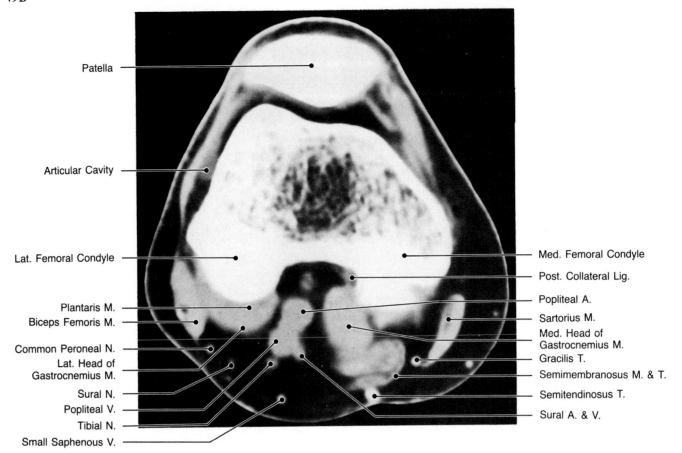

Patella

Articular Cavity

Lat. Femoral Condyle

Plantaris M.
Biceps Femoris M.

Common Peroneal N.
Lat. Head of
Gastrocnemius M.

Sural N.

Popliteal V.

Tibial N.

Small Saphenous V.

Med. Femoral Condyle
Post. Collateral Lig.
Popliteal A.
Sartorius M.
Med. Head of
Gastrocnemius M.
Gracilis T.
Semimembranosus M. & T.
Semitendinosus T.
Sural A. & V.

49C

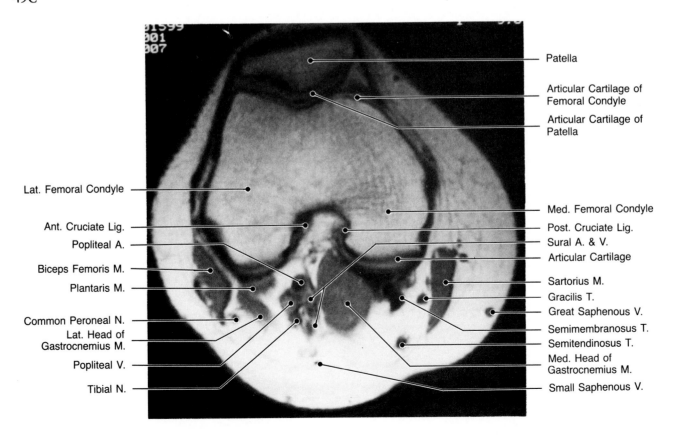

Patella

Articular Cartilage of
Femoral Condyle

Articular Cartilage of
Patella

Lat. Femoral Condyle

Ant. Cruciate Lig.

Popliteal A.

Biceps Femoris M.

Plantaris M.

Common Peroneal N.
Lat. Head of
Gastrocnemius M.

Popliteal V.

Tibial N.

Med. Femoral Condyle
Post. Cruciate Lig.
Sural A. & V.
Articular Cartilage
Sartorius M.
Gracilis T.
Great Saphenous V.
Semimembranosus T.
Semitendinosus T.
Med. Head of
Gastrocnemius M.
Small Saphenous V.

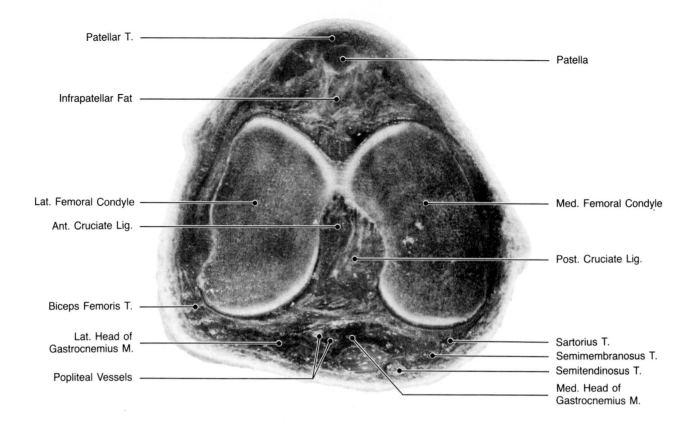

Patellar T.

Infrapatellar Fat

Lat. Femoral Condyle

Ant. Cruciate Lig.

Biceps Femoris T.

Lat. Head of
Gastrocnemius M.

Popliteal Vessels

Patella

Med. Femoral Condyle

Post. Cruciate Lig.

Sartorius T.
Semimembranosus T.
Semitendinosus T.
Med. Head of
Gastrocnemius M.

50A, 50B, 50C Axial/Anatomic Plate/CT/MRI/Mid Knee Joint. Axial sections through the distal aspect of the patella and femoral condyles. The popliteal vessels and tibial nerve are located between both medial and lateral heads of the gastrocnemius. Anteriorly the prominent infrapatellar fat pad is seen posterior to the patellar tendon.

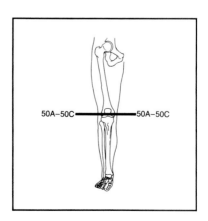

50A–50C   50A–50C

50B

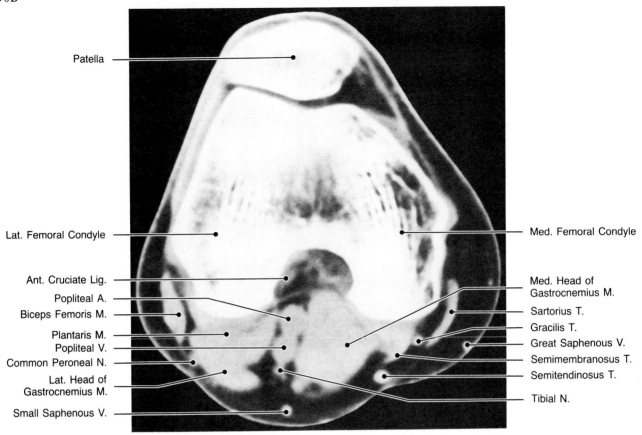

Patella

Lat. Femoral Condyle

Med. Femoral Condyle

Ant. Cruciate Lig.

Med. Head of
Gastrocnemius M.

Popliteal A.

Sartorius T.

Biceps Femoris M.

Gracilis T.

Plantaris M.

Great Saphenous V.

Popliteal V.

Semimembranosus T.

Common Peroneal N.

Semitendinosus T.

Lat. Head of
Gastrocnemius M.

Tibial N.

Small Saphenous V.

50C

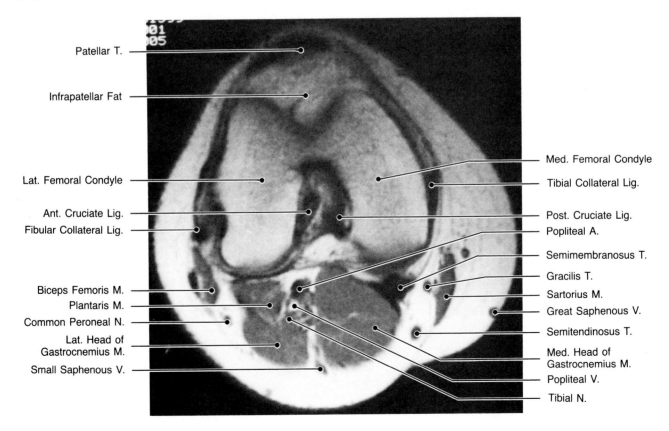

Patellar T.

Infrapatellar Fat

Med. Femoral Condyle

Lat. Femoral Condyle

Tibial Collateral Lig.

Ant. Cruciate Lig.

Post. Cruciate Lig.

Fibular Collateral Lig.

Popliteal A.

Semimembranosus T.

Gracilis T.

Biceps Femoris M.

Sartorius M.

Plantaris M.

Great Saphenous V.

Common Peroneal N.

Semitendinosus T.

Lat. Head of
Gastrocnemius M.

Med. Head of
Gastrocnemius M.

Small Saphenous V.

Popliteal V.

Tibial N.

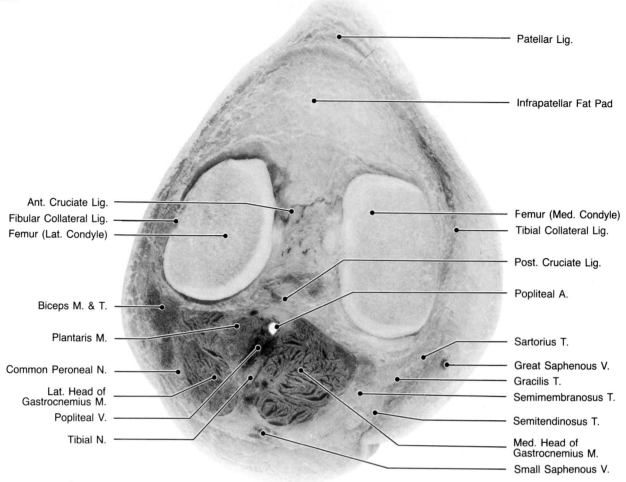

Patellar Lig.

Infrapatellar Fat Pad

Ant. Cruciate Lig.

Fibular Collateral Lig.

Femur (Lat. Condyle)

Femur (Med. Condyle)

Tibial Collateral Lig.

Post. Cruciate Lig.

Popliteal A.

Biceps M. & T.

Plantaris M.

Sartorius T.

Common Peroneal N.

Great Saphenous V.

Gracilis T.

Lat. Head of
Gastrocnemius M.

Semimembranosus T.

Popliteal V.

Semitendinosus T.

Tibial N.

Med. Head of
Gastrocnemius M.

Small Saphenous V.

51A, 51B, 51C Axial/Anatomic/CT/MRI/Mid Knee Joint.
Axial sections through the most distal portion of the femoral
condyles. In this image the intercondylar eminence is seen in
the midline. The posterior cruciate ligament is well seen both
in the CT (high intensity structure) (51B) and MRI (low signal
intensity structure) (51C) just posterior to the intercondylar
eminence. The plantaris muscle is now seen just anterior to the
lateral head of the gastrocnemius muscle. Both fibular and tibial collateral ligaments are also seen in the lateral and medial
aspect of the knee respectively.

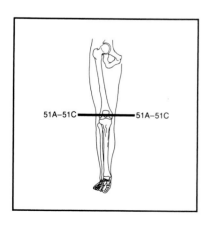

51A–51C   51A–51C

51B

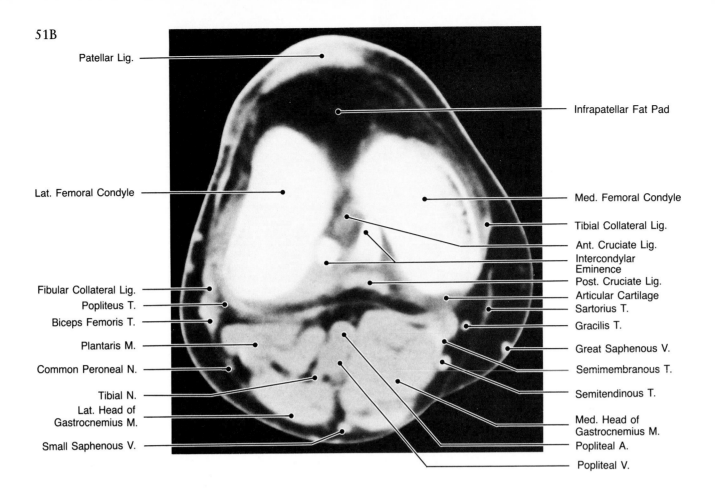

Patellar Lig.

Infrapatellar Fat Pad

Lat. Femoral Condyle

Med. Femoral Condyle

Tibial Collateral Lig.

Ant. Cruciate Lig.

Intercondylar Eminence

Post. Cruciate Lig.

Articular Cartilage

Sartorius T.

Fibular Collateral Lig.

Popliteus T.

Biceps Femoris T.

Gracilis T.

Plantaris M.

Great Saphenous V.

Common Peroneal N.

Semimembranous T.

Tibial N.

Semitendinous T.

Lat. Head of Gastrocnemius M.

Med. Head of Gastrocnemius M.

Small Saphenous V.

Popliteal A.

Popliteal V.

51C

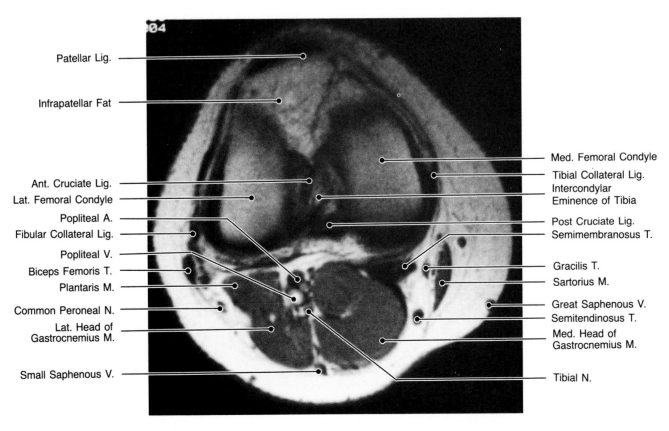

Patellar Lig.

Infrapatellar Fat

Med. Femoral Condyle

Tibial Collateral Lig.

Intercondylar Eminence of Tibia

Ant. Cruciate Lig.

Lat. Femoral Condyle

Popliteal A.

Post Cruciate Lig.

Fibular Collateral Lig.

Semimembranosus T.

Popliteal V.

Biceps Femoris T.

Gracilis T.

Plantaris M.

Sartorius M.

Common Peroneal N.

Great Saphenous V.

Semitendinosus T.

Lat. Head of Gastrocnemius M.

Med. Head of Gastrocnemius M.

Small Saphenous V.

Tibial N.

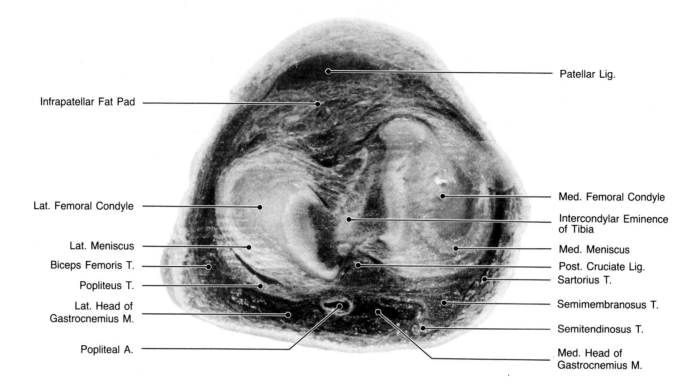

Patellar Lig.

Infrapatellar Fat Pad

Lat. Femoral Condyle

Med. Femoral Condyle

Intercondylar Eminence of Tibia

Lat. Meniscus

Med. Meniscus

Biceps Femoris T.

Post. Cruciate Lig.

Popliteus T.

Sartorius T.

Lat. Head of Gastrocnemius M.

Semimembranosus T.

Semitendinosus T.

Popliteal A.

Med. Head of Gastrocnemius M.

52A, 52B, 52C Axial/Anatomic/CT/MRI/Mid Knee Joint.
Axial sections through the knee joint. Both MRI and CT sections show structures corresponding to the femur and tibia as well as the menisci. On the CT (52B) the lateral and medial menisci are seen as high density structures surrounding the lateral and medial femoral condyles. The intercondylar eminence of the tibia is located in the midline. The posterior cruciate ligament is seen at the level of its tibial insertion. The popliteus tendon and muscle are seen posterior to the lateral meniscus. A small amount of fat separates the popliteus tendon from the posterior horn of the lateral meniscus. The anatomic section shows the recess of the popliteus tendon.

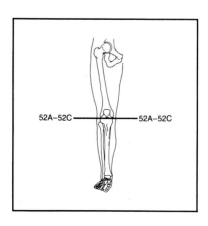

52A–52C   52A–52C

52B

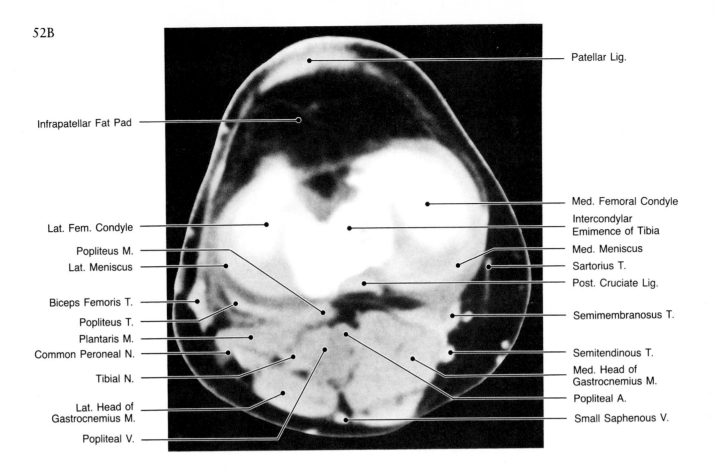

Patellar Lig.

Infrapatellar Fat Pad

Med. Femoral Condyle

Lat. Fem. Condyle

Intercondylar Emimence of Tibia

Popliteus M.

Med. Meniscus

Lat. Meniscus

Sartorius T.

Post. Cruciate Lig.

Biceps Femoris T.

Popliteus T.

Semimembranosus T.

Plantaris M.

Common Peroneal N.

Semitendinous T.

Tibial N.

Med. Head of Gastrocnemius M.

Lat. Head of Gastrocnemius M.

Popliteal A.

Popliteal V.

Small Saphenous V.

52C

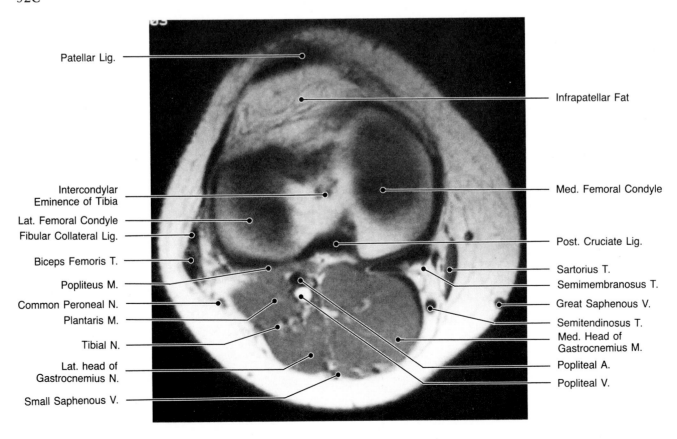

Patellar Lig.

Infrapatellar Fat

Med. Femoral Condyle

Intercondylar Eminence of Tibia

Lat. Femoral Condyle

Fibular Collateral Lig.

Post. Cruciate Lig.

Biceps Femoris T.

Sartorius T.

Popliteus M.

Semimembranosus T.

Common Peroneal N.

Great Saphenous V.

Plantaris M.

Semitendinosus T.

Tibial N.

Med. Head of Gastrocnemius M.

Lat. head of Gastrocnemius N.

Popliteal A.

Popliteal V.

Small Saphenous V.

53A

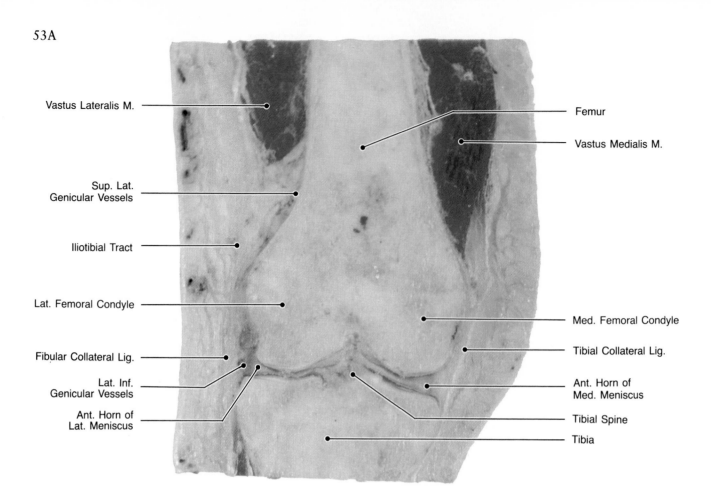

Vastus Lateralis M.

Sup. Lat. Genicular Vessels

Iliotibial Tract

Lat. Femoral Condyle

Fibular Collateral Lig.

Lat. Inf. Genicular Vessels

Ant. Horn of Lat. Meniscus

Femur

Vastus Medialis M.

Med. Femoral Condyle

Tibial Collateral Lig.

Ant. Horn of Med. Meniscus

Tibial Spine

Tibia

53A, 53B Coronal/Anatomic Plate/MRI/Anterior Knee Joint. Coronal sections through the anterior aspect of the knee joint. The lateral and medial meniscus can be seen in both anatomic and MRI sections. The lateral fibular collateral ligament and tibial collateral ligaments are seen on the MR image as thin dark bands laterally and medially, respectively. The fibular collateral ligament cannot be separated from the iliotibial band. Note the superior lateral and medial genicular vessels above the femoral condyles. The anterior and posterior cruciate ligaments are seen in the intercondylar notch.

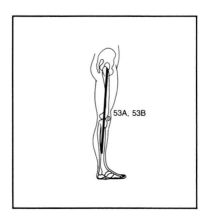

53A, 53B

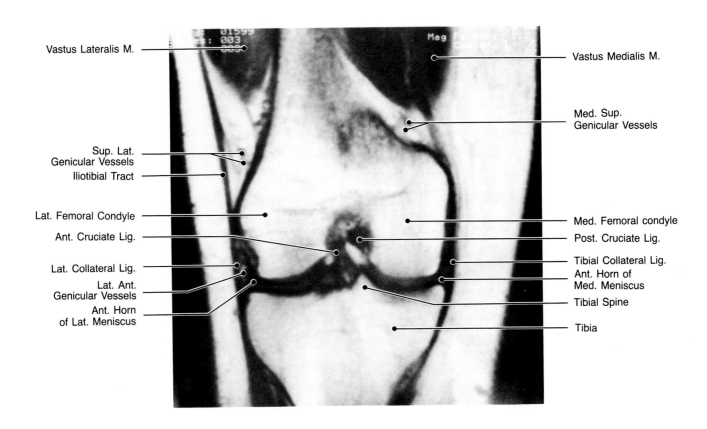

Vastus Lateralis M.

Vastus Medialis M.

Med. Sup.
Genicular Vessels

Sup. Lat.
Genicular Vessels

Iliotibial Tract

Lat. Femoral Condyle

Med. Femoral condyle

Ant. Cruciate Lig.

Post. Cruciate Lig.

Lat. Collateral Lig.

Tibial Collateral Lig.

Lat. Ant.
Genicular Vessels

Ant. Horn of
Med. Meniscus

Ant. Horn
of Lat. Meniscus

Tibial Spine

Tibia

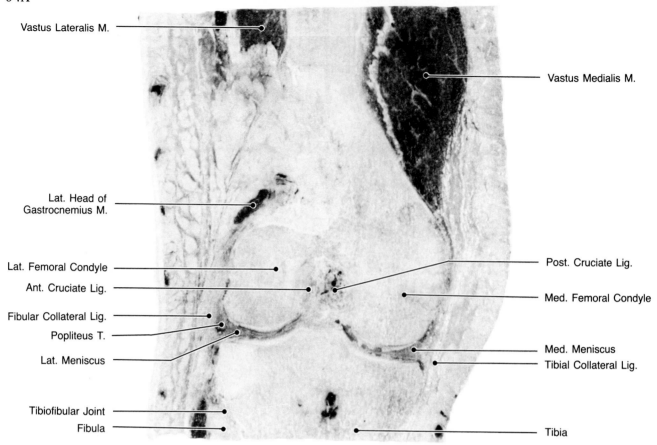

Vastus Lateralis M.

Vastus Medialis M.

Lat. Head of Gastrocnemius M.

Lat. Femoral Condyle

Post. Cruciate Lig.

Ant. Cruciate Lig.

Med. Femoral Condyle

Fibular Collateral Lig.

Popliteus T.

Med. Meniscus

Lat. Meniscus

Tibial Collateral Lig.

Tibiofibular Joint

Fibula

Tibia

54A, 54B Coronal/Anatomic Plate/MRI/Posterior Knee Joint.
Coronal sections through the posterior aspect of the knee joint. The plane of section crosses the femoral condyles vertically. The posterior horns of the medial and lateral menisci are well seen. Also note the anterior and posterior cruciate ligaments in the intercondylar notch as well as the insertions of the lateral and medial heads of the gastrocnemius muscle.

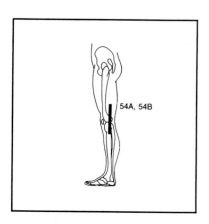

54A, 54B

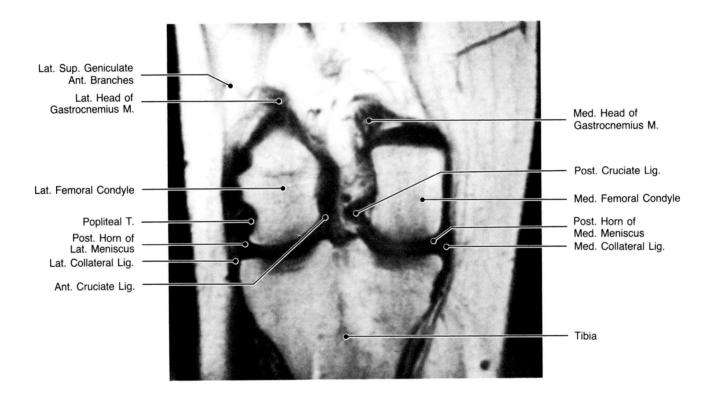

Lat. Sup. Geniculate Ant. Branches

Lat. Head of Gastrocnemius M.

Lat. Femoral Condyle

Popliteal T.

Post. Horn of Lat. Meniscus

Lat. Collateral Lig.

Ant. Cruciate Lig.

Med. Head of Gastrocnemius M.

Post. Cruciate Lig.

Med. Femoral Condyle

Post. Horn of Med. Meniscus

Med. Collateral Lig.

Tibia

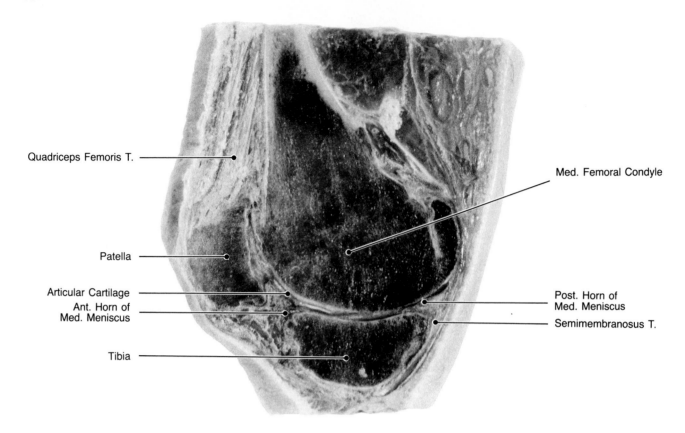

Quadriceps Femoris T. ——————

Patella ————

Articular Cartilage —————
Ant. Horn of
Med. Meniscus

Tibia ————

Med. Femoral Condyle

Post. Horn of
Med. Meniscus

Semimembranosus T.

55A, 55B Sagittal/Anatomic Plate/MRI/Medial Knee Joint. Sagittal sections through the medial aspect of the knee joint. The medial femoral condyle is seen surrounded by a high intensity structure corresponding to the articular cartilage on the MR image (55B). Anterior and posterior horns of the medial meniscus are well demonstrated and separated from the adjacent articular cartilage of the condyle and the proximal tibia.

55C Sagittal/MRI/Medial Knee Joint. Sagittal MRI of the knee. There is a horizontal line through the medial meniscus. This is a normal finding and should not be confused with a horizontal tear.

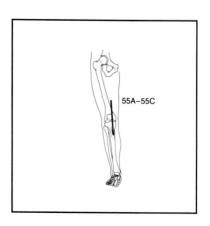

55A–55C

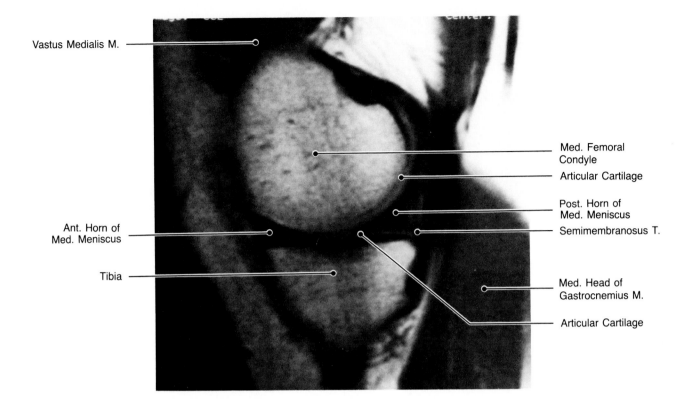

Vastus Medialis M.

Med. Femoral
Condyle

Articular Cartilage

Post. Horn of
Med. Meniscus

Ant. Horn of
Med. Meniscus

Semimembranosus T.

Tibia

Med. Head of
Gastrocnemius M.

Articular Cartilage

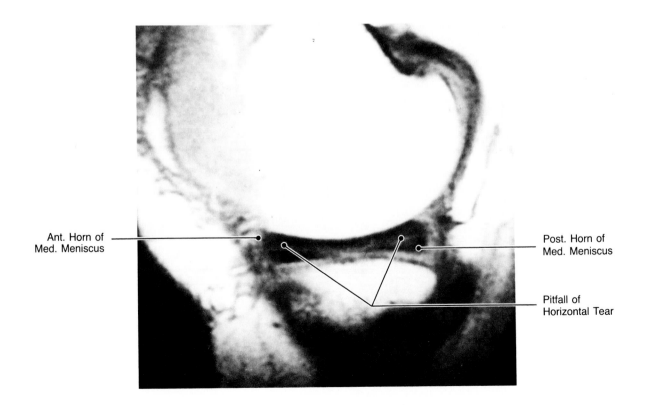

Ant. Horn of
Med. Meniscus

Post. Horn of
Med. Meniscus

Pitfall of
Horizontal Tear

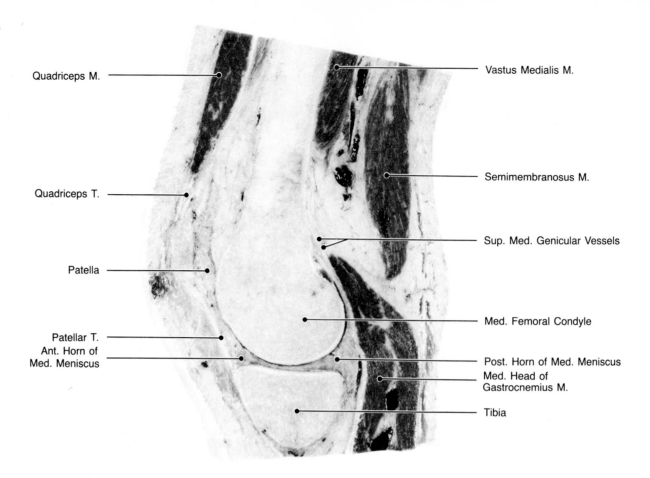

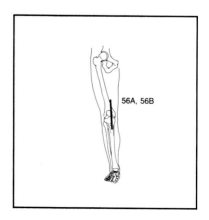

Quadriceps M.

Vastus Medialis M.

Quadriceps T.

Semimembranosus M.

Patella

Sup. Med. Genicular Vessels

Patellar T.
Ant. Horn of
Med. Meniscus

Med. Femoral Condyle

Post. Horn of Med. Meniscus
Med. Head of
Gastrocnemius M.

Tibia

56A, 56B Sagittal/Anatomic Plate/MRI/Medial Knee Joint.
Sagittal section through the medial femoral condyle. The anterior and posterior horns of the medial meniscus are again well visualized. In this section part of the femoropatellar joint is also noted. The large amount of infrapatellar fat provides excellent contrast for the low intensity signal of the anterior horn of the medial meniscus.

56A, 56B

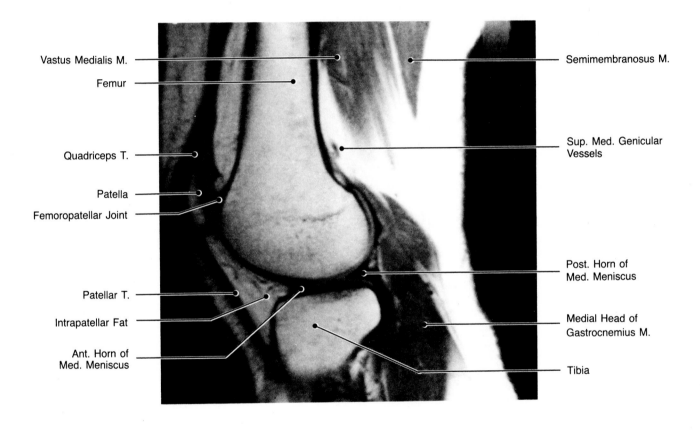

Vastus Medialis M.

Femur

Quadriceps T.

Patella

Femoropatellar Joint

Patellar T.

Intrapatellar Fat

Ant. Horn of
Med. Meniscus

Semimembranosus M.

Sup. Med. Genicular
Vessels

Post. Horn of
Med. Meniscus

Medial Head of
Gastrocnemius M.

Tibia

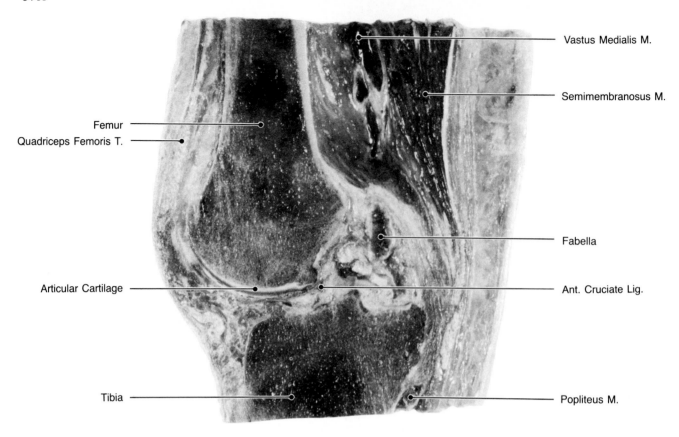

Vastus Medialis M.

Semimembranosus M.

Femur

Quadriceps Femoris T.

Fabella

Articular Cartilage

Ant. Cruciate Lig.

Tibia

Popliteus M.

57A, 57B, 57C Sagittal/Anatomic Plate/MRI/Knee Joint Midline. Sagittal sections through the midline of the knee joint. The cadaveric section is oriented obliquely to the sagittal plane. The posterior cruciate ligament can be discerned in (57B) as a low signal intensity band extending from the posterior aspects of the lateral femoral condyle to its insertion in the posterior aspect of the tibial epiphysis. Due to its oblique orientation with the image plane, the anterior cruciate ligament is seen only at the level of its insertion in the anterior aspect of the tibial spine. The popliteal artery and veins are surrounded anteriorly by fat in (57C). In this section the femoral patellar joint is also well defined, and the cartilaginous thickness of the posterior aspect of the patella and the femoral condyle can be assessed.

57D Sagittal/MRI/Mid Knee Joint. Sagittal MRI section through the intercondylar notch demonstrating the posterior cruciate ligament avulsed from its insertion in the posterior aspect of the tibia. Decreased signal intensity within the tibia corresponds to incomplete fracture and intraosseous hemorrhage.

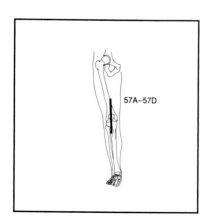

57A–57D

57B

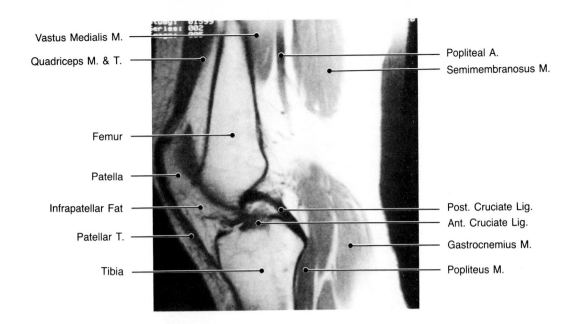

Vastus Medialis M.
Quadriceps M. & T.
Popliteal A.
Semimembranosus M.

Femur
Patella
Infrapatellar Fat
Patellar T.
Tibia

Post. Cruciate Lig.
Ant. Cruciate Lig.
Gastrocnemius M.
Popliteus M.

57C

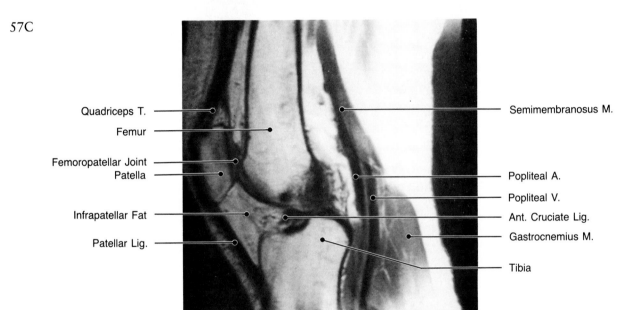

Quadriceps T.
Femur
Femoropatellar Joint
Patella
Infrapatellar Fat
Patellar Lig.

Semimembranosus M.

Popliteal A.
Popliteal V.
Ant. Cruciate Lig.
Gastrocnemius M.
Tibia

57D

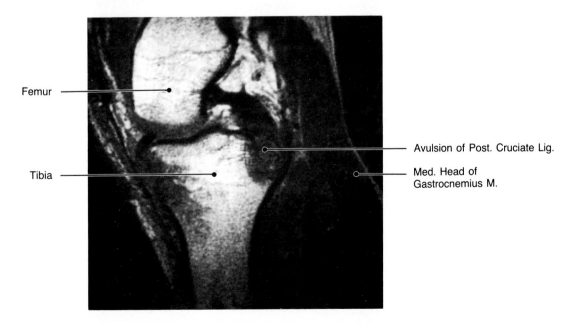

Femur

Tibia

Avulsion of Post. Cruciate Lig.
Med. Head of
Gastrocnemius M.

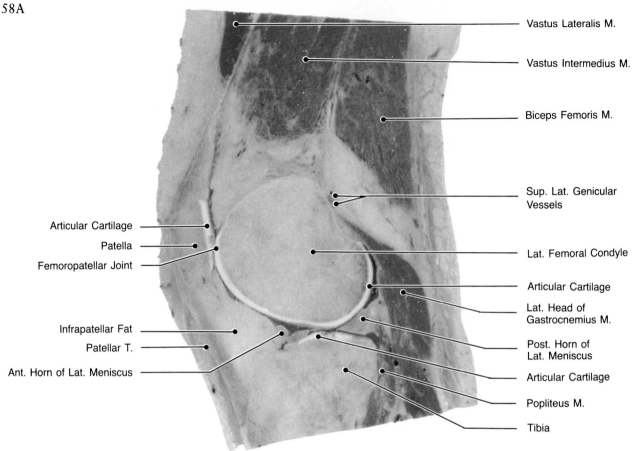

Vastus Lateralis M.

Vastus Intermedius M.

Biceps Femoris M.

Sup. Lat. Genicular Vessels

Articular Cartilage

Patella

Femoropatellar Joint

Lat. Femoral Condyle

Articular Cartilage

Lat. Head of Gastrocnemius M.

Infrapatellar Fat

Patellar T.

Post. Horn of Lat. Meniscus

Ant. Horn of Lat. Meniscus

Articular Cartilage

Popliteus M.

Tibia

58A, 58B Sagittal/Anatomic Plate/MRI/Lateral Knee Joint. Sagittal sections through the lateral aspect of the knee joint. The lateral femoral condyle, lateral meniscus, anterior and posterior horns, and the tibiofibular joint are clearly seen on the MR image (58B). The low signal intensity of the fibrocartilage of the meniscus is well demarcated from the high signal intensity of the articular, hyaline cartilage.

58C Sagittal/MRI/Lateral Knee Joint. Sagittal MRI section through the lateral aspect of the knee. There is a horizontal tear of the anterior horn of the lateral meniscus.

58A–58C

58B

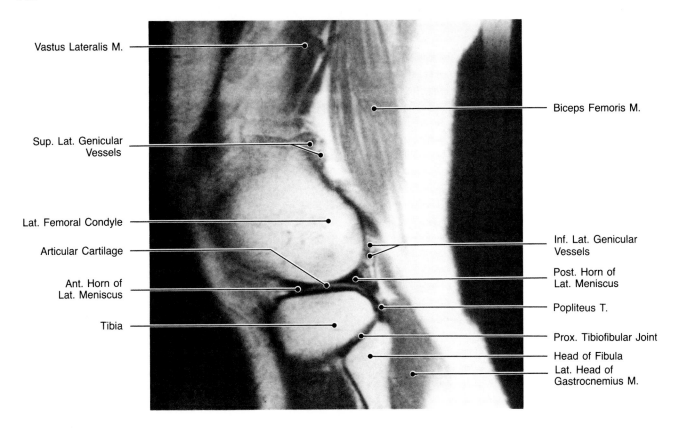

Vastus Lateralis M.

Biceps Femoris M.

Sup. Lat. Genicular Vessels

Lat. Femoral Condyle

Articular Cartilage

Inf. Lat. Genicular Vessels

Ant. Horn of Lat. Meniscus

Post. Horn of Lat. Meniscus

Popliteus T.

Tibia

Prox. Tibiofibular Joint

Head of Fibula

Lat. Head of Gastrocnemius M.

58C

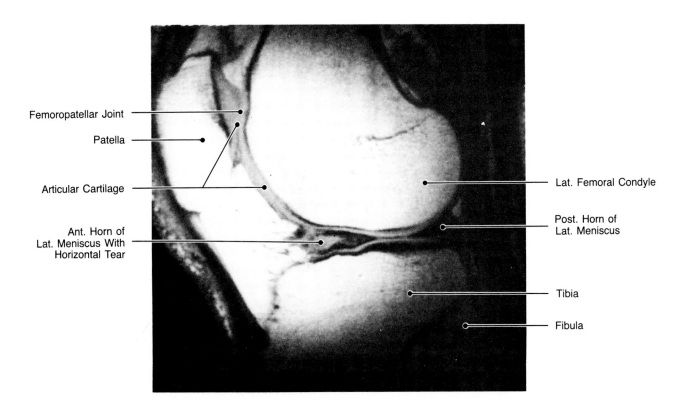

Femoropatellar Joint

Patella

Articular Cartilage

Lat. Femoral Condyle

Post. Horn of Lat. Meniscus

Ant. Horn of Lat. Meniscus With Horizontal Tear

Tibia

Fibula

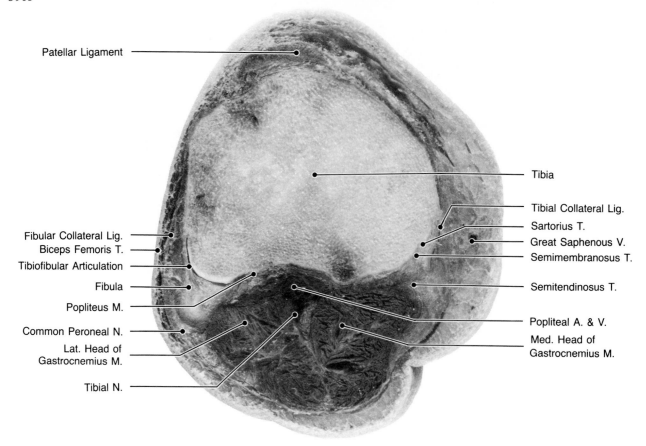

Patellar Ligament

Tibia

Tibial Collateral Lig.

Fibular Collateral Lig.
Biceps Femoris T.
Tibiofibular Articulation
Fibula
Popliteus M.
Common Peroneal N.
Lat. Head of
Gastrocnemius M.

Tibial N.

Sartorius T.
Great Saphenous V.
Semimembranosus T.

Semitendinosus T.

Popliteal A. & V.
Med. Head of
Gastrocnemius M.

59A, 59B, 59C Axial/Anatomic Plate/CT/MRI/Proximal Leg. Axial sections through the proximal tibia. The lower aspect of the patellar ligament is seen as a dense structure anterior to the tibia on the CT image (59B) and as a low signal intensity structure in the MR image (59C). The head of the fibula is barely seen lateral to the tibial plateau, with the insertions of both the fibular collateral ligament and biceps femoris tendon. The MR image shows high intensity in the lumen of the popliteal vein due to its low flow. On the CT image there is no difference between muscles and vessels, due to the lack of intravascular contrast.

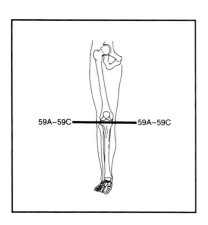

59A–59C   59A–59C

59B

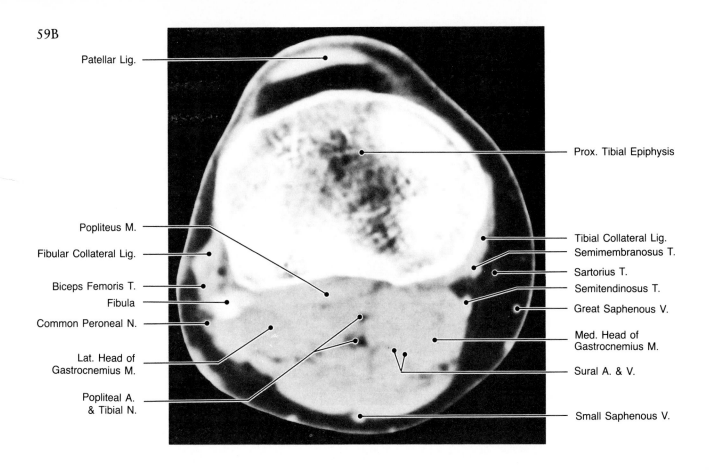

Patellar Lig.

Prox. Tibial Epiphysis

Popliteus M.

Fibular Collateral Lig.

Biceps Femoris T.

Fibula

Common Peroneal N.

Lat. Head of
Gastrocnemius M.

Popliteal A.
& Tibial N.

Tibial Collateral Lig.

Semimembranosus T.

Sartorius T.

Semitendinosus T.

Great Saphenous V.

Med. Head of
Gastrocnemius M.

Sural A. & V.

Small Saphenous V.

59C

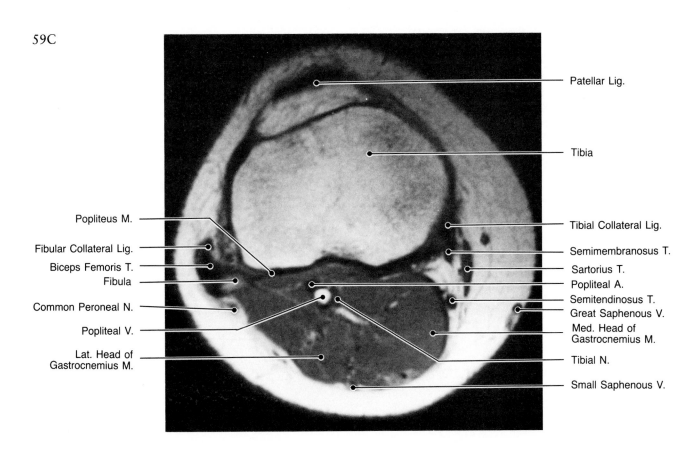

Patellar Lig.

Tibia

Popliteus M.

Fibular Collateral Lig.

Biceps Femoris T.

Fibula

Common Peroneal N.

Popliteal V.

Lat. Head of
Gastrocnemius M.

Tibial Collateral Lig.

Semimembranosus T.

Sartorius T.

Popliteal A.

Semitendinosus T.

Great Saphenous V.

Med. Head of
Gastrocnemius M.

Tibial N.

Small Saphenous V.

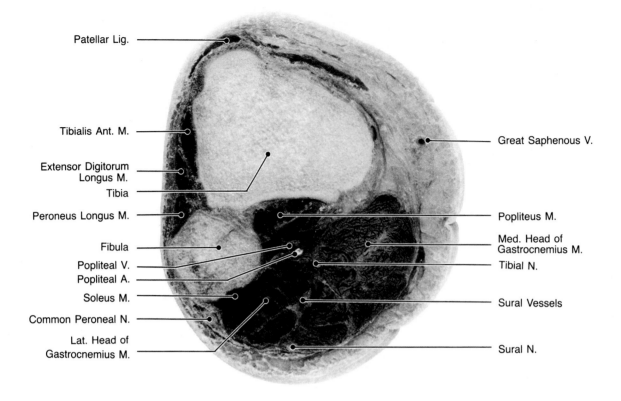

Patellar Lig.

Tibialis Ant. M.

Extensor Digitorum
Longus M.

Tibia

Peroneus Longus M.

Fibula

Popliteal V.

Popliteal A.

Soleus M.

Common Peroneal N.

Lat. Head of
Gastrocnemius M.

Great Saphenous V.

Popliteus M.

Med. Head of
Gastrocnemius M.

Tibial N.

Sural Vessels

Sural N.

60A, 60B, 60C Axial/Anatomic Plate/CT/MRI/Proximal Leg. Axial sections through the proximal tibia at the level of the proximal tibiofibular joint. The inferior half of the popliteal space is totally filled by the popliteus, soleus, and gastrocnemius muscles. The popliteal vessels are seen only on the MR image. The patellar ligament is seen in (60B), inserting on the anterior tibial tuberosity.

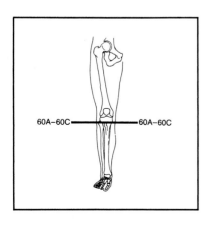

60A–60C   60A–60C

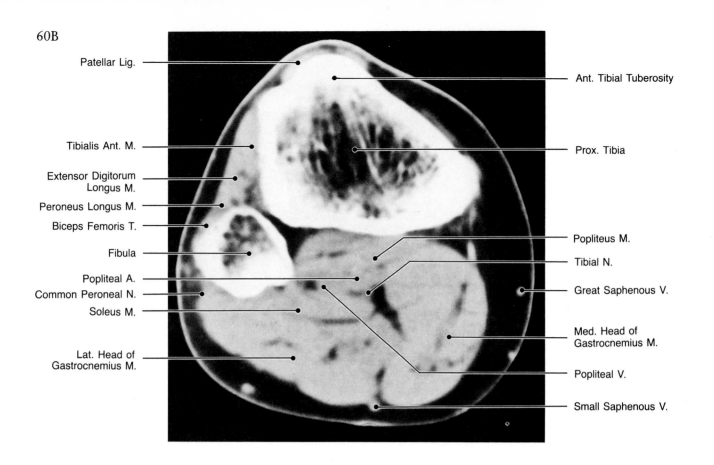

Patellar Lig.

Ant. Tibial Tuberosity

Tibialis Ant. M.

Prox. Tibia

Extensor Digitorum
Longus M.

Peroneus Longus M.

Biceps Femoris T.

Fibula

Popliteus M.

Tibial N.

Popliteal A.

Common Peroneal N.

Soleus M.

Great Saphenous V.

Med. Head of
Gastrocnemius M.

Lat. Head of
Gastrocnemius M.

Popliteal V.

Small Saphenous V.

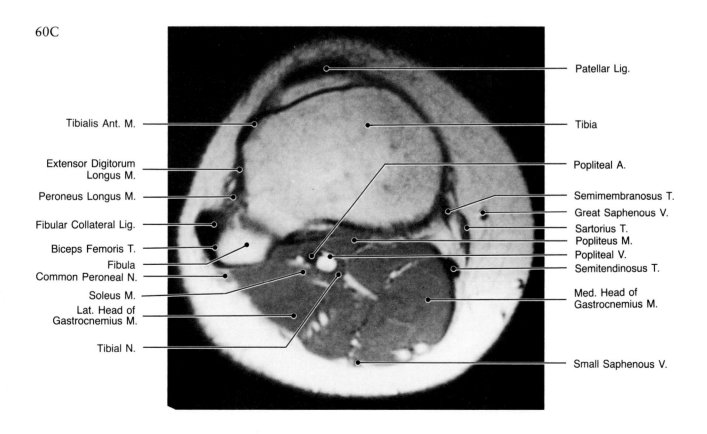

Patellar Lig.

Tibialis Ant. M.

Tibia

Extensor Digitorum
Longus M.

Popliteal A.

Peroneus Longus M.

Semimembranosus T.

Great Saphenous V.

Fibular Collateral Lig.

Sartorius T.

Popliteus M.

Biceps Femoris T.

Fibula

Popliteal V.

Common Peroneal N.

Semitendinosus T.

Soleus M.

Med. Head of
Gastrocnemius M.

Lat. Head of
Gastrocnemius M.

Tibial N.

Small Saphenous V.

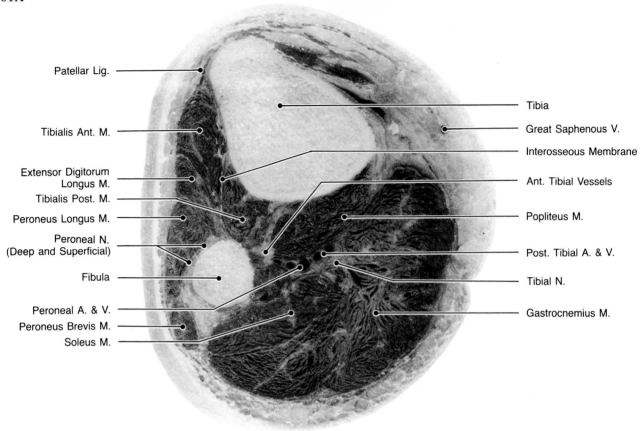

Patellar Lig.

Tibialis Ant. M.

Extensor Digitorum Longus M.

Tibialis Post. M.

Peroneus Longus M.

Peroneal N. (Deep and Superficial)

Fibula

Peroneal A. & V.

Peroneus Brevis M.

Soleus M.

Tibia

Great Saphenous V.

Interosseous Membrane

Ant. Tibial Vessels

Popliteus M.

Post. Tibial A. & V.

Tibial N.

Gastrocnemius M.

61A, 61B, 61C Axial/Anatomic Plate/CT/MRI/Proximal Calf. Axial sections through the proximal calf. The posterior compartment is occupied by the popliteus, soleus, and gastrocnemius muscles. The posterior tibial artery and vein and tibial nerve are located within the posterior compartment. The anterior compartment is occupied by the tibialis anterior, extensor digitorum longus, and peroneus longus muscles as well as the peroneal vessels. Both compartments are separated by the interosseous membrane, which is seen only in the anatomic section.

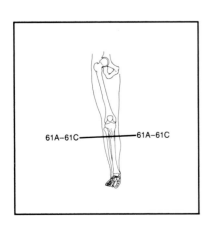

61A–61C   61A–61C

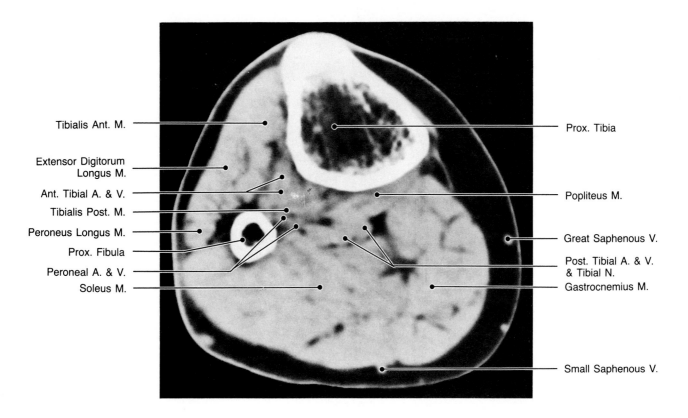

Tibialis Ant. M.

Extensor Digitorum Longus M.

Ant. Tibial A. & V.

Tibialis Post. M.

Peroneus Longus M.

Prox. Fibula

Peroneal A. & V.

Soleus M.

Prox. Tibia

Popliteus M.

Great Saphenous V.

Post. Tibial A. & V. & Tibial N.

Gastrocnemius M.

Small Saphenous V.

61C

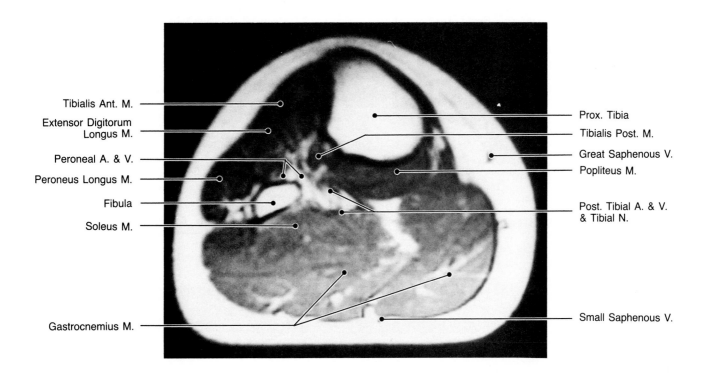

Tibialis Ant. M.

Extensor Digitorum Longus M.

Peroneal A. & V.

Peroneus Longus M.

Fibula

Soleus M.

Gastrocnemius M.

Prox. Tibia

Tibialis Post. M.

Great Saphenous V.

Popliteus M.

Post. Tibial A. & V. & Tibial N.

Small Saphenous V.

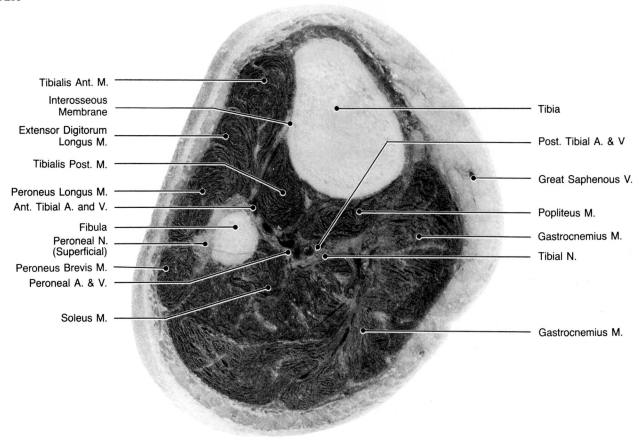

Tibialis Ant. M.

Interosseous Membrane

Extensor Digitorum Longus M.

Tibialis Post. M.

Peroneus Longus M.

Ant. Tibial A. and V.

Fibula

Peroneal N. (Superficial)

Peroneus Brevis M.

Peroneal A. & V.

Soleus M.

Tibia

Post. Tibial A. & V

Great Saphenous V.

Popliteus M.

Gastrocnemius M.

Tibial N.

Gastrocnemius M.

62A, 62B, 62C Axial/Anatomic Plate/CT/MRI/Proximal Calf. Axial sections through the proximal calf, about 5 cm below the previous section. More than 75 per cent of the posterior compartment is occupied by the soleus muscle. The gastrocnemius muscle has become smaller and is located more posteromedially.

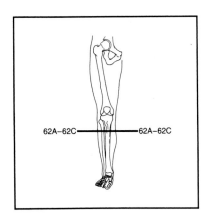

62A–62C   62A–62C

62B

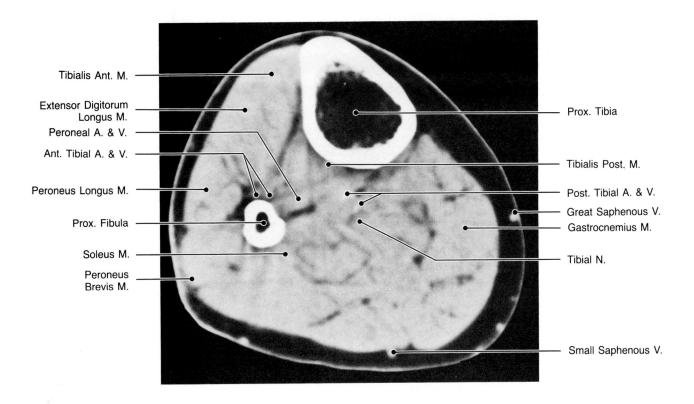

Tibialis Ant. M.

Extensor Digitorum Longus M.

Peroneal A. & V.

Ant. Tibial A. & V.

Peroneus Longus M.

Prox. Fibula

Soleus M.

Peroneus Brevis M.

Prox. Tibia

Tibialis Post. M.

Post. Tibial A. & V.

Great Saphenous V.

Gastrocnemius M.

Tibial N.

Small Saphenous V.

62C

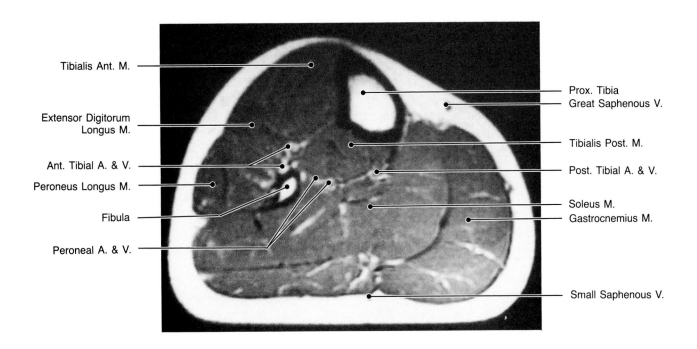

Tibialis Ant. M.

Extensor Digitorum Longus M.

Ant. Tibial A. & V.

Peroneus Longus M.

Fibula

Peroneal A. & V.

Prox. Tibia

Great Saphenous V.

Tibialis Post. M.

Post. Tibial A. & V.

Soleus M.

Gastrocnemius M.

Small Saphenous V.

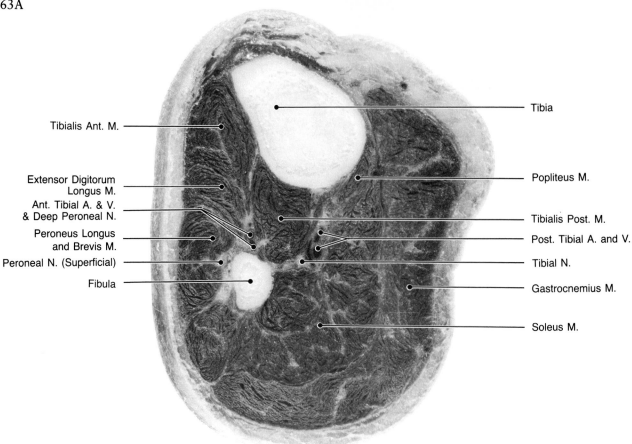

Tibialis Ant. M.

Extensor Digitorum
Longus M.

Ant. Tibial A. & V.
& Deep Peroneal N.

Peroneus Longus
and Brevis M.

Peroneal N. (Superficial)

Fibula

Tibia

Popliteus M.

Tibialis Post. M.

Post. Tibial A. and V.

Tibial N.

Gastrocnemius M.

Soleus M.

63A, 63B, 63C Axial/Anatomic Plate/CT/MRI/Mid Calf.
Axial sections through the mid calf. The peroneus longus and
peroneus brevis muscles occupy the lateral aspect of the anterior compartment, along with the extensor digitorum longus
and tibialis anterior muscles. The anterior tibial vessels are
seen immediately posterior to the tibialis anterior muscle.

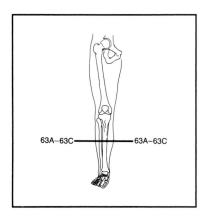

63A–63C   63A–63C

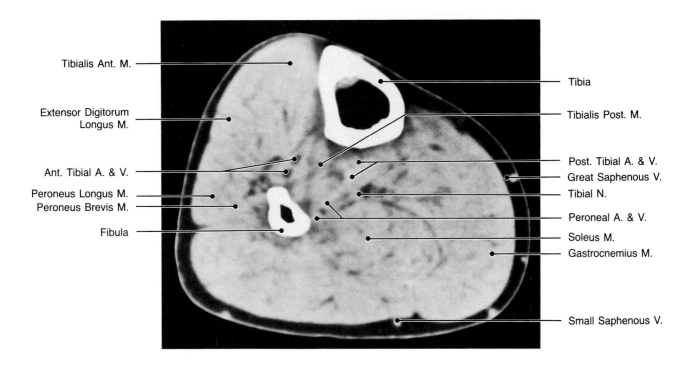

Tibialis Ant. M.

Extensor Digitorum
Longus M.

Ant. Tibial A. & V.

Peroneus Longus M.
Peroneus Brevis M.

Fibula

Tibia

Tibialis Post. M.

Post. Tibial A. & V.
Great Saphenous V.

Tibial N.

Peroneal A. & V.

Soleus M.
Gastrocnemius M.

Small Saphenous V.

63C

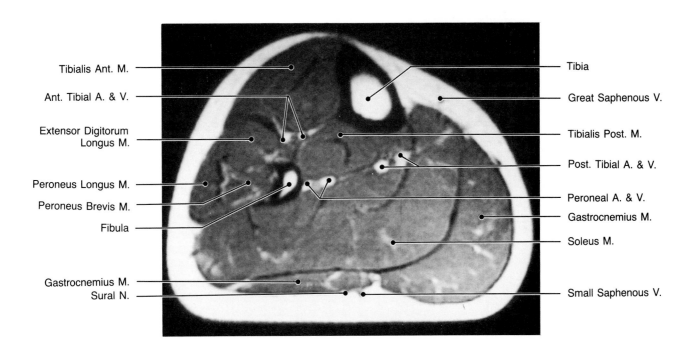

Tibialis Ant. M.

Ant. Tibial A. & V.

Extensor Digitorum
Longus M.

Peroneus Longus M.

Peroneus Brevis M.

Fibula

Gastrocnemius M.
Sural N.

Tibia

Great Saphenous V.

Tibialis Post. M.

Post. Tibial A. & V.

Peroneal A. & V.

Gastrocnemius M.

Soleus M.

Small Saphenous V.

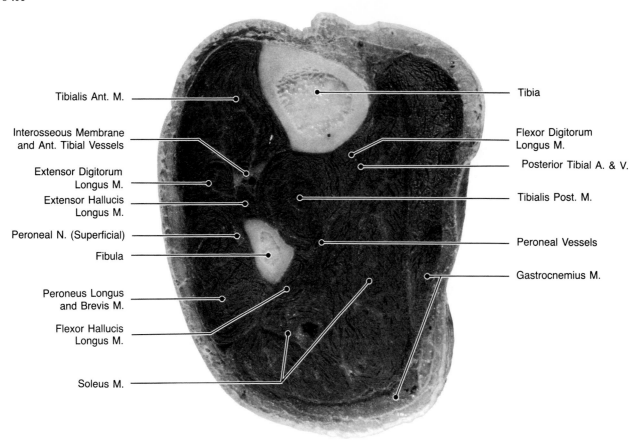

Tibialis Ant. M.

Interosseous Membrane and Ant. Tibial Vessels

Extensor Digitorum Longus M.

Extensor Hallucis Longus M.

Peroneal N. (Superficial)

Fibula

Peroneus Longus and Brevis M.

Flexor Hallucis Longus M.

Soleus M.

Tibia

Flexor Digitorum Longus M.

Posterior Tibial A. & V.

Tibialis Post. M.

Peroneal Vessels

Gastrocnemius M.

64A, 64B, 64C Axial/Anatomic Plate/CT/MRI/Mid Calf.
Axial sections through the mid calf about 5 cm below the previous section. The gastrocnemius muscle is seen only as a very thin band in the posterior medial aspect of the calf. Most of the posterior compartment is occupied by the soleus muscle.

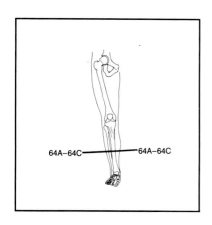

64A–64C   64A–64C

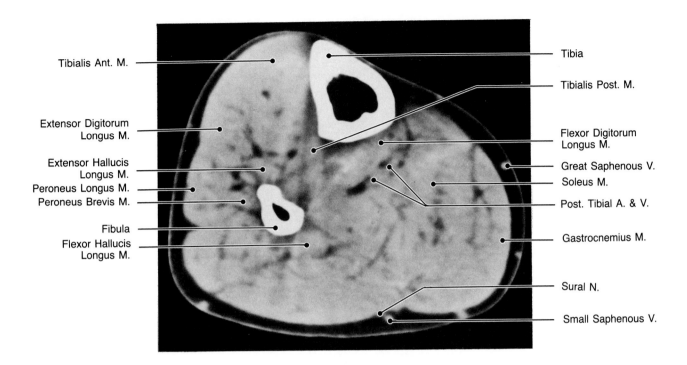

Tibialis Ant. M.

Extensor Digitorum
Longus M.

Extensor Hallucis
Longus M.
Peroneus Longus M.
Peroneus Brevis M.

Fibula
Flexor Hallucis
Longus M.

Tibia

Tibialis Post. M.

Flexor Digitorum
Longus M.

Great Saphenous V.
Soleus M.

Post. Tibial A. & V.

Gastrocnemius M.

Sural N.

Small Saphenous V.

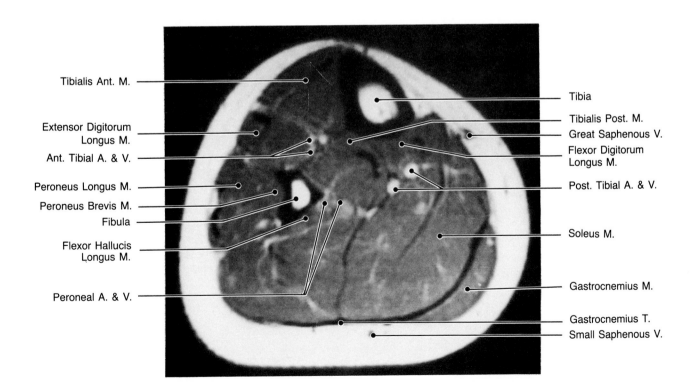

Tibialis Ant. M.

Extensor Digitorum
Longus M.
Ant. Tibial A. & V.

Peroneus Longus M.
Peroneus Brevis M.
Fibula

Flexor Hallucis
Longus M.

Peroneal A. & V.

Tibia

Tibialis Post. M.
Great Saphenous V.
Flexor Digitorum
Longus M.

Post. Tibial A. & V.

Soleus M.

Gastrocnemius M.

Gastrocnemius T.
Small Saphenous V.

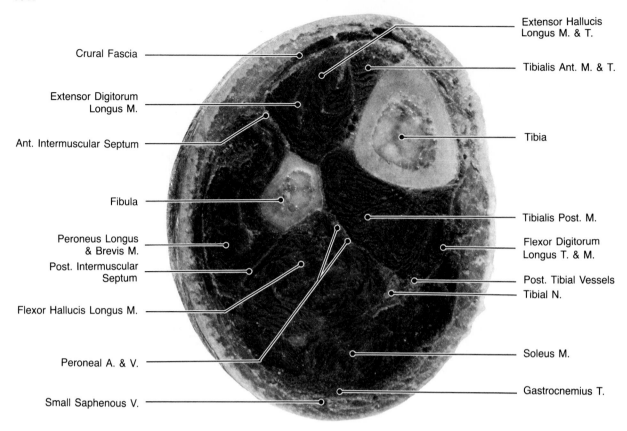

Crural Fascia

Extensor Digitorum
Longus M.

Ant. Intermuscular Septum

Fibula

Peroneus Longus
& Brevis M.

Post. Intermuscular
Septum

Flexor Hallucis Longus M.

Peroneal A. & V.

Small Saphenous V.

Extensor Hallucis
Longus M. & T.

Tibialis Ant. M. & T.

Tibia

Tibialis Post. M.

Flexor Digitorum
Longus T. & M.

Post. Tibial Vessels

Tibial N.

Soleus M.

Gastrocnemius T.

65A, 65B, 65C Axial/Anatomic Plate/CT/MRI/Lower Leg.
Axial sections through the lower third of the leg, about 5 cm
below the previous section. The gastrocnemius tendon is now
clearly identified in both CT (65B) and MR sections (65C).
The posterior compartment is now occupied by the flexor hallucis longus, tibialis posterior, flexor digitorum longus, and
soleus muscles. The anterior compartment is occupied by the
tibialis anterior, extensor hallucis longus, extensor digitorum
longus and peroneus longus and brevis muscles.

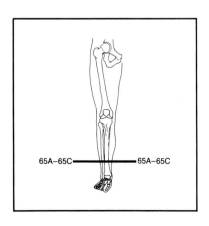

65A–65C   65A–65C

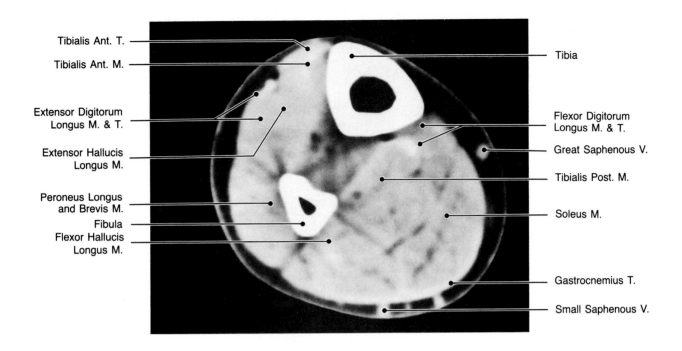

Tibialis Ant. T.

Tibialis Ant. M.

Extensor Digitorum Longus M. & T.

Extensor Hallucis Longus M.

Peroneus Longus and Brevis M.

Fibula

Flexor Hallucis Longus M.

Tibia

Flexor Digitorum Longus M. & T.

Great Saphenous V.

Tibialis Post. M.

Soleus M.

Gastrocnemius T.

Small Saphenous V.

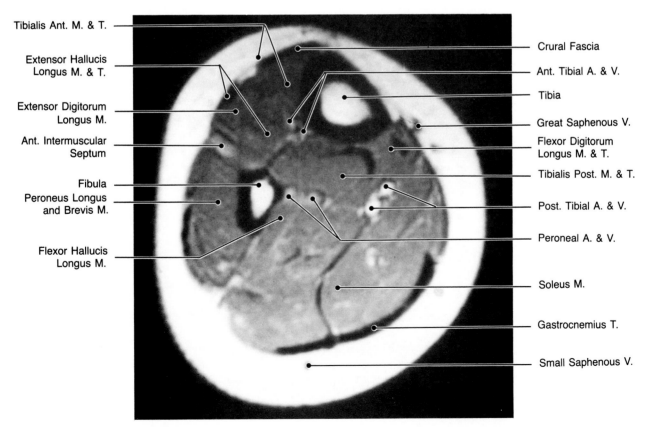

Tibialis Ant. M. & T.

Extensor Hallucis Longus M. & T.

Extensor Digitorum Longus M.

Ant. Intermuscular Septum

Fibula

Peroneus Longus and Brevis M.

Flexor Hallucis Longus M.

Crural Fascia

Ant. Tibial A. & V.

Tibia

Great Saphenous V.

Flexor Digitorum Longus M. & T.

Tibialis Post. M. & T.

Post. Tibial A. & V.

Peroneal A. & V.

Soleus M.

Gastrocnemius T.

Small Saphenous V.

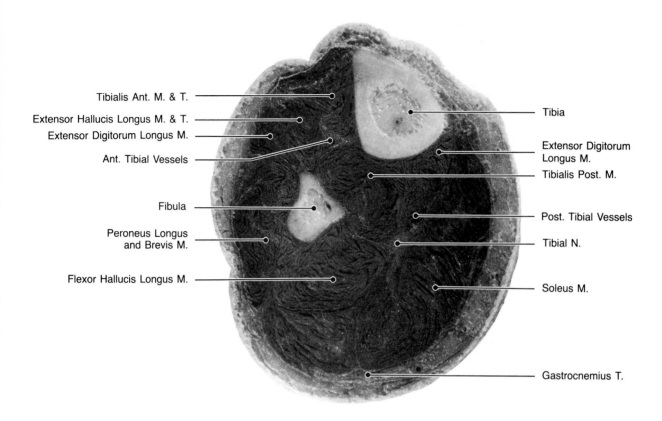

Tibialis Ant. M. & T.
Extensor Hallucis Longus M. & T.
Extensor Digitorum Longus M.
Ant. Tibial Vessels
Fibula
Peroneus Longus and Brevis M.
Flexor Hallucis Longus M.

Tibia
Extensor Digitorum Longus M.
Tibialis Post. M.
Post. Tibial Vessels
Tibial N.
Soleus M.
Gastrocnemius T.

66A, 66B, 66C Axial/Anatomic Plate/CT/MRI/Lower Calf.
Axial sections through the lower third of the calf, about 5 cm below the previous section. The gastrocnemius muscle is no longer seen. Instead the gastrocnemius tendon is seen as a high density band surrounding the soleus muscle posteriorly on the CT image (66B) and as a low intensity structure in the same location on the MR image (66C). The tibialis posterior muscle is also seen posterior to the tibia and medial to the flexor digitorum longus muscle and tendon in the medial aspect of the calf. The tendons of the tibialis anterior, extensor hallucis longus, extensor digitorum longus, and peroneus longus and brevis muscles can be seen in these sections, showing high density in the CT and low signal intensity in the MR images.

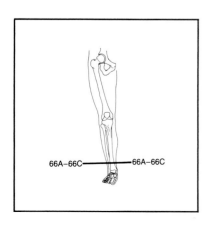

66A–66C   66A–66C

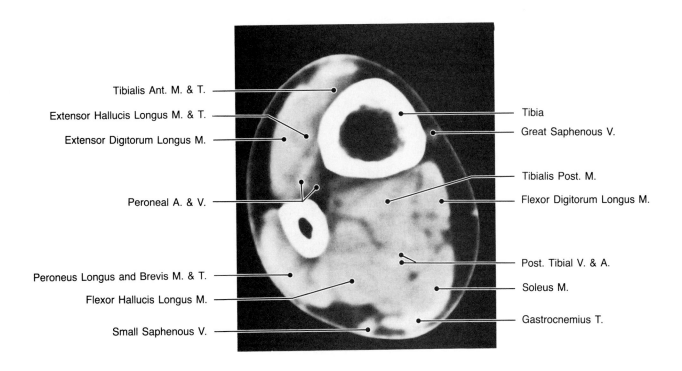

Tibialis Ant. M. & T.

Extensor Hallucis Longus M. & T.

Extensor Digitorum Longus M.

Peroneal A. & V.

Peroneus Longus and Brevis M. & T.

Flexor Hallucis Longus M.

Small Saphenous V.

Tibia

Great Saphenous V.

Tibialis Post. M.

Flexor Digitorum Longus M.

Post. Tibial V. & A.

Soleus M.

Gastrocnemius T.

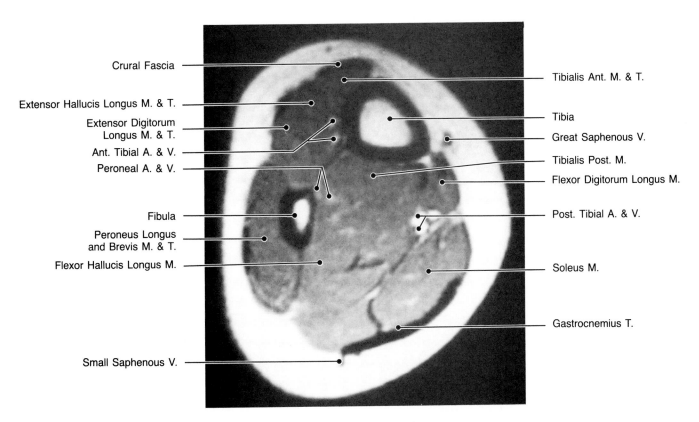

Crural Fascia

Extensor Hallucis Longus M. & T.

Extensor Digitorum Longus M. & T.

Ant. Tibial A. & V.

Peroneal A. & V.

Fibula

Peroneus Longus and Brevis M. & T.

Flexor Hallucis Longus M.

Small Saphenous V.

Tibialis Ant. M. & T.

Tibia

Great Saphenous V.

Tibialis Post. M.

Flexor Digitorum Longus M.

Post. Tibial A. & V.

Soleus M.

Gastrocnemius T.

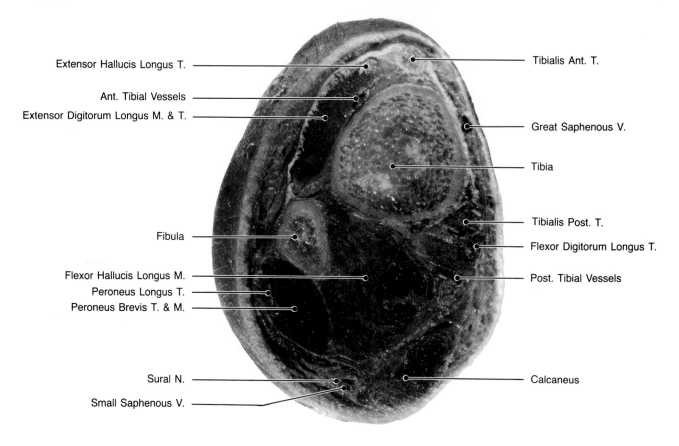

Extensor Hallucis Longus T. — Tibialis Ant. T.

Ant. Tibial Vessels —

Extensor Digitorum Longus M. & T. — Great Saphenous V.

Tibia

Tibialis Post. T.

Fibula — Flexor Digitorum Longus T.

Flexor Hallucis Longus M. — Post. Tibial Vessels
Peroneus Longus T. —
Peroneus Brevis T. & M. —

Sural N. — Calcaneus
Small Saphenous V. —

67A, 67B, 67C Axial/Anatomic Plate/CT/MRI/Lower Leg.
Axial section through the distal tibiofibular joint, just above
the lateral malleolus. The gastrocnemius tendon has become
thicker and smaller and at this level is named calcaneal (Achil-
les) tendon. The CT and MR sections correspond to different
patients. On the MRI the soleus muscle extends to a lower
level. The anterior compartment is occupied by tendons of the
tibialis anterior, extensor hallucis longus, and extensor digitorum longus. The peroneus longus and brevis are located
immediately posterior to the fibula.

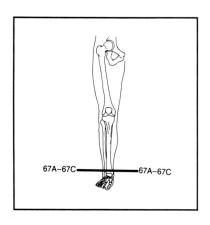

67A–67C   67A–67C

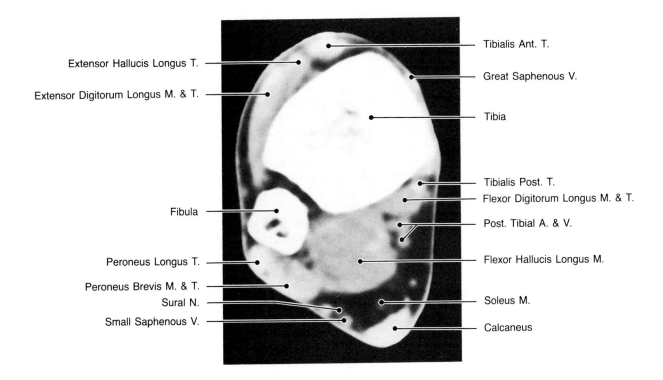

Extensor Hallucis Longus T.

Extensor Digitorum Longus M. & T.

Fibula

Peroneus Longus T.

Peroneus Brevis M. & T.

Sural N.

Small Saphenous V.

Tibialis Ant. T.

Great Saphenous V.

Tibia

Tibialis Post. T.

Flexor Digitorum Longus M. & T.

Post. Tibial A. & V.

Flexor Hallucis Longus M.

Soleus M.

Calcaneus

67C

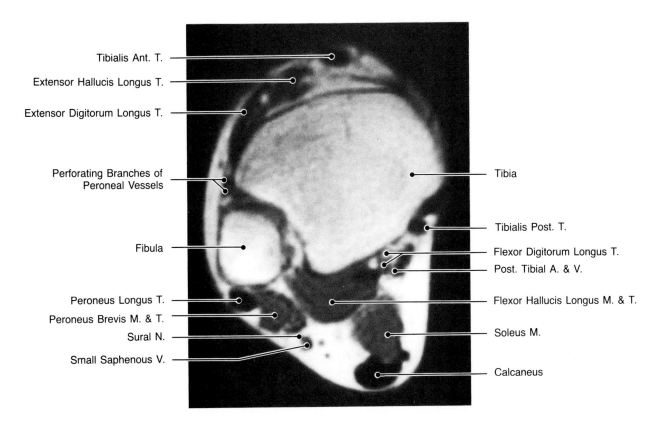

Tibialis Ant. T.

Extensor Hallucis Longus T.

Extensor Digitorum Longus T.

Perforating Branches of
Peroneal Vessels

Fibula

Peroneus Longus T.

Peroneus Brevis M. & T.

Sural N.

Small Saphenous V.

Tibia

Tibialis Post. T.

Flexor Digitorum Longus T.

Post. Tibial A. & V.

Flexor Hallucis Longus M. & T.

Soleus M.

Calcaneus

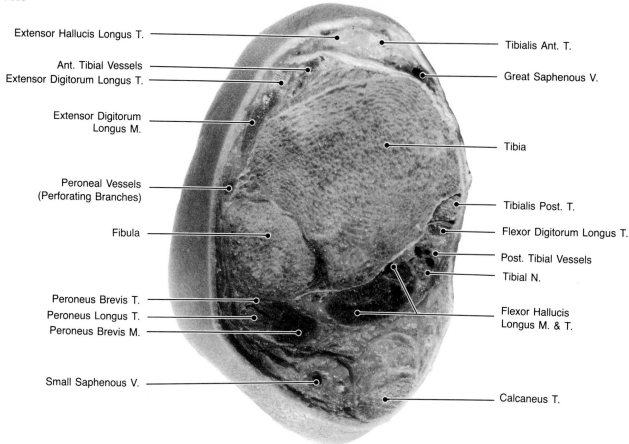

Extensor Hallucis Longus T. — Tibialis Ant. T.

Ant. Tibial Vessels — Great Saphenous V.
Extensor Digitorum Longus T.

Extensor Digitorum Longus M. — Tibia

Peroneal Vessels (Perforating Branches) — Tibialis Post. T.

Fibula — Flexor Digitorum Longus T.

Post. Tibial Vessels

Tibial N.

Peroneus Brevis T. — Flexor Hallucis Longus M. & T.
Peroneus Longus T.
Peroneus Brevis M.

Small Saphenous V. — Calcaneus T.

68A, 68B, 68C Axial/Anatomic Plate/CT/MRI/Upper Ankle Joint. Axial sections through the distal tibiofibular joint at the level of the lateral malleolus, about 3 cm below the previous section. The posteriomedial aspect of the ankle is occupied by the flexor hallucis longus muscle laterally and the flexor digitorum longus and tibialis posterior tendons medially and anteriorly. Immediately posterior to the lateral malleolus, the peroneus longus tendon is seen anterior to the peroneus brevis tendon.

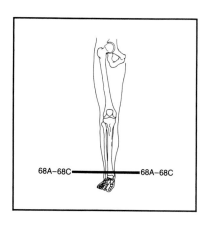

68A–68C   68A–68C

68B

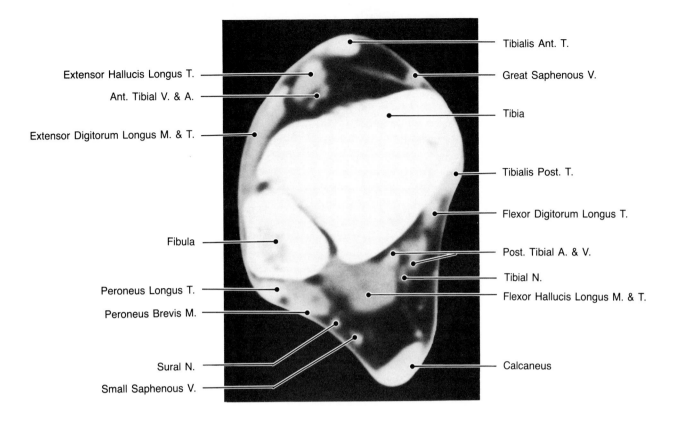

Extensor Hallucis Longus T.

Ant. Tibial V. & A.

Extensor Digitorum Longus M. & T.

Fibula

Peroneus Longus T.

Peroneus Brevis M.

Sural N.

Small Saphenous V.

Tibialis Ant. T.

Great Saphenous V.

Tibia

Tibialis Post. T.

Flexor Digitorum Longus T.

Post. Tibial A. & V.

Tibial N.

Flexor Hallucis Longus M. & T.

Calcaneus

68C

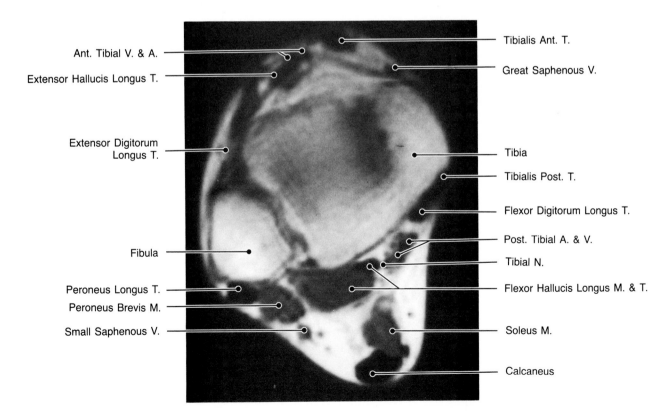

Ant. Tibial V. & A.

Extensor Hallucis Longus T.

Extensor Digitorum
Longus T.

Fibula

Peroneus Longus T.

Peroneus Brevis M.

Small Saphenous V.

Tibialis Ant. T.

Great Saphenous V.

Tibia

Tibialis Post. T.

Flexor Digitorum Longus T.

Post. Tibial A. & V.

Tibial N.

Flexor Hallucis Longus M. & T.

Soleus M.

Calcaneus

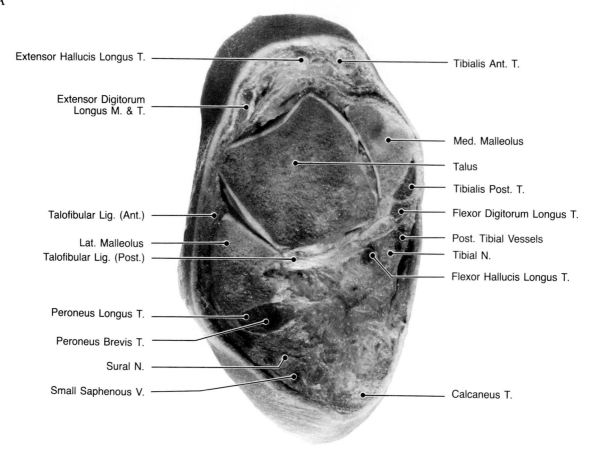

Extensor Hallucis Longus T. ——— ——— Tibialis Ant. T.

Extensor Digitorum
Longus M. & T. ———

——— Med. Malleolus

——— Talus

Talofibular Lig. (Ant.) ——— ——— Tibialis Post. T.

——— Flexor Digitorum Longus T.

Lat. Malleolus ——— ——— Post. Tibial Vessels
Talofibular Lig. (Post.) ——— ——— Tibial N.

——— Flexor Hallucis Longus T.

Peroneus Longus T. ———

Peroneus Brevis T. ———

Sural N. ———

Small Saphenous V. ——— ——— Calcaneus T.

69A, 69B, 69C Axial/Anatomic Plate/CT/MRI/Mid Ankle Joint. Axial sections through the ankle joint, at the level of the lateral and medial malleoli. The talus is now seen between the lateral malleolus of the fibula and medial malleolus of the tibia. The posterior talofibular ligament is seen in both CT (69B) and MRI (69C) sections. The tendons occupying the anterior aspect of the ankle are, from lateral to medial, extensor digitorium longus, extensor hallucis longus, and tibialis anterior. The posterior aspect of the ankle, from lateral to medial, is occupied by the peroneus longus, peroneus brevis, flexor hallucis longus, flexor digitorum longus, and tibialis posterior tendons.

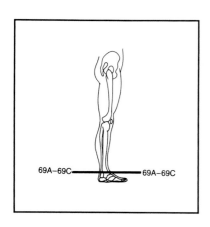

69A–69C ——— ——— 69A–69C

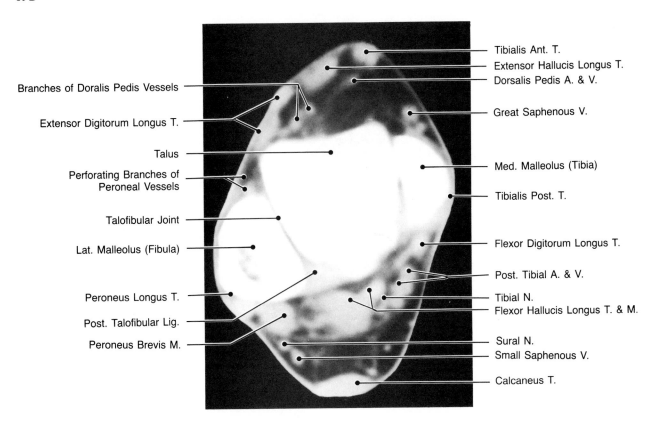

Branches of Doralis Pedis Vessels

Extensor Digitorum Longus T.

Talus

Perforating Branches of
Peroneal Vessels

Talofibular Joint

Lat. Malleolus (Fibula)

Peroneus Longus T.

Post. Talofibular Lig.

Peroneus Brevis M.

Tibialis Ant. T.
Extensor Hallucis Longus T.
Dorsalis Pedis A. & V.

Great Saphenous V.

Med. Malleolus (Tibia)

Tibialis Post. T.

Flexor Digitorum Longus T.

Post. Tibial A. & V.

Tibial N.
Flexor Hallucis Longus T. & M.

Sural N.
Small Saphenous V.

Calcaneus T.

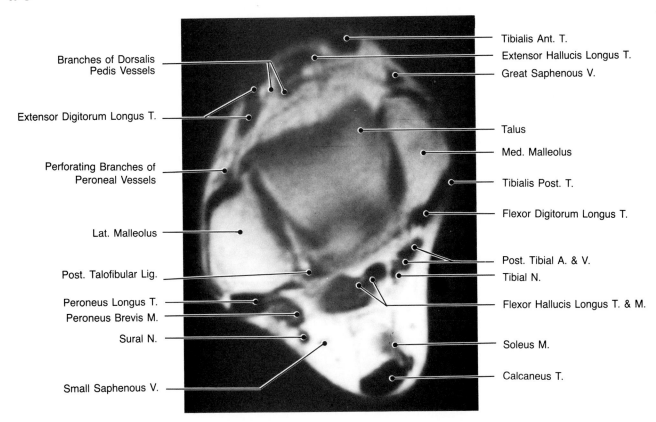

Branches of Dorsalis
Pedis Vessels

Extensor Digitorum Longus T.

Perforating Branches of
Peroneal Vessels

Lat. Malleolus

Post. Talofibular Lig.

Peroneus Longus T.
Peroneus Brevis M.

Sural N.

Small Saphenous V.

Tibialis Ant. T.
Extensor Hallucis Longus T.
Great Saphenous V.

Talus
Med. Malleolus

Tibialis Post. T.

Flexor Digitorum Longus T.

Post. Tibial A. & V.
Tibial N.

Flexor Hallucis Longus T. & M.

Soleus M.

Calcaneus T.

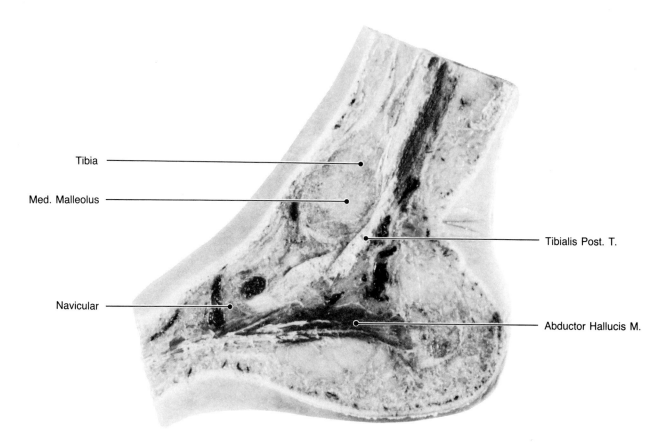

Tibia

Med. Malleolus

Tibialis Post. T.

Navicular

Abductor Hallucis M.

70A, 70B Sagittal/Anatomic Plate/MRI/Medial Malleolus.
Sagittal sections through the medial malleolus. The tibialis
posterior and flexor digitorum longus tendons are seen located
immediately posterior to the medial malleolus. The deltoid ligament is seen as a low intensity signal structure between the
distal tibia, medial malleolus, and talus on MRI (70B).

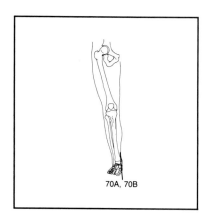

70A, 70B

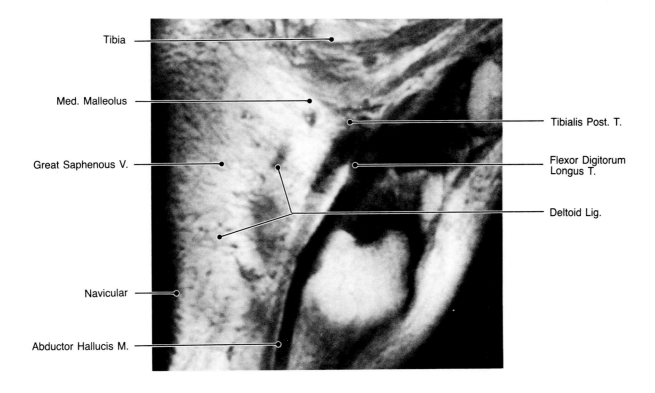

Tibia

Med. Malleolus

Great Saphenous V.

Navicular

Abductor Hallucis M.

Tibialis Post. T.

Flexor Digitorum Longus T.

Deltoid Lig.

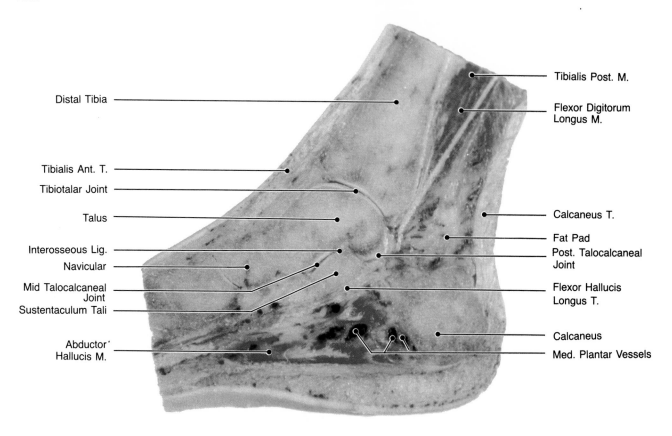

Tibialis Post. M.

Flexor Digitorum Longus M.

Distal Tibia

Tibialis Ant. T.

Tibiotalar Joint

Talus

Calcaneus T.

Fat Pad

Interosseous Lig.

Post. Talocalcaneal Joint

Navicular

Mid Talocalcaneal Joint

Sustentaculum Tali

Flexor Hallucis Longus T.

Abductor Hallucis M.

Calcaneus

Med. Plantar Vessels

71A, 71B Sagittal/Anatomic Plate/MRI/Mid Ankle Joint. Sagittal section through the mid ankle joint. The tibiotalar joint is clearly demonstrated. Note the separation between the articular cartilage of the tibia and the articular cartilage of the dome of the talus on the MRI section (71B). The thickness of both cartilages can be assessed separately. The sustentaculum tali is well demonstrated. The flexor hallucis longus tendon is located below the sustentaculum tali.

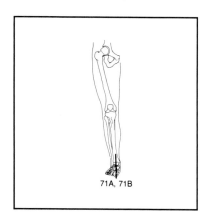

71A, 71B

71B

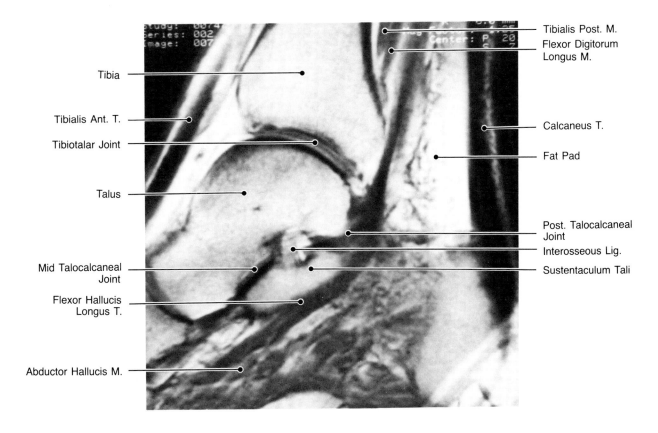

Tibialis Post. M.

Flexor Digitorum
Longus M.

Tibia

Tibialis Ant. T.

Tibiotalar Joint

Talus

Calcaneus T.

Fat Pad

Post. Talocalcaneal
Joint

Interosseous Lig.

Mid Talocalcaneal
Joint

Sustentaculum Tali

Flexor Hallucis
Longus T.

Abductor Hallucis M.

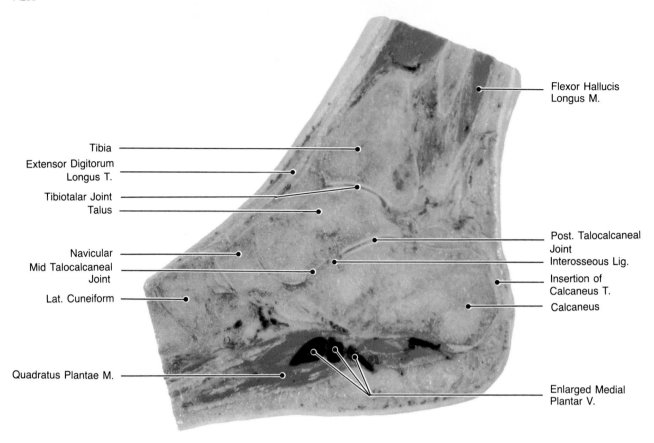

Flexor Hallucis
Longus M.

Tibia

Extensor Digitorum
Longus T.

Tibiotalar Joint

Talus

Navicular

Mid Talocalcaneal
Joint

Lat. Cuneiform

Quadratus Plantae M.

Post. Talocalcaneal
Joint
Interosseous Lig.

Insertion of
Calcaneus T.

Calcaneus

Enlarged Medial
Plantar V.

72A, 72B Sagittal/Anatomic Plate/MRI/Mid Ankle Joint. Sagittal section through the mid portion of the ankle joint about 2 cm lateral to the previous section. The subtalar joint is well defined in this view. Both mid and posterior talocalcaneal joints as well as the interosseous ligament are shown.

72C Sagittal/MRI/Mid Ankle Joint. Sagittal MRI section through the mid portion of the ankle joint in a patient with history of repeated ankle trauma. The posterior aspect of the tibiotalar joint is narrowed due to chondromalacia of the articular cartilage. A subchondral cyst and a loose intra-articular body are also seen.

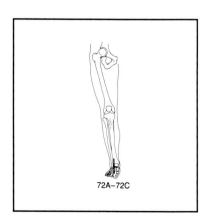

72A–72C

72B

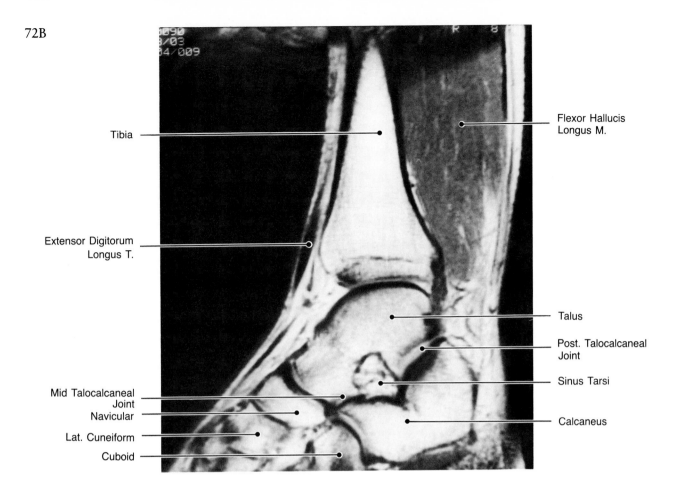

Tibia

Extensor Digitorum
Longus T.

Mid Talocalcaneal
Joint
Navicular

Lat. Cuneiform

Cuboid

Flexor Hallucis
Longus M.

Talus

Post. Talocalcaneal
Joint

Sinus Tarsi

Calcaneus

72C

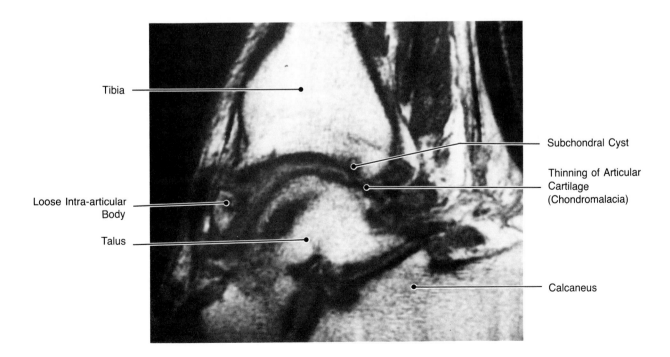

Tibia

Loose Intra-articular
Body

Talus

Subchondral Cyst

Thinning of Articular
Cartilage
(Chondromalacia)

Calcaneus

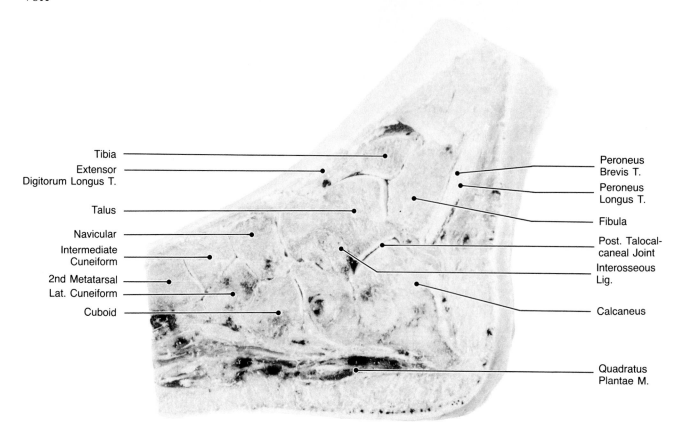

Tibia

Extensor Digitorum Longus T.

Talus

Navicular

Intermediate Cuneiform

2nd Metatarsal

Lat. Cuneiform

Cuboid

Peroneus Brevis T.

Peroneus Longus T.

Fibula

Post. Talocalcaneal Joint

Interosseous Lig.

Calcaneus

Quadratus Plantae M.

73A, 73B, 73C Sagittal/Anatomic Plate/MRI/Lateral Ankle Joint. Sagittal section through the lateral aspect of the tibiotalar joint (73A, 73B) and through the lateral malleolus of the fibula (73C). The subtalar joint and the interosseous ligament are again clearly demonstrated on 73B. The peroneus brevis and longus tendons are seen on 73C, immediately posterior to the lateral malleolus.

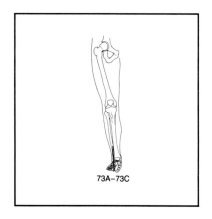

73A–73C

73B

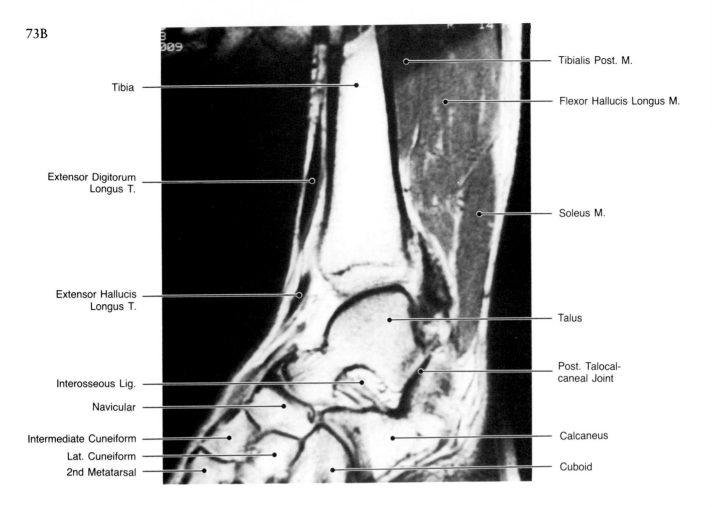

Tibia

Extensor Digitorum
Longus T.

Extensor Hallucis
Longus T.

Interosseous Lig.

Navicular

Intermediate Cuneiform

Lat. Cuneiform

2nd Metatarsal

Tibialis Post. M.

Flexor Hallucis Longus M.

Soleus M.

Talus

Post. Talocal-
caneal Joint

Calcaneus

Cuboid

73C

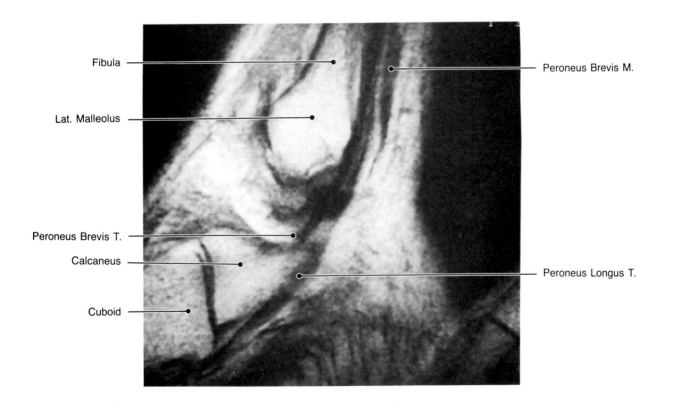

Fibula

Lat. Malleolus

Peroneus Brevis T.

Calcaneus

Cuboid

Peroneus Brevis M.

Peroneus Longus T.

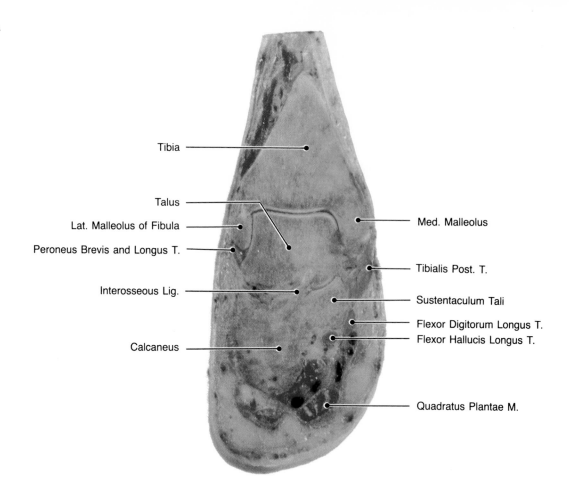

Tibia

Talus

Lat. Malleolus of Fibula

Peroneus Brevis and Longus T.

Interosseous Lig.

Calcaneus

Med. Malleolus

Tibialis Post. T.

Sustentaculum Tali

Flexor Digitorum Longus T.

Flexor Hallucis Longus T.

Quadratus Plantae M.

74A, 74B Coronal/Anatomic Plate/MRI/Mid Ankle Joint.
Coronal section through the tibiotalar joint, including both
lateral and medial malleoli. Note the tibiotalar joint and the
subtalar joint with the interosseous ligament crossing obliquely from the talus to the calcaneus. Note the sustentaculum
tali with the flexor hallucis longus immediately below it.

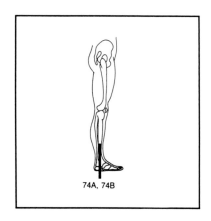

74A, 74B

Extensor Hallucis Longus M.

Tibia

Tibiotalar Joint

Talus

Fibula

Lat. Malleolus

Interosseous Lig.

Peroneus Brevis T.

Peroneus Longus T.

Calcaneus

Med. Malleolus

Deltoid Lig.

Tibialis Post. T.

Flexor Digitorum Longus T.

Sustentaculum Tali

Flexor Hallucis Longus T.

Abductor Hallucis Longus M.

Quadratus Plantae M.

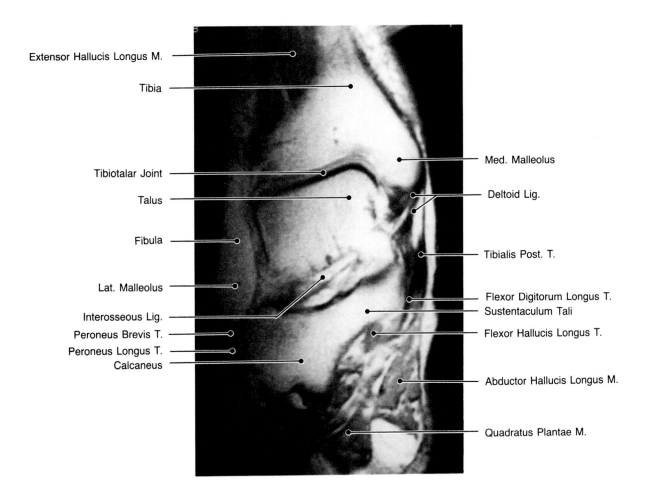

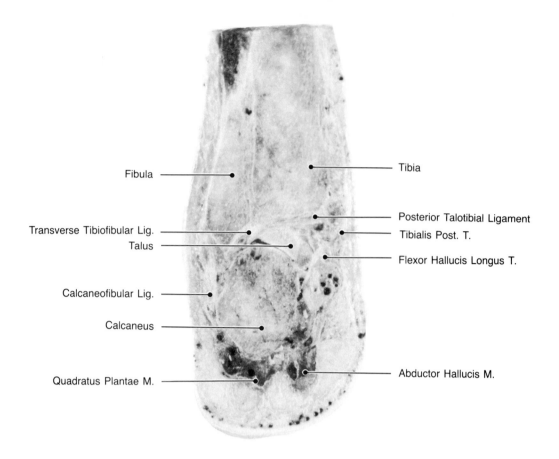

Fibula — Tibia

Transverse Tibiofibular Lig. — — Posterior Talotibial Ligament
Talus — — Tibialis Post. T.
— Flexor Hallucis Longus T.

Calcaneofibular Lig. —

Calcaneus —

Quadratus Plantae M. — — Abductor Hallucis M.

75A, 75B Coronal/Anatomic Plate/MRI/Posterior Ankle Joint. Coronal section through the posterior aspect of the tibiotalar joint about 2 cm posterior to the previous section. The calcaneofibular ligament in the lateral aspect of the joint as well as the transverse tibiofibular ligament can be identified.

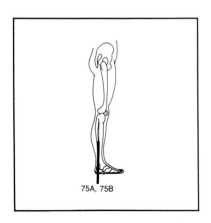

75A, 75B

75B

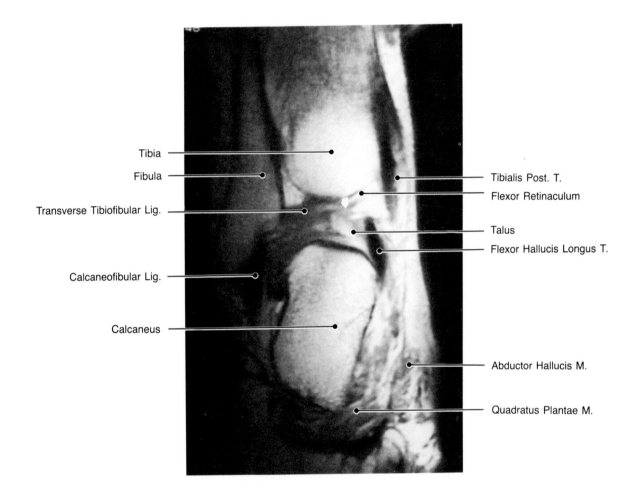

Tibia

Fibula

Transverse Tibiofibular Lig.

Calcaneofibular Lig.

Calcaneus

Tibialis Post. T.

Flexor Retinaculum

Talus

Flexor Hallucis Longus T.

Abductor Hallucis M.

Quadratus Plantae M.

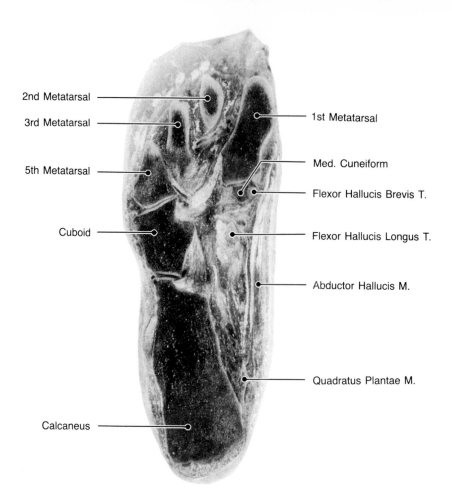

2nd Metatarsal

3rd Metatarsal

5th Metatarsal

Cuboid

Calcaneus

1st Metatarsal

Med. Cuneiform

Flexor Hallucis Brevis T.

Flexor Hallucis Longus T.

Abductor Hallucis M.

Quadratus Plantae M.

76A, 76B, 76C Axial/Anatomic Plate/CT/MRI/Foot. Axial
section through the foot. The flexor hallucis brevis and flexor
hallucis longus tendons are seen posterior to the base of the
first metatarsal and medial cuneiform. Parts of the quadratus
plantae and abductor hallucis muscles are also seen.

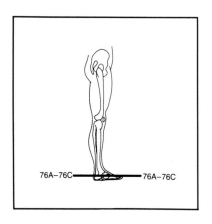

76A–76C   76A–76C

76B

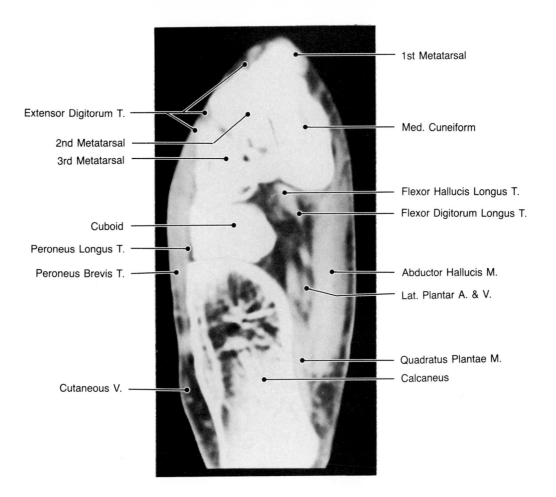

1st Metatarsal

Extensor Digitorum T.

2nd Metatarsal

3rd Metatarsal

Med. Cuneiform

Flexor Hallucis Longus T.

Flexor Digitorum Longus T.

Cuboid

Peroneus Longus T.

Peroneus Brevis T.

Abductor Hallucis M.

Lat. Plantar A. & V.

Quadratus Plantae M.

Calcaneus

Cutaneous V.

76C

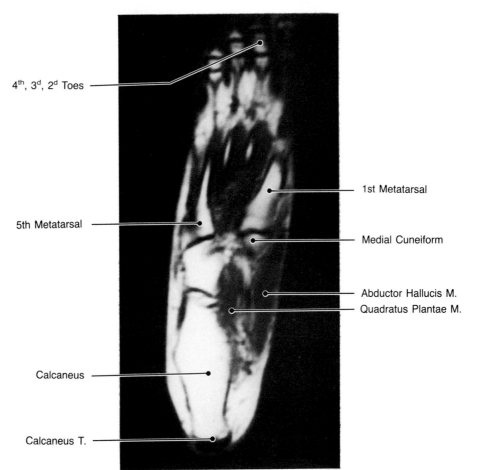

4th, 3^d, 2^d Toes

1st Metatarsal

5th Metatarsal

Medial Cuneiform

Abductor Hallucis M.

Quadratus Plantae M.

Calcaneus

Calcaneus T.

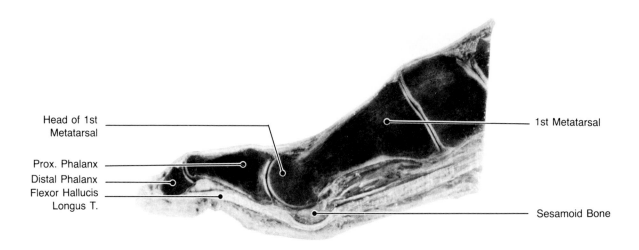

Head of 1st Metatarsal

1st Metatarsal

Prox. Phalanx

Distal Phalanx

Flexor Hallucis Longus T.

Sesamoid Bone

77A, 77B Sagittal/Anatomic Plate/MRI/Great Toe. Sagittal section through the great toe. This section demonstrates the insertion of the flexor hallucis longus on the plantar aspect of the distal phalanx. The integrity of this tendon can be assessed on the MRI section.

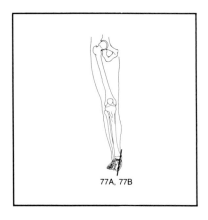

77A, 77B

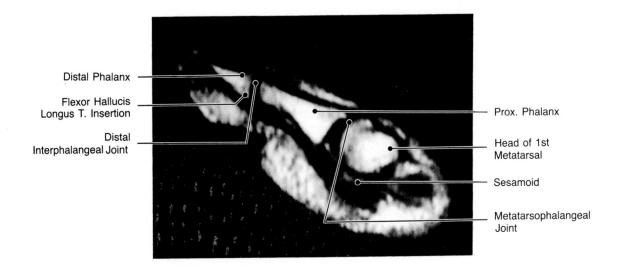

Distal Phalanx

Flexor Hallucis
Longus T. Insertion

Distal
Interphalangeal Joint

Prox. Phalanx

Head of 1st
Metatarsal

Sesamoid

Metatarsophalangeal
Joint

Index